# Thermal Plasmas

## Fundamentals and Applications

## Volume 1

# Thermal Plasmas

## Fundamentals and Applications

## Volume 1

Maher I. Boulos

*University of Sherbrooke*
*Sherbrooke, Quebec, Canada*

Pierre Fauchais

*University of Limoges*
*Limoges, France*

and

Emil Pfender

*University of Minnesota*
*Minneapolis, Minnesota*

**Springer Science+Business Media, LLC**

Library of Congress Cataloging in Publication Data

Boulos. Maher I.
Thermal plasmas: fundamentals and applications, Volume 1 / Maher I. Boulos, Pierre Fauchais, and Emil Pfender.
p. cm.
Includes bibliographical references and index.

1. High temperature plasmas. 2. High temperature plasmas—Industrial applications. I. Fauchais, Pierre. II. Pfender. Emil, 1925- . III. Title.
QC718.5.H5B68 1994
530.4'4—dc20 94-6326
CIP

DOI 10.1007/978-1-4899-1337-1


Originally published by Plenum Press, New York in 1994.
MyCopy version of the original edition 1994

# Preface

Thermal plasma technology has evolved over the past decade into an advanced interdisciplinary science that is attracting increasing attention. Its principal applications are in materials processing, including extractive metallurgy, melting and refining of metals and alloys, plasma chemical synthesis, plasma chemical vapor deposition, plasma and arc spraying, plasma waste destruction, and plasma synthesis of advanced ceramics.

Although some of these applications are presently well-established technologies such as arc plasma welding, plasma cutting, arc spraying, and atmospheric and vacuum plasma spraying, present research activities indicate that refinement and optimization of these processes and especially intelligent processing and automation are still in the forefront of research endeavors. Plasma synthesis of fine particles down to the nanometer size range, the plasma chemical vapor deposition of thin films, and the plasma destruction of toxic waste materials are among the more recent applications of thermal plasmas, which are still in their early stages of development.

As engineers or research scientists undertake a design or an R & D project in any of these areas, they are faced with the major difficulty of having to be acquainted with a wide range of scientific disciplines varying from plasma physics, statistical thermodynamics, high-temperature chemical kinetics, advanced transport phenomena, and materials science. Publications dealing with thermal plasma technology tend to be dispersed over a wide range of scientific journals, which makes it rather difficult for the expert as well as the newcomer to follow progress in this field.

Having maintained an active involvement in this field over the past 20 years, and having taught graduate level and continuing education courses covering different aspects of thermal plasma technology, we have decided to put our knowledge and diverse experiences and backgrounds together in a single reference textbook devoted to the fundamentals and applications of thermal plasma technology. We choose to address the book to practicing engineers and research scientists looking for a simple and clear review of the principal fundamental concepts involved rather than an exhaustive survey of the subject. But this book may also serve as an introductory text for graduate students entering the field of thermal plasma technology. Ample references are given in each chapter for a more in-depth study of the topics covered.

Because of the diversity of the topics to be covered, it was not possible to cover all of them in a single volume. Our choice was, therefore, to present the subject matter of this book in two complementary volumes, the first devoted to the more fundamental concepts of plasma physics and gaseous electronics, thermodynamics and transport properties of plasmas, while the second deals more with the engineering aspects of plasma generation, transport phenomena under plasma conditions, diagnostic techniques, and industrial applications of thermal plasma technology.

This first volume is divided into eight chapters preceded by a brief introduction. In the first chapter, the plasma state is defined and various methods of thermal plasma generation are introduced along with a brief discussion of thermal plasma properties and applications of thermal plasmas. Chapters 2, 3, and 4 cover the essential elements of atomic and molecular theory, kinetic theory, and gaseous electronics, respectively, which are needed for an understanding of material presented in the following chapters. After deriving the plasma equations in Chapter 5, thermodynamic and transport properties are considered in the following two chapters. The most up-to-date values of these properties for selected gases and gas mixtures are presented in the Appendix in the form of tables. The last chapter (8) is devoted to radiation transport.

This book has grown out of work done by the authors and their co-workers at their respective universities. Thanks are due to many present and former students and co-workers of the authors who contributed directly or indirectly to this book. Mrs. Pierrette Robidoux painstakingly typed the manuscript and Mrs. P. McMurry provided assistance in technical editing.

Very special thanks are due to the wives of the authors—Alice, Paulette, and Maja—for their patience, understanding, and support.

M. Boulos
P. Fauchais
E. Pfender

# Contents

# Thermal Plasmas

## Fundamentals and Applications

## Volume 1

Chapter 1

# The Plasma State

## 1.1. PRELIMINARY DEFINITION OF THE PLASMA STATE

The plasma state is frequently referred to as the fourth state of matter in the sequence: solid, liquid, gas, and *plasma*. This classification as a state of matter is justified by the fact that more than 99% of the known universe is in the plasma state. A typical example is the sun, whose interior temperatures exceed $10^7$ K. The high energy content of a plasma compared to that of solids, liquids, or ordinary gases lends itself to a number of important applications.

### 1.1.1. What Is a Plasma?

The following preliminary definition of the plasma state will be restricted to *gaseous* plasmas, which consist of a mixture of electrons, ions, and neutral particles. Since the masses of ions and neutrals are much higher than the electron mass ($m_H/m_e = 1840$, where $m_H$ is the mass of the H atom and $m_e$ is the electron mass), neutrals and ions are classified as the *heavy particles* or the *heavy component* in a plasma. Some of these heavy particles may be in an *excited state* due to the high energy content of a plasma. Particles in an excited state can return to their ordinary or *ground state* by *photon emission.* The latter process is at least partially responsible for the *luminosity* of a plasma. In addition to ions and neutral particles in the ground state, a plasma also contains excited species and photons, i.e., in general a plasma consists of electrons, ions, and neutrals in the ground state, excited species, and photons. Such a mixture, however, qualifies as a plasma only if the negative and positive charges balance each other, i.e., overall a plasma must be electrically neutral. This property is known as *quasi-neutrality*.

In contrast to an ordinary gas, plasmas are electrically conducting due to the presence of free charge carriers. In fact, plasmas may reach electrical conductivities exceeding those of metals at room temperature. For example, a hydrogen plasma at one atmosphere heated to a temperature of $10^6$ K has approximately the same electrical conductivity as copper at room temperature.

A more rigorous definition of the plasma state, taking second-order effects into account, will be given in Chapter 4.

### 1.1.2. Temperature in a Plasma

Kinetic temperatures in a plasma, as in any gaseous medium, are defined by the average kinetic energy of a particle (molecule, atom, ion, or electron), i.e.,

$$\tfrac{1}{2}m\overline{v^2} = \tfrac{3}{2}kT \tag{1.1}$$

where $m$ is the mass of the particle, $(\overline{v^2})^{1/2}$ is its rms or effective velocity, $k$ is the Boltzmann constant, and $T$ represents the absolute temperature (K). Equation (1.1) implies that the particles follow a Maxwell–Boltzmann distribution, which can be expressed by

$$dn_v = nf(v)\,dv \tag{1.2}$$

with

$$f(v) = \frac{4}{\sqrt{\pi}}\left(\frac{2kT}{m}\right)^{3/2} v^2 \exp\left(-\frac{mv^2}{2kT}\right) \tag{1.3}$$

The distribution function $f(v)$ is shown in Fig. 1.1 and reaches a

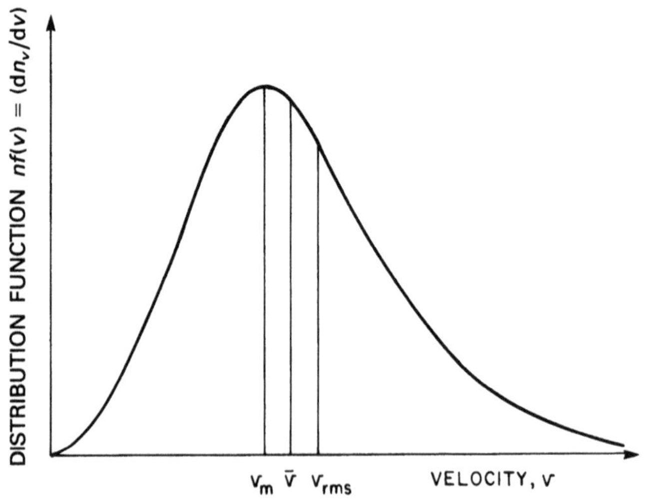


**FIG. 1.1.** Maxwell–Boltzmann distribution of velocities, where $v_m = (2kT/m)^{1/2}$ is the most probable velocity, $\bar{v} = (8kT/\pi m)^{1/2}$ is the mean velocity, $v_{rms} = (3kT/m)^{1/2}$ is the root-mean-square velocity.

maximum at $v_m = (2kT/m)^{1/2}$. The number density of molecules with velocities between $v$ and $v + dv$ is given by $dn_v$. From this distribution it follows that the average velocity is

$$\bar{v} = \int_0^\infty v f(v)\, dv = (8kT/\pi m)^{1/2} \tag{1.4}$$

and the mean-square velocity is

$$\overline{v^2} = \int_0^\infty v^2 f(v)\, dv = \frac{3kT}{m} \tag{1.5}$$

The establishment of a Maxwell–Boltzmann distribution among the particles in a plasma or in an ordinary gas depends strongly on the interaction among the particles, i.e., on the collisional frequency and on the energy exchange during a collision. By applying the conservation equations to an elastic binary collision of particles with mass $m$ and $m'$, one can show that, on the average, exchange of kinetic energy is given by[1]

$$\Delta E_{\text{kin}} = \frac{2mm'}{(m + m')^2} \tag{1.6}$$

This result implies that for particles of the same mass ($m = m'$), $\Delta E_{\text{kin}} = 1/2$, and therefore any distortion of the Maxwell–Boltzmann distribution among particles of the same mass will be eliminated by fewer than 10 collisions.

These considerations demonstrate that in a collision-dominated plasma, we can assume that heavy species and electrons among themselves will have a Maxwell–Boltzmann distribution that permits the definition of a corresponding temperature for these species.

If the subscript $r$ designates the various components (such as electrons, ions, and neutrals) in a plasma, then the Maxwell–Boltzmann distribution for each of these components can be written in terms of their kinetic energy $E_r = \frac{1}{2}mv_r^2$ as

$$dn_{E_r} = \frac{2n_r}{\sqrt{\pi}}(kT_r)^{-3/2} \exp\left(-\frac{E_r}{kT_r}\right) dE_r \tag{1.7}$$

where $T_r$ represents the temperature of the component $r$. As the following discussion will show, the temperatures of the various components of a plasma may or may not be the same.

Let us consider the energy exchange between electrons and heavy species. With $m' = m_e$ (electron mass) and $m = m_h$ (mass of the heavy species) we find from Eq. (1.6) that

$$\Delta E_{\text{kin}} = \frac{2m_e}{m_h} \tag{1.8}$$

because $m_e \ll m_h$. Thus many collisions ($>10^3$) are required to eliminate energy (or temperature) differences between electrons and heavy species.

The most common way to generate and maintain a plasma is by means of an electric discharge. In such a discharge the high-mobility electrons pick up energy from the applied electric field and then transfer part of this energy to the heavy particles through elastic collisions. But even with an excellent collisional coupling (high collision frequency) between electrons and heavy particles, there will always be a difference between the electron temperature and the temperature of the heavy species in the plasma. The energy transferred from an electron to a heavy particle in a single elastic collision may be expressed by

$$\tfrac{3}{2}k(T_e - T_h)\frac{2m_e}{m_h} \tag{1.9}$$

where $T_e$ and $T_h$ represent the electron and the heavy particle temperatures, respectively. The energy that an electron acquires from the electric field ($E$) between collisions is given by

$$eE\overline{v_d\tau_e} \tag{1.10}$$

where $\overline{v_d}$ is the average drift velocity and $\overline{\tau_e}$ the average free flight time between collisions. With $\overline{\tau_e} = l_e/\overline{v_e}$, where $\overline{v_e} = (8kT_e/\pi m_e)^{1/2}$ and $l_e$ is the mean free path (mfp) of the electrons, it follows that for a steady-state situation

$$\frac{T_e - T_h}{T_e} = \frac{3\pi m_h}{32 m_e}\left(\frac{el_eE}{\frac{3}{2}kT_e}\right)^2 \tag{1.11}$$

According to Eq. (1.11), kinetic equilibrium ($T_e = T_h$) requires that the energy acquired by the electrons in an electric field between collisions must be very small compared to the average kinetic energy of the electrons. Another interpretation of Eq. (1.11) considers the fact that

$$l_e \sim \frac{1}{p} \qquad (p \text{ is pressure})$$

and thus

$$\frac{T_e - T_h}{T_e} = \frac{\Delta T}{T_e} \sim \left(\frac{E}{p}\right)^2 \tag{1.12}$$

This relation shows that the parameter $E/p$ plays a governing role for determining the kinetic equilibrium situation in a plasma. For small values of $E/p$, the electron temperature approaches the heavy particle temperature—this is one of the basic requirements for the existence of *Local Thermodynamic Equilibrium* (LTE) in a plasma. Additional conditions for LTE include excitation and chemical equilibrium as well as certain limitations on the gradients in the plasma. Details of all LTE requirements for a plasma will be discussed in Chapter 4.

A plasma that is in kinetic equilibrium and simultaneously meets all other LTE requirements is classified as a *thermal plasma.* In contrast, plasmas with strong deviations from kinetic equilibrium ($T_e \gg T_h$) are classified as *nonthermal* or *nonequilibrium* plasmas. Both types of plasmas will be further discussed in the following section.

### 1.1.3. Different Types of Plasmas

Sometimes plasmas are categorized as *natural* or *man-made plasmas.* As mentioned previously, natural plasmas comprise more than 99% of the universe known today. Two of the earliest known plasma phenomena are *lightning strikes* and the *aurora borealis.* These two plasma phenomena occur at relatively high and extremely low pressures, respectively, leading to the drastic differences in their appearance. In the case of lightning, narrow, high-luminosity channels with numerous dying-off side branches are observed, while the aurora borealis appears as a diffuse, widespread (of astronomic dimensions), low-luminosity event. The pressure of a plasma affects not only its luminosity (relatively low in tenuous plasmas) but also the energy (or temperature) of the various plasma components and their thermodynamic state.

Plasmas occur over such a wide range of pressures, so it is customary to classify them in terms of *electron temperatures* and *electron densities.* Figure 1.2 shows such a classification for some natural and man-made plasmas. The temperature in Fig. 1.2 is plotted in units of eV (1 eV corresponds to 7740 K if there is a M–B distribution). Extremely tenuous plasmas (found, for example, in the *solar corona*) may assume temperatures exceeding $10^6$ K, while plasmas of similar densities in the *ionosphere* may have temperatures of $10^3$ K or even lower. *Flames,* which may also be classified as plasmas, show somewhat higher electron densities and temperatures. In a flame at atmospheric pressure, the degree of ionization $\xi = n_e/(n_e + n)$ is, however, on the order of $10^{-10}$.

*Glow discharges,* which are typically operated in a pressure range from $10^{-4}$ to 1 kPa, reveal electron temperatures on the order of $10^4$ K

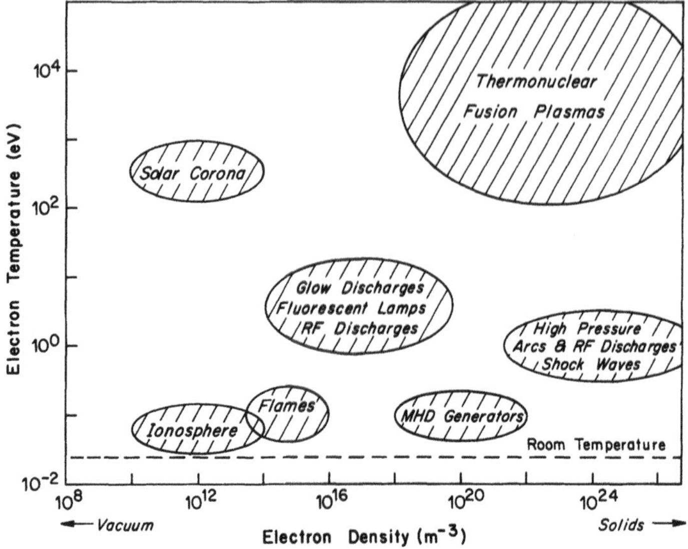


FIG. 1.2. Classification of plasmas.

and heavy-particle temperatures close to room temperature. In a fluorescent lamp, for example, the electron temperature may reach $2.5 \times 10^4$ K, but the heavy-particle temperature remains close to 300 K.

Extreme conditions in terms of electron density and temperature exist in *thermonuclear fusion plasmas.* In the case of inertial confinement fusion plasmas, electron and ion densities may exceed $10^{26}\,m^{-3}$, and temperatures for both magnetically confined and inertially confined plasmas are typically above $10^6$ K and may be as high as $10^8$ K.

Thermal plasmas, i.e., plasmas that approach a state of LTE and that are of primary interest in this text, reveal temperatures around $10^4$ K with electron densities ranging from $10^{21}$ to $10^{26}\,m^{-3}$.

In order to establish a clear distinction between *thermal* and *nonthermal* plasmas, some of the typical features of these two types of plasma will be discussed in the following paragraphs.

#### 1.1.3.1. Thermal Plasmas

Thermal plasmas, which are classified as "hot" plasmas in the American and European literature and as "low temperature" plasmas (to distinguish them from thermonuclear fusion plasmas) in the Russian literature, are by definition in or close to LTE.

Over the past years, it has become increasingly clear that the existence of LTE in a plasma is the exception rather than the rule. Many plasmas that are classified as thermal plasmas do not meet all requirements for LTE, i.e., they are not in *complete local thermodynamic equilibrium* (*CLTE*). As will be discussed later on in more detail, one of the main reasons for deviations from CLTE is the lack of the excitation equilibrium (Boltzmann distribution). In particular, the lower-lying energy levels of atoms may be underpopulated due to the high radiative transition probabilities of these levels, resulting in a corresponding overpopulation of the ground state. Because of the small contribution of excited species to the enthalpy of a plasma, this type of deviation from CLTE is immaterial for most engineering applications. For this reason, such plasmas are still treated as thermal plasmas or, more accurately, as plasmas in *partial local thermodynamic equilibrium* (*PLTE*). However, caution must be exercised if emission spectroscopy is used for diagnostics in such plasmas. Substantial errors may be incurred if energy levels that deviate from excitation equilibrium are used.

More serious deviations from LTE may be expected in the fringes of a plasma or in the vicinity of walls or electrodes. Deviations from both kinetic ($T_e \neq T_h$) and chemical (composition) equilibrium may be found in such regimes.[2] In high-speed plasma flows, deviations from chemical equilibrium are likely because chemical reactions cannot follow the rapid macroscopic motion of the species; a chemically "frozen" situation results. In this case electron densities may be substantially higher than one would expect from the prevailing temperatures.[3] A more detailed discussion of such deviations from LTE will follow in Chapter 4.

As mentioned in the previous section, the parameter $E/p$ plays a crucial role in attaining kinetic equilibrium. Kinetic equilibrium becomes feasible for small values of $E/p$, i.e., high pressures and/or small values of $E$. Typically, pressures in LTE plasmas exceed 10 kPa (≈0.1 atm), as illustrated for an arc plasma in Fig. 1.3; for pressures below 10 kPa, the electron and heavy-particle temperatures separate ($T_e > T_h$).

The magnitude of the electric field is associated with the electrical conductivity ($\sigma_e$) of the plasma through Ohm's law:

$$j = \sigma_e E \tag{1.13}$$

where $j$ is the current density. For a given current density, the field strength decreases as $\sigma_e$ increases.

The following example shows the order of magnitude of $E/p$ required for kinetic equilibrium in a high-intensity argon arc operated at a pressure of 100 kPa. In this case the order of magnitude of this parameter is $E/p = 1\ \text{V/m} \cdot \text{kPa}$.

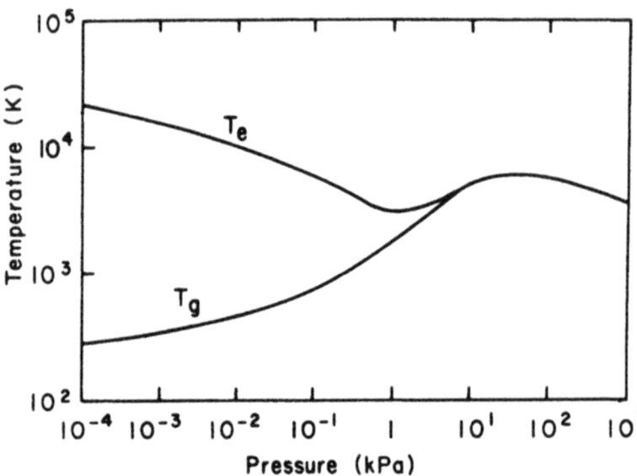


**FIG. 1.3.** Behavior of electron temperature ($T_e$) and heavy-particle temperature ($T_h$) in an arc plasma.

The particle number density can be used instead of the pressure in this parameter. According to Dalton's law

$$p = \sum_r n_r kT \tag{1.14}$$

provided that LTE exists in the plasma. The subscript $r$ stands for electrons, ions, and neutral particles. If the degree of ionization is very small ($\xi \ll 1$) Eq. (1.14) reduces to

$$p = nkT \tag{1.15}$$

where $n$ is the number density of the neutral particles.

#### 1.1.3.2. Nonthermal Plasmas

Nonequilibrium plasmas are frequently classified as "cold" plasmas, because of the low temperature of the heavy species ($T_h \ll T_e$). in contrast to thermal plasmas, nonequilibrium plasma systems are in most cases operated at pressures $p < 10\,\text{kPa}$. According to the previously discussed $E/p$ criterion, substantial deviations from kinetic equilibrium are expected for large values of $E/p$. Typical values of $E/p$ or $E/n$ are several orders of magnitude higher for nonthermal plasmas than for thermal plasmas. A typical value for a glow discharge operated at a pressure of 0.1 Pa is on the order or $E/p = 10^7\,\text{V/m kPa}$.

## 1.2. GENERATION OF THERMAL PLASMAS

Plasmas can be generated by passing an electric current through a gas. Since gases at room temperature are excellent insulators, a sufficient number of charge carriers must be generated to make the gas electrically conducting. This process is known as electrical breakdown, and there are many possible ways to accomplish this breakdown. Breakdown of the originally nonconducting gas establishes a conducting path between a pair of electrodes. The passage of an electrical current through the ionized gas leads to an array of phenomena known as gaseous discharges. Such gaseous discharges are the most common, but not the only, means for producing plasmas. For various applications plasmas are produced by electrodeless RF discharges, by microwaves, by shock waves, and by laser or high-energy particle beams. Finally, plasmas can also be produced by heating gases (vapors) in a high-temperature furnace. Because of inherent temperature limitations, this method is restricted to metal vapors with low ionization potentials.

Owing to space limitations, the discussion in this section will be restricted to the most common electrically generated, steady-state thermal plasmas. Transient plasmas and plasmas that are not produced by electrical discharges (i.e., those produced by laser beams, high-energy particle beams, shock waves, or heating in a furnace) will not be included in this survey.

The most widely used electrical methods for producing thermal plasmas employ *high-intensity arcs* or *inductively coupled high-frequency discharges*; more recently, *microwave discharges* have been considered a viable method for producing such plasmas. These three types of discharges will be discussed in the following sections.

### 1.2.1. High-Intensity Arcs

For the sake of simplicity, only DC arcs will be considered here, although AC arcs are in wide use in actual applications.

The potential distribution in high- as well as in low-intensity arcs shows a peculiar behavior, as indicated in Fig. 1.4. Steep potential drops in front of the electrodes and relatively small potential gradients in the arc column suggest dividing the arc into three parts: the cathode region, the anode region, and the arc column. The latter is a true plasma that will approach a state of LTE in a high-intensity arc.

A high-intensity arc is defined as a discharge operated at current levels $I > 50\,\mathrm{A}$ and pressures $p > 10\,\mathrm{kPa}$. In contrast to low-intensity arcs, high-intensity arcs are characterized by strong macroscopic flows induced by the arc itself.[4,5] Any variation of the current-carrying cross section of the arc leads, via the interaction of the arc current with its own magnetic field, to a pumping action of the type sketched in Fig. 1.5. At sufficiently high currents ($I > 100\,\mathrm{A}$) and axial current density

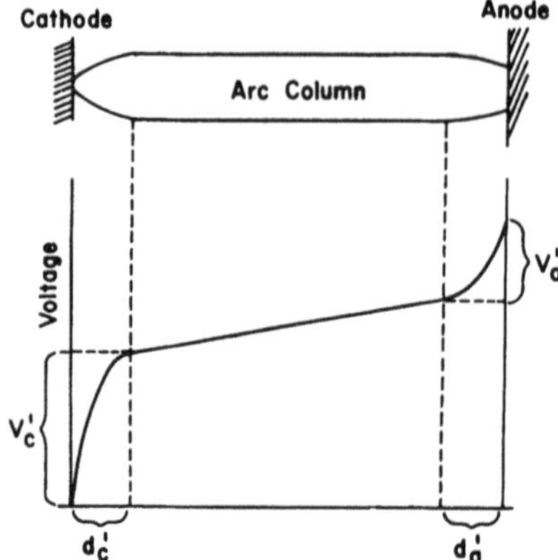


FIG. 1.4. Typical potential distribution along an arc.

variations, flow velocities of the order of 100 m/s are produced. The cathode jet phenomenon, also known as the Maecker effect,[4] is a typical example.

Temperatures and charged particle densities, which are the most important properties of an arc plasma, can vary over a wide range. These properties are determined by the arc parameters, including the arc geometry. Figure 1.6 shows the temperatures and electron densities of different types of arcs, some of which will be discussed in the following sections.

For arc applications, it is useful to classify arc columns in terms of their methods of stabilization. There is a direct link between the method of stabilizing the arc column and the options available for the design of arc devices.

For stable operation, most electric arcs require some kind of stabilizing mechanism that must be either provided externally or produced by the arc itself. Here the term *stabilization* refers to a particular

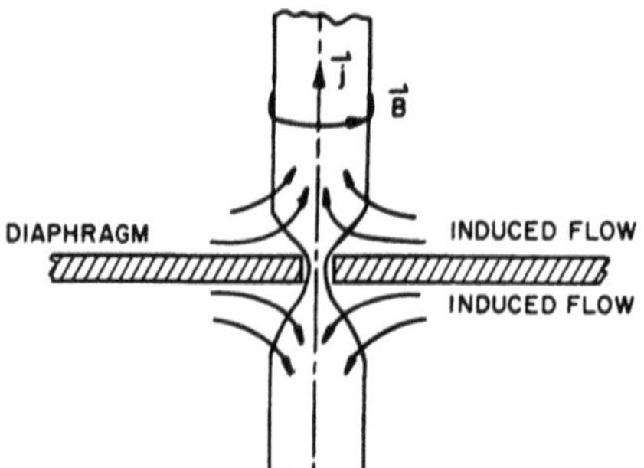


FIG. 1.5. The pumping action induced by arc constriction.

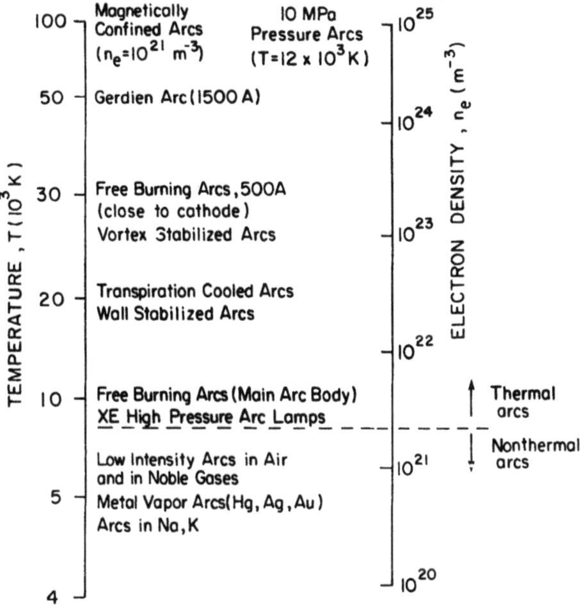


**FIG. 1.6.** Survey of arc temperatures and electron densities.

mechanism that keeps the arc column in a given, stable position, i.e., any accidental excursion of the arc from its equilibrium position causes an interaction with the stabilizing mechanism such that the arc column is forced to return to its equilibrium position. This stable position is not necessarily a stationary one; the arc may, for example, rotate or move along rail electrodes with a certain velocity. Stabilization implies in this situation that the arc column can only move in a well-defined pattern controlled by the stabilizing mechanism.[5]

### 1.2.1.1. Free-Burning Arcs

As the name implies, no external stabilizing mechanism is imposed on the arc in this case; but that does not exclude the possibility that this arc generates its own stabilizing mechanism. Although high-intensity arcs may be operated in the free-burning arc mode, they are frequently classified as self-stabilized arcs (see Section 1.2.1.5) if the induced gas flow due to the interaction of the self-magnetic field and the arc current is the dominating stabilizing mechanism. Free-burning, high-intensity arcs to which these conditions apply will be discussed in Section 1.2.1.5.

Arcs operated at extremely high currents (up to 100 kA) are known as ultrahigh current arcs and should also be mentioned in this category. Although most experiments in this current range utilize pulsed discharges, the relatively long duration (≈10 ms) of the discharge justifies classifying them as arcs. There is considerable interest in such arcs for applications involving melting and steelmaking, chemical arc furnaces, and high-power switchgear. Visual observations of ultrahigh current arcs in arc furnaces reveal a rather complex picture of large, grossly turbulent plasma volumes, vapor jets emanating from the electrodes, and parallel current paths with multiple, highly mobile electrode spots. In this situation there is no evidence for any dominating stabilizing mechanism. Induced gas flows and vapor jets exist simultaneously, interacting with each other in a complicated way. For certain polarities of the arc and certain electrode materials, stable vapor jets that are able to stabilize the arc column have been observed. Thus, the generation of vapor jets by the arc represents another possible mechanism for self-stabilization of arcs.

In a comprehensive survey, Edels[6] gives a description of the characteristic features and properties of ultrahigh current arcs, including 120 pertinent references. Continuing studies in this area are mainly concerned with radiation properties and flow fields in such arcs.

#### 1.2.1.2. Wall-Stabilized Arcs

The principle of wall-stabilization of arcs has been known for more than 80 years, having been introduced in connection with arc lamps. A long arc enclosed in a narrow tube with circular cross section will assume a rotationally symmetric, coaxial position within the tube. Any accidental excursion of the arc column toward the wall will be compensated for by the increased heat conduction to the wall, which reduces the temperature and therefore the electrical conductivity at this location. In short, the arc will be forced to return to its equilibrium position. In this situation, increased thermal conduction and the associated secondary effects provide the stabilizing mechanism.

In order to cope with the extremely high wall heat fluxes experienced with high-intensity arcs enclosed in small diameter tubes, metal tubes composed of a stack of insulated, water-cooled disks (usually Cu) have been introduced as arc vessels. This arrangement is known as the wall-stabilized, cascade arc, and it has been extensively used as a basic research tool.

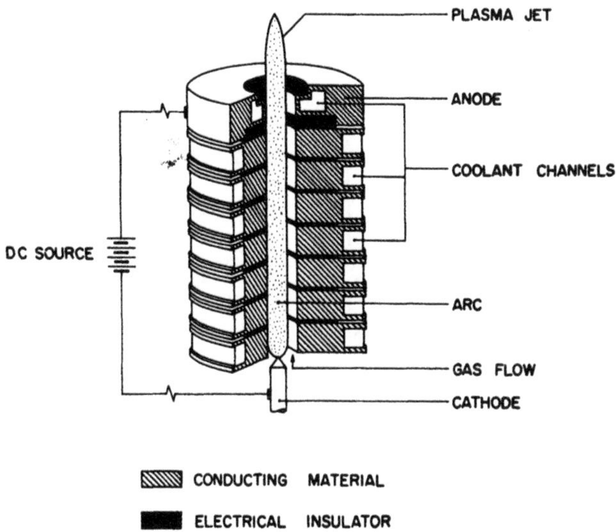


**FIG. 1.7.** Cutaway view of a wall-stabilized, cascaded arc.

Figure 1.7 shows a cutaway view of a typical wall-stabilized arc. The much higher electrical conductivity of metals compared with that of the arc column requires segmentation of the tube enclosing the arc; a continuous metal tube would cause a double arc (arcing from the cathode to the metal tube and from the metal tube to the anode) seeking the path of least resistance.

The maximum possible temperature or enthalpy attainable in a constricted, wall-stabilized arc is limited by the highest permissible heat flux the wall is able to withstand. Sophisticated water-cooling arrangements permit wall heat fluxes up to $2 \times 10^5$ kW/m$^2$.

### 1.2.1.3. Convection-Stabilized Arcs

Among the various possible ways of stabilizing an arc by a superimposed convective flow, vortex stabilization has played a particularly important role.

The principle of vortex stabilization of arcs was reported around the turn of the last century.[7] In the case of vortex or whirl stabilization, the arc is confined to the center of a tube in which an intense vortex of a gas or liquid is maintained. Centrifugal forces drive the cold fluid toward the walls of the arc chamber, which is thus thermally well protected. In addition to the circumferential component of the vortex flow, there is also a superimposed axial component that continuously supplies cold fluid.

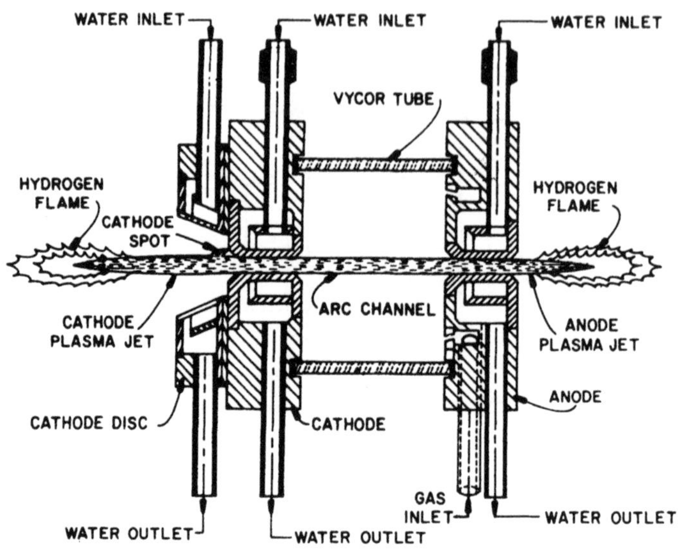


**FIG. 1.8.** Schematic of a gas vortex-stabilized arc.

Various gases and gas mixtures are used as working fluids in actual applications of vortex-stabilized arcs. Figure 1.8 shows a schematic of a gas vortex-stabilized arc arrangement developed for the generation of fully ionized, atmospheric-pressure hydrogen plasmas. Both electrodes in this arrangement are water-cooled. The working fluid enters tangentially at the anode through small orifices. The vortex generated in this way confines the arc to the center of the arc chamber, reducing the arc diameter to approximately 2–3 mm. The intense convective cooling of the arc fringes due to the vortex flow around the arc enhances the power dissipation per unit length of the arc column which, in turn, results in high axis temperatures. In this type of vortex-stabilized hydrogen arc, axis temperatures close to 25,000 K have been reached.

By superimposing an axial flow on an otherwise unstable arc, stability can be achieved. Convective heat transfer from the arc to the cold gas shroud surrounding the arc plays essentially the same role as conduction in the case of a wall-stabilized arc.

#### 1.2.1.4. Magnetically Stabilized Arcs

Since an arc is an electrically conducting medium, it will interact not only with its own magnetic field but also with externally applied magnetic fields. This interaction has attracted increasing interest over the past 20 years because of its potential for many arc applications. Magnetically influenced or stabilized arcs are used extensively in the development of arc gas heaters for material processing, in circuit breakers, in arc furnaces, etc. According to the existing literature, the interaction of arcs with magnetic fields can be divided into the following categories:

- Magnetic stabilization of arcs in cross flow
- Magnetically deflected arcs
- Magnetically driven arcs

The first category refers to arcs exposed to strong cross flows in such a way that the arc column bends in the downstream direction if the electrode roots are fixed. At the same time, arc length and arc voltage drop increase, and for a sufficiently strong flow the required arc voltage may surpass the available voltage, i.e., the arc extinguishes. In order to stabilize an arc in this situation, a magnetic field may be applied so that the drag force exerted on the arc by the flow is balanced by the $\vec{j} \times \vec{B}$ force. Experimental studies on magnetically balanced arcs in cross flow indicate that the magnetic field strength required for balancing the arc is proportional to $v^2$, where $v$ is the gas speed; i.e., the arc behaves like a solid body as far as aerodynamic drag is concerned. The cross section of the arc assumes the shape of an ellipse with the major axis normal to the flow. Without the balancing $\vec{j} \times \vec{B}$ force, the major axis of the ellipsoidal cross section of the arc is in the direction of the flow. These findings have been confirmed by a number of other studies that have included detailed analyses. Many refined experimental studies have also discussed balanced arcs in supersonic flow.

The second category deals with magnetically deflected arcs and the secondary effects induced by this deflection. Although the applied magnetic fields exert a strong influence on the arc, the primary stabilizing effects of an arc enclosed in a tube is due to confining walls. Thus this type of arc may also be classified as a magnetically influenced, wall-stabilized arc.

The interaction of curved arcs with their own magnetic field can produce effects similar to those observed in magnetically deflected arcs.

Magnetically driven arcs are also classified as magnetically stabilized arcs, as previously explained. Such arcs have been extensively studied from a more basic point of view in connection with the phenomenon of retrograde motion. With regard to applications involving arc gas heaters, the coaxial, magnetically rotated arc has been of great interest for efficient heating of gases to well-controlled temperature levels in the range from $3 \times 10^3$ to $6 \times 10^3$ K.

### 1.1.2.5. Self-Stabilized Arcs

The transition from a low-intensity to a high-intensity arc, which occurs above currents of 50 A at atmospheric pressure, manifests itself in a drastic change in the stability of the arc column. Below 50 A the arc column is subject to irregular motion induced by free convection effects. For arc currents in the range from 50 to 100 A the column becomes suddenly motionless and stiff with a visually well-defined boundary. The previously mentioned cathode jet phenomenon gives rise to this transition. The maximum velocity in the jet is related to total current and current density at the cathode, as follows:

$$v_{max} \sim (Ij)^{1/2} \tag{1.16}$$

As soon as this velocity substantially exceeds those induced by free convection effects (on the order of 1 m/s), the described transition will occur. The current at which this transition takes place depends on the conditions in the cathode region (current density and variation of current density in the axial direction). The transition is usually reversible, i.e., by lowering the current, a transition to the free-convection-dominated low-intensity arc occurs.

The well-known bell shape of the free-burning high-intensity arc is observed when the cathode jet impinges on an anode normal to the cathode axis. Figure 1.9 shows calculated and measured isotherms of such an arc.[8]

Another type of self-stabilization is found in extremely short arcs (arcs with electrode gaps on the order of 1 mm). The behavior of such arcs is determined by the proximity of the electrodes and the prevailing steep axial gradients. The arc column as such does not exist in this situation. The remaining arc consists of the nonuniform electrode regions which in contrast to a fully developed arc column, reveal strong axial gradients of the plasma properties.[9] These regions can be considered thermal boundary layers and in this situation may even partially overlap. The contour of the remaining part of the arc approaches the shape of an ellipse with the electrode roots as focus points.

## 1.2.2. Thermal RF Discharges

An RF discharge can be maintained either by capacitive or inductive coupling with the power source. In capacitive coupling the high-frequency electric field is responsible for maintaining the discharge; thus this type of discharge is known as the E discharge. In contrast, an inductively coupled discharge is maintained by the time-varying magnetic field and is denoted as an H discharge.

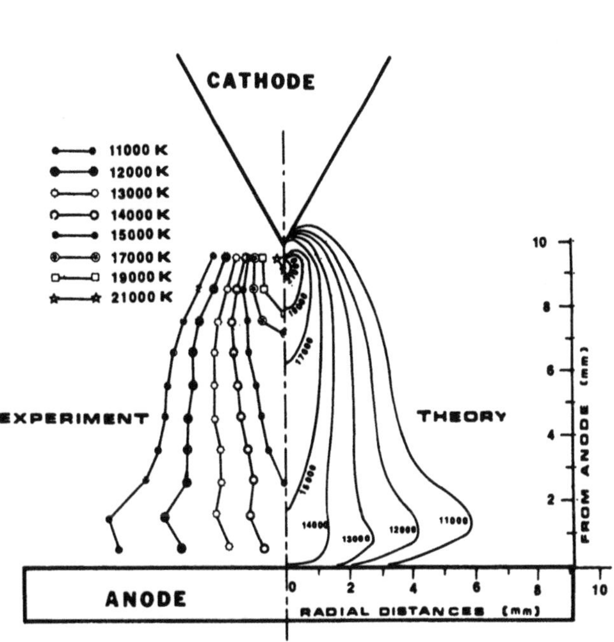


FIG. 1.9. Measured and calculated isotherms in a high-intensity arc.

For the generation of thermal plasmas, the H discharge is by far more important, because the E discharge, which relies on displacement currents for establishing a closed electrical circuit, requires extremely high frequencies for producing a thermal plasma.

In the following discussion it will be assumed that the plasma in the tube sketched in Fig. 1.10 is already established. The time-varying magnetic field caused by the RF current flowing in the coil induces an electric ring field, which in turn drives a current of density $\vec{j}$, since the plasma is a conducting medium. This induced current consists of closed loops directed opposite to the primary current in the coil. The current

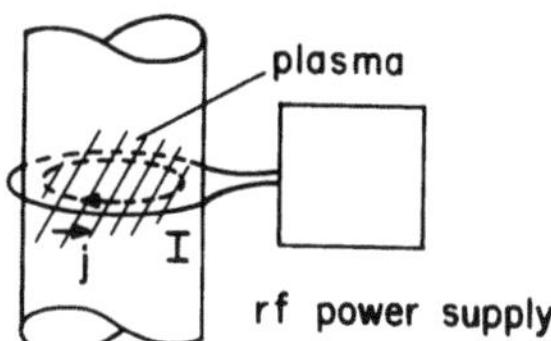


FIG. 1.10. Schematic of an inductively coupled RF discharge.

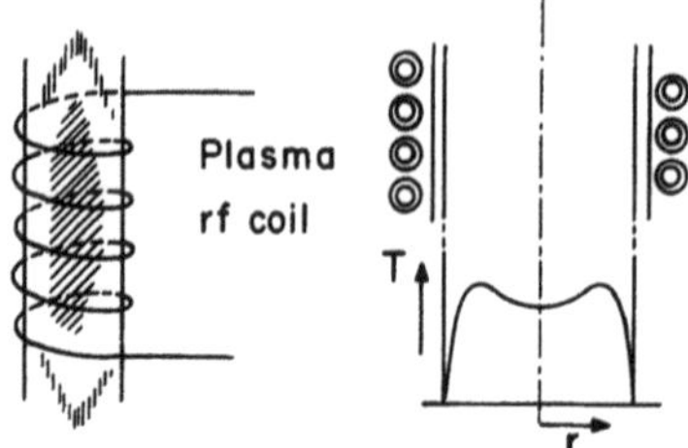


**FIG. 1.11.** Plasma appearance and schematic temperature distribution in a RF discharge.

distribution and the associated temperature distribution show off-axis peaks as indicated in Fig. 1.11.

The off-axis peak of the current density is primarily due to the skin effect. Since the plasma has a relatively high electrical conductivity, the alternating magnetic field cannot penetrate the plasma, especially at very high frequencies. This phenomenon is usually quantified in terms of the skin depth $\delta$, which is a measure for the penetration of the RF field into the plasmas. The skin depth is given by

$$\delta = \left(\frac{1}{\pi f \sigma_e \mu}\right)^{1/2} \tag{1.17}$$

where $\sigma_e$ represents the electrical conductivity of the plasma, $f$ is the frequency of the power supply, and $\mu$ is the permeability of the plasma. Although Eq. (1.17) holds strictly only for plane materials with uniform electrical conductivity, this equation is useful for examining trends. An increase in the frequency reduces the skin depth, as does an increase in the electrical conductivity.

The nonuniform current density distribution with an off-axis peak close to the walls of the discharge vessel gives rise to enhanced heat dissipation ($j^2/\sigma_e$) close to the wall. This effect, in combination with radiative cooling of the core of the plasma, causes an off-peak in the temperature distribution as indicated in Fig. 1.11.

The H discharge can be operated in a wide frequency range from approximately 100 kHz to almost 100 MHz. It has been suggested that this frequency range (or wavelength range: $\lambda = c/f$ where $c$ is the speed of light) be divided into the following intervals, assuming a characteristic length of $L = 1$ m.

$$\frac{\lambda}{L} \gg 100 \quad (f \ll 3\,\text{MHz}) \text{ low-frequency discharge}$$

$$100 > \frac{\lambda}{L} > 10 \quad (3\,\text{MHz} < f < 30\,\text{MHz}) \text{ high-frequency discharge}$$

$$\frac{\lambda}{L} \ll 10 \quad (f \gg 30\,\text{MHz}) \text{ ultrahigh frequency discharge}$$

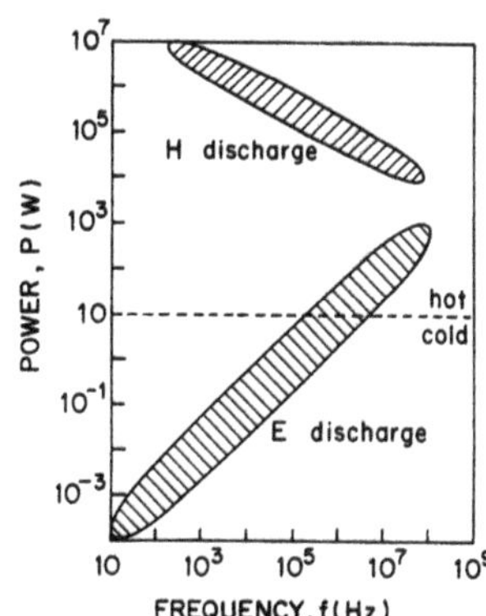


**FIG. 1.12.** Power dissipation in the H and E discharge.

As the dissipated power in the plasma increases, the plasma approaches LTE, i.e., the state of a thermal plasma. Figure 1.12 shows qualitatively how the power dissipation per unit volume varies as a function of the frequency. The figure also shows a comparison with the E discharge.

Stability considerations discussed in a review by Eckert[10] seem to indicate that the skin effect has a stabilizing influence on the H discharge provided that the ratio $R/\delta \leq 1.75$, where $R$ is the discharge radius and $\delta$ the skin depth [Eq. (1.17)]. More details on RF inductively coupled plasmas can be found in the general bibliography at the end of this chapter.

### 1.2.3. Microwave Discharges

Although electric arcs and inductively coupled RF discharges play a dominant role as thermal plasma sources, microwave discharges are also considered to be viable plasma sources. Conventional microwave discharges require that the discharge be an integral part of the microwave circuit. This requirement imposes inherent limitations on the flexibility of the discharge parameters, in particular on the configuration and sizc of the plasma volume.

The past 15 years have seen the development of more flexible microwave devices that make use of *electromagnetic surface waves or traveling wave discharges* (TWD) for sustaining a plasma. According to a recent survey titled "Plasma Sources Based on the Propagation of Electromagnetic Surface Waves,"[11] plasmas are sustainable with this approach for frequencies ranging from 1 MHz to 10 GHz, pressures from $10^{-3}$ Pa to several hundred kPa, and discharge tube diameters from 0.5 to 150 mm. Plasmas produced by TWD are stable, reproducible, and quiescent, with low fluctuations of electron density. As in the case of

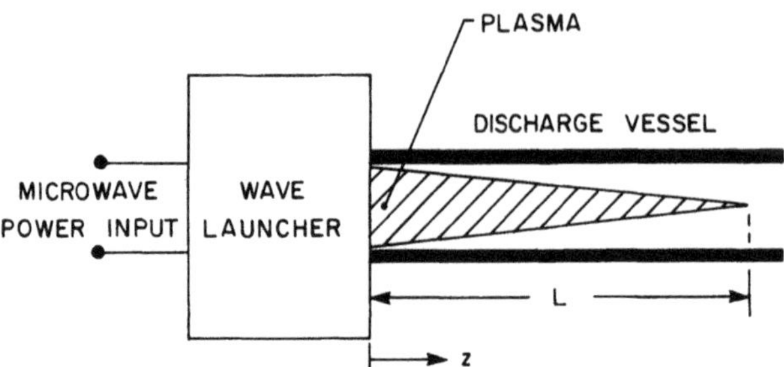


**FIG. 1.13.** A traveling wave discharge (TWD).

electric arcs and inductively coupled RF discharges, the electrons in a microwave discharge are primarily responsible for the absorption of energy from the electric field. This implies that operation at low pressures ($p < 10\,\text{kPa}$) will result in strong deviations from kinetic equilibrium ($T_e \gg T_h$) because of the poor collisional coupling between electrons and heavy particles at reduced pressures.

The following discussion will consider only higher pressure levels ($p > 10\,\text{kPa}$), at which plasmas produced by TWD may approach a state of LTE. Figure 1.13 shows a schematic diagram for a TWD. The key component is the wave launcher, which includes an impedance matching network and field applicator. The surface wave is excited by the wave launcher, which surrounds a small part of the discharge tube. After breakdown of the initially nonconducting gas in the discharge tube (accomplished by an RF spark or other means of preionization), the surface wave initiated by the wave launcher travels along the interface between the plasma and the discharge tube, supplying power for sustaining the discharge. The length over which microwave power is transferred to the plasma is usually large compared to the plasma diameter. Since the traveling wave loses energy continuously as it moves along the tube, less and less energy is supplied to the plasma with increasing distance from the wave launcher, i.e., the plasma shows strong nonuniformities that are typical of plasmas produced by TWD. Figure 1.14 shows the qualitative behavior of the wave power flux and the electron density as a function of the distance from the wave launcher. At $z = L$, the power flux drops to zero, resulting in a rapid decay of the electron density at this location.

It should be pointed out that there is an intimate interaction between the wave and the plasma. The wave supplies energy to the plasma, but without the plasma, the wave could not exist. In a steady-state situation there is a local balance between the power supplied by the wave and the power lost from the plasma.

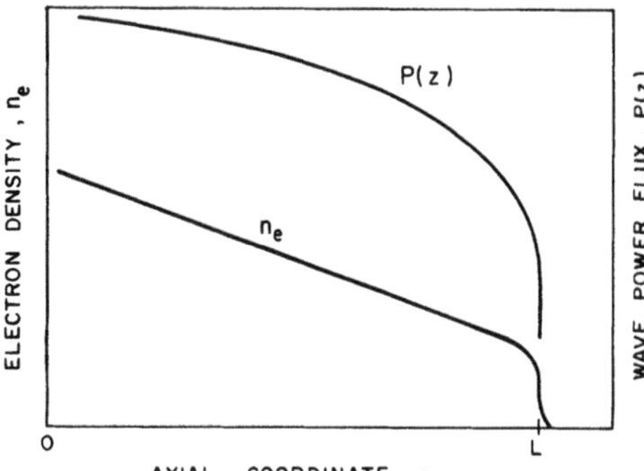


**FIG. 1.14.** Power flux and electron density (schematically) in a TWD.

Recently, a simplified model has been developed for an atmospheric-pressure TWD,[12] and this model has been applied to a discharge in nitrogen. Figure 1.15 shows in normalized form the behavior of the heat flux potential

$$S(r, z) = \int \kappa \, dT \tag{1.18}$$

and the electron density, $n_e(r, z)$; $r$ and $z$ are radial and axial coordinates, respectively, and $\kappa$ represents the thermal conductivity. The hatched areas ($n_e$) correspond approximately to the luminous region of the plasma.

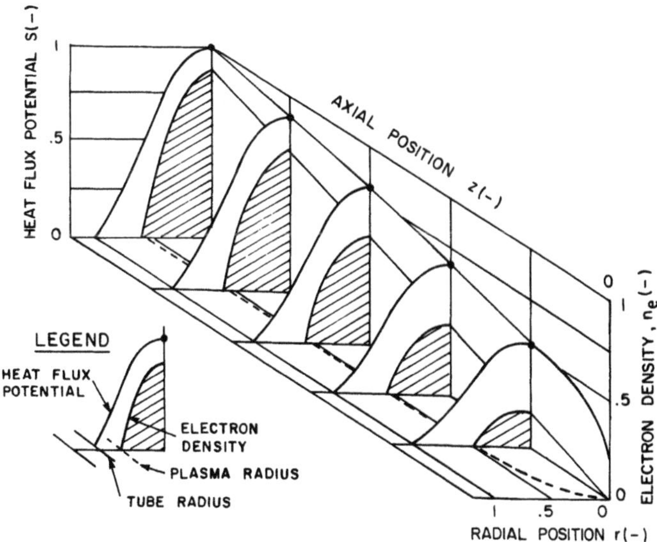


**FIG. 1.15.** Heat flux potential and electron density in a TWD.

Owing to the relatively high losses of atmospheric-pressure plasmas (mainly by conduction to the surrounding vessel), a steady-state TWD at atmospheric pressure requires a correspondingly high power input, which is limited, however, by the heat flux that the discharge vessel can tolerate. The maximum wall heat flux occurs close to the wave launcher.

In terms of wall heat fluxes, there is a certain similarity between a wall-stabilized arc and a TWD. This similarity, however, does not apply to the plasma, which shows axial uniformity in the fully developed regime of a wall-stabilized arc but would be classified as a traveling-wave-supported, decaying plasma in the case of the TWD.

## 1.3. PROPERTIES OF THERMAL PLASMAS

With the availability of high-speed computers, the modeling of thermal plasmas, plasma reactors, and thermal plasma processing became an extremely important research tool. The prerequisite for any modeling work, however, is a data base of thermodynamic and transport properties.

In this section, a brief overview of the most important properties of thermal plasmas will be given, starting with the *plasma composition.* As will be shown later in this section, *thermodynamic and transport properties* depend directly on the plasma composition.

A comprehensive discussion of plasma properties and of the basis for calculating such properties will be reserved for Chapters 6, 7, and 8.

### 1.3.1. Plasma Composition

For the sake of simplicity, the following considerations will be based on thermal plasmas that contain only one type of ion, namely, singly ionized atoms. If such a plasma is generated from a monatomic gas (for example, argon), then only three species compose the plasma—electrons, neutral argon atoms (some of them may be in an excited state), and positive argon ions (again, some may be in an excited state).

$$\mathrm{Ar} \leftrightarrow \mathrm{Ar}^{+} + e$$

The plasma composition in this situation is described by a set of equations: the Eggert–Saha equation, Dalton's law, and the condition for quasineutrality of the plasma:

$$\frac{n_e n_i}{n} = \frac{2Q_i}{Q}\left(\frac{2\pi m_e kT}{h^2}\right)^{3/2} \exp\left(-\frac{E_i}{kT}\right) \tag{1.19}$$

$$p = (n_e + n_i + n)kT \tag{1.20}$$

$$n_e = n_i \tag{1.21}$$

In the Eggert–Saha equation (1.19), $n_e$ is the electron number density, while $n_i$ and $n$ represent ion and neutral number densities, respectively, regardless of whether the ions and neutrals are in an excited state or in the ground state; $Q_i$ and $Q$ are the partition functions of the ions and neutrals, respectively, $h$ is Planck's constant, and $E_i$ represents the ionization energy. The partition functions (or sum over all states) are given by

$$\begin{aligned} Q_i &= \sum_s g_{i,s} \exp(-E_{i,s}/kT) \\ Q &= \sum_s g_s \exp(-E_s/kT) \end{aligned} \tag{1.22}$$

where $g_{i,s}$ and $g_s$ are the statistical weights of the energy levels of the ions and neutrals, respectively, while $E_{i,s}$ and $E_s$ are the corresponding energy levels of their excited states. The equations for the partition functions imply that the populations of the excited states follow a Boltzmann distribution.

The Eggert–Saha equation can be derived from thermodynamic principles (minimization of Gibbs free energy), and therefore it can be considered a "mass action law" for the ionization process. It should be pointed out that the ionization energy $E_I$ requires a correction term, $-\delta E_I$, which accounts for the lowering of the ionization energy due to the electric microfields in a plasma. These microfields are primarily a function of the charged particle densities.

For a given pressure, Eqs. (1.19)–(1.21) permit the calculation of the plasma composition as a function of temperature. Since the previously mentioned ionization-energy correction term is primarily a function of the electron (or ion) density, a few iterations are necessary to calculate $n_e(T) = n_i(T)$ and $n(T)$. Figure 1.16 shows, as an example, the composition of a thermal argon plasma at a pressure of 100 kPa. Since the pressure is kept constant, the total particle number density $n_t = n_e + n_i + n$ decreases with increasing temperature.

If a plasma is generated from a molecular gas (for example, nitrogen), the number of possible species comprising the plasma will be increased due to the presence of molecular species. The chemical processes that may occur in the plasma will include dissociation of molecules into atoms and ionization of some atoms. The formation of molecular ions will be neglected. The dissociation process in a nitrogen plasma

$$N_2 \leftrightarrow N + N$$

can be described by an equation similar to the Eggert–Saha equation,

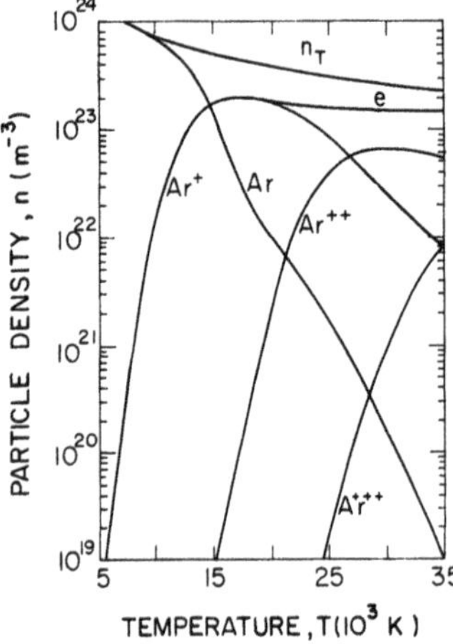


**FIG. 1.16.** Composition of an argon plasma at 100 kPa.

i.e., the mass action law for the dissociation process. Taking dissociation, ionization, and the presence of additional species into account, the composition of the nitrogen plasma can be calculated. The results for a nitrogen plasma at $p = 100\,\text{kPa}$ are shown in Fig. 1.17. For $T > 10^4\,\text{K}$, nitrogen molecules are no longer present due to dissociation, and ionization of nitrogen atoms reaches a peak around $T = 1.5 \times 10^4\,\text{K}$. For temperatures $T > 2 \times 10^4\,\text{K}$, the plasma is, in practical terms, fully ionized, i.e., the number density of atoms becomes negligible.

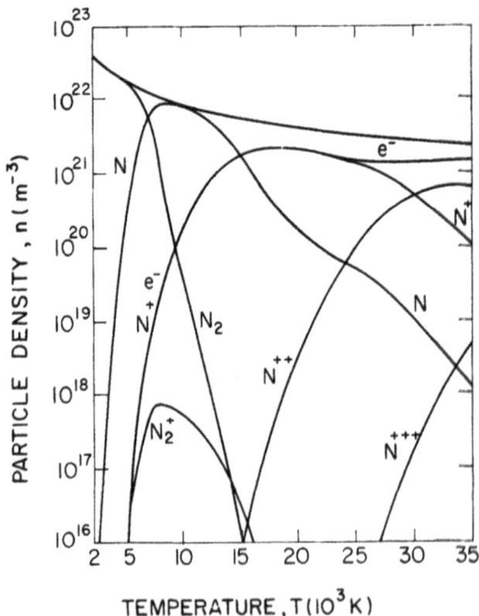


**FIG. 1.17.** Composition of a nitrogen plasma at 100 kPa.

Similar calculations are feasible for plasmas generated from more complex molecules and from gas mixtures. They will be discussed further in Chapter 6.

### 1.3.2. Thermodynamic Properties

The thermodynamic properties of plasmas include the mass density, the internal energy, the enthalpy, the specific heat, and the entropy. In addition, there are derived thermodynamic functions: the Helmholtz function (free energy) and the Gibbs function (free enthalpy or chemical potential).

The mass density $\rho$ follows directly from the plasma composition as

$$\rho = \sum_i n_i m_i \tag{1.23}$$

where $n_i$ refers to the number density of the various species present in the plasma and $m_i$ represents the corresponding mass.

As an example, Fig. 1.18 shows the mass density of a nitrogen plasma at $p = 100$ kPa. Similar calculations can be conducted for more complex plasmas, including plasmas produced from gas mixtures (see Chapter 6).

The other thermodynamic functions, including the derived functions, can be calculated from the *partition functions*, which play a crucial role

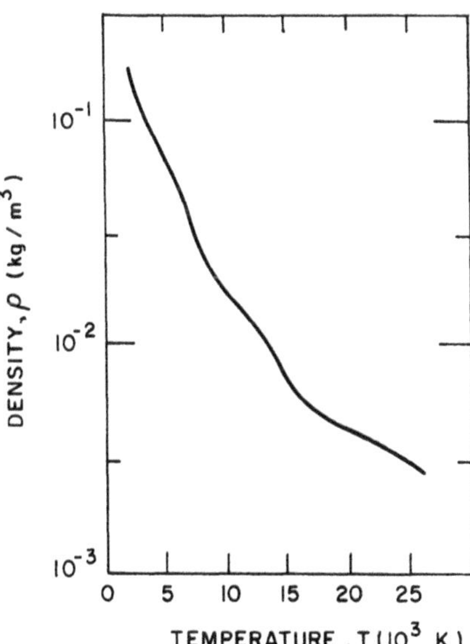


**FIG. 1.18.** Mass density of a nitrogen plasma at 100 kPa.

in the evaluation of thermodynamic functions. For this reason, the evaluation of partition functions will be briefly discussed, along with the underlying basic assumptions for their derivation.

#### 1.3.2.1. Partition Functions

The partition functions establish the link between the coordinates of microscopic systems and macroscopic thermodynamic properties. In general, the partition function of a particle can be expressed as

$$Q = \sum_s g_s \exp(-E_s/kT) \tag{1.24}$$

where $E_s$ represents all forms of energy that a particle can assume and $g_s$ accounts for the degeneracy or statistical weight of each energy level.

It is customary to divide the energy of a particle into translational energy ($E_{s,\mathrm{tr}}$) and internal energy ($E_{s,\mathrm{int}}$), i.e.,

$$E_s = E_{s,\mathrm{tr}} + E_{s,\mathrm{int}} \tag{1.25}$$

These energies are associated with the corresponding translational and internal degrees of freedom of a molecule. The latter include electronic excitation, rotation, vibration, nuclear spins, and chemical reactions. In the Born–Oppenheimer approximation that is valid for gases and plasmas, the total internal energy of a molecule can be expressed as the sum of all the previously mentioned energies. Thus the total partition function $Q_t$ of a molecule can be expressed by a simple product

$$Q_t = Q_{\mathrm{tr}} \cdot Q_{\mathrm{rot}} \cdot Q_{\mathrm{vib}} \cdot Q_{\mathrm{el}} \cdot Q_{\mathrm{nucl}} \cdot Q_{\mathrm{ch}} \tag{1.26}$$

where the individual partition functions represent translational, rotational, vibrational, electronic, nuclear and chemical contributions, respectively.

The translational partition function can be obtained by integration over all spatial and momentum coordinates of a molecule to give

$$Q_{\mathrm{tr}} = \frac{V}{h^3}(2\pi mkT)^{3/2} \tag{1.27}$$

where $V$ is the volume of the system and $m$ the mass of the molecule.

The evaluation of the internal partition function for atoms is rather straightforward, because atoms do not have rotational and vibrational

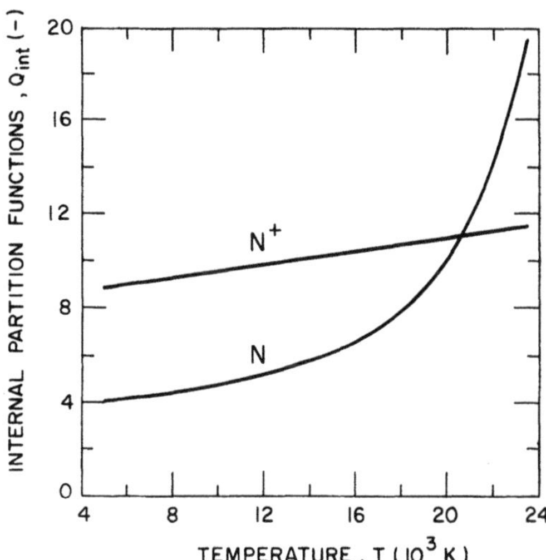


**FIG. 1.19.** Partition functions of nitrogen atoms and ions.

degrees of freedom and the chemical contribution is the ionization process with a single energy level $E_I - \Delta E_I$. Therefore, the partition function of an atom can be expressed as

$$Q_t = \frac{V}{h^3}(2\pi mkT)^{3/2} \cdot Q_{\text{el}} \cdot Q_{\text{nucl}} \cdot \exp[-(E_I - \Delta E_I)/kT] \quad (1.28)$$

where

$$Q_{\text{el}} = \sum_s g_s \exp(-E_s/kT) \quad (1.29)$$

accounts for electronic excitation; $Q_{\text{nucl}}$ is associated with the spin of the atomic nucleus. As an example Fig. 1.19 shows the partition function of nitrogen atoms and ions.

The evaluation of rotational and vibrational contributions to the total partition function of molecules will be reserved for Chapter 6.

In general, the translational contribution to the total partition function of a molecule or atom is by far the largest contribution. Among the contributions to the internal partition function of a molecule, the rotational and vibrational contributions dominate.

#### 1.3.2.2. Thermodynamic Functions

As previously mentioned, the internal energy, enthalpy, specific heat, entropy, Helmholtz function, and Gibbs function of a plasma can be calculated from the corresponding partition functions and the previously

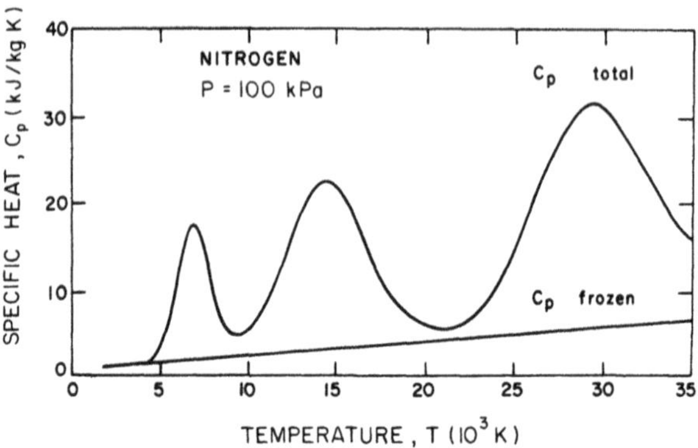


**FIG. 1.20.** Specific heat of a nitrogen plasma at 100 kPa.

discussed plasma composition. The derivation of these expressions will be given in Chapter 6.

As an example of a typical thermodynamic function, Fig. 1.20 shows the specific heat of a nitrogen plasma at constant pressure ($p = 100$ kPa). The pronounced peaks around $7 \times 10^3$ K and $1.5 \times 10^4$ K are associated with chemical reactions in the nitrogen plasma. In the process of dissociation, which occurs between $5 \times 10^3$ and $10^4$ K, the capacity of the plasma for storing energy (dissociation energy) is substantially enhanced. Similarly, the energy storage capacity of a nitrogen plasma is enhanced between $10^4$ and $2 \times 10^4$ K due to ionization.

As a result of the strong contributions of chemical reactions to the specific heat, it is customary to separate these contributions from all other components of the specific heat ("frozen" chemistry) so that the total specific heat can be expressed by

$$c_p = c_{pf} + c_{pr} \tag{1.30}$$

where $c_{pf}$ is the "frozen" part and $c_{pr}$ the reactional contribution to the specific heat.

Similar considerations apply to the other thermodynamic functions.

### 1.3.3. Fluxes and Transport Properties

Many theoretical considerations in plasma physics are based on the assumption of uniform plasmas. It is, however, very difficult—if not impossible—to produce such uniform plasmas. Actual plasmas will reveal gradients in such characteristics as particle number densities ($n$), applied electrical potentials ($V$), temperatures ($T$), and velocity components ($v_x$). These gradients can be considered "driving forces" that give rise to fluxes. If the magnitude of these gradients remains within certain limits, there will be linear relationships between the driving forces and the fluxes. Examples of such relationships are:

| | | |
|---|---|---|
| Fick's law | $\vec{\Gamma} = -D \operatorname{grad} n$ | (1.31) |
| Ohm's law | $\vec{j} = -\sigma_e \operatorname{grad} V$ | (1.32) |
| Fourier's law | $\vec{q} = -\kappa \operatorname{grad} T$ | (1.33) |
| and | $\vec{f}_x = -\mu \operatorname{grad} v_x$ | (1.34) |

where $\vec{\Gamma}, \vec{j}$, and $\vec{q}$ refer to fluxes due to diffusion, electrical conduction, and thermal conduction, respectively. The term $\vec{f}_x$ represents the frictional force in the $x$ direction. These linear relationships between fluxes and driving forces incorporate the so-called transport coefficients $D$, $\sigma_e$, $\kappa$, and $\mu$, which are known as the diffusion coefficient, the electrical conductivity, the thermal conductivity, and the viscosity, respectively.

Energy and momentum, for example, are transferred between particles by collisions. Thus sufficient details of the collision processes between particles must be known in order to determine transport coefficients, because these coefficients depend on the collision cross sections between particles. Since molecules and atoms have complex electronic structures, a theoretical description of the collisional interaction between particles becomes a formidable task. In many cases, highly simplified models (considering, for example, a molecule or atom as a classic sphere) have been developed for determining collision cross sections, and reference is frequently made to measurements of such cross sections.[5] Unfortunately, the experimental data base is still rather limited. For this reason, transport coefficients, especially in more complex mixtures, are frequently unknown or suffer from a high degree of uncertainty. The problem is further exacerbated by deviations from LTE, which are frequently experienced in actual plasmas. At this point it seems that the establishment of a reliable data base to cover a wide spectrum of gas mixtures including deviations from LTE will require many years.

Transport processes and transport coefficients will be further discussed in Chapter 5, and in Chapter 7 it will be shown that transport coefficients for pure gases and for some simple gas mixtures are available today. A listing of some of them is provided in the appendix at the end of this volume.

The thermal conductivity of nitrogen as a function of temperature is a typical example of a transport coefficient. Figure 1.21 shows the contributions of molecules, atoms, ions, electrons, and chemical reactions to the total value of $\kappa$. As in the case of the specific heat (Fig. 1.20), chemical reactions give rise to pronounced peaks around $7 \times 10^3$ K and

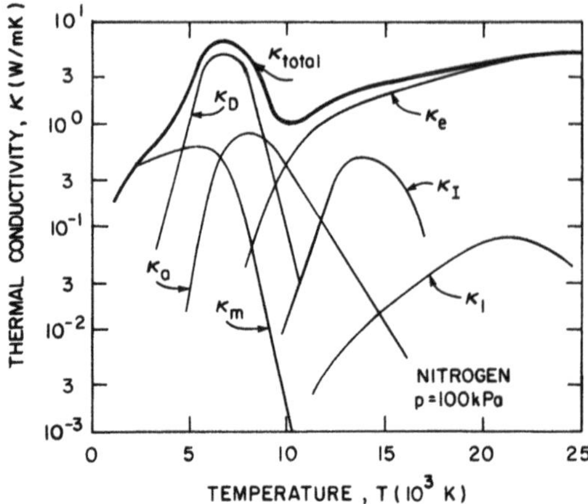


**FIG. 1.21.** Contribution of molecules, $\kappa_m$, atoms, $\kappa_a$, electrons, $\kappa_e$, ions, $\kappa_i$, and chemical reactions; dissociation, $\kappa_D$, and ionization, $\kappa_I$, to the total heat conductivity of a nitrogen plasma at 100 kPa.

$1.5 \times 10^4$ K due to dissociation and ionization, respectively. In an acutal nitrogen plasma, these peaks explain the transport of dissociation and ionization energy from regimes of higher to regimes of lower temperature.

Finally, since a thermal plasma is a highly luminous body, radiative transport must be considered. To determine radiative transport coefficients, the various mechanisms responsible for the emission and absorption of radiation in such plasmas must be taken into account. The spectrum from a typical thermal plasma generated from a monatomic gas reveals continuous as well as line radiation. Electronic transitions of excited atoms or ions from higher to lower energy states cause the emission of spectral lines. Since the electron involved in the radiation process remains in a bound state, radiation of this type is also referred to as *bound–bound* radiation. The total energy transport by line radiation is frequently only a small fraction of the total radiation energy from a plasma; the energy transport depends on the number and wavelength of the emitted lines, which in turn depend on the nature of the plasma fluid, in particular, on the number of possible species at a given temperature. The plasma of a given gas may be a "strong" or "weak" line radiator, depending on the plasma density and composition, which are functions of pressure and temperature.

Continuous radiation in a plasma under the previously specified conditions results from recombination of ions with electrons (*free–bound* radiation) and from bremsstrahlung (*free–free* radiation). In the process of radiative recombination, a free electron is captured by a positive ion into a certain bound energy state and the excess energy is converted into radiation. Recombination may occur into all possible energy levels of an ion; thus the number of continuous spectra for a particular species coincides with the number of electronic energy states of this ion. The entire free–bound continuum consists, therefore, of a superposition of all continuous spectra emitted by the different species that exist in the plasma.

Bremsstrahlung has its origin in the interaction of free electrons with other charged particles, i.e., a free electron may lose kinetic energy in the Coulomb field of an ion, and this energy is readily converted into radiation. Since both the initial state and the final state of the electrons are free states in which the electrons may assume arbitrary energies within the Maxwellian distribution, the emitted radiation is of the continuum type.

The total radiation continuum consisting of free–free and free–bound radiation frequently dominates the radiative balance in thermal plasmas ($p \geq 100\,\text{kPa}$).

If molecular species are present in the plasma, the spectrum will also contain radiation bands due to the excitation of vibrational and rotational energy modes of molecules.

The total radiation originating from the various emission mechanisms just described leaves the plasma without appreciable attenuation as long as the plasma can be considered optically thin. This assumption may fail for line and band radiation, as well as for continuum radiation. Very strong absorption occurs, for example, for resonance lines. In general, absorption effects become more pronounced as the pressure increases. Plasmas at very high pressures become optically thick and may approach the radiation intensity of a blackbody radiator if the temperature is sufficiently high. An argon arc, for example, will behave as a blackbody radiator in a certain wavelength range for pressures $p \geq 10^4\,\text{kPa}$ and temperatures $T > 2 \times 10^4\,K$.[13]

Radiation emitted by plasmas has been extensively used for diagnostic purposes. Over the past 30 years plasma spectroscopy has become a highly sophisticated and powerful diagnostic tool that plays an extremely important role in plasma physics.[14–19]

For many years thermal plasmas have been treated as optically thin, even at higher pressures. Only in recent years has the significance of reabsorption of radiation in the plasma been recognized as an important mechanism that may affect other plasma transport properties, particularly the thermal conductivity. Figure 1.22, which shows a schematic diagram of the radiative balance in high-pressure argon arcs,[20] illustrates these effects. At low pressure and/or temperatures, the contribution of radiation to the energy balance is negligible (region I). Since only ohmic heating and heat conduction are involved, the resulting temperature

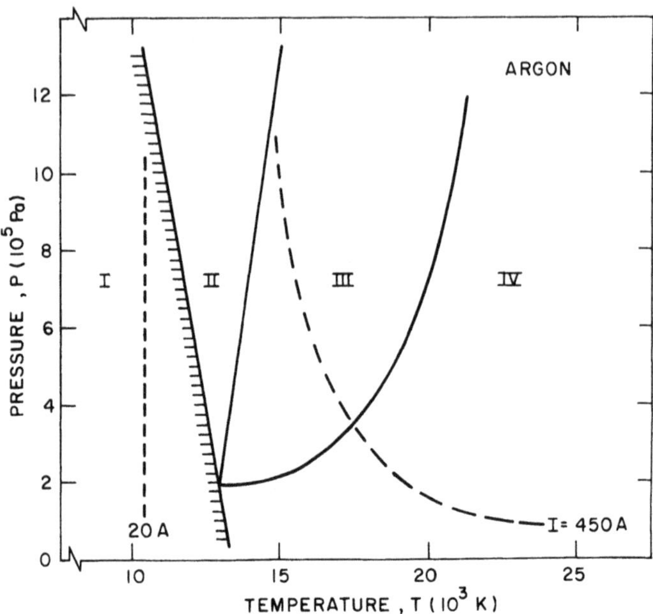


FIG. 1.22. Radiative properties of high-pressure argon arcs.

profiles are relatively narrow. As the temperature increases, radiation can no longer be neglected, particularly at higher pressures.

In general, radiation for wavelengths $\lambda > 2000$ Å can be considered optically thin, while radiation for $\lambda < 2000$ Å (UV) will be partially or totally absorbed in the plasma, depending on the mean free path length of the photons ($\lambda_{ph}$). For UV radiation, $\lambda_{ph}$ is inversely proportional to the absorption coefficient which, in turn, is proportional to the number density of the atoms. Therefore, $\lambda_{ph}$ decreases with increasing pressure for a given temperature. At pressures $p > 300$ kPa and temperatures $T < 15{,}000$ K, $\lambda_{ph} \ll R$ (arc radius), i.e., UV radiation is immediately reabsorbed without contributing to the energy transport in the arc (region II in Fig. 1.22). In this situation the energy balance of the arc is determined by ohmic heating, heat conduction, and optically thin radiation.

At higher temperatures ($T > 15{,}000$ K) in the same pressure region ($p \geq 300$ kPa), $\lambda_{ph} < R$ and the transport of UV radiation can be described by ordinary diffusion of radiation (region III). In this region, the energy balance is governed by ohmic heating, optically thin radiation, and a modified heat conduction term that includes radiative transport (emission and reabsorption of UV radiation).

As the temperature further increases (or at relatively low pressures), $\lambda_{ph} \sim R$ and radiative transport can no longer be described by local properties in the arc. An integral expression is needed that depends not only on local temperature and pressure in the arc but also on the field strength. The resulting radiation term is denoted as "far-reaching diffusion of radiation" (region IV), in contrast to the ordinary diffusion of radiation. Optically thin radiation as well as the previously described far-reaching radiation exert a strong influence on the temperature profiles of arcs. Radiative energy transport toward the arc fringes enlarges the arc diameter, and in the case of confined (wall-stabilized) arcs this energy transport leads to almost rectangular shapes for the temperature profiles.

It is interesting to note that for a given pressure, energy transport by radiation increases sharply with temperature. At axis temperatures of 26,000 K, for example, approximately 95% of the energy input to the arc core of an atmospheric-pressure nitrogen arc is dissipated by radiation.[21] For temperatures above 13,000 K, the emission and reabsorption of radiation play a governing role in the energy transport within atmospheric-pressure nitrogen arcs.[22] A similar situation is to be expected for other working gases or gas mixtures, especially at higher pressure levels.[23]

## 1.4. THERMAL PLASMA TECHNOLOGY

Materials and materials processing will be one of the more important technical issues for the remainder of this and the beginning of the next century. This issue will not be restricted to the development of new materials but will also include the refining of materials, the conservation of materials (by hard facing, coating, etc.), and the development of new processing routes that are more energy efficient, more productive, and less damaging to our environment. Thermal plasma technology will play an important role in these developments. Its potential for developing new materials-related technologies is increasingly recognized, and many research laboratories all over the world are now advancing the frontiers of our knowledge in this exciting field. An interesting example of the utility of thermal plasma processing has been recently demonstrated in connection with the high rate deposition of diamond films.[24]

Since the second volume of this treatise will be entirely devoted to thermal plasma technology, this section will merely summarize known thermal plasma processes that are already in use or close to becoming commercialized.

### 1.4.1. Plasma Deposition

Today, plasma deposition of coatings and films is probably the fastest growing area in thermal plasma technology. Plasma deposition includes *plasma spraying, thermal plasma chemical vapor deposition* (*TPCVD*), and *thermal plasma physical vapor deposition* (*TPPVD*).

#### 1.4.1.1. Plasma Spraying

Figure 1.23 shows a schematic of an atmospheric-pressure plasma spray (APS) arrangement using a DC plasma torch. A high-intensity arc is operated between a stick-type cathode and a nozzle-shaped water-cooled anode. Plasma gas introduced along the cathode is heated by the arc to plasma temperatures, and leaves the anode nozzle as a plasma jet or plasma flame. Fine powder suspended in a carrier gas is injected into the plasma jet, where the powder particles are accelerated and heated.

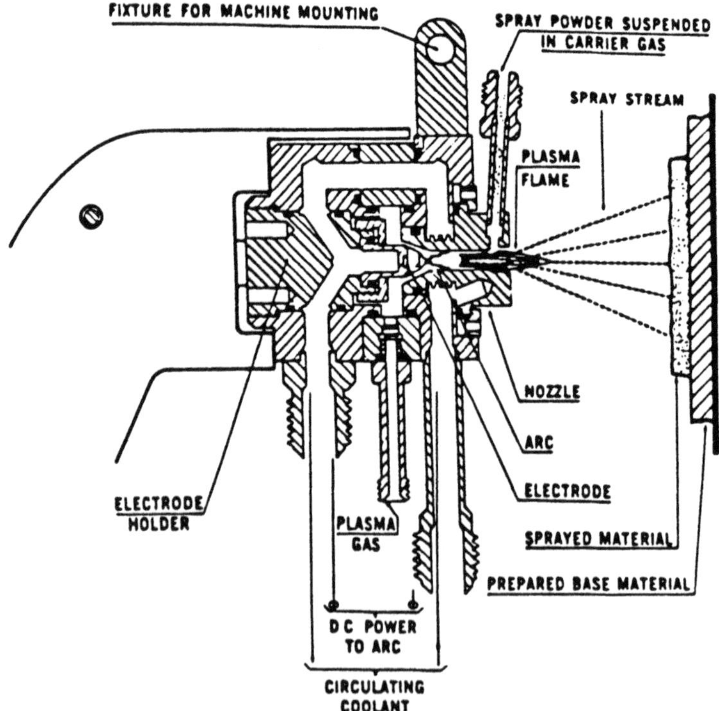


**FIG. 1.23.** An atmospheric-pressure plasma spray arrangement.

As the molten powder particles impinge with high velocities on a substrate, they form a more or less dense coating.

Besides DC torches, RF plasma torches may also be used for this spray process. In addition, plasma spraying at reduced pressures ($\approx$10 kPa) has attracted considerable attention over the past 10 years, especially for coating of aircraft engine parts.

Today, applications of plasma spraying include corrosion-, temperature-, and abrasion-resistant coatings and production of monolithic and near net shapes that also take advantage of the rapid solidification process. Powders of glassy metals can be plasma sprayed without changing their amorphous characteristics. Recently, high-temperature superconductive materials have been deposited by the plasma spray process.

Although plasma spraying has been known for more than 30 years, our knowledge of the underlying fundamentals of the plasma spray process is still incomplete, particularly with regard to the behavior of the plasma jet, the interaction of the plasma with the injected powder particles, and the coating formation.

#### 1.4.1.2. Thermal Plasma Chemical Vapor Deposition (TPCVD)

TPCVD is a relatively new technology with great potential for future applications. In the TPCVD process, the high energy densities of thermal plasmas are used to generate high-density vapor-phase precursors for the deposition of thick films. The cooled substrate is brought into intimate contact with the plasma, and the deposition precursors traverse a nonequilibrium boundary layer to nucleate at the substrate surface. In this way high-quality films can be obtained with high densities and high uniformity of the crystal orientation in the different grains. Compared to conventional CVD or plasma-enhanced CVD deposition techniques, TPCVD offers considerably higher deposition rates (orders of magnitude higher) because of the higher density of the precursor species. Compared to plasma spraying, TPCVD has the advantage of offering better control over the film quality at the expense of a lower deposition rate. In addition, TPCVD allows the deposition of materials without a liquid phase, such as carbon and many carbides, and adjustment of the film stoichiometry in the deposition of chemical compounds. In principle, normal plasma spray equipment can be used for TPCVD; liquid- or vapor-phase injection of the reactants or precursors is preferred to solid particle injection. Injection of particles requires sufficiently long precursor residence times in the hot plasma region to allow complete evaporation of the particle.

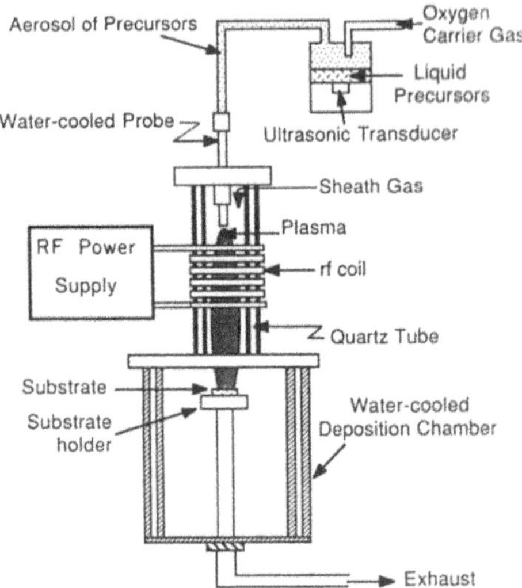


**FIG. 1.24.** Setup for rf induction plasma deposition of superconducting films.

As an example, Fig. 1.24 shows a schematic arrangement for TPCVD of superconducting films using an RF-generated plasma.[25] The system consists essentially of three parts; (1) an RF thermal plasma reactor, (2) a liquid precursor atomizing and feeding system that includes a water-cooled stainless-steel aerosol feeding probe and an ultrasonic nebulizer, and (3) a helium-cooled stainless-steel substrate holder inside the deposition chamber. Yttrium, barium, and copper nitrates dissolved in distilled water were used as the starting reagents. The liquid precursors were atomized with the ultrasonic nebulizer and were introduced into the RF thermal plasma using oxygen as the carrier gas. Once in the plasma, the liquid precursors evaporated, decomposed, and reacted in the boundary layer over the substrate or on the substrate to form the oxide films. In addition to superconducting and other refractory oxide films, TPCVD lends itself to the formation of carbide, nitride, boride, and even diamond films. The equipment required for TPCVD is essentially the same as that for plasma spraying.

#### 1.4.1.3. Thermal Plasma Physical Vapor Deposition (TPPVD)

In this process the plasma is merely the heat source for evaporating materials to be deposited on cooled substrates. It is used, for example, in vacuum arcs where the metallic vapor generated from the electrodes serves as the arc operating fluid and also deposits as a film on the cooled substrates.

### 1.4.2. Plasma Synthesis of Fine Powders

Synthesis of fine powders in thermal plasmas is closely related to the previously discussed TPCVD processes. In powder synthesis, however, rapid quenching of the vapor is necessary in order to nucleate particles before the vapor impinges on the cooled walls of the plasma reactor. The supersaturation of the vapor species due to high quench rates provides the driving force for particle nucleation. High quench rates lead to the production of ultrafine particles (down to nm size) by homogeneous nucleation. Ceramic powders such as carbides, nitrides, oxides, and solid solutions have been successfully synthesized in thermal plasma reactors.

The equipment used for this purpose, including the plasma reactors, is similar to that employed for TPCVD. Recent novel advances in reactor and process designs have enhanced the quality of powders produced in thermal plasmas. These novel designs include the RF–DC hybrid plasma reactors, the reactive submerged arc (RSA) reactor, the multiple plasma jet reactor, and the counterflow liquid-injection plasma reactor. These new techniques aim to maximize the heating, the mixing, and the residence time of materials in the plasma. Either the discharge itself or the plasma flame downstream of the discharge can be used for synthesizing the powders. In thermal plasma synthesis the reactants can be gases, liquids, or solids before injection into the plasma; however, the availability of gas-phase precursors for metals is severely limited, so the most commonly used reactants for plasma synthesis have therefore been solids. In this case the injection process may constitute a severe problem because of the high viscosity of the plasma. Recently, a liquid-injection method has been developed to overcome the problems associated with solid injection and to capitalize on the benefits of gaseous reactants.

For more details, the reader is referred to a survey on the plasma synthesis of ceramics.[26]

### 1.4.3. Thermal Plasma Decomposition

Any material exposed to thermal plasmas at their typical temperatures ($T \geq 10^4$ K) will be decomposed, possibly into its elemental constituents. Although this decomposition process may be of interest in breaking down extremely stable chemical compounds, its primary attraction seems to be in the field of toxic waste destruction. Compared to competing technologies, thermal plasma processing offers a number of advantages:

- The high temperatures cause fast and complete pyrolysis of organic hazardous wastes, as well as melting and possible vitrification of inorganic wastes, resulting in a volume reduction and encapsulation of nondestructible wastes.
- The high energy density obtainable in a plasma reactor allows the use of smaller installations for comparable waste throughputs, thus reducing capital costs and favoring construction of small mobile units.
- the use of electric arcs to generate the high-temperature gas reduces the total gas throughput and with it the need to heat excess gas for combustion; the consequence is a reduction in the required capacity for off-gas treatment systems. A wider choice of process gases is also possible, thus improving control over the process chemistry.
- The reduced system size and the high energy density allow rapid start-up and shut-down times.
- The ultraviolet radiation emitted by the plasma can lead to process enhancement in some waste-destruction processes, such as pyrolysis of organic chlorides.

In view of all these potential advantages of the plasma-based processes, it is not surprising that numerous development efforts are being pursued all over the world for utilization of thermal plasmas in waste-processing applications. Some large-scale industrial developments have been reported over recent years[27–29] and Fig. 1.25, for example, shows a plasma furnace for medical-waste pyrolysis.[30] The furnace heat source consists of a 300-kW plasma arc torch operated in the transferred mode, with the second electrode being a pool of molten iron maintained at about 2000 °C by the plasma torch. The furnace throughput is approximately 250 kg of hospital wastes per hour. The medical waste is charged into the top of the furnace, and the down-moving waste in the vicinity of the plasma arc column is pyrolyzed to form stable gases such as hydrogen, carbon monoxide, carbon dioxide, methane, etc. The metallic and siliceous fractions of the waste are melted, with the light fraction floating on top and the metallic fraction joining the molten iron pool. At predetermined time intervals, the plasma torch is moved to melt a path for tapping the molten materials, which consist of metal and slag.

### 1.4.4. Plasma Metallurgy

The application of thermal plasmas to metallurgical processing has a long and colorful history. The extremely high temperatures feasible in electric arcs were recognized almost 200 years ago, and in 1815 the first attempts to use arcs for melting refractory materials were reported.

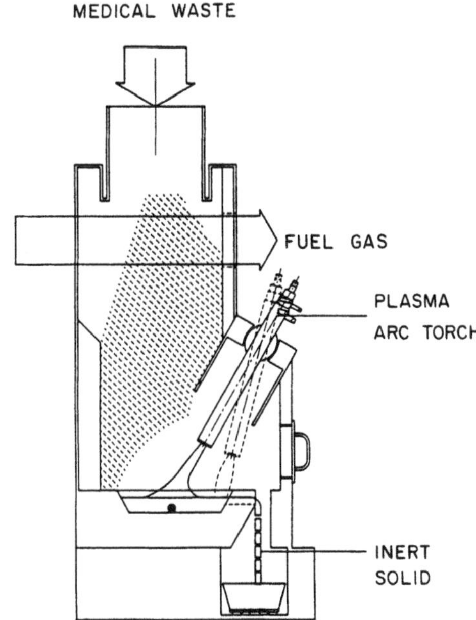


**FIG. 1.25.** Plasma furnace for medical-waste pyrolysis.

In contrast to conventional arc furnaces with consumable cathodes, which have been known for almost 100 years, newer developments—known as arc-plasma or plasma furnaces—employ plasma torches with nonconsummable cathodes in a controlled furnace atmosphere. There are several advantages to this approach, including reduced noise levels, elimination of graphite electrode consumption, and higher yields, which are particularly important when processing valuable alloying elements such as nickel and molybdenum.

Besides melting and remelting applications, plasma metallurgy also includes extractive metallurgy. Both types of applications will be briefly discussed here.

### 1.4.4.1. Melting and Remelting

A wide variety of plasma torches are in use today or projected for future use for scrap melting, alloying (ladle heating), iron melting in cupolas, tundish heating, and other melting/remelting applications. These plasma torches operate with DC or AC either in the nontransferred or in the transferred mode, and at power levels up to almost 10 MW. In the case of nontransferred arcs, the plasma torch is essentially an arc gas heater producing extremely hot gases that emanate from the torch in the form of a plasma jet. The more common approach, however, makes use

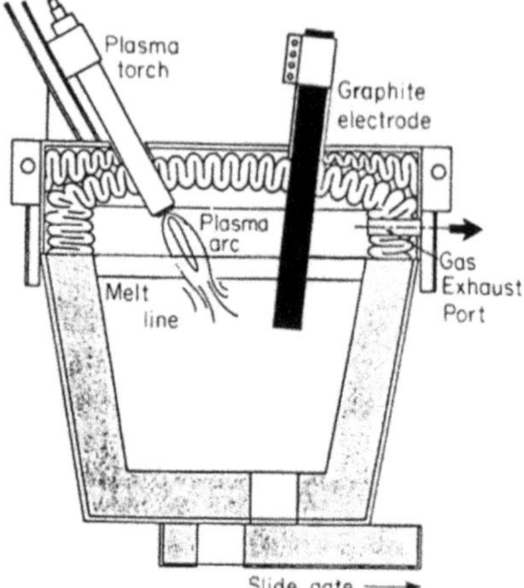


**FIG. 1.26.** High-power plasma-torch arrangement for ladle heating.

of transferred arcs, in which the molten pool serves as one of the electrodes and the major energy input is at the arc root of the molten bath surface.

A typical example of such a torch used for ladle heating is shown in Fig. 1.26. The DC arc is operated in the transferred mode, requiring a return electrode (graphite rod). The main objective in this application is to hold steel in the ladle at constant or slightly increasing temperature levels.

an interesting development based on plasma furnaces operated with AC arcs has been reported by Krupp.[31] These furnaces have been designed for melting scrap, production of alloy steels, and reheating steel in tundishes or ladles. In contrast to the previously discussed plasma surface, the frequently troublesome bottom electrode is no longer required and, compared with conventional arc furnaces, these furnaces have higher yields, lower noise levels, and reduced flicker. A schematic of the plasma torch used in such furnaces is shown in Fig. 1.27. The thermionically emitting hot electrode consists of tungsten and has been designed to withstand AC operation.

#### 1.4.4.2. Extractive Metallurgy

Extractive metallurgy refers to the extraction of metals in either pure or alloyed form from their respective ores. The increasing world demand for metals and alloys over the past 30 years has exerted a strong influence on processing technologies, in particular on plasma processing.

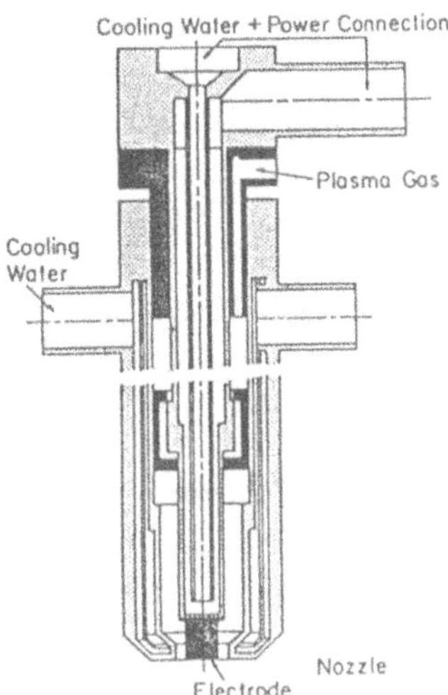


**FIG. 1.27.** High-power plasma for AC operation.

Thermal plasma reactors for smelting, melting, or refining operations represent highly concentrated heat sources that allow for high processing rates per unit reactor volume. In spite of this interesting feature, there have been very few industrial-scale applications of plasma reactors in extractive metallurgy. This fact may be explained by the required production size of smelting systems. A modern economic steel smelter, for example, must operate at approximately 30 ton/h; this rate requires a plasma furnace capacity of close to 100 MW.[32] Even with multiple torch operation, this power level is beyond present technology. Nevertheless, a number of potential applications have attracted considerable interest, including the smelting of virgin ores, calcination, preheating of gases and feed stock to augument existing processes, and specialized systems for recovering metals from baghouse dust and from other waste materials. Some of the applications have been extensively researched and, in some instances, pursued to the pilot-plant state or even to full-scale industrial installations.[32]

### 1.4.5. Plasma Densification

Various materials, in particular refractories, can be densified in thermal plasmas. The most important plasma densification processes are *spheroidization* and *sintering.*

### 1.4.5.1. Spheroidization

Heating irregularly shaped powder particles in a thermal plasma results in melting of those particles. Surface tension in the liquid phase causes the molten particles to form spherical droplets and, upon quenching, the droplets freeze into spherical or nearly spherical powder particles that may be denser than the starting material.

Fine particles are spheroidized commercially for a variety of applications. A wide range of materials have been spheroidized, including oxides and carbides,[32] to produce materials with controlled porosity, catalysts, abrasives, and free-flowing powders for spray applications.

Plasma densification and spheroidization of presintered agglomerates (derived from spary-drying) have been used to produce dense, spherical powders of metals such as W and Mo and carbide–metal mixtures such as W–Co.

### 1.4.5.2. Plasma Sintering

In contrast to plasma spheroidization and densification of powders, plasma sintering is at this time still in the laboratory stage.

Compared with conventional technology, sintering of high-tech ceramics in thermal plasmas has the potential to drastically reduce the time period required for this process. In addition, plasma sintering offers the opportunity to restrain grain growth and to tailor heat transfer during the sintering process to obtain desirable structures and properties in the sintered materials.

Plasma sintering is a pressureless sintering process that can cover a pressure range from 100 kPa to a few kPa. At pressures below 10 kPa, the plasma may no longer be classified as a thermal plasma because of substantial deviations from LTE.

The essential characteristic of plasma sintering and of any other sintering process is an increase in density and strength of a cold pressed powder compacted sample upon heating.

Rapid sintering of ceramics has been observed by various investigators using gaseous plasmas produced by microwaves, glow discharges, or RF discharges. Extensive work by Johnson *et al.*[34] using a 5-MHz RF argon plasma shows rapid sintering of various aluminas. They also achieved rapid sintering of MgO and doped alumina using three different plasma devices, namely, a microwave-induced plasma, a hollow cathode discharge, and an RF inductively coupled plasma. The observed product was fine-grained structure of sintered oxide.

The effects of gas composition and pressure on RF plasma sintering of MgO without sintering aids have also been reported. Only a few reports about plasma sintering of nonoxide ceramics such as silicon carbide have been published. Kijima[35] has successfully used an argon RF plasma to sinter silicon carbide, a most difficult ceramic to sinter, to near theoretical density without substantial grain growth.

### 1.4.6. Plasma Welding and Cutting

Using arc-generated thermal plasmas for welding is a well-established technology. Both TIG (tungsten inert gas) and MIG (metal inert gas) welding processes are in wide use today. In TIG welding, a nonconsumable tungsten electrode serves as the cathode and the workpiece is the anode. Inert gas or gas mixtures (Ar, He) are blown along the cathode to prevent contamination from the surroundings. In MIG welding, the arc is maintained between the workpiece and a consumable wire electrode that is fed continuously through the torch at controlled speeds. Inert gas is fed simultaneously through the torch into the weld zone, protecting the weld from the contaminating effects of the atmosphere. Other welding methods make use of submerged arcs. Since these technologies are well covered in the welding literature, they will not be further discussed in this treatise.

Plasma cutting is also a well-established technology but it has seen some new developments over the past years, including air-operated and air-cooled low amperage cutting torches that are extensively used in automotive repair shops. Another more recent development considers high-current cutting torches for underwater cutting. This technology may become important in the dismantling of nuclear power plants that are beyond their useful life span.

In arc cutting, the specific heat fluxes at the workpiece are at least one order of magnitude higher (typically in the range from 100 to 200 $kW/cm^2$) than for arc welding. This difference implies that the arc for arc cutting must be extremely constricted, resulting in high current densities and correspondingly high temperatures in the axis of the arc, even at relatively low amperages ($\leq$20 A).

## LIST OF SYMBOLS

$c_p$ specific heat at constant pressure
$D$ diffusion coefficient

| | |
|---|---|
| $E$ | energy (eV) |
| $E_i$ | ionization energy (eV) |
| $f_x$ | frictional force in $x$ direction (N/m$^2$) |
| $g$ | statistical weight |
| $h$ | Planck's constant ($6.6 \times 10^{-34}$ W s$^2$) |
| $j$ | current density (A/m$^2$) |
| $k$ | Boltzmann constant ($1.38 \times 10^{-23}$ J/K) |
| $l$ | mean free path length (m) |
| $m, m'$ | mass of a particle (kg) |
| $n$ | number density of particles (m$^{-3}$) |
| $p$ | pressure (Pa) |
| $Q$ | partition function |
| $q$ | heat flux (W/m$^2$) |
| $R$ | arc radius (m) |
| $r$ | radial coordinate (m) |
| $S$ | heat flux potential (W/m) |
| $T$ | absolute temperature (K) |
| $V$ | electric potential (V) |
| $v$ | thermal velocity of a particle (m/s) |
| $\bar{v}$ | average thermal velocity (m/s) |
| $v_d$ | drift velocity of a particle (m/s) |
| $v_m$ | most probable thermal velocity (m/s) |
| $v_{\text{rms}}$ | root-mean-square thermal velocity (m/s) |

## Greek Symbols

| | |
|---|---|
| $\Gamma$ | particle flux (m$^{-2}$ s$^{-1}$) |
| $\delta$ | change of quantity |
| $\kappa$ | thermal conductivity (W/mK) |
| $\kappa_a$ | contribution of atoms to the thermal conductivity (W/mK) |
| $\kappa_D$ | contribution of dissociation to the thermal conductivity (W/mK) |
| $\kappa_e$ | contribution of electrons to the thermal conductivity (W/mK) |
| $\kappa_i$ | contribution of ions to the thermal conductivity (W/mK) |
| $\kappa_I$ | contribution of ionization to the thermal conductivity (W/mK) |
| $\kappa_m$ | contribution of molecules to the thermal conductivity (W/mK) |
| $\lambda$ | wavelength of radiation (nm) |
| $\lambda_{\text{ph}}$ | mean free path length of photons (nm) |
| $\mu$ | dynamic viscosity (kg/ms) |
| $\xi$ | degree of ionization |
| $\rho$ | mass density (kg/m$^3$) |
| $\sigma_e$ | electrical conductivity (ohm$^{-1}$ m$^{-1}$) |

$\tau$ free flight time (s)

## Subscripts

| | |
|---|---|
| ch | chemical |
| $e$ | electrons |
| el | electronic |
| $f$ | frozen |
| $h$ | heavy particles |
| $i$ | ions |
| int | internal |
| kin | kinetic |
| nucl | nuclear |
| $p$ | at constant pressure |
| $r$ | refers to component $r$; reactional |
| rot | rotational |
| $s$ | excited state |
| tr | translational |
| $v$ | between $v$ and $v + dv$ |
| vib | vibrational |
| $x$ | in $x$ direction |

## GENERAL BIBLIOGRAPHY

Boulos, M. I., "The Inductively Coupled R.F. (Radio Frequency) Plasma," *Pure Appl. Chem.* **57,** 1321 (1985).
Brown, S. C., Jr., *Basic Data of Plasma Physics,* New York: Wiley, 1959.
Cambel, A. B., *Plasma Physics and Magneto Fluid Mechanics,* New York: McGraw-Hill, 1963.
Chen, F. C., *Introduction to Plasma Physics,* New York: Plenum Press, 1974.
Cobine, J. D., *Gaseous Conductors,* New York: Dover, 1958.
Eckert, H. U., "The Induction Arc: A State-of-the-Art Review," *High Temp.* **6,** 99–134 (1974).
Engle, A. V., *Ionized Gases,* 2nd ed., Oxford: Clarendon Press, 1965.
Griem, H. R., *Plasma Spectroscopy,* New York: McGraw-Hill, 1964.
*Handbook of Physics,* Vol. XXII, Berlin: Springer-Verlag, 1956.
Huddlestone, R. H. and S. L. Leonard, eds., *Plasma Diagnostic Techniques,* New York: Academic Press, 1965.
Lochte-Holtgreven, W., ed., *Plasma Diagnostics,* Amsterdam: North-Holland, 1968.
Mitchner, M. and C. H. Kruger, *Partially Ionized Gases,* New York: Wiley, 1973.
Pfender, E., "Electric Arcs and Arc Gas Heaters," Chapter 5 in *Gaseous Electronics, Vol. 1,* New York: Academic Press, 1978.
Somerville, J. M., *The Electric Arc,* New York: Wiley, 1959.

## REFERENCES

1. A. M. Howatson, *An Introduction to Gas Discharges* (Oxford: Pergamon Press, 1965): 22.
2. H. A. Dinulescu and E. Pfender, *J. Appl. Phys.* 51 (1980): 3149.
3. C. H. Chang and E. Pfender, *Plasma Chem. and Plasma Process.* 10 (1990): 473.
4. H. Maecker, *Z. Phys.* 141 (1955): 198.
5. E. Pfender, "Electric Arcs and Arc Gas Heaters," Chapter 5 in *Gaseous Electronics,* Vol. 1, M. N. Hirsh and H. J. Oskam, Eds. (New York: Academic Press, 1978): 291–398.
6. H. Edels, "Properties of the High Pressure Ultra High Current Arc" (Proc. of the Eleventh Int. Conf. on Phenomena in Ionized Gases, Prague, Czechoslovakia, Czechoslovak Academy of Sciences, Institute of Physics, 18040 Prague 8, Na Slovance 2, CSSR, 1973).
7. O. Schoenherr, *Elektrotechn. A.* 30 (1909); 365.
8. K. C. Hsu, K. Etemadi, and E. Pfender, *J. Appl. Phys.* **54** (1983): 1293.
9. G. Ecker, "Electrode Components of the Arc Discharge," *Erg. D. ExaktenNaturwiss.* (1961): 1.
10. H. U. Eckert, *High Temp. Sci.* 6 (1974): 99.
11. M. Moisan and Z. Zakrzewski, "Plasma Sources Based on the Propagation of Electromagnetic Surface Waves," *J. Phys. D: Appl. Phys.* **24,** 1025 (1993).
12. H. Nowakowska, Z. Zakrzewski, and M. Moisan, *J. Phys. D: Appl. Phys.* 22 (1990): 789.
13. W. Finkelnburg and Th. Peters, "*Kontinuierliche spektren,*" in *Encyclopaedia of Physics,* Vol. 28 (Berlin: Springer-Verlag, 1957).
14. W. Finkelnburg and H. Maecker, "*Electric Arcs and Thermal Plasmas,*" in *Encyclopaedia of Physics,* Vol. 22 (Berlin: Springer-Verlag 1956): 254.
15. H. R. Griem, *Plasma Spectroscopy* (New York: McCraw-Hill, 1964).
16. R. H. Huddlestone and S. L. Leonard, eds., *Plasma Diagnostic Techniques* (New York: Academic Press, 1965).
17. W. Lochte-Holtgreven, ed., *Plasma Diagnostics* (Amsterdam: North-Holland, 1968).
18. G. V. Marr, *Plasma Spectroscopy* (Amsterdam: Elsevier, 1968).
19. H. R. Griem, *Spectral Line Broadening by Plasmas* (New York: Academic Press, 1974).
20. K. Kopainsky, *Z. Phys.* 248 (1971): 417.
21. W. Hermann and E. Schade, *Z. Phys.* 233 (1970): 333.
22. K. A. Ernst, J. Kopainsky, and J. Mentel, *Z. Phys.* 265 (1973): 253.
23. U. H. Bauder and D. L. Bartelheimer, *IEEE Trans. Plasma Sci. PS-1* 4 (1971): 23.
24. N. Ohtake and M. Yoshikawa, *J. Electrochem. Soc.* 137 (1990): 717.
25. H. Zhu, Y. C. Lau and E. Pfender, *J. Appl. Phys.* 69 (1991): 3404.
26. P. C. Kong and E. Pfender, "Plasma Synthesis of Ceramics—A Review," in *Materials Processing—Theory and Practices* (Amsterdam: Elsevier, 1992).
27. T. G. Barton and J. A. Mordy, *Can. J. Physiol. Pharmacol.* 62 (1984): 976.
28. J. V. R. Heberlein, W. G. Melilli, S. V. Dighe, and W. H. Reed, "Adaptation of Non-Transferred Plasma Torches to New Applications of Plasma Systems" (Proc. of Workshop on Industrial Plasma Applications, M. Boulos, ed., Pugnochiuso, Italy, 1989): 1.
29. R. C. Eschenbach, "Use of Plasma Torches for Melting Special Metals and for Destroying and Stabilizing Hazardous Waste" (Proc. of Workshop on Industrial Plasma Applications, M. Boulos, ed., Pugnochiuso, Italy, 1989): 127.
30. S. L. Camacho, "Plasma Pyrolysis of Medical Waste" (Proc. of First Int. EPRI Plasma Symp., 1990).

31. D. Neuschuetz, *Iron and Steel Engineer,* 23 (May 1985).
32. National Materials Advisory Board, "Plasma Processing of Materials," (National Academy Press, USA, 1985).
33. V. M. Slepstov, A. M. Proshedromirskaya, and A. M. Taranets, *Russ. Metall. Fuels* 7 (1967): 113.
34. D. L. Johnson, V. A. Kramb, and D. Lynch, *Emergent Process Methods for High Technology Ceramics,* MRS 17, R. F. Davis, H. Palmour III, and R. L. Porter, eds. (New York: Plenum Press, 1982).
35. K. Kijima, (Proc. 8th Int. Symp. on Plasma Chemistry, Tokyo, Japan, 1987): 1632.

Chapter 2

# Basic Atomic and Molecular Theory

This chapter will cover only the basic aspects of atomic and molecular theory that are necessary for an understanding of the following chapters. It begins with a description of Bohr's atomic model, the hydrogen atom, and its eigenfunctions. The structure of more complex atoms is discussed next. A brief presentation on diatomic molecules is given at the end of the chapter.

## 2.1. ATOMIC MODELS

Before atomic theories were developed, it was assumed that an atom was an elementary particle that represented the smallest unit of mass (in Greek, *atomos* means "not divisible"). Today, we know that atoms consist of other elementary particles such as protons, neutrons, and electrons.

Atomic spectra were already known, long before atomic models were introduced, and there was no doubt that there must be a close relationship between atomic structure and atomic spectra. The first atomic model, conceived by Rutherford around the end of the nineteenth century, assumed that the mass of an atom is concentrated in a positively charged nucleus, with electrons revolving around this nucleus. Rutherford's model was based on the results of his scattering experiments, which involved $\alpha$-rays on thin foils, and on independent studies by Lennard, who used electrons for such scattering experiments. These experiments clearly demonstrated that atoms consist mostly of empty space, with nuclear diameters typically on the order of $10^{-14}$ m and atomic diameters (diameters of the electron orbits) on the order of $10^{-10}$ m.

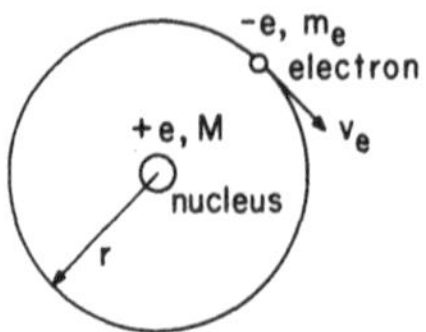


FIG. 2.1. Rutherford model of a hydrogen atom.

For simplicity, we will first consider the hydrogen atom, (Fig. 2.1). According to Rutherford, the centrifugal force of the revolving electron is balanced by the Coulomb attraction (Fig. 2.1), i.e.,

$$\frac{m_e v_e^2}{r_1} = \frac{e^2}{4\pi\varepsilon_0 r_1^2} \tag{2.1}$$

where $m_e = 9.1 \times 10^{-31}$ kg is the mass of an electron, and $v_e$ and $e$ are its velocity and electric charge ($e = 1.6 \times 10^{-19}$ C), respectively. The electron orbit is assumed to be circular with a radius $r_1$; $\varepsilon_0$ is the dielectric constant of a vacuum ($\varepsilon_0 = 8.86 \times 10^{-12}$ As/Vm).

According to classical concepts, the positive nucleus and the revolving electron would be thought of as a time-varying dipole that should radiate constantly, emitting a continuous spectrum. As a result, the atom would lose energy continuously and, at the same time, $r_1$ would have to decrease until the electron finally fell into the nucleus. It was obvious that classical physics could explain neither the typical line spectra of atoms nor the existence of stable atoms. It was not until 1913 that Niels Bohr found the key to understanding the line spectra of atoms and the existence of stable atoms.

### 2.1.1. Bohr's Model

Bohr established a number of ingenious postulates that severely limited the validity of classical physics for atomic systems. Twelve years later, the validity of these postulates was theoretically confirmed by Heisenberg and Schrödinger through quantum mechanics. Bohr could show that his postulates, listed below, eliminated the shortcomings of the classical atomic model.

(a) An electron moves in a circular orbit around the nucleus (Fig. 2.1), in such a way that the Coulomb attraction and the centrifugal force are balanced (classical mechanics).

(b) "Quantum orbits" correspond to definite energy states ($E$); there is no radiation given off in these states.
(c) The innermost quantum orbit with the smallest possible radius ($r_1$) belongs to the normal atomic state (ground state).
(d) To bring an electron into a higher orbit, a certain amount of energy, the "excitation energy" for that particular orbit, must be expended. After an average time of approximately $10^{-8}$ s in the higher orbit, the electron "jumps" spontaneously to an orbit of lower energy and eventually back to the ground state.

If the energy that belongs to an upper level (higher orbit) is $E_u$ and the corresponding energy of a lower orbit is $E_l$, then the transition from the higher to the lower orbit results in the emission of a photon of energy $h\nu_{ul}$, i.e.,

$$E_u - E_l = h\nu_{ul} = \frac{hc}{\lambda_{ul}} \tag{2.2}$$

where $h$ is Planck's constant ($h = 6.6 \times 10^{-34}$ Ws$^2$), $\nu_{ul}$ is the frequency and $\lambda_{ul}$ the wavelength of the emitted photon, and $c$ is the velocity of light in a vacuum ($c = 2.998 \times 10^8$ m/s). For the sake of simplicity, the subscript $ul$ will be deleted in the followng discussion.

According to Bohr's postulates, the allowed orbits are distinguished by a quantum condition, namely,

$$2\pi r(m_e v_e) = nh \qquad \text{with } n = 1, 2, 3, \ldots \text{ (integer)} \tag{2.3}$$

This quantum condition states that the angular momentum [on the left side of Eq. (2.3)] can change only in multiples of $h$, which has the dimension of energy $\times$ time $=$ "action." As a generalization of this finding, we know today that all physical quantities that have the dimension of an action can only change in discrete steps, namely, in multiples of $h$.

### 2.1.2. Line Emission

The consequences of Bohr's theory will now be considered for one-electron systems, i.e., atoms that have only one electron in the outermost electron shell. In such systems the charge and mass of the nucleus and electron are represented by

nucleus: $+Z'e, M$
electron: $-e, m_e$

$Z'$ represents the number of protons in the nucleus and $M$ the mass of such an atom; $Z' = 1$ for a hydrogen atom, $Z' = 2$ for a singly ionized helium atom, $Z' = 3$ for a doubly ionized lithium atom, and so forth.

From classical mechanics we have (Bohr's first postulate)

$$\frac{Z'e^2}{4\pi\varepsilon_0 r^2} = \frac{m_e v_e^2}{r} \tag{2.4}$$

This equation is a generalization of Eq. (2.1). By combining Eq. (2.4) with the quantum condition, Eq. (2.3), we obtain

$$r_n = \frac{4\pi\varepsilon_0 \hbar^2}{Z'e^2 m_e} \cdot n^2 \qquad \text{with } n = 1, 2, 3, \ldots \tag{2.5}$$

for the radii of allowed Bohr orbits where $\hbar = h/2\pi$. The smallest Bohr orbit, $r_1$, belongs to the ground state. Combining Eqs. (2.3) and (2.4) gives the velocity of an electron in a quantum orbit:

$$v_e = \frac{n\hbar}{m_e r_n} = \frac{Z'e^2}{4\pi\varepsilon_0 \hbar} \cdot \frac{1}{n} \qquad \text{with } n = 1, 2, 3, \ldots \tag{2.6}$$

According to Eq. (2.6), electrons can only assume discrete velocities in their corresponding quantum orbits.

For the hydrogen atom in the ground state ($Z' = 1$ and $n = 1$), Eqs. (2.5) and (2.6) give

$$r_1 = 5.3 \times 10^{-11}\ \text{m}$$

$$v_e = 2.2 \times 10^6\ \text{m/s}$$

This velocity is the highest velocity an electron can have in a hydrogen atom ($\approx$1% of the velocity of light).

The total energy of an electron in a quantum orbit consists of two parts: its potential energy and its kinetic energy. The potential energy, $V_P$, is defined as the energy required to bring the electron from infinity to an orbit of radius $r$, i.e.,

$$V_P = \int_\infty^r \frac{Z'e^2}{4\pi\varepsilon_0 r^2}\, dr = -\frac{Z'e^2}{4\pi\varepsilon_0 r} \tag{2.7}$$

The potential energy appears with a negative sign because the Coulomb force between the positive nucleus and the electron is an attractive force.

The corresponding kinetic energy, $T_K$, of an electron in a quantum orbit is given by

$$T_K = \tfrac{1}{2}m_e v_e^2 = \frac{Z'e^2}{8\pi\varepsilon_0 r} \tag{2.8}$$

and the total energy, $E$, is then given by

$$E = V_P + T_K = -\frac{Z'e^2}{8\pi\varepsilon_0 r} \tag{2.9}$$

or

$$E_n = -\frac{(Z')^2 e^4 m_e}{2\hbar^2(4\pi\varepsilon_0)^2}\frac{1}{n^2} \qquad \text{with } n = 1, 2, 3, \ldots \tag{2.10}$$

Equation (2.10) represents the so-called energy eigenvalues of a one-electron system. It indicates that not only the angular momentum but also the total energy of the electron is governed by a quantum condition.

According to Eq. (2.10) we can draw the energy level diagram for the hydrogen atom ($Z' = 1$) as shown in Fig. 2.2. The energy eigenvalues of the various levels are expressed in terms of electron volts with ($1\ \text{eV} = 1.6 \times 10^{-19}\ \text{Ws}$). This figure follows the generally accepted practice of assigning the ground state ($n = 1$) an energy of zero and the ionized state ($n = \infty$) an energy of $(Z')^2 e^4 m_e/2\hbar^2(4\pi\varepsilon_0)^2$. This practice is equivalent to rewriting Eq. (2.10) as

$$E_n' = \frac{(Z')^2 e^4 m_e}{2\hbar^2(4\pi\varepsilon_0)^2}\left(1 - \frac{1}{n^2}\right) \qquad \text{with } n = 1, 2, 3, \ldots \tag{2.11}$$

The lowest energy state of the hydrogen atom ($n = 1$) is the most stable state and is known as the ground state ($E_l = 0$ according to Eq. (2.11)). As $n \to \infty$, the electron is no longer bound to the nucleus—it is liberated

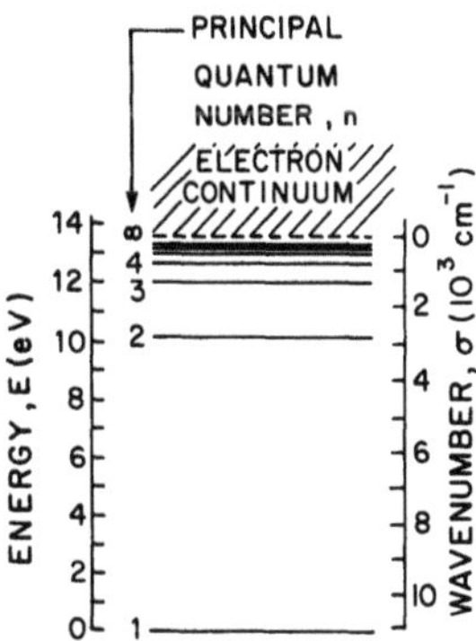


FIG. 2.2. Energy level diagram for the hydrogen atom.

from the atom or ionized. Ionization corresponds to the energy level $E_\infty = 13.6\,\text{eV}$, while levels for which $n > 1$ are defined as *excited states* of the atom.

A transition from a higher quantum orbit ($n_u$) to a lower quantum orbit ($n_l$) leads, according to one of Bohr's postulates, to the emission of a photon of frequency $\nu$, i.e.,

$$\nu = \frac{E_u - E_l}{h} = -\frac{m_e(Z')^2e^4}{(4\pi\varepsilon_0)^2 4\pi\hbar^3}\left(\frac{1}{n_u^2} - \frac{1}{n_l^2}\right) \tag{2.12}$$

where $n_u > n_l$ and both $n_u$ and $n_l$ are integers. In terms of the wave number $\sigma = 1/\lambda = \nu/c$ one finds

$$\sigma = -R_\infty(Z')^2\left(\frac{1}{n_u^2} - \frac{1}{n_l^2}\right) \tag{2.13}$$

where $R_\infty$ is the Rydberg constant given by

$$R_\infty = \frac{m_e e^4}{(4\pi\varepsilon_0)^2 4\pi c\hbar^3} \tag{2.14}$$

The essential predictions of Bohr's theory can be summarized as follows:

(a) The lowest energy state that an atom or ion (particle) can assume is the state for which $n = 1$, known as the ground state.
(b) Excitation of a particle to a level where ($n > 1$) requires energy, the particle can acquire in an electrical discharge.
(c) The excess energy of a particle is generally emitted in the form of radiation. An excited particle returns to the ground state by a single transition or by a series of transitions in which the electron assumes states of successively lower energies.
(d) A very large number of such energy transitions comprises an atomic and/or ionic spectrum, i.e., all possible transitions will occur during observation of such a spectrum. The wave number, $\sigma$, of the emitted spectral lines is given by Eq. (2.13) for all possible combinations of $n_u$ and $n_l$, where $n_u > n_l$.

For the sake of simplicity, the hydrogen atom ($Z' = 1$) will be considered in the following discussion. For a fixed value of $n_l = 2$, Eq. (2.13) becomes

$$\sigma = R_\infty\left(\frac{1}{2^2} - \frac{1}{n_u^2}\right) \qquad \text{with } n_u = 3, 4, 5, 6, \ldots \tag{2.15}$$

This expression describes the Balmer spectrum of the hydrogen atom, provided that $R_\infty$ is identified as $R_H$, the Rydberg constant of the hydrogen atom; $R_H$ is obtained by replacing the electron mass in Eq. (2.14) by the reduced mass $\mu = m_e/(1 + m_e/M)$. This extremely small correction of $R_\infty$ (about 0.05%) accounts for the finite mass of the hydrogen nucleus compared to that of the electron, and brings predictions and experiments into perfect agreement.

In addition to the Balmer spectrum, Bohr's theory predicted other spectral series for the hydrogen-atom:

| | | |
|---|---|---|
| $n_l = 1$, | $n_u = 2, 3, 4, \ldots$ | Lyman series (ultraviolet) |
| $n_l = 2$, | $n_u = 3, 4, 5, \ldots$ | Balmer series (visible) |
| $n_l = 3$, | $n_u = 4, 5, 6, \ldots$ | Paschen series (near-infrared) |
| $n_l = 4$, | $n_u = 5, 6, 7, \ldots$ | Brackett series (infrared) |
| $n_l = 5$, | $n_u = 6, 7, 8, \ldots$ | Pfund series (far-infrared) |

The success of Bohr's theory was particularly impressive because none of the spectral series other than the Balmer and Paschen series were known at the time Bohr developed his theory. The existence of the other spectral series was experimentally confirmed afterward.

The predictions of Bohr's theory work equally well for other one-electron systems such as $He^+$, $Li^{++}$, etc., which have spectra similar to that of the hydrogen atom but with larger wave numbers [as given by Eq. (2.13)].

### 2.1.3. Line Absorption

The previous discussion referred to spectra produced by emission of radiation (*emission spectra*). In *absorption spectra,* absorption of radiation is considered. Such spectra are also easy to understand in terms of Bohr's theory. Atoms or ions can only absorb radiation with appropriate wavelengths or energies. The absorption of radiation of frequency $\nu_{ul}$ will increase the energy level of the atom from its lower excited level $E_l$ to an upper level $E_u$ according to Eq. (2.2). Equation (2.2) can be rewritten in terms of the wave number, $\sigma$, as

$$\sigma = \frac{1}{\lambda} = \frac{E_u - E_l}{hc} \tag{2.16}$$

At room temperature, almost all atoms are in the ground state, so absorption can occur only from this state. This implies that for hydrogen, only the Lyman series will appear in absorption at room temperature. Since absorption occurs from one level only (the ground state), absorption spectra show many fewer lines than emission spectra.

### 2.1.4. Franck–Hertz Experiment

A basic experiment performed by J. Franck and G. Hertz in 1914 proved beyond any doubt the validity of Bohr's theory. Their experimental arrangement, shown in Fig. 2.3, consisted of a discharge tube (triode) with a thermionically emitting cathode, a grid, and an anode. The tube is filled with mercury (Hg) vapor at a pressure of approximately 0.1 Pa ($10^{-3}$ Torr). A variable potential between the cathode and the grid accelerates the thermionically emitted electrons toward the grid. The anode is negatively biased (0.5 V) with respect to the grid.

As the electrons travel toward the grid, collisions with Hg atoms will occur. These collisions, however, will be elastic with negligible energy exchange as long as the energy of the electrons remains below the energy of the first excited level of the Hg atoms. Electrons with energies exceeding 0.5 eV will be able to overcome the negative-bias potential of the anode and will be collected by the anode. An increasing number of electrons will reach the anode as the potential of the grid is raised until the grid potential reaches the potential of the first excited state of the Hg atom (4.9 V). At this point, some of the electrons will suffer inelastic collisions with the Hg atoms, leaving these electrons with energies below 0.5 eV, i.e., these electrons will be repelled by the anode, resulting in a substantial reduction of the electron current collected by the anode.

The measurements made by Franck and Hertz, shown in Fig. 2.4, confirm this behavior. By further increasing the potential of the grid, the electrons will be accelerated until they suffer a second inelastic collision at a grid potential of $2 \times 4.9 = 9.8$ V, accompanied by a corresponding drop of the current collected by the anode. A third drop in the current is observed at $3 \times 4.9 = 14.7$ V (Fig. 2.4), which corresponds to a third inelastic collision between electrons and Hg atoms. The excited Hg atoms should, according to Bohr's postulates, return to the ground state emitting photons of a fixed wavelength

$$\lambda = \frac{hc}{\Delta E} = 253.7 \text{ nm} \tag{2.17}$$

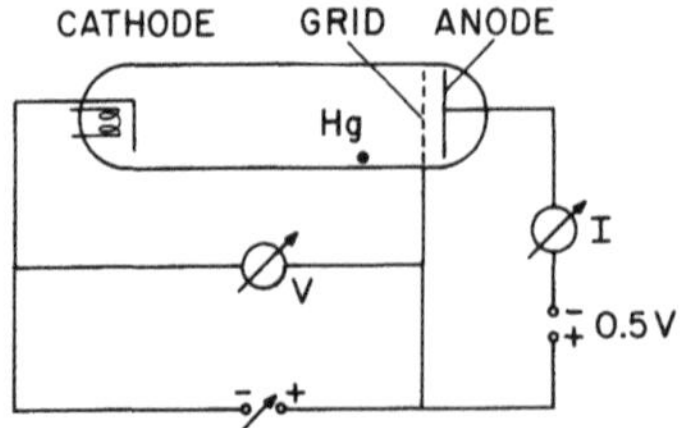


**FIG. 2.3.** Schematic arrangement of the Franck/Hertz experiment.

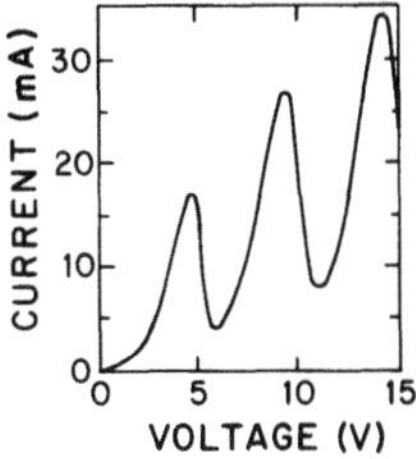


**FIG. 2.4.** Results of the Franck/Hertz experiment.

where $\Delta E = 4.9\,\text{eV}$ represents the energy difference between the first excited state of the Hg atom and its ground state. By observing the light emitted from the space between the cathode and the grid with a spectrometer, Franck and Hertz found indeed a single spectral line at a wavelength of 253.7 nm. This impressive confirmation of Bohr's theory was later extended to higher excitation levels of atoms.

## 2.2. THE HYDROGEN ATOM AND ITS EIGENFUNCTIONS

### 2.2.1. The Schrödinger Equation

The general equation describing the propagation of a wave can be written as

$$\nabla^2\Psi - \frac{1}{u^2}\frac{\partial^2\Psi}{\partial t^2} = 0 \tag{2.18}$$

where $\nabla^2$ represents the Laplace operator, $\Psi$ the wave function, and $u$ the phase velocity of the wave. The last term in Eq. (2.18) is the second derivative of the wave function with respect to time. In Cartesian coordinates the operator $\nabla^2$ can be expressed by

$$\nabla^2 = \frac{\partial^2}{\partial x^2} + \frac{\partial^2}{\partial y^2} + \frac{\partial^2}{\partial z^2} \tag{2.19}$$

Equation (2.18), which is valid for macroscopic systems, will be transformed into an equation (the Schrödinger equation) that holds for microscopic (atomic-scale) systems. For this transformation, quantum mechanical concepts such as the wavelength of matter (de Broglie wavelength) must be introduced. In terms of the de Broglie wavelength

$$\lambda = \frac{h}{mv} \tag{2.20}$$

where $h$ is Planck's constant, $m$ is the mass of a particle, and $v$ its velocity, the phase velocity of a wave can be expressed by

$$u = \lambda \nu = \frac{h\nu}{mv} \tag{2.21}$$

The numerator in Eq. (2.21) represents not only the well-known energy of a photon but also, in terms of de Broglie's postulate, the energy, $E$, of any particle. Since the total energy of a particle is comprised of kinetic and potential energy, one can write the kinetic energy as

$$\tfrac{1}{2}mv^2 = E - V_P \tag{2.22}$$

or

$$m^2v^2 = 2m(E - V_P) \tag{2.23}$$

Substituting Eq. (2.23) into Eq. (2.21) gives

$$u = \frac{h\nu}{[2m(E - V_P)]^{1/2}} \tag{2.24}$$

and substituting Eq. (2.24) into the general wave equation (2.18) leads to

$$\nabla^2\Psi - \frac{2m(E - V_P)}{h^2\nu^2}\frac{\partial^2\Psi}{\partial t^2} = 0 \tag{2.25}$$

In general, the wave function $\Psi$ is a function of location and time, i.e., $\Psi = \Psi(x, y, z, t)$. The wave function can also be expressed as a product of an amplitude $\psi$ and a periodic function in time

$$\Psi(x, y, z, t) = \psi(x, y, z)\exp(-2\pi i\nu t) \tag{2.26}$$

where $i = \sqrt{-1}$ is the imaginary unit.

Substituting Eq. (2.26) into Eq. (2.25) results in the famous Schrödinger equation

$$\nabla^2\psi + \frac{8\pi^2 m}{h^2}(E - V_P)\psi = 0 \tag{2.27}$$

This equation is a homogeneous second-order partial differential equation that describes the steady-state properties of atomic systems.

### 2.2.2. Solution of the Schrödinger Equation

If the potential energy of an atomic system, $V_P(x, y, z)$, is specified, solutions of the Schrödinger equation can be obtained in terms of the so-called "eigenfunctions" $\psi(x, y, z)$ that are a consequence of the boundary conditions of the system. The parameters that appear in these eigenfunctions are the quantum numbers.

Solutions of the Schrödinger equation also define the possible energy states of atomic systems, known as the eigenvalues. For an atomic system the eigenvalues can assume only quantized values, as discussed in the previous paragraph.

The mathematical formalism required for the solution of the Schrödinger equation, even for the simplest atomic system, is rather extensive. In general, the choice of an appropriate coordinate system depends on the geometry of the problem. For example, in the case of the H atom, a spherical coordinate system $(r, \theta, \phi)$ is used since the potential energy depends only on $r$. Substituting the value of $V_P$ given by Eq. (2.7) in Eq. (2.27), the Schrödinger equation for the H atom can be written in spherical coordinates as follows:

$$\frac{\partial^2\psi}{\partial r^2} + \frac{2\partial\psi}{r\partial r} - \frac{1}{r^2}\left[\frac{1}{\sin\theta}\frac{\partial}{\partial\theta}\left(\sin\theta\frac{\partial\psi}{\partial\theta}\right) + \frac{1}{\sin^2\theta}\frac{\partial^2\psi}{\partial\phi^2}\right] + \frac{8\pi^2 m_e}{h^2}\left(E + \frac{e^2}{4\pi\varepsilon_0 r}\right)\Psi = 0 \quad (2.28)$$

It is customary to solve this type of partial differential equation by separation of variables. In this process it is assumed that the solution of Eq. (2.28) can be expressed as a product of three functions $(R, \Theta, \Phi)$, where each function depends on only a single variable, i.e.,

$$\psi(r, \theta, \phi) = R(r) \cdot \Theta(\theta) \cdot \Phi(\phi) \quad (2.29)$$

By introducing Eq. (2.29) into Eq. (2.28), the partial differential equation with the variables $r$, $\theta$, and $\phi$ transforms into three ordinary differential equations, i.e.,

$$\frac{d^2R}{dr^2} + \frac{2}{r}\frac{dR}{dr} + \left[\frac{8\pi^2 m_e}{h^2}\left(E + \frac{e^2}{4\pi\varepsilon_0 r}\right) - \frac{A}{r^2}\right]R = 0 \quad (2.30)$$

$$\frac{1}{\sin\theta}\frac{d}{d\theta}\left(\sin\theta\frac{d\Theta}{d\theta}\right) + \left(A - \frac{m^2}{\sin^2\theta}\right)\Theta = 0 \quad (2.31)$$

$$\frac{d^2\Phi}{d\phi^2} + m^2\Phi = 0 \quad (2.32)$$

with $A$ and $m^2$ as separation constants.

Mathematical considerations show that:

The differential equation for $R(r)$ has only discrete solutions in which the quantum number $l$ takes on integer values between 0 and $n - 1$, i.e.,

$$0 \leq l \leq n - 1 \tag{2.33}$$

The solution of the differential equation for $\Theta(\theta)$ is valid for discrete values of the quantum number $m$ over the range between 0 and $\pm l$, i.e.,

$$m = 0, \pm 1, \pm 2, \ldots, \pm l$$

The solution of the equation for $\Phi(\phi)$ is valid only for the $n$ and $l$ quantum numbers as defined above.

The solutions $\psi(r, \theta, \phi)$, which depend on $n$, $l$, and $m$, are known as the eigenfunctions of the hydrogen atom. The complete solution in nonnormalized form can be expressed as

$$\psi(\rho, \theta, \phi) = \rho^{l} \cdot u_{n-l-1}(\rho) \exp(-\rho/2) P_l^{(m)}(\cos\theta) \exp(im\phi) \tag{2.34}$$

with $\rho = 2r/r_1 n$; $P_l^{(m)}(\cos\theta)$ is a polynomial of degree $l$ calculated according to Rodrigues formula, and $u_{n-l-1}$ is a polynomial of degree $n - l - 1$.

This equation provides a complete description of the behavior of the electron in the stable energy states of the H atom.

### 2.2.3. Quantum Numbers

As we saw in the last section, the eigenfunctions and energy states of the H atom are characterized by the three quantum numbers $n$, $l$, and $m_l$ with

$$n \geq l + 1 \qquad \text{and} \qquad 0 \leq |m_l| \leq l \tag{2.35}$$

It is customary to add a subscript $l$ to the quantum number $m$ to avoid confusing it with the particle mass.

A fourth quantum number of the electron spin, $s$, does not follow from this calculation. According to the Pauli exclusion principle, the states of two electrons in an atom cannot be identical. If $n$, $l$, and $m_l$ are identical for two electrons, then $s$ has to be different. The spin quantum number, $s$, can assume only two values:

$$s = \pm\tfrac{1}{2} \tag{2.36}$$

The equation for the energy eigenvalues of the H atom is governed by the quantum numbers $m_l$ and $l$. For every $l$, there are $2l + 1$ different eigenfunctions. With the restriction that $l \leq n - 1$, one finds that the total number of eigenfunctions for a given energy state is

$$\sum_{l=0}^{n-1} (2l + 1) = n^2 \tag{2.37}$$

This result shows that there can be more than one eigenfunction for one energy state, a property known as degeneracy. Disregarding the electron spin, the energy eigenvalues of the H atom are $(n^2 - 1)$-fold degenerate. Including the electron spin adds a factor of two to the number of eigenfunctions per energy state because of the two possible spin orientations of the electron.

### 2.2.4. Probability Distribution

Before considering the probability distribution for various combinations of quantum numbers, a physical interpretation of the quantum numbers is useful:

- $n$ Principal quantum number, describes the energy state of an atomic system.
- $l$ Quantum number for the angular momentum of the electron, usually called the azimuthal quantum number, describes the shape of electron "orbits."
- $m_l$ Magnetic quantum number, describes the orientation of the electron "orbits."
- $s$ Spin quantum number, assumes the values $\pm\frac{1}{2}$, depending on the orientation of the electron spin.

Considering only the $r$-dependence of the wave function, a probability distribution function $P$ can be established:

$$dP = |\psi^2|\, d\tau = R^2 \cdot 4\pi r^2\, dr \tag{2.38}$$

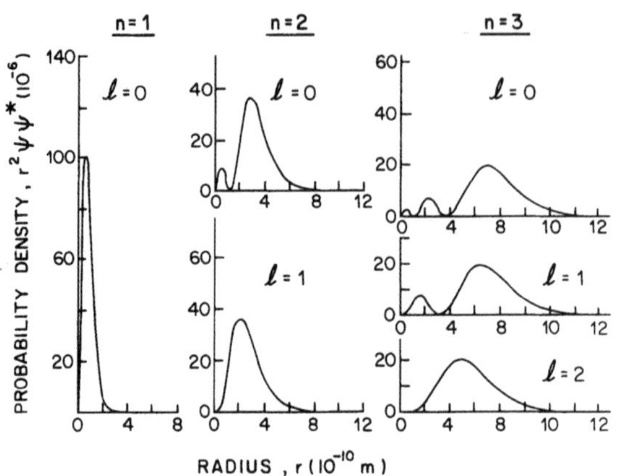


FIG. 2.5. Examples of probability density distributions for finding the electron in a hydrogen atom.

and $P = 4\pi r^2 R^2$ is the probability density. Figure 2.5 shows examples of such probability density distributions on a relative scale for different combinations of $n$ and $l$.

As we have discussed, the possible electronic states in an H atom are characterized by a set of four quantum numbers $(n, l, m_l, s)$. The first two quantum numbers in this set are of particular interest for describing electronic states. It is customary to assign letters to the values of the second quantum number ($l$), i.e.,

$$\begin{aligned} l &= 0 \rightarrow s \\ l &= 1 \rightarrow p \\ l &= 2 \rightarrow d \\ l &= 3 \rightarrow f \\ l &= 4 \rightarrow g \\ l &= 5 \rightarrow h \\ &\vdots \end{aligned}$$

Starting with $l = 3(f)$, the letters follow the alphabet. The first three letters are derived from the appearance of the spectra of the more complicated alkaline atoms, which show four characteristic spectral series referred to as the sharp series ($s$), the principal series ($p$), the diffuse series ($d$), and the Bergmann series ($f$). The alkaline atoms show an important similarity to the H atom. As will be shown later in this chapter, the inner electron shells of the alkaline atoms are completely filled and the outermost shell has only one electron. As a first approximation, such atoms can be treated as one-electron systems.

Some examples of the notations used for various combinations of the first and second quantum numbers are listed below:

$$\left.\begin{matrix} n = 1 \\ l = 0 \end{matrix}\right\} 1s \text{ electron} \qquad \left.\begin{matrix} n = 4 \\ l = 0 \end{matrix}\right\} 4s \text{ electron}$$

$$\left.\begin{matrix} n = 2 \\ l = 0 \end{matrix}\right\} 2s \text{ electron} \qquad \left.\begin{matrix} n = 4 \\ l = 1 \end{matrix}\right\} 4p \text{ electron}$$

$$\left.\begin{matrix} n = 2 \\ l = 1 \end{matrix}\right\} 2p \text{ electron} \qquad \left.\begin{matrix} n = 4 \\ l = 2 \end{matrix}\right\} 4d \text{ electron}$$

$$\left.\begin{matrix} n = 3 \\ l = 0 \end{matrix}\right\} 3s \text{ electron} \qquad \left.\begin{matrix} n = 4 \\ l = 3 \end{matrix}\right\} 4f \text{ electron}$$

$$\left.\begin{matrix} n = 3 \\ l = 1 \end{matrix}\right\} 3p \text{ electron} \qquad \vdots$$

$$\left.\begin{matrix} n = 3 \\ l = 2 \end{matrix}\right\} 3d \text{ electron} \qquad \vdots$$

An interesting wave-mechanical picture of the electronic states of a hydrogen atom can be obtained by considering the vibrational modes of an elastic sphere (Fig. 2.6).

## 2.3. THE STRUCTURE OF MORE COMPLEX ATOMS

### 2.3.1. Atomic Structure

In general, an atom can have $Z'$ protons in the nucleus, resulting in a total nuclear charge of $+Z'e$. In addition, a nucleus contains a certain number of neutrons which, together with the protons, make up the mass of the nucleus. Since an atom is electrically neutral, the total charge of the electrons surrounding the nucleus must be $-Z'e$.

The nuclear mass of an atom in atomic mass units is indicated by a superscript after the atomic symbol, and the nuclear charge is indicated by a subscript in front of the atomic symbol, for example,

$$_1H^1, {}_2He^4, {}_3Li^6, {}_{18}Ar^{40}, {}_{22}Ti^{48}, {}_{82}Pb^{207}$$

Isotopes have the same nuclear charge, but the number of neutrons in the nucleus differs. Typical examples of pairs of isotopes are $_3Li^6$ and $_3Li^7$ and $_{92}U^{235}$ and $_{92}U^{238}$. The chemical properties of the elements are determined by the arrangement of the electrons around the nucleus and not by the mass of the nucleus, i.e., isotopes cannot be distinguished chemically.

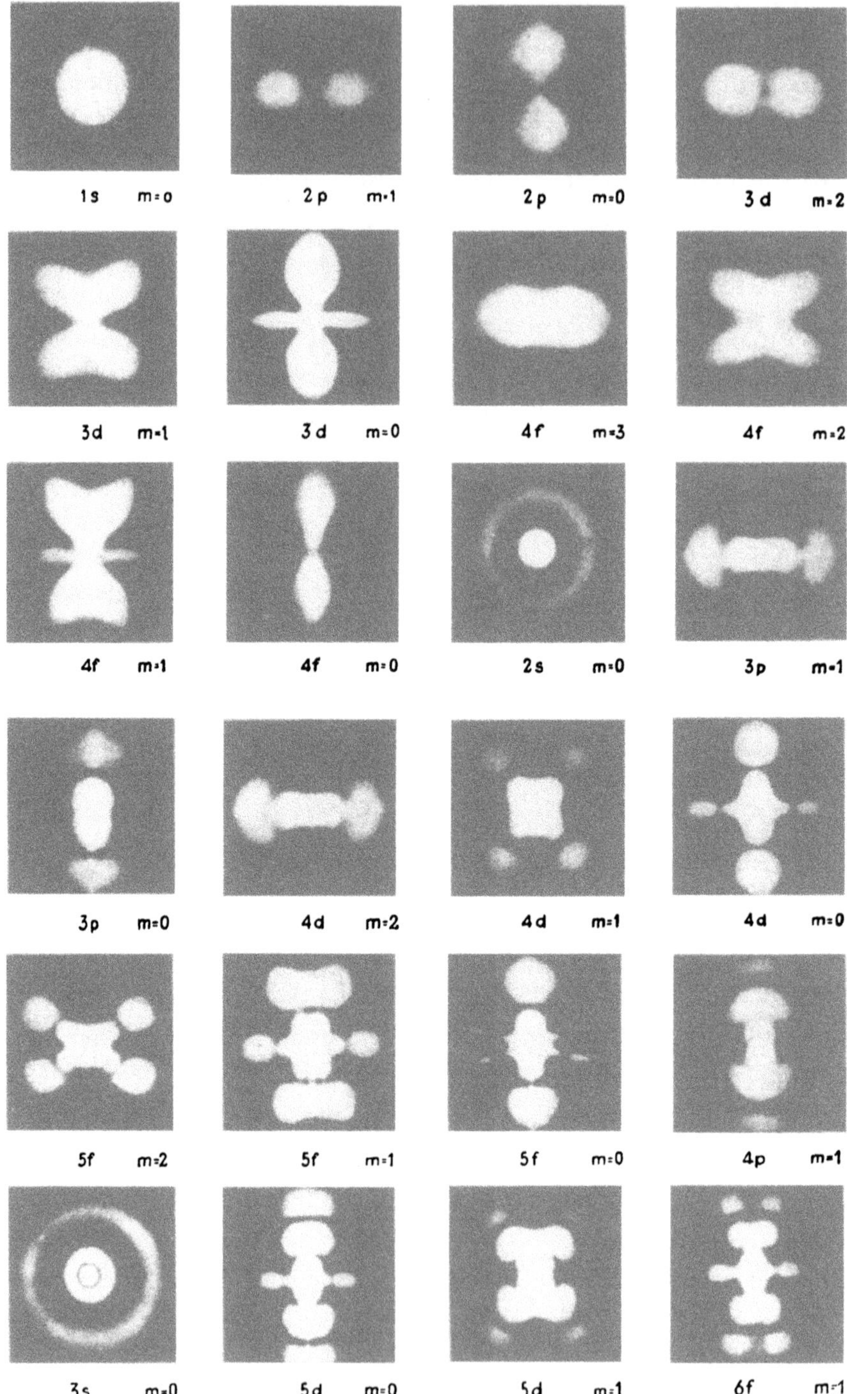


**FIG. 2.6** Wave-mechanical picture of the electronic states in a hydrogen atom [after W. Finkelnburg, *Structure of Matter* (Berlin: Springer-Verlag, 1964)].

**TABLE 2.1.** Schematic Outline of Atomic Structure

| Shell | $n$ | $l$ | Designation of electrons | Number of electrons in subshell | Total number of electrons in shell | Configuration of completed shell |
|---|---|---|---|---|---|---|
| $K$ | 1 | 0 | $1s$ | 2 | 2 | $1s^2$ |
| $L$ | 2 | 0 | $2s$ | 2 | 8 | $2s^2 2p^6$ |
| | 2 | 1 | $2p$ | 6 | | |
| $M$ | 3 | 0 | $3s$ | 2 | 18 | $3s^2 3p^6 3d^{10}$ |
| | 3 | 1 | $3p$ | 6 | | |
| | 3 | 2 | $3d$ | 10 | | |
| $N$ | 4 | 0 | $4s$ | 2 | 32 | $4s^2 4p^6 4d^{10} 4f^{14}$ |
| | 4 | 1 | $4p$ | 6 | | |
| | 4 | 2 | $4d$ | 10 | | |
| | 4 | 3 | $4f$ | 14 | | |

In atomic models of more complex atoms it is assumed that the electrons are arranged in shells. The first two quantum numbers of an electronic state defined the electron shells, as shown below:

| | | | | |
|---|---|---|---|---|
| $n =$ | 1 | 2 | 3 | 4 |
| main shells: | $K$ | $L$ | $M$ | $N$ |
| $l =$ | 0 | 1 | 2 | 3 |
| subshells: | $s$ | $p$ | $d$ | $f$ |

These definitions provide a schematic outline of atomic structure, which is shown in Table 2.1 for the first four electron shells. The number of electrons in each subshell is indicated by superscripts following the designation of the electron state (last column of Table 2.1).

Using this schematic, a table of the electronic structures of the elements can be established (Table 2.2). Table 2.2 outlines the electronic structures of the elements up to completion of the $4s$ shell. The horizontal double lines in Table 2.2 indicate completed main shells, and the single lines correspond to completed subshells.

Potassium (K) reveals an irregularity. A new shell is started even though the $3d$ subshell is not yet completed. This irregular behavior has been confirmed by spectroscopic observations as well as by energetic considerations. For Cu the opposite effect is observed: two electrons move into the $3d$ subshell leaving only one electron in the $4s$ subshell. The irregular behavior occurs several times throughout the schematic electron arrangement of the elements.

From an extended Table 2.2, the electron configuration for every

**TABLE 2.2.** Electronic Structure of the Elements

| $Z'$ | Element | Number of electrons in subshells | | | | | | | | | |
|---|---|---|---|---|---|---|---|---|---|---|---|
| | | 1*s* | 2*s* | 2*p* | 3*s* | 3*p* | 3*d* | 4*s* | 4*p* | 4*d* | 4*f* |
| 1 | H | 1 | | | | | | | | | |
| 2 | He | 2 | | | | | | | | | |
| 3 | Li | 2 | 1 | | | | | | | | |
| 4 | Be | 2 | 2 | | | | | | | | |
| 5 | B | 2 | 2 | 1 | | | | | | | |
| 6 | C | 2 | 2 | 2 | | | | | | | |
| • | • | • | • | • | | | | | | | |
| • | • | • | • | • | | | | | | | |
| • | • | • | • | • | | | | | | | |
| 10 | Ne | 2 | 2 | 6 | | | | | | | |
| 11 | Na | 2 | 2 | 6 | 1 | | | | | | |
| 12 | Mg | 2 | 2 | 6 | 2 | | | | | | |
| 13 | Al | 2 | 2 | 6 | 2 | 1 | | | | | |
| • | • | • | • | • | • | • | | | | | |
| • | • | • | • | • | • | • | | | | | |
| • | • | • | • | • | • | • | | | | | |
| 18 | Ar | 2 | 2 | 6 | 2 | 6 | | | | | |
| 19 | K | 2 | 2 | 6 | 2 | 6 | | 1 | | | |
| 20 | Ca | 2 | 2 | 6 | 2 | 6 | | 2 | | | |
| 21 | Sc | 2 | 2 | 6 | 2 | 6 | 1 | 2 | | | |
| • | • | • | • | • | • | • | • | • | | | |
| • | • | • | • | • | • | • | • | • | | | |
| • | • | • | • | • | • | • | • | • | | | |
| 28 | Ni | 2 | 2 | 6 | 2 | 6 | 8 | 2 | | | |
| 29 | Cu | 2 | 2 | 6 | 2 | 6 | 10 | 1 | | | |
| 30 | Zn | 2 | 2 | 6 | 2 | 6 | 10 | 2 | | | |
| • | • | • | • | • | • | • | • | • | | | |
| • | • | • | • | • | • | • | • | • | | | |
| • | • | • | • | • | • | • | • | • | | | |

element can be derived. Copper, for example, has the electronic configuration $1s^2 2s^2 2p^6 3s^2 3p^6 3d^{10} 4s^1$.

Knowledge of the electron configuration of atoms is very important because the chamical properties of atoms are determined mainly by their electron configurations. For example, elements such as the alkali metals

(Li, Na, K, Ru, and Cs), all of which have one outer electron, show strong similarities in their chemical behavior.

### 2.3.2. Electronic States of Atoms

#### 2.3.2.1. Momentum

Like atomic energy levels, the angular momentum of an atom, which is characterized by $\vec{l}$ in wave mechanics, can assume only discrete values of energy levels. For a bound electron, the magnitude of $\vec{l}$ is $\sqrt{l(l+1)} \cdot \hbar$, where $l$ is the azimuthal quantum number. Thus, a $d$ electron ($l = 2$) has $\sqrt{6}$ units of angular momentum, where the unit of the angular momentum is $\hbar$. Similarly, a $5p$ electron [an electron in the fifth (O) shell] has $\sqrt{2}$ units of angular momentum.

Electrons with identical values of $n$ and $l$ are called equivalent electrons. The number of equivalent electrons ($r$) in a subshell is written as a superscript to the right of the subshell: $nl^r$. Thus, $6d^2$ denotes 2 electrons in the 6th ($P$) shell, each electron having $\sqrt{6}$ units of angular momentum.

According to classical field theory, the angular momentum vector $\vec{l}$ of an atom in a magnetic or electric field $\vec{F}$ will prescribe a cone with the field direction as axis and will have a constant component $m_l$ as shown in Fig. 2.7a. Quantum theory allows only discrete values of $m_l$ (between $-l$ and $+l$), as shown in Fig. 2.7b. The quantum number $l$ can assume all integer values from 0 to $n - 1$.

The remaining quantum number is the quantum number $s$ (not to be confused with the subshells designated by $l = 0$), which is associated

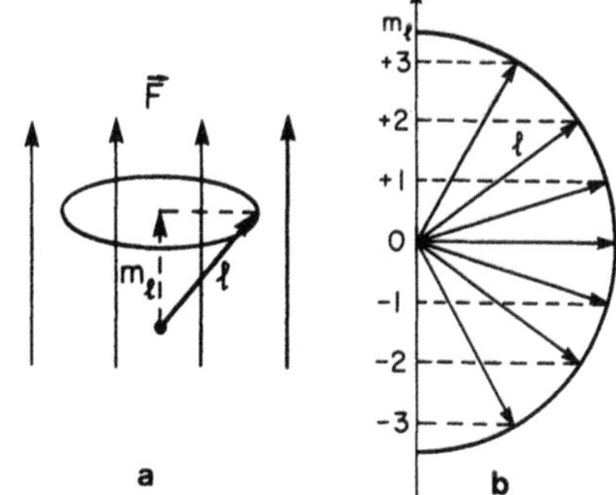


**FIG. 2.7.** (a) Precession of the angular momentum $\vec{l}$ of an electron in a magnetic or an electric field. (b) Space quantization of $\vec{l}$ in a field $\vec{F}$ for $\vec{l} = 3$ (after Herzberg[1]).

with the electron spin. The quantum number $s$ can only assume two values, $s = \pm 1/2$. Therefore, there can be only two $s$ electrons ($l = 0$) in the first shell ($n = 1, l = 0, m_l = 0, s = \pm 1/2$). The second shell ($n = 2$) can contain two $s$-electrons ($n = 2$, $l = 0$, $m_l = 0$, $s = \pm 1/2$) and six $p$ electrons ($n = 2, l = 1, m_l = -1, 0, 1, s = \pm 1/2$), and so on.

Since each electron is characterized by an angular momentum quantum number, $l$, and a spin quantum number, $s$, these individual quantum numbers can be added to describe the total angular momentum $\vec{j} = \vec{l} + \vec{s}$. The total angular momentum quantum number $j = l \pm 1/2$ is associated with the total angular momentum $\sqrt{j(j+1)} \cdot \hbar$. When a magnetic field is applied, the total angular momentum vector of the electron orbit can assume $(2j + 1)$ different orientations in the magnetic field, i.e., $m_j = j, (j - 1), \ldots, -(j - 1), -j$.

### 2.3.2.2. Energy Transitions

As we saw in Section 2.1.1, the transition of an electron from one energy level to another is associated with the emission or absorption of electromagnetic radiation, i.e.,

$$h\nu_{ul} = E_u - E_l$$

Only transitions between certain energy levels are permitted. The permitted transitions are determined by selection rules that are based on the observed spectra of atoms.

The lowest energy state of the atom is the *ground state.* The other energy states (in which the electron is still bound to the particle) are the *electronically excited states.* Their lifetime is generally short ($10^{-8}$ to $10^{-6}$ s). The resonance excited state is a special type of excited state in which the probability of radiative transition to the ground state is very high (the lifetime of electrons in this state is extremely small, $\sim 10^{-8}$ s). The resonance state is the first excited state and thus it can appear in both emission and absorption. Photons with the corresponding energy for transition to the resonance excited state will be absorbed by particles in the ground state with very high efficiency.

Radiative de-excitation or electronic excitation can cause the transition of an electron to an energy level from which the transition rules forbid a radiative transition to a lower energy level. The lifetime of such a state, called a *metastable state,* can be orders of magnitude higher than for spontaneous transitions. such states can be de-excited only by collisions with other particles or by absorption of radiative energy. Thus, such particles act as an energy reservoir.

When one or more electrons are completely removed from a particle, the particle is said to be in an *ionized state.* Electrons removed by ionization processes are crucial for maintaining an electrical discharge. The ionization energy depends on the atomic number $Z'$. For the noble gases (He, Ne, Ar, . . .), in which each shell contains the maximum number of electrons allowed by the Pauli principle (completed shells), higher energies are required to remove a bound electron than for other elements. The higher the atomic number $Z'$ (and principal quantum number $n$), the greater the number of completed shells shielding the positive charges of the nucleus. It is, therefore, easier to remove one electron from Xe than from Kr. Ionization energies vary from 24.5 eV for He down to 14 for Kr and 13 for Xe (Table 2.3).

The alkali metals (Li to Fr) have the lowest ionization energies, because the binding energy acting on their single outermost electron is low due to shielding by the electrons in the inner, completed shells. The chemical activity of the alkali metals is, correspondingly, very high and they are known as electropositive elements.

The increase of the ionization potential is gradual from the alkali metals toward elements with the next completed shell. In general, the energy level of the first excited state (resonance state) follows a pattern similar to the variation of the ionization energies with $Z'$ or $n$ (Table 2.3).

Negative ions are formed by the attachment of an additional electron to a neutral atom. Of course, elements that need only one electron to form a completed shell (for example, F, which has only seven electrons in the $L$ shell compared with 8 for Ne) will very easily attach an extra electron due to the field resulting from the nucleus and the electronic charges. Thus, the halogens (from F to At) are the most strongly electronegative elements. Elements with two electrons missing from a completed shell (for example, O) also tend to favor electron attachment. Table 2.4 lists the electron affinity of various atoms.

### 2.3.3. Designation of Electron Configurations

The nomenclature adopted for describing the transitions between atomic energy levels generally refers to the nature of the coupling between the electrons that are effective in producing the spectrum. *Because the core of completed atomic shells has zero angular momentum, only electrons in partially filled outer shells need to be considered.* We do, however, need to include those situations in which an atom with completed shells, such as argon, is excited so that an electron from a completed shell is moved to an outer, incomplete shell.

**TABLE 2.3.** Electronic Data for Selected Atoms

| Z | Element | Atomic weight ($10^{-3}$ kg) | Ionization energy (eV) | Energy and designation of the first excited states (eV) | (−) | Energy and designation of the metastable states (eV) | (−) | Radiative lifetime of metastable states (s) |
|---|---|---|---|---|---|---|---|---|
| 1 | H | 1.008 | 13.659 | 10.198 | $2p\ ^2P^0_{1/2}$ | 10.198 | $2s\ ^2S_{1/2}$ | 0.12 |
| | | | | 10.198 | $2p\ ^2P^0_{3/2}$ | | | |
| 2 | He | 4.004 | 24.481 | 21.216 | $2p\ ^1P^0_1$ | 19.818 | $2s\ ^3S_1$ | 0.0197 |
| | | | | | | 20.614 | $2s\ ^1S_0$ | 9000 |
| 3 | Li | 6.941 | 5.391 | 1.848 | $2p\ ^2P^0_{1/2}$ | — | — | — |
| | | | | 1.848 | $2p\ ^2P^0_{3/2}$ | | | |
| 7 | N | 14.007 | 14.534 | 10.335 | $3s\ ^4P_{1/2}$ | 2.384 | $2p^3\ ^2D_{3/2}$ | 61,200 |
| | | | | 10.329 | $3s\ ^4P_{3/2}$ | 2.384 | $2p^3\ ^2D_{5/2}$ | 144,000 |
| | | | | 10.325 | $3s\ ^4P_{5/2}$ | 3.575 | $2p^3\ 2P^0_{1/2}$ | 40 |
| | | | | | | 3.575 | $2p^3\ 2P^0_{3/2}$ | 166 |
| 8 | 0 | 15.999 | 13.618 | 9.521 | $3s\ ^3S^0_1$ | 0.020 | $2p^4\ ^3P_1$ | — |
| — | — | — | — | — | — | 0.028 | $2p^4\ ^3P_0$ | — |
| — | — | — | — | — | — | 1.967 | $2p^4\ ^1D_2$ | 110–147 |
| — | — | — | — | — | — | 4.189 | $2p^4\ ^1S_0$ | 0·76–0·90 |
| — | — | — | — | — | — | 9.146 | $3s\ ^5S^0_2$ | — |
| 10 | Ne | 20.183 | 21.564 | 16.847 | $3s\ ^1P^0_1$ | 16.618 | $4s\ ^3P_2$ | 0.8–430 |
| — | — | — | — | 18.380 | $3p\ ^3S_1$ | 16.670 | $4s\ ^3P_1$ | — |
| | | | | | | 16.714 | $4s\ ^3P_0$ | 0.8–24.8 |
| 11 | Na | 22.99 | 5.139 | 2.102 | $3p\ ^2P^0_{1/2}$ | — | — | — |
| | | | | 2.104 | $3p\ ^2P^0_{3/2}$ | — | — | — |
| | | | | 3.191 | $4p\ ^2S_{1/2}$ | — | — | — |
| 17 | Cl | 35.453 | 12.967 | 9.202 | $4s\ ^2P_{3/2}$ | 0.109 | $3p^5\ ^2P^0_{1/2}$ | 81 |
| | | | | 9.281 | $4s\ ^2P_{1/2}$ | 8.421 | $4s\ ^4P_{5/2}$ | — |
| | | | | | | 8.987 | $4s\ ^4P_{3/2}$ | — |
| | | | | | | 9.029 | $4s\ ^4P_{1/2}$ | — |
| 18 | Ar | 39.944 | 15.755 | 11.623 | $4s\ ^2P^0_1$ | 11.548 | $4s\ ^2P^0_0$ | 1.3–55.9 |
| | | | | 11.821 | $4s\ ^2P^0_3$ | 11.723 | $4s\ ^2P^0_2$ | 1.3–44.9 |
| 19 | K | — | 4.341 | 1.609 | $4p\ ^2P^0_{1/2}$ | — | — | — |
| | | | | 1.616 | $4p\ ^2P^0_{3/2}$ | — | — | — |
| 36 | Kr | 83.80 | 13.999 | 10.03 | $5s\ ^3P^0_1$ | 9.91 | $5s\ ^3P_2$ | 1.85 |
| | | | | 10.64 | $5s\ ^1P^0_1$ | 10.56 | $5s\ ^3P_0$ | 0.49–1.0 |
| | | | | 11.30 | $5p\ ^3S_1$ | | | |

Electrons interact with each other through coupling forces arising from:

electrostatic repulsions between electrons,
magnetic fields resulting from both the orbital motions and the spins of the electrons,

**TABLE 2.4.** Electron Affinities of Various Atoms

| Ion formed | (eV) |
|---|---|
| $H^-$ ($1s^2$) | 0.746 |
| $H^-$ ($2s^2$) from $2s$ | 0.434 |
| $H^-$ ($2s2p$) from $2s$ | 0.460 |
| $C^-$ | 1.249 |
| $O^-$ | 1.466 |
| $F^-$ | 3.448 |
| $Cl^-$ | 3.612 |
| $NO^-$ | 0.911 |
| $O_2^-$ | 0.438 |
| $OH^-$ | 1.830 |

exchange forces between electron spins (these forces can be understood only on a quantum-mechanical basis)

Both the energy levels of the atom and the probability of transitions between electronic levels depend on the nature and magnitude of these interactions.

The potential energy of an atom is determined by the energies of its electrons in their various orbits. Each individual electron $i$ can be assigned an orbital quantum number, $l_i$, and a spin quantum number, $s_i$ ($s_i = \pm 1/2$). Most of the various interactions are accounted for by either $L$–$S$ coupling or $j$–$j$ coupling;[1] $j$–$j$ coupling occurs when the magnetic interaction between the orbital angular momentum and the spin angular momentum of each electron becomes dominant compared to the electrostatic and exchange interactions between different electrons, which is generally the case when the charge $Z'$ of the nucleus increases. It can be used for intermediate coupling (as in the case of noble gases).

### 2.3.3.1. $L$–$S$ Coupling

In $L$–$S$ coupling, the orbital angular momenta of the individual electrons are strongly coupled among themselves. Therefore, the total orbital angular momentum $\vec{L}$ is formed by combining the orbital angular momenta $\vec{l}_i$ of the various electrons. The magnitude of the orbital angular momentum $\vec{L}$ is $\sqrt{L(L+1)} \cdot \hbar$, where $L$ is the associated quantum number.[1]

Because the spins $(\vec{s}_i)_i$ of the individual electrons can be regarded as strongly coupled, the total spin momentum $\vec{S}$ is formed by combining the spin momenta $\vec{s}_i$ of the separate electrons. The magnitude of the spin angular momentum $\vec{S}$ is $\sqrt{S(S+1)} \cdot \hbar$, where $S$ is the associated quantum number.[1]

In $L$–$S$ coupling the total orbital angular momentum $\vec{L}$ and the total spin momentum $\vec{S}$ are coupled by weak magnetic forces. A total angular momentum $\vec{J}$, characterized by the quantum number $J$, can be expressed by

$$\vec{J} = \vec{L} + \vec{S} \tag{2.39}$$

The magnitude of the total momentum $\vec{J}$ corresponding to the quantum number $J$ is $\sqrt{J(J+1)} \cdot \hbar$, where $J$ assumes values between

$$J = L + S, L + S - 1, L + S - 2, \ldots, L - S \qquad \text{when } S < L$$

and (2.40)

$$J = L + S, L + S - 1, L + S - 2, \ldots, S - L \qquad \text{when } S > L$$

In the first case ($S < L$) there are $2S + 1$ possible values of $J$, while in the second case there are $2L + 1$ possible values of $J$; $2S + 1$ is called the *multiplicity*. A given $L$ value together with the corresponding multiplicity defines a *spectral term*. A given $J$ for a given term defines a *spectral level*, and spectral lines arise from transitions between spectral levels. A *multiplet* consists of all possible levels of a given term.

*2.3.3.1.a. Designation of L-Values.* The designations used for the total angular momentum quantum numbers are analogous to the designations used for orbital angular momenta of single electrons, except that capital letters are used instead of small letters. The total orbital angular momentum designations are $L = 0, 1, 2, 3, 4, \ldots$, corresponding to the $S, P, D, F, G, \ldots$, levels, respectively.

*2.3.3.1.b. Designation of Terms.* A term is defined by the $L$ value and the multiplicity. It is formed by placing the number denoting the multiplicity ($2S + 1$) as a left-hand superscript on the $L$-value designation. For example, if $S = 1/2$ and $L = 2$, the multiplicity is given by $2S + 1 = 2$, and the designation appropriate to $L = 2$ is $D$. Therefore, the term is designated by $^2D$.

*2.3.3.1.c. Designation of Levels.* A level corresponding to a given term and a given $J$ value is designated by adding the $J$ value as a right-hand subscript to the term designation. In the example above where $S = 1/2$ and $L = 2$, there are two possible $J$ values, $J = 5/2$ and $J = 3/2$, according to Eq. (2.40). The corresponding two levels of the $D$ term are

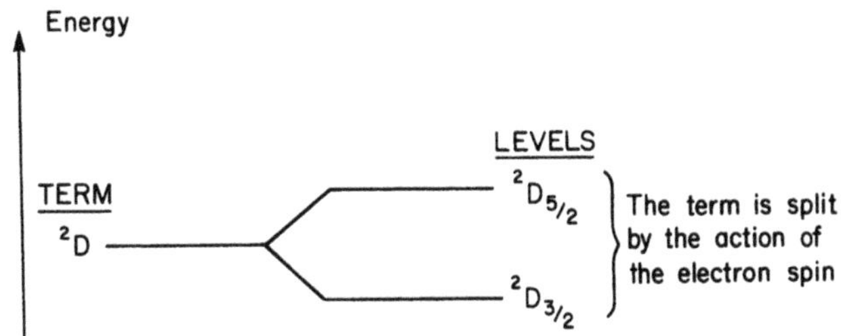


**FIG. 2.8.** $^2D$ term split into two levels by electron spin.

designated by $^2D_{5/2}$ and $^2D_{3/2}$ and shown schematically in Fig. 2.8. Splitting into $2J + 1$ states may occur, for example, in the presence of an electric or a magnetic field. $2J + 1$ is often referred to as the statistical weight of the level and is important for deriving the radiative power of a spectral line or for evaluating partition functions, as will be discussed in Section 6.2.4.

*2.3.3.1.d. Designation of Parity.* The parity is either odd or even depending on whether $\Sigma_i l_i$ has an odd or even value. The expression $\Sigma_i l_i$ denotes the sum of the angular momentum quantum numbers for the electrons in the atom. For completed shells, the total number of electrons is even and so is the parity. Thus, we need to evaluate $\Sigma_i l_i$ only for electrons in partially filled shells.

Odd parity is designated by a superscript "o" to the right of the term symbol, e.g., $^2P^o$, while an even parity is designated by omitting the superscript "o."

*2.3.3.1.e. Example.* According to the preceding definitions, the ground state of a copper atom will be designated by

$$1s^2\,2s^2\,2p^6\,3s^2\,3p^6\,3d^{10}\,4s\;(^2S_{1/2})$$

Since the shells up to $3p$ do not play any significant role in the formation of the spectrum, this designation is abbreviated by

$$3d^{10}\,4s\;(^2S_{1/2})$$

Because all the shells up to and including $3d$ are complete, only the outer $s$ electron contributes to the total angular momentum and to the resultant spin. $L = 0$ and $S = 1/2$ result in multiplicity of $2S + 1 = 2$, leading to the term designation $^2S_{1/2}$.

Excitation of the $4s$ electron to other orbits such as $4p$ or $5d$ leads, according to the preceding arguments, to $^2P^o$ and $^2D$ terms, respectively. The $^2P^o$ term is split into two levels, $^2P^o_{1/2}$ and $^2P^o_{3/2}$, and the $^2D$ term into the levels $^2D_{5/2}$ and $^2D_{3/2}$.

On the other hand, if one of the electrons of the copper $3d$ shell is excited, one obtains the configuration $3d^9\,4s\,nx$, where $n > 4$ and $x$ corresponds to $s, p, d, f, \ldots,$. The core now has a residual spin of 1/2, which can couple with the 1/2 spin of both the $4s$ and the $nx$ electrons to give a total resultant spin of:

$$S = 1/2 + 1/2 + 1/2 = 3/2; \qquad 2S + 1 = 4$$

or

$$S = 1/2 + 1/2 - 1/2 = 1/2; \qquad 2S + 1 = 2$$

The corresponding multiplicities are 4 and 2, known as quartets and doublets, respectively.

*2.3.3.1.f. Energy of the Spectral Levels.* The energy of a bound electron increases with an increase in the principal quantum number $n$ and, in general, with the value of the angular momentum quantum number $l$. Usually, one uses some simple rules applicable for ground-state terms (these rules do not necessarily apply to higher configurations).

Triplet terms lie below singlet terms in an energy diagram (the energy is smaller for higher $S$).
The difference in term values between two levels $J + 1$ and $J$ of a multiplet with the same $L$ and $S$ is proportional to $J + 1$, e.g., in the case of a $^3P_{2,1,0}$ multiplet, the term difference between the $^3P_2$ and $^3P_1$ is twice that between the $^3P_1$ and $^3P_0$.

The tables of Moore,[4–6] for example, list the energies of the different excited states corresponding to the spectral terms for most of the atoms and their ions.

*2.3.3.1.g. Selection Rules for Dipole Radiation in the Case of L–S Coupling.* The transition probability for the emission or absorption of the radiation between any two term values is governed by the square of the appropriate matrix element (see the book by Cohen–Tannoudiji *et al.* cited in the general bibliography at the end of this chapter). It is possible to provide a set of simple rules (selection rules) that indicate whether electric dipole transitions are allowed or forbidden. In strict $L$–$S$ coupling, these rules are

1. The parity must change.
2. The multiplicity must remain unchanged, i.e., intercombination lines are forbidden.
3. $J$ must change by $\pm 1$ or 0, however, the transition $J = 0$ to $J = 0$ is *not* allowed.

4. $L$ must change by $\pm 1$ or 0, however, the transiton $L = 0$ to $L = 0$ is *not* allowed.

Now consider the excitation of the outer 4*s* electron of copper to 4*p* and 5*d* orbits. As we have already seen, this leads to $^2P^o$ and $^2D$ terms, respectively. The $^2P^o$ term is then split into the two levels $^2P^o_{3/2}$ and $^2P^o_{1/2}$, and the $^2D$ term is split into the two levels $^2D_{5/2}$ and $^2D_{3/2}$. Applying the selection rule gives rise to three possible transitions, as shown in Fig. 2.9a. If doublet transitions involve an $S$ term, only two transitions are possible, as shown in Fig. 2.9b. The maximum number of transitions between two quartet terms is 9, as shown in Fig. 2.9c.

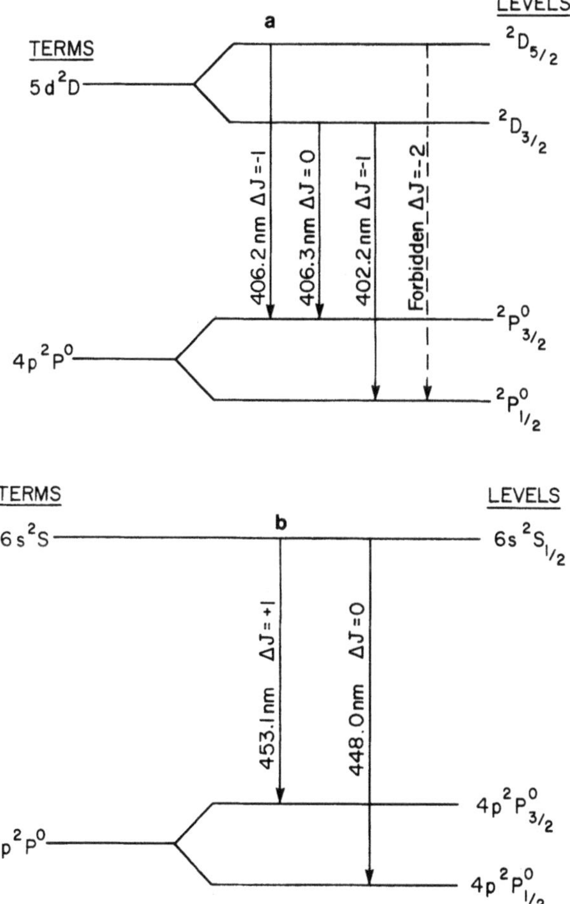


FIG. 2.9. Example of transitions in the copper spectrum: (a) Three lines in a doublet system. (b) Two lines in a doublet system. (c) Nine lines in a quartet system.

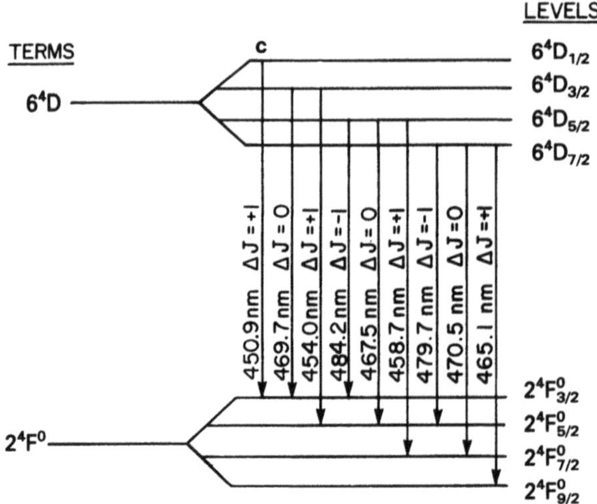


**FIG. 2.9.** (Continued).

#### 2.3.3.2. *j–j* Coupling

With increasing $Z'$ and the corresponding increase in the distance between the outermost electrons, their coupling among themselves ($L$–$S$ coupling) decreases. The spin–orbit interaction becomes predominant, resulting in ($j$–$j$) coupling for the heavy species and noble gases.

The coupling of $\vec{s}_i$ and $\vec{l}_i$ results in a total angular momentum $\vec{j}_i$ for the $i$th bound electron ($\vec{j}_i = \vec{l}_i + \vec{s}_i$). The total angular momenta $\vec{j}_i$ of the individual electrons combine to form a total angular momentum $\vec{J}$.

The correspondence between the two designations, ($j$–$j$) and ($L$–$S$), is rather simple and will be illustrated for the He atom with the terms related to an $s, p$ configuration:

$$s, p \left| \begin{array}{ll} l_1 = 0 & s_1 = 1/2 \\ l_2 = 1 & s_2 = 1/2 \end{array} \right.$$

In $L$–$S$ coupling, $L = 1$ and $S = 1$ or $S = 0$ give $^3P_{2,1,0}$ or $^1P_1$. With $j$–$j$ coupling

$$(l_1, s_1) = j_1 = 1/2$$
$$(l_2, s_2) = j_2 = 3/2, 1/2$$

corresponding to two terms

$$(1/2, 1/2) \quad \text{or} \quad J = 1, 0$$
$$(1/2, 3/2) \quad \text{or} \quad J = 2, 1$$

The four terms obtained from the two types of coupling correspond to each other in the following way:

$$\left.\begin{array}{l} {}^1P_1 \cdots J = 1 \\ {}^3P_2 \cdots J = 2 \end{array}\right| (1/2, 3/2)$$

$$\left.\begin{array}{l} {}^3P_1 \cdots J = 1 \\ {}^3P_0 \cdots J = 0 \end{array}\right| (1/2, 3/2)$$

## 2.4. EXCITED STATES OF DIATOMIC MOLECULES

When a particle consists of more than one atom, the determination of its energy states becomes a rather complex problem compared to the situation encountered with atoms. The discussion in this section will be restricted to diatomic molecules.

### 2.4.1. Energy States

Figure 2.10 gives a schematic representation of a diatomic molecule in which each of the two atoms (X and Y) is surrounded by a cloud of electrons and the two atoms are joined by outer, shared electron orbits. For example, in the $H_2$ molecule the two H atoms are held together by forces that arise from the two shared electrons, which move around the two nuclei in an orbit resembling a three-dimensional 8. The balance of

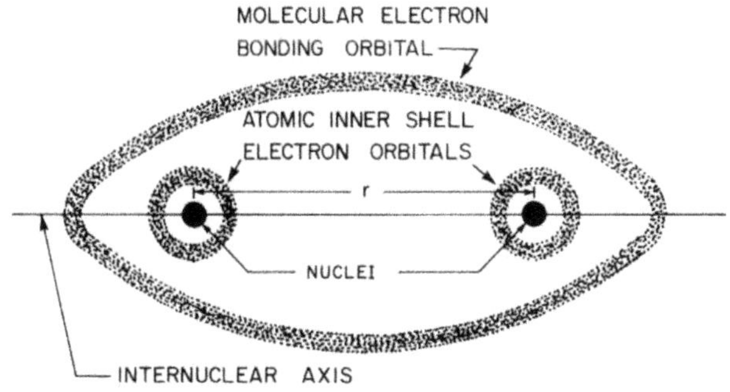


**FIG. 2.10.** Schematic representation of a diatomic molecule.

the electrostatic forces involved results in a finite separation $r$ between the two atomic centers. The atoms can, however, vibrate in an $r$-dependent force field about an equilibrium internuclear distance. The nuclear "dumbbell" can also rotate about two axes orthogonal to the internuclear *axis*. Therefore, *vibrationally* and *rotationally* excited energy states are possible in addition to the electronically excited states of X and Y. The solution of the Schrödinger equation shows, as it did for atoms, that only discrete energy levels are allowed, thus defining quantum numbers for each of the electronically, the vibrationally, and the rotationally excited states. This solution is much more complex than the solution for atoms because the electrostatic forces vary with $r$ and because potential energies must be determined for each electronic orbit. The results given by this solution are very sensitive to the choice of potential energies in the ground state as well as in the excited states.

Figure 2.11, for example, is a typical energy level diagram used for diatomic molecules, approximated by a Morse potential

$$V(r) = D_e\{1 - \exp[-a(r - r_e)]\}^2 \tag{2.41}$$

where $a$ is a constant for the particular molecular state and $r_e$ the equilibrium distance between the nuclei. Quantity $D_e$ is the energy difference between the equilibrium position of the nuclei ($r = r_e$) and the free atoms ($r = \infty$); it is often referred to as the dissociation energy with respect to the minimum energy. (This energy is not the real dissociation energy, which is given by $D_0 = D_e - \frac{1}{2}h\nu_0$; at rest, when the vibrational quantum number $v = 0$, the vibrational energy of the molecule is not zero but $\frac{1}{2}h\nu_0$, where $\nu_0$ is the oscillation frequency for the classical vibration of two masses separated by a distance $r_e$.) The part of the potential energy curve for $r > r_e$ corresponds to an attractive potential,

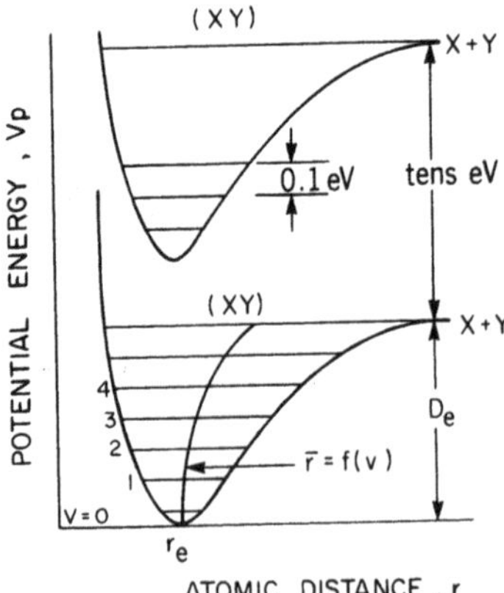


**FIG. 2.11.** Potential level diagram of a diatomic molecule XY in the ground state and in an electronically excited state.

and the part for $r < r_e$ to a repulsive potential. It should be noted that, except for low values of the vibrational quantum number $v$, the vibrations are asymmetrical, as shown by the line $\bar{r} = f(v)$ passing through the middle of the classical vibrational amplitudes (see Fig. 2.11).

Taking into account the different quantum numbers involved [electronic, vibrational ($v$), and rotational ($J$)], the solution of the wave equation in the simplest case when the various excitation modes (electronic, vibrational, rotational) may be considered as independent[2] (the Born–Oppenheimer approximation) leads to energy eigenvalues of the molecule (with the Morse potential) described by

$$E = E_e + \frac{ha}{\pi}\sqrt{\frac{D_e}{2\mu}}(v + \tfrac{1}{2}) + \frac{h^2}{8\pi^2\mu r_e^2}J(J + 1) \tag{2.42}$$

where the first term represents the noninteracting electronic energy. The second term contains the vibrational energy, which is assumed to be simply harmonic (this is true for vibrational levels close to the bottom of the Morse potential, which can be approximated in this region by a simple parabola; see Fig. 2.11). The third term accounts for the rotational energy of the molecule (using the rigid rotator approximation without coupling). In this expression, $\mu$ is the reduced mass of the molecule.

As already mentioned, spectral energies are proportional to $1/\lambda$, and it is therefore customary to express them in units of $\text{cm}^{-1}$ $(E/hc)$. With the classical notation (see Herzberg[2]), Eq. (2.42) becomes

$$T = T(e) + G(v) + F(J)$$

with

$$G(v) = \omega_e \cdot (v + 1/2), \qquad T(e) = \frac{E_e}{hc}, \qquad \text{and} \qquad F(J) = B_e \cdot J(J + 1)$$

where

$$\omega_e = \frac{a}{c\pi}\sqrt{\frac{D_e}{2\mu}} \qquad \text{and} \qquad B_e = \frac{h}{8\pi^2\mu c r_e^2} \tag{2.43}$$

$\omega_e$ corresponds to the vibrational term when the vibration is assumed to be harmonic, and $B_e$ is the rotational constant of the molecule at rest (the separation distance $r_e$ between the two nuclei is the equilibrium distance; the moment of inertia is $I = \mu r_e^2$).

The differences between the various energy modes (see Fig. 2.11) are very significant: a few eV between electronically excited states, approximately 0.1 eV between vibrational levels, and approximately 0.01 eV between rotational levels.

Of course expression (2.42) is oversimplified because, in solving the Schrödinger equation, one has to take into account:

the degree of anharmonicity of the actual molecule (Morse potential instead of a simple parabola),
the change in the internuclear distance in the higher vibrational state resulting in modification of the rotation,
the centrifugal correction arising from the increase in the internuclear distance during higher-energy rotation, and
the spin effect.

Taking these effects into account results in more complex expressions for the energy. We will try to explain these expressions through the momentum description of the molecule.

### 2.4.2. Classification of the Electronic States of Diatomic Molecules

The motion of the electrons and nuclei in the molecule, the resultant spin, and even the spins of the individual electrons are not independent of each other. This is rather evident, because the motions of the nuclei and the electrons create electric currents, which produce magnetic fields. An interaction is therefore to be expected, because an intrinsic property of the spin is its magnetic momentum. The different relevant vectors in the momentum description of the molecule are

the orbital electronic angular momentum $\vec{L}$,
the spin electronic angular momentum $\vec{S}$, and
the angular momentum of rotation of the nuclei $\vec{N}$

(The latter is orthogonal to the internuclear axis.)

#### 2.4.2.1. Orbital Angular Momentum

In a diatomic molecule the charges of the nuclei create a strong electric field in the direction of the internuclear axis, around which the angular momentum $\vec{L}$ precesses with a constant component $M_L \cdot h$, where $M_L$ can assume only the values $L, L-1, \ldots, -L$. In an electric field, unlike in a magnetic field, reversing the direction of motion of all

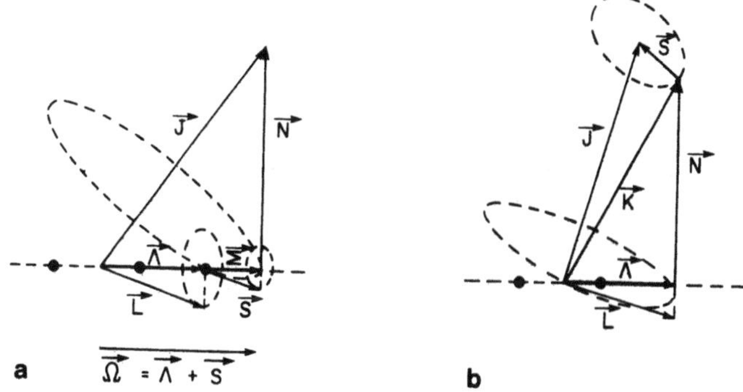


**FIG. 2.12.** Coupling of angular momentum vectors for a homonuclear molecule in Hund's coupling in case (a) and case (b).

the electrons does not change the energy of the system but changes $M_L$ into $-M_L$. Therefore, the energy is a function only of $M_L$, and states with different $M_L$ have, in general, widely differing energies, since the electric field that causes this splitting is very strong. Therefore, the electronic states are classified according to $M_L$, written as $\Lambda$. The corresponding angular momentum vector $\vec{\Lambda}$ represents the component of the electronic angular momentum along the internuclear axis (see Fig. 2.12).

The state of a molecule is designated according to its value of $\Lambda$, as shown in the following table:

| $\Lambda$ | 0 | 1 | 2 | 3 |
|---|---|---|---|---|
| State | $\Sigma$ | $\Pi$ | $\Delta$ | $\Theta$ |

When different electronic states of the same molecule have the same value of $\Lambda$, the Greek symbols ($\Sigma, \Pi, \Delta, \Theta, \ldots$) are preceded by a Roman capital letter. The ground state for electronic levels is designated by $X$, and the others by $A$, $B$, $C, \ldots$, in order of increasing energy. $\Pi, \Delta, \Theta, \ldots$ states are twofold degenerate since $M_L$ can have two values, $+\Lambda$ and $-\Lambda$, with the same energy.

### 2.4.2.2. Spin

The spins of individual electrons combine to form a total spin momentum $\vec{S}$. The corresponding quantum number $S$ can be an integer or a fraction, depending on whether the total number of electrons is even or odd. In $\Sigma$ states, $\vec{S}$, which is not affected by an electric field, is fixed in space as long as the molecule does not rotate. On the other hand, if $\Lambda \neq 0$, there will be an internal magnetic field in the direction of the internuclear axis as a result of the orbital motion of the electrons. This magnetic field causes a precession of $\vec{S}$ about the internuclear axis with a constant component $M_s \cdot (\hbar)$ (see Fig. 2.12). For molecules, $M_s$ is denoted by $\Sigma$ (this is not the same as the one $\Sigma$ corresponding to $\Lambda = 0$) with quantum theory allowing the values $\Sigma = S, S - 1, \ldots, -S$, i.e., $2S + 1$ values. The multiplicity of the electronic states is indicated by a superscript (corresponding to $2S + 1$) to the left of the Greek letter denoting the value of $\Lambda$. For example, $B^3\pi$ denotes the electronic excited state $B$ with $\Lambda = 1$ and $S = 1$. In contrast, for $\Sigma$ states ($\Lambda = 0$) $M_s = 0$ even if the spin quantum number $S \neq 0$. In this case one gets a singlet state even if the value of $2S + 1$ is given, as, for example, in the case of the $^2\Sigma$ state of $N_2^+$.

### 2.4.2.3. Total Angular Momentum of the Electrons

The total electronic angular momentum about the internuclear axis, designated by $\vec{\Omega}$, is obtained by adding $\vec{\Lambda}$ and $\vec{\Sigma}$ (see Fig. 2.12a). Because $\vec{\Lambda}$ and $\vec{\Sigma}$ are collinear (along the line joining the nuclei), an algebraic addition is sufficient; thus the quantum number $\Omega$ is given by

$$\Omega = |\Lambda + \Sigma| \tag{2.44}$$

For $\Lambda \neq 0$ there are $2S + 1$ different values of $\Omega$, corresponding to somewhat different energies of the resulting molecular states. For a multiplet, the electronic energy can be written as

$$T_e = T_0 + A \cdot \vec{\Lambda} \cdot \vec{\Sigma} \tag{2.45}$$

where $T_0$ is the term value when the spin is neglected and $A$ is a constant for a given multiplet.

### 2.4.2.4. Angular Momenta for the Rotation of the Molecule

According to classical mechanics, the rotation of a diatomic molecule surrounded by a "rigid" cloud of electrons leads to identical moments of inertia $I_B$ for both axes perpendicular to the intermolecular axis and passing through the center of gravity. The moment of inertia $I_A$ about the internuclear axis is much smaller than $I_B$, but the corresponding angular momenta are of the same order of magnitude $\vec{J}$, since the electrons rotate much faster than the heavy particles. The total angular momentum is the sum of the angular momentum taken at right angles to the internuclear axis (designated by $\vec{N}$) and the total electronic angular momentum along the axis $\vec{\Omega}$ (see Fig. 2.12a). Quantum mechanics indicates that $\vec{J}$ is quantized, i.e.,

$$|\vec{J}| = \sqrt{J(J+1)} \cdot \hbar \tag{2.46}$$

The rotational quantum number $J$ can only be an integer $(0, 1, 2, \ldots)$.

Thus, the energy levels for rotation will be given by

$$F(J) = B_e \cdot J \cdot (J+1) + (A - B_e)\Lambda^2 \tag{2.47}$$

and

$$B_e = \frac{h}{8\pi^2 c I_B} = \frac{h}{8\pi^2 \mu c r_e^2} \quad \text{and} \quad A = \frac{h}{8\pi^2 c I_A} \tag{2.48}$$

where $c$ is the velocity of light.

Note that, because of the small values of $I_A$, $A > B_e$.

### 2.4.2.5. Coupling of Rotation and Electronic Motion

In an actual molecule, rotational vibrational and electronic transitions occur simultaneously, and the influence of the corresponding interactions is very important for the energy states for the molecule. The interaction of vibrational and electronic transitions is accounted for by the interaction potential $V(r)$. Therefore, the interactions of rotational and electronic transitions need closer examination.

The simplest case is, of course, when $S$ and $\Lambda$ are zero (these are the $\Sigma$ states for which the angular momentum of nuclear rotation $\vec{N}$ is identical with the total angular momentum $J$). This case corresponds to a simple rigid rotator. The other cases have been classified by Hund as cases (a), (b), (c), and intermediate cases. Only cases (a) and (b) will be briefly described here.

*2.4.2.5.a. Hund's case (a).* For this case, the interaction of the vectors $\vec{S}$ and $\vec{\Lambda}$ is strong. $\vec{S}$ is coupled very strongly with the internuclear axis, precessing around it at a constant angle in such a way that the axial component of $\vec{S}$ is quantized ($\Sigma$ takes $2S + 1$ values). $\vec{\Omega}$ is defined and combined vectorially with the angular momentum of rotation of the nuclei $\vec{N}$ (not nuclear spin) to form a resultant angular momentum for the molecule (see Fig. 2.12a), i.e.,

$$\vec{J} = \vec{\Omega} + \vec{N} \tag{2.49}$$

$\vec{J}$, which is of constant amplitude, has a fixed direction around which $\vec{N}$ and $\vec{\Omega}$ (as well as the internuclear axis itself) precess much more slowly than do $\vec{L}$ and $\vec{S}$ about the internuclear axis.

For a given $\Omega, J$ takes the values

$$J = \Omega, \Omega + 1, \Omega + 2, \ldots$$

Levels for $J < \Omega$ do not exist.

*2.4.2.5.b. Hund's case (b).* As in case (a), $\vec{L}$ precesses rapidly around the internuclear axis of the molecule and $\vec{\Lambda}$ is quantized, but in this case the magnetic field associated with $\vec{\Lambda}$ is so weak that the interaction between $\vec{\Lambda}$ and $\vec{S}$ is small compared with the effect of the rotation of the molecule on the spin. $\vec{S}$ is no longer coupled with the axis; therefore, $\vec{\Sigma}$ does not exist and $\vec{\Omega}$ is not defined.

$\vec{\Lambda}$ and the orbital nuclear moment $\vec{N}$, which are parallel and perpendicular to the internuclear axis, respectively, form a resultant $\vec{K}$,

$$\vec{K} = \vec{\Lambda} + \vec{N} \tag{2.50}$$

around which $\vec{\Lambda}$ and $\vec{N}$ precess (see Fig. 2.12b).

The corresponding quantum number $K$ is an integer that can assume the values

$$K = \Lambda, \Lambda + 1, \Lambda + 2, \ldots$$

$\vec{K}$ and the resultant spin $\vec{S}$ form the total angular momentum $\vec{J}$. $\vec{J}$ has a fixed direction around which $\vec{K}$ and $\vec{S}$ precess slowly compared with the molecular rotation:

$$\vec{J} = \vec{K} + \vec{S} \tag{2.51}$$

For given values of $K$, $J$ can assume the values

$$J = K + S, K + S - 1, \ldots, (K - S)$$

Thus, each level $K$ is composed of $2S + 1$ sublevels. Figure 2.12b indicates the position of these vectors.

The coupling case depends on the excitation of the molecule, particularly on its rotational excitation. As soon as $J$ increases sufficiently, the rotational speed of the molecule, which is usually small compared with the speed of precession of $\vec{S}$ around $\vec{\Lambda}$ [case (a)], becomes comparable with it; in this situation the influence of rotation becomes predominant [case (b)]. For example, Shemansky and Jones[7] assert that a state belongs to Hund's case (a) as long as $A/B_e \gg J$.

### 2.4.3. General Remarks about Molecular Spectra

As already mentioned in the Born–Oppenheimer approximation [see Eq. (2.42)], the total energy can be expressed by the sum of the electronic, vibrational, and rotational energy [$T = T_e + G_e(v) + F_v(J)$].

The electronic energy, including the coupling of the spin and the angular momentum of the electrons, is given by Eq. (2.45).

The vibrational energy depends on the interaction potential (see Fig. 2.11), which in the simple case of the harmonic oscillator is reduced to a parabola [$V(r) = f \cdot (r - r_e)^2$]. In the actual situation a cubic term or even a fourth-order term has to be added to this expression. In this case the solution of the Schrödinger equation leads to

$$G(v) = \omega_e(v + \tfrac{1}{2}) - \omega_e x_e(v + \tfrac{1}{2})^2 + \omega_e y_e(v + \tfrac{1}{2})^3 + \cdots \qquad (2.52)$$

with $\omega_e y_e \ll \omega_e x_e \ll \omega_e$.

The values of $x_e$, $y_e$ (constants for each electronic state) can be found, for example, in Herzberg.[2]

The rotational energy, as we have already seen, must be corrected to take into account not only the interaction between the angular momentum of the nuclei (at a right angle to the internuclear axis) and the electronic angular momentum (see Fig. 2.12) but also the interaction between the rotation and the vibration. During vibration, the internuclear distance and consequently the moment of inertia and the rotational constant $B_e$ are changing. Since the period of vibration is very small compared to the period of rotation, one uses a mean value for $B_e$ in the vibrational state considered and $B_v$ is written as

$$B_v = B_e - \alpha_e(v + \tfrac{1}{2}) \qquad (2.53)$$

where $\alpha_e$ is a constant (for a given electronic state) that is small compared to $B_e$ (see Herzberg[2]). The detailed quantum-mechanical equations lead finally to

$$F(J) = B_v J(J + 1) - D_v J^2(J + 1)^2 + (A - B_v)\Lambda^2 \qquad (2.54)$$

with

$$D_v = D_e + \beta_e(v + \tfrac{1}{2}) \qquad (2.55)$$

and

$$D_e = \frac{4B_e^3}{\omega_e^2} \tag{2.56}$$

Table 2.5 gives some energy values for various excited states of selected molecules.

The spectral emissions of excited molecules consist of spectral bands due to electronic, vibrational, and rotational levels. Each line within a band corresponds to a transition between rotational levels, transitions between vibrational levels determine the band structure, and the location of possible potential curves in the energy level diagram is determined by electronic excitations.

The details of molecular transition rules are rather complex; therefore, only the main rules will be discussed here, and the complexity of molecular spectra will be demonstrated with a relatively simple example.

The wave numbers of spectral lines corresponding to transitions between two electronic states (in emission or absorption) are given by

$$\sigma = (T_e' - T_e'') + (G' - G'') + (F' - F'') \tag{2.57}$$

where the single-primed letters refer to the upper state and the double-primed letters to the lower state. The upper and lower states are determined by the transition rules.

The selection rules for transitions between electronic terms $\Delta\Lambda = \Lambda' - \Lambda''$ and between electronic levels are: $\Delta\Lambda = 0, \pm 1$, $\Delta\Omega = 0, \pm 1$, and $\Delta\Sigma = 0$ for the spin; there are also some rules for symmetry properties (see Herzberg[2]). There are no selection rules for vibrational transitions.

The selection rules for rotational transitions $\Delta J = J' - J''$ are given by $\Delta J = 0, \pm 1$, with the restriction that transitions from $J = 0$ to $J = 0$ or from $\Omega = 0$ to $\Omega = 0$ are not allowed. The line series for which $\Delta J = 0$ is called the $Q$ branch, the series with $\Delta J = -1$ is the negative or $P$ branch.

A band represents the collection of all lines (of the $Q$, $R$, $P$ branches) that belong to one electron transition and one specific vibrational transition $v$.

### 2.4.4. The $N_2^+(1-)$ Spectra

As an example, let us consider the simple case of the (1−) system of $N_2^+$, which is often seen in thermal plasmas where nitrogen is the plasma gas or when nitrogen is entrained during the operation of a plasma jet in

**TABLE 2.5.** Electronic Data for Selected Molecules

| Molecule | Molecular weight $10^{-3}$ kg | Thermal dissociation energy (eV) | Ground states | | | Metastable states | | | First ionization | | Dissociative ionization | |
|---|---|---|---|---|---|---|---|---|---|---|---|---|
| | | | Designation | $E_v$ (eV) | $E_r$ (eV) | Designation | $E_e$ (eV) | $\tau_m$ (s) | Energy (eV) | State | Energy (eV) | State |
| $H_2$ | 2.02 | 4.588 | ${}^1\Sigma_g^+$ | 0.545 | 1.50E-2 | $C^3\Pi_u$ | 11.87 | 1.02–1.76 | 15.426 | $H_2^+$ | 18.0 | $H^+$ |
| $H_2^+$ | 2.02 | 2.648 | ${}^2\Sigma_g^+$ | 0.285 | 7.40E-3 | — | — | — | — | — | — | — |
| $H_2O$ | 18.02 | 9.509 | — | 0.198 | 1.00E-3 | — | — | — | 13.0 | $H_2O^+$ | 18.7 | $OH^+$ |
| $N_2$ | 28.01 | 9.756 | ${}^1\Sigma_g^+$ | 0.293 | 4-98E-4 | $A^3\Sigma_u^+$ | 6.224 | 1.36 | 15.58 | $N_2^+$ | 23.4 | $N^+$ |
| | | | | | | $a^1\Sigma_u^-$ | 8.400 | 0.017–0.5 | | | | |
| | | | | | | $a^1\Pi_g$ | 8.590 | — | | | | |

atmospheric air.[8] In this system transitions occur between the electronically excited state $B^2\Sigma_u^+$ and the fundamental state $X^2\Sigma_g^+$ of $N_2^+$.

#### 2.4.4.1. Rotational Structure

Since both the upper and lower states are $\Sigma$ states, Hund's case (b) always applies; with $\Lambda = 0$ and $\Sigma = 0$, the spin vector $\vec{S}$ is not coupled with the internuclear axis at all and $\vec{\Omega}$ is not defined. In this case $S = 1/2$ and we have

$$J = K \pm 1/2$$

The transition rule for $K$ requires that $\Delta K = \pm 1$, $\Delta K = 0$ being forbidden for $\Sigma$ levels. The separation of the two sublevels $J$, for a given $K$, will in general be very small compared to the separation of successive rotational levels. If we neglect the sublevels, each band consists of a $P$ branch ($\Delta K = -1$) and an $R$ branch ($\Delta K = +1$). Using the simplified formula (2.54) in Eq. (2.57) and writing $K'$ as a function of $K''$ (denoted as $K$ in the following), we obtain

$$\sigma_P(K) = \sigma_0' - (B_{v'}' + B_{v''}'')K + (B_{v'}' - B_{v''}'' - D_{v'}' + D_{v''}'')K^2 + \cdots \tag{2.58}$$

and

$$\sigma_R(K) = \sigma_0' + (2B_{v'}' - 4D_{v'}') + (3B_{v'}' - B_{v''}'' - 12D_{v'}')K + (B_{v'}' - B_{v''}'' - 13D_{v'}' + D_{v''}'')K^2 + \cdots \tag{2.59}$$

where $\sigma_0'$ is the wave number for $J' = J'' = 0$.

The selection rule $\Delta K = \pm 1$ gives $K = 1$ for the smallest value of the $P$ branch ($\Delta K = -1$) and $K = 0$ for the $R$ branch ($\Delta K = +1$) (see Fig. 2.13). In such a case the preceding formulas (2.58) and (2.59) show that there is no line at the position $\sigma = \sigma_0$.

In fact, if the fine structure of the band (the separation of the $J$

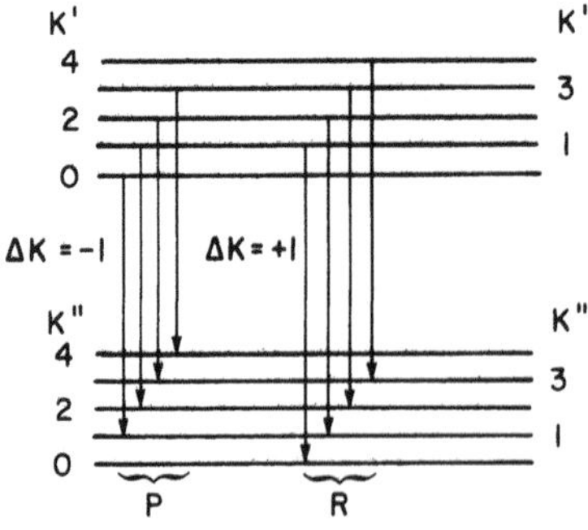


**FIG. 2.13.** Rotational transitions for the $P$ and $R$ branches of the transition $B^2\Sigma_u^+ \rightarrow X^2\Sigma_g^+$ of $N_2^+$ [(1−) transition].

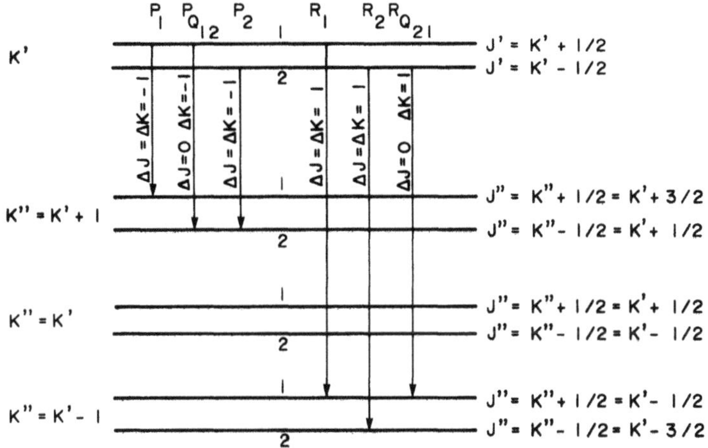


**FIG. 2.14.** Rotational transitions for the principal ($\Delta K = \Delta J$) and satellite branches ($\Delta K \neq \Delta J$) for the $N_2^+(1-)$.

lines) is considered the structure of the band is more complex: each band ($P$ and $R$) is divided into three branches because each line is separated into three components according to the values $\Delta J$ (see Fig. 2.14). However, the differences in wavelength between the three $J$ lines in the $P$ branch and the three $J$ lines in the $R$ branch are smaller than 0.1 Å and, unless a very high resolution monochromator (resolution power greater than 200,000) is used, only one line is seen in the spectra. In the following discussion we will consider only lines from the $P$ and $R$ branches, corresponding to $\Delta K = -1$ and $+1$, respectively, neglecting $\Delta J$.

The values of the corresponding coefficients for the $N_2^+$ system are given in Table 2.6. If each branch is represented by plotting the wave number of each line along the abscissa and the corresponding $K$ value along the ordinate, the Fortrat parabola [see Fig. 2.15 for the $N_2^+(1-)$ transition] is obtained; it can take two forms according to the values of $B_v'$ and $B_v''$ (the $D_v$ terms are negligible).

Owing to the quadratic term in $K^2$, some of the lines of the same branch run together to form a "band head" where only lines of the same branch are found. In the case of the $N_2^+$ 0–0(1−) transition, the first 26 lines of the $P$ branch form the band head shown in Fig. 2.16. The first peak of the band head is the integral of the first 17 $P$ lines. Of course the $R$ lines are found only for wave numbers higher than $\sigma_0'$ (see Fig. 2.15), and the $R$ branch corresponds to lines of the $P$ branch with $K_p > 26$. Because of the dispersion of the monochromator used to register the spectra shown in Fig. 2.16 and the small wavelength difference between

**TABLE 2.6.** Electronic, Vibrational, and Rotational Constants of the $B^2\Sigma_u^+$ and $X^2\Sigma_g^+$ states of $N_2^+$

| State | $T_e$ ($cm^{-1}$) | $\omega_e$ ($cm^{-1}$) | $\omega_e x_e$ ($cm^{-1}$) | $\omega_e y_e$ ($cm^{-1}$) | $B_e$ ($cm^{-1}$) | $\alpha_e$ ($cm^{-1}$) | $D_e$ ($cm^{-1}$) | $\beta_e$ ($cm^{-1}$) | $r_e$ ($cm^{-1}$) | $\sigma_{00}$ ($cm^{-1}$) |
|---|---|---|---|---|---|---|---|---|---|---|
| $B^2\Sigma_u^+$ | 25,461.5 | 2419.84 | 23.19 | −0.5375 | 2.083 | 0.0195 | — | — | 0.1075 | B X 25,566.0 |
| $X^2\Sigma_g^+$ | 0 | 2207.19 | 16.14 | −0.0400 | 1.9328 | 0.0208 | $5.75 \times 10^{-6}$ | $0.29 \times 10^{-6}$ | 0.1118 | — |

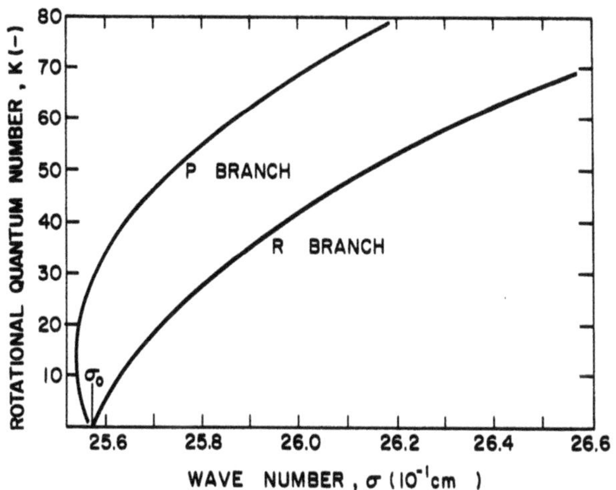


**FIG. 2.15.** Fortrat's parabola for the 0–0 band of the transition $B^2\Sigma_u^+ \rightarrow X^2\Sigma_g^+$ of $N_2^+$.

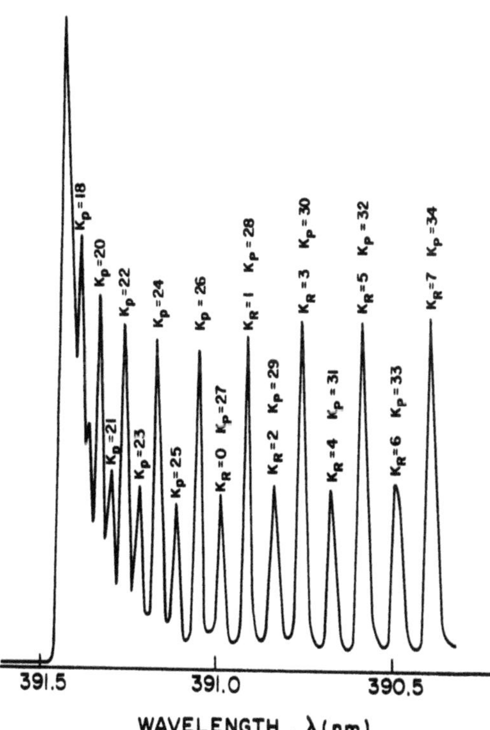


**FIG. 2.16.** (0–0) band head and first rotational lines of the system $B^2\Sigma_u^+ \rightarrow X^2\Sigma_g^+$ of $N_2^+$.

two adjacent lines of $P$ and $R$ branches (less than 0.2 Å), the line corresponding to $K_P = K_R + 27$ almost overlaps with the line $R$ corresponding to $K_R$, and both lines merge in a unique line, sometimes referred to as a global or total line. This is why the lines corresponding to wavelengths smaller than 391.0 nm in Fig. 2.16 are referred to by two $K$ numbers, one for the line from the $P$ branch and one for the line from the $R$ branch. Figure 2.16 (which is based on experimental results) also shows that the line intensities alternate. This alternation occurs because the rotational statistical weight includes the nuclear spin,[3] which is different for the symmetrical (index $s$) and for the antisymmetrical (index $a$) levels (see Herzberg[2] for symmetry properties):

$$g_s = \frac{I+1}{2I+1}, \qquad g_a = \frac{I}{2I+1}$$

where $I$ is the spin vector quantum number. In the (1−) spectra of $N_2^+$, $I = 1$, the symmetrical levels corresponding to odd values of the initial state of $K'$ will have their intensity multiplied by $\frac{2}{3}$, while those corresponding to even values will have their intensity multiplied by $\frac{1}{3}$. Thus in the line notation ($K = K''$), the lines with $K_P$ or $K_R$ even (corresponding to $K' = K \pm 1$, i.e., $K'$ odd) will have twice the intensities of the lines for which $K$ is odd ($K' = K \pm 1$ then being even). Thus the lines with even $K_P$ are about twice the intensity of the lines with odd $K_P$; this can be seen very clearly from Fig. 2.16 for $K_P < 27$ (where only $K_P$ lines are present). For "global" lines where an even $K_R$ is superimposed with an odd $K_P$, some sort of compensation takes place. But at $K_P = K_R + 27$ and beyond, the intensity of $K_P$, which is proportional to $2K + 1$ (see Section 6.2.4) dominates the intensity of $K_R$ and the global lines with even $K_P$ have higher intensities than those with odd $K_P$.

### 2.4.4.2. Vibrational Structure

If we consider only transitions between levels for which $J = 0$ and use the approximate expression for $F(J) = B_v J(J+1)$, the vibrational structure is given by

$$\sigma_0' = T_e' - T_e'' + G'(v') - G''(v'') = \sigma_e' + \sigma_v' \tag{2.60}$$

Using expression (2.52) and writing $\Delta v = v' - v''$ ($\Delta v$ = constant is called a sequence), $\sigma = \sigma_0' - \sigma_0''$ can be expressed as

$$\begin{aligned}\sigma = \sigma_{00} &+ \omega_0' \Delta v - \omega_0' x_0' (\Delta v)^2 \\ &- (\omega_0'' - \omega_0' + 2\omega_0' x_0' \Delta v) v'' - (\omega_0' x_0' - \omega_0'' x_0'') v''^2 \cdots\end{aligned} \tag{2.61}$$

**TABLE 2.7.** Deslandre's Table for the Bands of $B^2\Sigma_u^+ \rightarrow X^2\Sigma_g^+$ of $N_2^+$ [a]

| $v'$ \ $v''$ | 0 | 1 | 2 | 3 | 4 | 5 | 6 | 7 | 8 | 9 | 10 |
|---|---|---|---|---|---|---|---|---|---|---|---|
| 0 | (25,566)<br>391.44 | (23,391)<br>427.81 | (21,249)<br>470.92 | (19,140)<br>522.83 | (17,064)<br>586.47 | | | | | | |
| 1 | (27,937)<br>358.21 | (25,763)<br>388.43 | (23,620)<br>(423.65 | (21,511)<br>465.18 | (19,435)<br>514.88 | (17,392) | | | | | |
| 2 | (30,256) | (28,081)<br>356.39 | (25,939)<br>385.79 | (23,830)<br>419.91 | (21,754)<br>459.97 | (19,711)<br>507.66 | 565.31 | | | | |
| 3 | (32,517) | (30,342)<br>329.87 | (28,200)<br>354.89 | (26,091)<br>383.54 | (42,014)<br>416.68 | (21,972)<br>455.41 | (19,962)<br>501.27 | (17,987)<br>556.41 | | | |
| 4 | | (32,537) | (30,395)<br>329.34 | (28,286)<br>353.83 | (26,210)<br>381.81 | (24,167)<br>414.05 | (22,158)<br>451.59 | (20,182)<br>495.79 | | | |
| 5 | | | (32,518) | (30,409) | (28,332)<br>353.26 | (26,290) | (24,280)<br>412.10 | (22,305)<br>449.03 | (20,364)<br>491.32 | (18,457)<br>542.08 | (16,584) |
| 6 | | | | (32,449) | (30,373) | (28,330) | (26,321) | (24,345)<br>411.09 | (22,404)<br>446.66 | (20,497)<br>488.17 | (18,624)<br>532.73 |
| 7 | | | | | (32,320) | (30,278) | (28,268) | (26,293) | | (22,445)<br>(445.93) | (20,572)<br>486.44 |
| 8 | | | | | | (32,121) | (30,112) | (28,137) | (26,195) | | (22,416)<br>446.66 |
| 9 | | | | | | | (31,839) | (29,864) | (27,922) | (26,015) | |
| 10 | | | | | | | | (31,460) | (29,519) | (27,612) | (25,739) |

[a] Numbers in parenthesis are in $cm^{-1}$, all others in nm.

where $\sigma_{00}$ is the wavenumber of the transition taking place between the level $v' = 0$ of the upper electronic state and $v'' = 0$ of the lower electronic state. All transitions between two levels $v'$ and $v''$ are allowed, whatever the value of $\Delta v$.

For a given electronic system the wavelengths of each band are presented as a function of $v'$ and $v''$ in a table (Deslandre's Table) in such a manner that the bands of each $v''$ progression are found on a horizontal line and the bands of each $v'$ progression in a vertical column. The data from Table 2.6 have been summarized in Table 2.7 (Deslandre's Table) to give the wavelengths of the band heads, and the wave numbers of the zero lines ($K' = K'' = 0$) for some given transitions ($v' - v'' = 2, 1, 0, -1, -2, -3$) of the system $B^2\Sigma_u^+ \rightarrow X^2\Sigma_g^+$ for $N_2^+$. It can be seen[2] that the bands which lie on the principal diagonal of the table (from the upper left corner to the lower right corner) or on a parallel to this diagonal are generally close together in wavelength; such bands, which are characterized by $\Delta v$ = constant, are called sequences. The bands of a sequence follow a progression that is a quadratic function of $v''$ except when $\omega_0''x_0'' = \omega_0'x_0'$.

Figure 2.17 gives the relative intensity (in arbitray units) of a few vibrational transitions for the $N_2^+$ (1−) system at 8000 K. This DC nitrogen plasma jet spectrum, obtained with a large opening at the

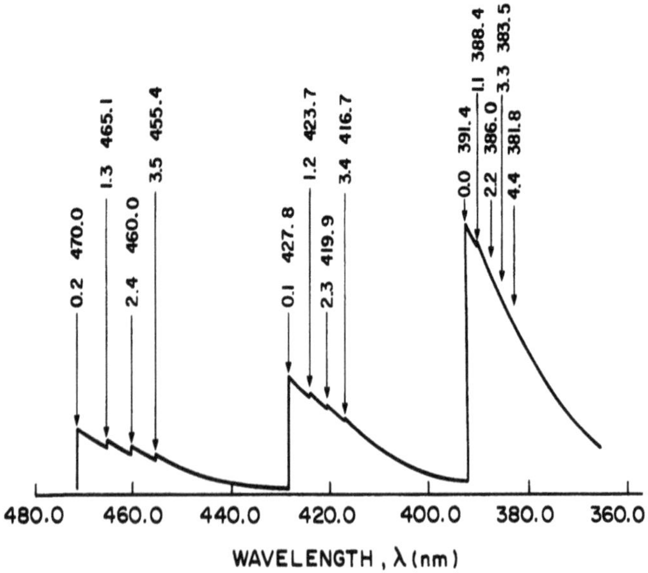


**FIG. 2.17.** Location of the various bands of the system $B^2\Sigma_u^+ \rightarrow X^2\Sigma_g^+$ of $N_2^+$ between 470.0 nm and 370.0 nm (the scale of the intensity is arbitrary).

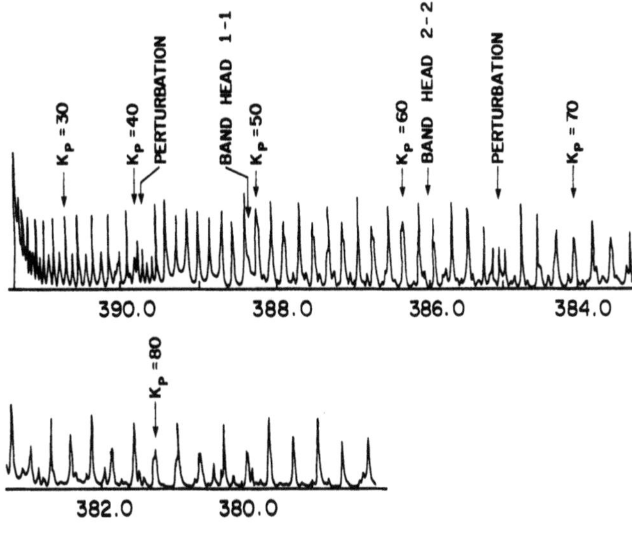


**FIG. 2.18.** Detailed structure of the 0–0, 1–1, and 2–2 vibrational transitions of the system $B^2\Sigma_u^+ \rightarrow X^2\Sigma_g^+$ of $N_2^+$.

entrance slit of the monochromator, represents an integration of the rotational lines like those shown in Fig. 2.16. The complexity of the spectra at high temperatures is due to the strong development of the bands (the rotational structure of the molecule is excited up to $J$ = 100–120). For example, for the 0–0 transition of (1−), the rotational lines for $K_p > 49$ of this band overlap with the 1–1 band, and so on (see Fig. 2.18 for a detailed spectrum of the 0–0 band). Thus, the interpretation of the molecular spectra is not straightforward, and a lot of care must be taken in evaluating rotational and vibrational temperatures from these spectra.

## LIST OF SYMBOLS

$a$ constant
$A$ separation constant
$a_p$ coefficients
$B_e$ rotational constant of a diatomic molecule at rest (cm$^{-1}$)
$c$ velocity of light ($2.998 \times 10^8$ m/s)
$D_e$ energy difference between the equilibrium position of the nuclei of a diatomic molecule and the free atom (cm$^{-1}$)
$e$ elementary charge (electron charge = $-1.6 \times 10^{-19}$ As)

| | |
|---|---|
| $E$ | Energy (eV or $cm^{-1}$) |
| $g$ | statistical weight |
| $h$ | Planck's constant ($\approx 6.6 \times 10^{-34}$ $Ws^2$) |
| $\hbar$ | $= h/2\pi$, reduced Planck's constant |
| $I$ | moment of inertia or spin vector quantum number |
| $i$ | imaginary unit ($=\sqrt{-1}$) |
| $\vec{J}$ | total momentum |
| $J$ | associated quantum number |
| $K$ | $\vec{K} = \vec{\Lambda} + \vec{N}$ |
| $k$ | Boltzmann constant ($k = 1.38 \times 10^{-23}\, J/K \cdot$ particle) |
| $\vec{L}$ | orbital angular momentum |
| $L$ | associated quantum number |
| $l$ | quantum number of the angular momentum |
| $m$ | mass of a particle (or separation constant) (kg) |
| $M$ | mass of an atom (kg) |
| $m_e$ | electron mass ($9.11 \times 10^{-31}$ kg) |
| $m_l$ | magnetic quantum number |
| $\vec{N}$ | angular momentum of rotation of the nuclei of a diatomic molecule |
| $n$ | principal quantum number |
| $P_l^{(m)}$ | polynomial of degree $l$ and of order $m$ |
| $P(r)$ | probability density function |
| $r$ | radial coordinate |
| $R(r)$ | radial part of the wave function |
| $R_\infty$ | Rydberg constant |
| $r_1$ | radius of first Bohr orbit (ground state) (m) |
| $\vec{S}$ | total spin momentum |
| $S$ | associated quantum number |
| $s$ | spin quantum number |
| $t$ | time (s) |
| $T(e)$ | equivalent electronic energy of a diatomic molecule ($cm^{-1}$) |
| $T_e$ | electron temperature (K) |
| $T_h$ | temperature of the heavy particles (K) |
| $T_K$ | kinetic energy (J) |
| $u$ | phase velocity of a wave |
| $u(\rho)$ | polynomial |
| $v$ | vibrational quantum number |
| $v_e$ | electron velocity (m/s) |
| $V_P$ | potential energy (J) |
| $X$ | symbol for an atomic chemical species in the ground state |
| $X^*$ | symbol for an atomic chemical species in an excited state |
| XY | diatomic molecule in its ground state |

XY* diatomic molecule in an excited state
$x$ coordinate
$y$ coordinate
$z$ coordinate
$Z'$ number of protons in a nucleus
$Z$ number of protons and neutrons in a nucleus

### Greek Symbols

$\Delta$ Laplace operator
$\varepsilon_0$ dielectric constant ($=8.86 \times 10^{-12}$ As/Vm)
$\theta$ cylindrical coordinate
$\lambda$ wavelength (nm)
$\lambda_n$ parameter ($= E_1/E_n$)
$\lambda_{ul}$ wavelength corresponding to a transition from an upper level $u$ to a lower level $l$ (nm)
$\Lambda$ associated quantum number
$\vec{\Lambda}$ angular momentum
$\mu$ reduced mass (kg)
$\nu$ frequency ($s^{-1}$)
$\rho$ reduced coordinate
$\sigma$ wave number ($cm^{-1}$)
$\sigma_e$ electrical conductivity ($ohm^{-1}\, m^{-1}$)
$\phi$ Cylindrical coordinate
$\psi$ amplitude of wave function
$\psi^*$ conjugate complex wave function
$\Psi$ wave function
$\omega_e$ vibrational term when vibration is assumed to be harmonic ($cm^{-1}$)
$\vec{\Omega}$ $\vec{\Omega} = \vec{\Lambda} + \vec{S}$

### Subscripts

$u$ upper energy level
$l$ lower energy level
$n$ refers to principal quantum number
$\infty$ infinite mass (or infinite distance)

## GENERAL BIBLIOGRAPHY

Bashkin, S. and J. Stoner Jr., eds., *Atomic Energy Level and Grotrian Diagrams,* Amsterdam: North-Holland, 1978.

Cohen-Tannoudiji, C., B. Diu, and F. Laloë, *Quantum Mechanics,* Volume 1, New York: Wiley, 1977.

Condon, E. U. and G. H. Shortley, *The Theory of Atomic Spectra,* Cambridge: Cambridge University Press, 1959.
Delcroix, J. L., *Physique des plasmas,* Paris: Monographic Dunod, 1966.
Finkelnburg, W., *Structure of Matter,* Berlin: Springer-Verlag, 1964.
*Handbook of Chemistry and Physics,* 49th Cleveland, Ohio: Chemical Rubber Co., 1968.
Kuhn, H. G., *Atomic Spectra,* New York: Academic Press, 1962.
Messiah, A., *Quantum Mechanics,* Volumes 1 and 2, Amsterdam: North-Holland, 1958.
Morrison, M. N., T. L. Estle, and N. F. Lane, *Quantum States of Atoms, Molecules, and Solids,* Englewood Cliffs, N.J.: Prentice-Hall, 1976.
Schiff, L. I., *Quantum Mechanics,* New York: McGraw-Hill, 1968.

## REFERENCES

1. G. Herzberg, *Atomic Spectra and Atomic Structure* (New York: Dover Publications, 1944).
2. G. Herzberg, *Spectra of Diatomic Molecules* (New York: D. van Nostrand, 1959).
3. P. Fauchais, K. Lapworth, and J. M. Baronnet, "First Report on Measurement of Temperature and Concentration of Excited Species in Optically Thin Plasmas" (IUPAC Subcommittee on Plasma Chemistry, P. Fauchais, ed., Limoges, France: Limoges University, 1980).
4. C. E. Moore, *Atomic Energy Levels,* NBS Circular 467, **1** (1949).
5. C. E. Moore, *Atomic Energy Levels,* NBS Circular 467, **2** (1952).
6. C. E. Moore, *Atomic Energy Levels,* NBS Circular 467, **3** (1958).
7. D. E. Shemansky and A. V. Jones, *Planet. Space Sci.* 16 (1968): 1115.
8. J. M. Baronnet, "Étude comparative de méthodes de mesure de la température de rotation-vibration de la molécule $N_2^+$ dans un plasma d'azote produit par un générateur à arc," Thèse de 3eme cycle (Université de Limoges, France, 1971).

## Chapter 3

# Kinetic Theory

## 3.1. PARTICLES AND COLLISIONS

In a plasma six basic types of particles can be identified:

- free electrons (designated as $e$);
- atoms and molecules in their fundamental or ground state (designated as X);
- excited atoms and molecules (generally designated as $X^*$);
- positive ions (atomic ions $X^+$, $X^{++}$, $X^{+++}, \ldots$, or molecular ions such as $X_2^+$);
- negative ions; certain atoms and molecules, particularly those with an almost-completed outer electron shell, form negative ions when an electron attaches itself to the neutral particle;
- photons; they have no mass, and their velocity $c$ is equal to the velocity of light.

When two particles initially separated by a large distance $d$ approach one another, they start to interact; if after the interaction some measurable (in principle) change has occurred, we say that a collision has taken place. According to the classical mechanical model in which actual particles (excluding photons) are replaced by rigid spheres without any electronic structure, a collision would occur only if such spheres made physical contact. In reality, particles "sense" each other as a result of their electronic structure long before they come into physical contact. The initial interaction as two particles approach each other is due to a mutual deformation of their electronic shells (for example, polarization), resulting in an attractive force (for the case of the Lenard-Jones potential, this force is $\sim d^{-7}$). As the outer electron shells of the two particles begin to penetrate each other, a strong repulsive force is experienced due to the positively charged nuclei (for the Lenard-Jones potential, this force is

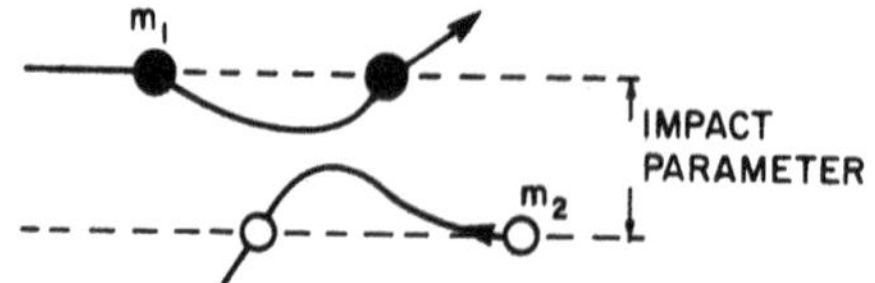


FIG. 3.1. Schematic of a collision process between heavy particles.

$\sim d^{-13}$). The interaction potential between two particles usually consists of a longer-range attractive part and a very-short-range repulsive part. Interactions of neutral particles following, for example, the Lenard-Jones potential description are classified as short-range interactions (see Fig. 3.1). In contrast, interactions between charged particles result in long-range interactions because of the far-reaching Coulomb force ($\sim d^{-2}$) between charged particles.

The result of this mutual influence or collision is that the particles will sensibly deflect each other's paths. A collision is defined as a resulting deflection greater than a certain minimum value. The kinetic and/or potential energy of the participating particles changes as a consequence of a collision, and the distinction between elastic and inelastic collisions is based on this energy change.

*Elastic collisions* are collisions in which the total kinetic energy is conserved. Practically all collisions in neutral gases at ambient temperatures are elastic. In the elastic collision process, the fraction of energy, $K$, transferred from one particle of mass $m$ to another of mass $M$ averaged over all angles is

$$K = \frac{2mM}{(m + M)^2} = \frac{2m}{M} \qquad \text{when } m \ll M \tag{3.1}$$

This form of energy transfer is of little importance for electron–atom collisions. In contrast, injecting a cold gas into a thermal plasma causes the mean kinetic energy (or temperature) of the injected atoms and molecules (heavy particles) to change drastically (a few collisions are sufficient), while the electron kinetic energy (electron temperature) is little affected (thousands of collisions are necessary).

*Inelastic collisions* are collisions in which the total kinetic energy of the particles changes as a modification of the internal energy of the particles occurs. Processes such as excitation, ionization, recombination, charge transfer, attachment, detachment, and dissociation are typical examples of inelastic collisions.

If $\Delta E_{\text{total}}$ is the total kinetic energy change suffered by the collision partners during an encounter, then $\Delta E_{\text{total}} < 0$ indicates that part of the kinetic energy of one of the particles was used to increase the internal energy of the other particle; $\Delta E_{\text{total}} > 0$ indicates a transformation of part of the internal energy of one of the particles into kinetic energy of the other particle (a superelastic collision).

## 3.2. CROSS SECTIONS AND COLLISION FREQUENCIES

For more details the reader can refer to the books of Delcroix (1966), Fowler (1956), Landau and Lifshitz (1967), and Reif (1988).

### 3.2.1. Collision Probabilities

Let $P(t)$ be the probability that a particle with relative velocity ($\vec{V} = \vec{v}_1 - \vec{v}_2$) *survives* a time $t$ without suffering a collision [of course, $P(0) = 1$ and $P(\infty) = 0$].

To describe the collision, let $w \cdot dt$ be the probability that a particle suffers a collision between $t$ and $t + dt$. Quantity $w$ is thus the probability per unit time that a particle suffers a collision; $w$ is also known as the *collision rate.*

The customary assumption is that $w$ is *independent* of the past history of the particle, i.e., it does not matter when the particle suffered its last collision. Assuming that $\vec{V}$ does change appreciably in times of the order of $w^{-1}$, it follows that

$$P(t + dt) = P(t)(1 - w\,dt) \tag{3.2}$$

Hence

$$\frac{1}{P}\frac{dP}{dt} = -w \tag{3.3}$$

or

$$P = C\exp\left(-\int_0^t w(t')\,dt'\right) \tag{3.4}$$

If $w$ does not vary with $\vec{V}$, it follows that

$$P = C\exp(-wt) \tag{3.5}$$

and because $P(0) = 1$, $C$ must be equal to one and hence

$$P = \exp(-wt) \tag{3.6}$$

If $P^*(t)\,dt$ is defined as the probability that a particle, *after surviving* without collisions for a time $t$, *suffers* a collision in the time interval from $t$ to $t + dt$,

$$P^*(t)\,dt = \exp(-wt)w\,dt \tag{3.7}$$

$P^*(t)\,dt$ is also normalized (asserting that one particle collides at some (time):

$$\int_0^\infty P^*(t)\,dt = \int_0^\infty \exp(-wt)w\,dt = \int_0^\infty \exp(-y)\,dy = 1 \tag{3.8}$$

The mean time between two collisions (referred to as the "collision time" or "relaxation time") is defined as

$$\tau = \bar{t} = \int_0^\infty tP^*(t)\,dt \tag{3.9}$$

$$\tau = \int_0^\infty \exp(-wt)tw\,dt \tag{3.10}$$

$$\tau = \frac{1}{w} \tag{3.11}$$

This expression will be used in Section 3.2.4 to derive an expression for the mean free path (mfp).

### 3.2.2. Collision Cross Sections

Consider two particles of respective masses $m_i$ and $m_j$, with position vectors $\vec{r}_i$ and $\vec{r}_j$ and velocities $\vec{v}_i$ and $\vec{v}_j$. From a frame of reference that is fixed with respect to particle $j$, their motion is described by their relative position $\vec{R}_{ij} = \vec{r}_i - \vec{r}_j$ and their relative velocity $\vec{V}_{ij} = \vec{v}_i - \vec{v}_j$. In this frame of reference it can be assumed that particle $j$ is the target (see Fig. 3.2a) at rest and that a uniform flux ($F_i$) of type $i$ particles per unit area and unit time impinges with a relative velocity $\vec{V}_{ij}$ on the target. A number $dN_i$ of particles of type $i$ will be scattered by the target particle per unit time and will have final velocities in the range $\vec{V}'_{ij}$ to $\vec{V}'_{ij} + d\vec{V}'_{ij}$. At large distances from the target particle we can define a small solid angle $d\Omega'$ about the direction $\theta$, $\varphi$ of the scattered beam (see Fig. 3.2b). If the collision process is elastic so that energy is conserved, then $|\vec{V}'| = |\vec{V}|$. $dN_i$ is proportional to the incident flux $F_i$ and to the solid angle:

$$dN_i = F_i \cdot \sigma(\vec{V}_{i,j}, \theta, \varphi)\,d\Omega' \tag{3.12}$$

$\sigma(\vec{V}_{i,j}, \theta, \varphi)$ is called the *differential scattering cross section*. It has the

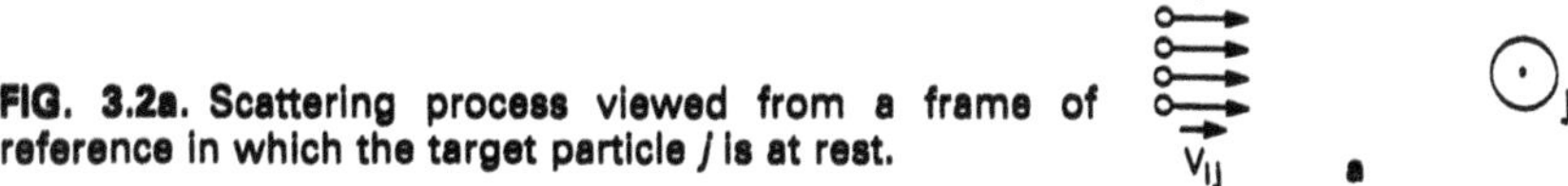


**FIG. 3.2a.** Scattering process viewed from a frame of reference in which the target particle $j$ is at rest.

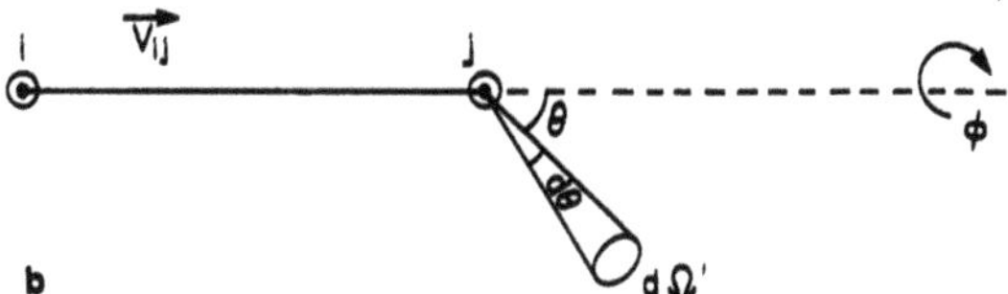


**FIG. 3.2b.** Collision process between two hard spheres of radii $r_i$ and $r_j$.

dimensions of area per unit solid angle since $F_i$ is expressed per unit area and unit time.

The total number $N_i$ of particles of type $i$ scattered per unit time in *all* directions is obtained by integrating Eq. (3.12) over the full solid angle:

$$N_i = \int_{\Omega'} F_i \sigma_{ij} \, d\Omega' = F_i \sigma_0(V_{ij}) \tag{3.13}$$

where

$$\sigma_0(V_{ij}) = \int_{\Omega'} \sigma(\vec{V}_{ij}, \theta, \varphi) \, d\Omega' \tag{3.14}$$

is called the total scattering cross section; it has units of area ($m^2$) and depends on the magnitude of the relative velocity of the two particles.

The differential cross section is often written as

$$\sigma_{ij}(\vec{V}_{ij}, \theta, \varphi) \, d\Omega' = p_{ij}(\vec{V}_{ij}, \theta, \varphi) \sigma_{ij}(V_{ij}) \, d\Omega' \tag{3.15}$$

where $p_{ij}(V_{ij}, \theta, \varphi)$ is a proportionality factor. Since

$$\sigma_0(V_{ij}) = \int_0^{\pi} \int_0^{2\pi} \sigma_{ij}(\vec{V}_{ij}, \theta) \sin\theta \, d\theta \, d\varphi \tag{3.16}$$

we have

$$\sigma_0(V_{ij}) = 2\pi \int_0^{\pi} \sigma_{ij}(\vec{V}_{ij}, \theta) \sin\theta \, d\theta \tag{3.17}$$

if there is no dependence on the azimuthal angle $\varphi$. The differential cross section is primarily of importance for transport phenomena, where the average scattering angle is an important parameter.

The total collision cross section $\sigma_0(V_{ij})$ can be interpreted as the effective geometrical blocking area that the field particles present to the beam of incident particles (see Fig. 3.2a), and thus for hard, sperical particles

$$\sigma_0 = \pi(r_i + r_j)^2 \tag{3.18}$$

which reduces to

$$\sigma_0 = 4\pi r_i^2 \tag{3.19}$$

for two identical particles.

Although atoms do not have any well-defined dimension but interact with other atoms over a certain range of distances, it is still useful to associate a radius $r$ with an atom. In the simple model of Bohr, $r = n^2 r_1/Z'$, where $n$ is the principal quantum number, $r_1$ the radius of the first Bohr orbit, and $Z'$ the charge of the nucleus. For hydrogen, $r_1 = 0.529 \times 10^{-10}$ m, and if we use the area presented by a sphere of radius $r_1$ as an estimate of the order of magnitude of the total cross section, we obtain

$$\sigma_0 = 4\pi \times (5.3 \times 10^{-11})^2 = 3.53 \times 10^{-20}\,\text{m}^2 \tag{3.20}$$

Thus, depending on the atoms considered, a first approximation indicates a range from $10^{-20}$ to $10^{-19}$ m$^2$ for the total collision cross section $\sigma_0$. The Bohr radius gives generally the right order of magnitude for $\sigma_0$. The actual value of $\sigma_0$ is usually greater than $4\pi r_1^2$.

This definition of the total cross section makes no distinction between the types of collisions that may occur. In fact, it represents an averaging over the elastic and all other relevant nonelastic collision cross sections. Of course, $\sigma_0$ also depends on the relative velocity $\vec{V}_{ij}$ of the target particle $j$ and on the colliding particle $i$. It is also clear that for inelastic collisions $\sigma_0$ will be zero if the relative kinetic energy of the colliding particles does not exceed a threshold energy, that is, the energy of the corresponding excited state. For example, for the first excited state of the H atom, this value is 10.2 eV; for the N atom, 10.34 eV; for the O atom, 9.15 eV; and for the Ar atom, 11.5 eV. On the other hand, if the relative velocity is very high, the time for interactions becomes short and the probability of the collision occurring is reduced.

### 3.2.3. Collision Frequencies and Scattering Cross Sections

The collision frequency or collision rate is defined as $\tau^{-1}$ (see Section 3.2.1), i.e., the inverse of the mean time between collisions which in turn can be related to the total cross section. If $n_i$ is the density of

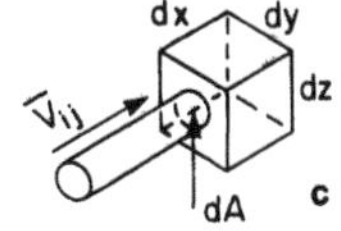


**FIG. 3.2c.** Incident flux of particles of type *i* scattered by particles of type *j*.

particles of type $i$, their relative flux incident on any one particle $j$ situated in a volume $d^3r$ is given by (see Fig. 3.2c)

$$F_i = \frac{n_i(\bar{V}_{ij} \cdot dt \cdot dA)}{dt \cdot dA} = n_i \bar{V}_{ij} \tag{3.21}$$

where $\bar{V}_{ij}$ is the mean relative velocity, which will be defined in Section 3.5.5. A fraction $n_i \bar{V}_{ij} \sigma_0$ of these incident particles is scattered per unit time in all possible directions by one target particle. The total number of particles $i$ scattered by all particles $j$ in volume $d^3r$ is then given by

$$(n_i \bar{V}_{ij} \sigma_0)(n_j d^3r) \tag{3.22}$$

Dividing this by the number $n_i d^3r$ of type $i$ particles in $d^3r$ gives the collision probability or collision frequency $\nu = \tau^{-1}$ (see Section 3.2.1) per unit time for one particle of this type:

$$\tau_{ij}^{-1} = \nu_{ij} = \bar{V}_{ij} \sigma_{ij}^0 n_j \tag{3.23}$$

This collision probability may be enhanced by a large density of target particles $j$, a large relative particle speed (for example, electrons versus heavy particles), and a large total scattering cross section.

### 3.2.4. Mean Free Paths

The mean free path (mfp) $l_{ij}$ is the average path length traversed by a particle of type $i$ between two successive collisions with particles of type $j$. In time $\tau_{ij}$ the particle $i$ covers a distance $\tau_{ij} \cdot \bar{v}_i$, where $\bar{v}_i$ is the mean velocity of particle $i$ (see Section 3.5.4) and thus

$$l_{ij} = \frac{\bar{v}_i}{\bar{V}_{ij}} \cdot \frac{1}{n_j \sigma_0} \tag{3.24}$$

For example, if one considers collisions between electrons $i$ and heavy species $j$, $\bar{V}_{ij} = \bar{v}_i - \bar{v}_j \approx \bar{v}_i$ and thus

$$l_e = \frac{1}{n\sigma_0} \tag{3.25}$$

where $n$ is the number density of neutral particles (neglecting ion density).

In these calculations $\tau_{ij}$, $l_{ij}$, and $\sigma_0$ have been assumed to be independent of $V_{ij}$, but this dependence can be taken into account as shown in Section 3.5.5. In this case it will be shown that the ratio $\bar{v}_i/\bar{V}_{ij} = 1/\sqrt{2}$. For a mixture of different chemical species $j$, the mean free path for particles of type $i$ (for example, electrons) is defined as

$$l_i = \frac{1}{\sqrt{2}\sum_j n_j \sigma_0 \bar{V}_{ij}} \tag{3.26}$$

It is useful to evaluate the mean free path $l$ in a cold gas at atmospheric pressure. Since the density of the molecules in a gas at 0 °C and $10^5$ Pa is $n = 2.69 \times 10^{25}\,\text{m}^{-3}$, with $\sigma = 10^{-19}\,\text{m}^2$ one obtains

$$l = \frac{1}{n\sigma} = 3.7 \times 10^{-7}\,\text{m} \tag{3.27}$$

In an ideal gas (see von Engel[1]) for $100 < p < 10\,\text{MPa}$, one finds generally

$$l \sim T/p \tag{3.28}$$

For $T < 600\,\text{K}$ an equation of the form

$$l(T) = \frac{l_{300}}{1 + C/T} \tag{3.29}$$

can be used. But in gases at very high temperatures such as in thermal plasmas, the relationship has the form

$$l(T) = \text{const} \cdot T^{5/4} \tag{3.30}$$

To obtain the distribution of mean free paths, consider a group of $N_0$ particles, initially moving through the gas with a velocity $v$. Let $N(x)$ be the number of particles reaching $x$ without undergoing any collisions. The number of particles that will undergo collisions between $x$ and $x + dx$ and will leave the initial group of particles will be $dN = N(x) \cdot dw$. Hence, with $dw = n \cdot \sigma \cdot dx$ [see Eqs. (3.11) and (3.23)]

$$P(x) = \frac{N(x)}{N_0} = \exp(-n\sigma x) = \exp(-x/l) \tag{3.31}$$

where $P(x)$ is the probability of a free path exceeding a distance $x$, i.e., 37% of the free path lengths are shorter than $l$ and 63% exceed $l$.

### 3.2.5. Total Effective Cross Section $Q_l(v)$ for Collision Processes

The result of an encounter is then defined as the number of collisions per unit path length for test particles $i$ with particles of type $j$:

$$Q_l = n_j \sigma_{ij} = \frac{1}{l_{ij}} (\text{m}^{-1}) \tag{3.32}$$

Note that due to the presence of particles of density $n_j$ in Eq. (3.32), the pressure and temperature at which $Q_l$ is calculated must be specified. Usually $Q_l$, also called the effective cross section per unit volume, is given at 133 Pa (1 Torr). The total effective cross section for all collisions is the sum of the particular cross sections for particular types of collisions:

$$Q = \sum_l Q_l \tag{3.33}$$

## 3.3. ELEMENTARY PROCESSES FOR ELASTIC COLLISIONS

It will be shown later that electron interactions are the most frequent processes in plasmas. Therefore, special attention will be given to interactions with electrons.

The concept of differential cross section is very important, because at energies above 1 eV, electrons are not scattered isotropically by gas molecules. Instead, they show preferred scattering directions with pronounced forward scattering. In light gases, these extrema are most noticeable over a relatively small range of electron energies (up to 6 eV in $H_2$, and 15 eV in He), but in heavier gases, they are observed at energies up to 800 eV.

The total cross section $\sigma_0$ generally varies strongly with the incident electron energy (see Fig. 3.3a from Brown[2] for noble gases) and reveals the same general behavior for atoms in the same column of the periodic table of elements. For noble gases below 1 eV, the cross section is very small due to diffraction effects (the Ramsauer effect, see Present[2]). For electron energies exceeding 20 eV, the cross section decreases monotonically. Electrons with such energies may gain more and more energy from the electric field in a discharge and simultaneously loose less and less energy by collision (runaway electrons).

Figure 3.3b from Brown[2] shows the behavior of the total cross section for some selected diatomic molecules.

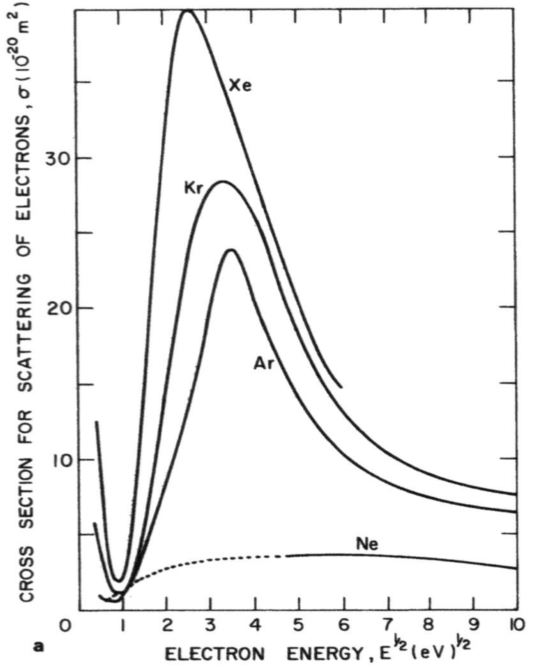


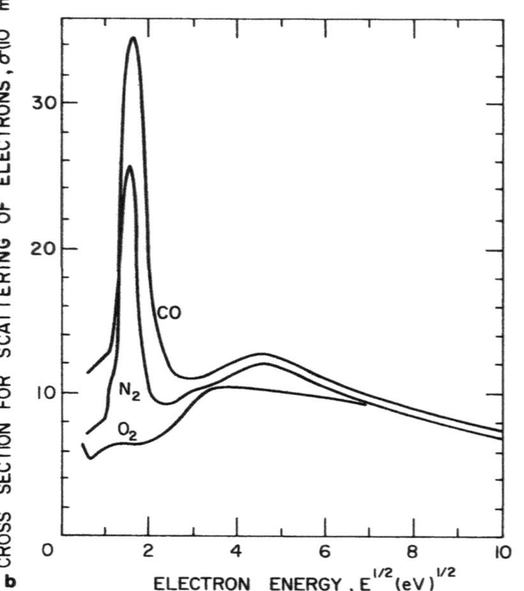


**FIG. 3.3.** Cross sections, $\sigma_0$, for total scattering of electrons (after Brown[2]): (a) in various noble gases, (b) in various diatomic gases.

## 3.4. ELEMENTARY PROCESSES FOR INELASTIC COLLISIONS

Although inelastic collisions belong to the next chapter on gaseous electronics, some of the basic features of inelastic collisions will be discussed in this section. During every collision, a substantial fraction of the kinetic and/or internal energy of the colliding particles will be exchanged or converted into other forms of energy (internal energy, chemical energy).

There are numerous types of inelastic collision processes (more than 200). In this book, the discussion will be restricted, however, to those that are of particular importance for thermal plasmas. For more details on other types of inelastic collisions see Massey and Burhop (1969), Massey (1969), Massey (1971), and Massey and Gilbody (1974).

### 3.4.1. Excitation

As already mentioned, when an atom or a molecule of type X in the ground state absorbs sufficient energy during a collision (with an electron or a heavy particle) or through absorption of a photon, one of its bound electrons reaches a higher energy level; this atom or molecule is said to be in an excited state, denoted by X*, In the following discussion the subscript $i$ will refer to the chemical species and the index $s$ to the excited states (of chemical species $i$).

#### 3.4.1.1. Excitation by Photons

Schematically, the process of excitation by a photon can be written as

$$X + h\nu \rightarrow X^*$$

Photons can produce excited states if the energy of the incident photons, $h\nu$, is at least equal to the energy difference between the upper ($E_u$) and lower ($E_l$) states ($E_u - E_l = E^*$). The probability of the process depends on the selection rules governing the reverse process. *Thus, a metastable state* (from which an electron cannot revert to the ground state by emission of radiation) *cannot be excited from the ground state by absorption of radiation.* The cross sections for photoexcitation are usually very small ($<10^{-22}$ m$^2$). The most probable photoexcitation occurs from the ground state (the reverse process generates the resonance line), and this phenomenon gives rise to the trapping of radiation. For more information see Mitchner and Kruger (1973).

### 3.4.1.2. Excitation by Electron Impact

Electrons will produce excited states if their energy is greater than the energy difference $E^*$ between the upper and lower states:

$$X + e(E_1) \rightarrow X^* + e(E_2), \qquad E_1 - E_2 > E^*$$

In a collision of this type, linear momentum and angular momentum about the center of mass must be conserved. Thus the difference in the angular momentum of the atom between its initial and final states, $\Delta J$ ($J = L + S$), must be balanced by the change in the angular momentum, $\Delta P$, during collisions, i.e.,

$$\Delta P = \hbar \, \Delta J \tag{3.34}$$

Thus the probability of excitation becomes very small if the energy of the electron is just equal to the energy difference between the upper and lower states, since the electron would then have to remain stationary after the collision. The maximum value of $\sigma$ ($\sigma_{\max} \sim 10^{-20}\,\text{m}^2$) is approached with increasing electron energy for energies several times the energy gap for singlet–singlet allowed transitions. In the singlet–triplet (forbidden) transition, the total spin number $S$ changes from 0 to 1, and hence one of the electrons must have its spin vector reversed. For most atoms, this process occurs if one atomic electron is replaced by an impacting electron having the correct spin orientation. In this case $\sigma$ rises rapidly with increasing electron energy, and the maximum is reached a few eV above the threshold energy. Figure 3.4 from Francis[4] shows a

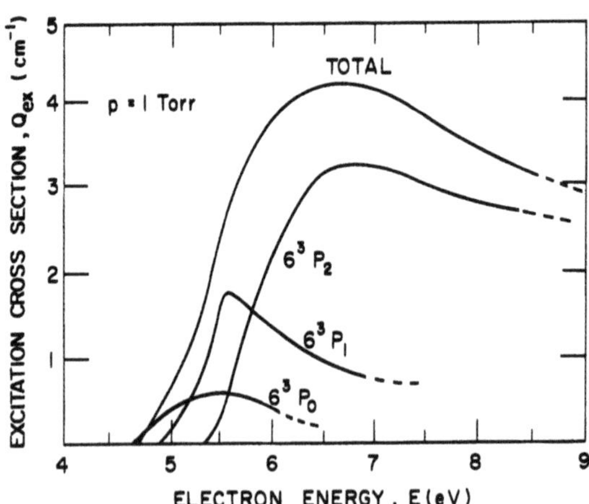


**FIG. 3.4.** Calculated effective cross sections, $Q_{ex}$, for the excitation of the mercury atom from an initial state $7^3S_1$ by electron impact ($p$ = 1 Torr) (after Francis[4]).

typical curve for excitation cross sections. After the maximum is reached, $\sigma$ decreases with a $(\log E)/E$ dependence for allowed transitions and more rapidly (with $1/E$) for forbidden transitions.

For molecules (the differences between rotational energies at 300 K are a few $10^{-3}$ eV), the interaction between the free electrons and the permanent quadrupole moment of the bound electrons (for a certain type of molecule) will excite higher rotational states (provided the selection rule $\Delta J = \pm 2$ for quadrupoles is satisfied). Usually, $\sigma_{rot}$ rises steeply with energy.

The vibrational cross sections (provided the energy of the electron is sufficient, $\varepsilon_{vib} > 0.1$ eV) are generally much larger than the rotational ones.

#### 3.4.1.3. Excitation by Impact of Atoms or Ions

Atoms can be excited by collisions with fast atoms or ions, but the energies, required to reach $\sigma_{max}$ are two orders of magnitude higher than the corresponding energies for electrons. Therefore, this process is of little importance in thermal plasmas (see Francis[4] and McDaniel[5]).

### 3.4.2. Ionization

When an atom or a molecule of type $r$ has absorbed sufficient energy ($E_I$) to release one of its outer electrons, the particle is said to be ionized. Ionization can occur through absorption of a photon, electron impact, or impact of a heavy particle.

#### 3.4.2.1. Ionization by Photons

In the following discussion the photon energies will be restricted to the optical regime. For the ionization of an atom or molecule in the ground state, the photon energy $h\nu$ must be equal to or greater than $E_I$ ($h\nu > E_I$ for the process $h\nu + X \rightarrow X^+ + e$). In terms of the wavelength of the incident photon one finds

$$\lambda_I < \frac{1240}{E_I} \text{(nm)} \tag{3.35}$$

where $E_I$ is in eV. For the alkali elements, $\lambda_I$ must be between 200 and

300 nm, and for noble gases, smaller than 50 nm. The corresponding cross sections rises sharply from zero at the threshold ($E_I$) to a value of about $10^{-21}$ m$^2$ as the photon energy is increased and then passes through successive steps as other electrons are removed.

Ionization of an already excited atom by photons can also occur:

$$h\nu + X^* \rightarrow X^+ + e$$

The energy of the photon, $h\nu$, must satisfy the condition

$$h\nu \geq E_I - E^*$$

### 3.4.2.2. Ionization by Electron Impact

For particles in the ground state, ionization by electron impact occurs when the energy of the electron exceeds the ionization energy of the particle, i.e.,

$$X + e(E_1) \rightarrow X^+ + e(E_2) + e, \qquad E_1 > E_I$$

Figure 3.5 from von Engel[1] shows the effective electron collision cross sections for ionization of various atoms and molecules. The cross section increases rapidly once the threshold energy has been surpassed. The maximum values are on the order of $10^{-20}$ m$^2$ for $\sigma(Q_i \sim 10$ to $20\ \text{cm}^{-1}$ for $p = 133$ Pa) and correspond, in general, to electron energies around 100 eV for most gases (except for alkaline metals, for which they are about 20 eV).

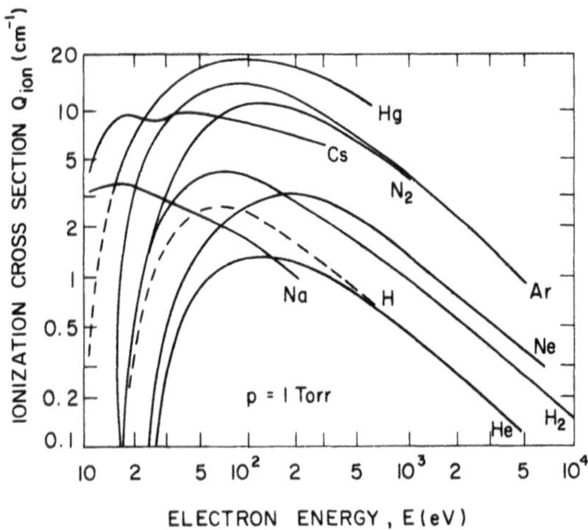


**FIG. 3.5.** Effective cross section, $Q_{ion}$, for ionization by impact in selected gases ($p$ = 1 Torr) (after von Engel[4]).

High-energy electrons can give rise to double ionization of atoms. For example, electrons with $E > 80\,\text{eV}$ can lead to double ionization of helium:

$$e + \text{He} \rightarrow \text{He}^{++} + 3e$$

Since electron energies of this magnitude are practically nonexistent in the thermal plasmas considered here, such processes will be neglected.

Collisions between electrons and molecules may give rise to molecular ions, for example,

$$\text{H}_2 + e \rightarrow \text{H}_2^+ + 2e$$

and also to dissociative ionization:

$$\text{H}_2 + e \rightarrow \text{H}^+ + \text{H} + 2e$$

For the hydrogen molecule, $\sigma_{max}$ occurs at approximately the same energy for both processes, but the value of $\sigma_{max}$ for dissociative ionization is one order of magnitude smaller than the value for molecular ion formation.

### 3.4.2.3. Ionization by Impact of Atoms or Molecules

If the target particle of mass $m_2$ is assumed to be at rest, then the incident particle (mass $m_1$) must have at least a kinetic energy $E_s$ (threshold) given by

$$E_s = \frac{m_1 + m_2}{m_2} E_I \tag{3.36}$$

derived from momentum and energy conservation. In general, $E_s$ is substantially higher than $E_I$. Usually, the increase in the cross section after passing the threshold energy is lower than for electrons, and $\sigma_{max}$ occurs for energies in the keV range. Again, such particle energies are beyond the range considered in the context of this book.

For metastable atoms, ionization may occur by collisions between excited atoms, i.e.,

$$\text{X}^m + \text{X}^m \rightarrow \text{X}_2^+ + e$$

$$\text{X}^m + \text{X}^m \rightarrow \text{X}^+ + \text{X} + e$$

The probability of the first process is supposed to be much smaller than that of the second process.[5]

### 3.4.3. Inelastic Collisions of the Second Kind

Collisions of the second kind are those in which internal energy (excitation energy) from one particle is transferred to another particle. The excitation energy of the colliding particle may be transferred as kinetic energy to the resultant particles, or it may cause excitation or ionization of the receiving particle. In the case of ionization, most of the excess energy is carried off by the electron removed from the particle. As a rule, the cross section for this process is small when the energy difference ($\Delta E$) is large and the velocity of the impacting particle is small.

#### 3.4.3.1. Associative Ionization

The production of molecular ions can occur in the following way:

$$X^* + X \rightarrow X_2^+ + e$$

provided that the energy condition

$$E^* \geq E_b$$

is met, where $E_b$ is the binding energy of $X_2^+$.

#### 3.4.3.2. Ionization of Already Excited Atoms by Electron Impact

$$e(E_1) + X^* \rightarrow X^+ + e + e(E_2)$$

This process can occur when $E_1 \geq E_I - E^*$. The excited atom involved in this process is usually a metastable atom, because of the short lifetime of regular excited states.

#### 3.4.3.3. Charge Exchange Processes

In this type of process there are ionized states both before and after the encounter:

$$X^+ + Y \rightarrow X + Y^+$$

but there is no production of free electrons. If X = Y, one has a so-called resonance phenomenon and the collision probability assumes a maximum. In this case one has a charge exchange process, especially when one of the particles is fast (an ion accelerated in an electric field, subscript $f$) and the other relatively slow (subscript $s$). Schematically, this process can be written

$$X_f^+ + X_s \rightarrow X_f^* + X_s^+$$

#### 3.4.3.4. The Penning Effect

Of special interest for low-pressure electrical discharges is the process in which ionization of one particle occurs by impact on a metastable particle with an excitation energy $E_m \geq E_I$ (the Penning effect). A typical example is ionization of argon ($E_I = 15.76$ eV) by the metastable atom of Ne $^3P_2$ ($E_m = 16.53$ eV). The probability for ionization in this case is large (almost unity per collision) because $\Delta E = 0.77$ eV is small, and the effect is further enhanced by the long lifetime of the metastable state.

For example, Ne–Ar and He–Ar mixtures exhibit the Penning effect, while He–Ne and Ar–Kr do not have the necessary favorable energy-level combinations, as shown in the following table.

| Atom | First metastable state $E_m$ (eV) | Ionization energy $E_i$ (eV) |
|---|---|---|
| Helium | 19.8 | 24.6 |
| Neon | 16.6 | 21.6 |
| Argon | 11.5 | 15.7 |
| Krypton | 9.9 | 14 |

## 3.5. DISTRIBUTION FUNCTIONS

The quantities we have just discussed are related to the relative velocities (energies) of individual particles on impact. But to describe a plasma on a macroscopic scale, microscopic properties have to be averaged, and for this purpose the velocity distribution of the particles is needed. For more details the reader can refer to Fowler (1956), Hirschfelder (1954), Landau and Lifshitz (1967), Mayer and Mayer (1940), Munster (1969), and Reif (1988).

### 3.5.1. Definition

The velocity distribution function $f(\vec{v}, \vec{r}, t)$ depends, in general, on the velocity (components of the velocity vector $\vec{v}$ in velocity space), on time, and on the space coordinate (components of the position vector $\vec{r}$). The number of particles dwelling at time $t$ in the volume element $dx\,dy\,dz$ (which, for convenience, is denoted as $d\vec{r}$) and having velocities in the range from $\vec{v}$ to $\vec{v} + d\vec{v}$ is given by

$$d^6N = f(\vec{v}, \vec{r}, t) \cdot d\vec{v} \cdot d\vec{r} \tag{3.37}$$

The total particle number density $n(\vec{r}, t)$ is then given by

$$n(\vec{r}, t) = \int f(\vec{v}, \vec{r}, t)\, d\vec{v} \tag{3.38}$$

and the integration is carried out over the total velocity space [according to the presence of exponential terms in $f(\vec{v}, \vec{r}, t)$, in most cases integration may be performed for velocity vectors $\vec{v}$ varying between $-\infty$ and $+\infty$].

The distribution function is normalized by the requirement that

$$\int_{-\infty}^{+\infty} f_n(\vec{v}, \vec{r}, t)\, d\vec{v} = 1 \tag{3.39}$$

A *steady state* is characterized by a distribution function $f(\vec{v}, \vec{r})$ independent of $t$; *a uniform distribution* function is characterized by a function $f(\vec{v}, t)$ independent of $\vec{r}$; *an isotropic distribution* exists when $f(\vec{v}, \vec{r}, t)$ depends on $\vec{r}$, $t$, and $|\vec{v}| = v$ (the absolute value of $\vec{v}$ has no directional dependence). For an isotropic distribution, the fraction of particles with velocities between $v$ and $v + dv$ can be expressed by

$$4\pi v^2 f(v, \vec{r}, t)\, dv \tag{3.40}$$

where $4\pi v^2\, dv$ is the volume between the spheres of radii $v$ and $v + dv$ in velocity space.

### 3.5.2. Particle Fluxes

The fundamental kinetic description of a partially ionized gas is provided by the velocity distribution function for each species $i$. Particle fluxes in such a gas may result from the random motion of the particles.

There is, however, no net flux of particles of any type, unless there are gradients that are able to give rise to net particle fluxes.

Any function $\chi(\vec{r}, \vec{v}, t)$ that describes particle properties in terms of position $\vec{r}$ and velocity $\vec{v}$ at time $t$ has a mean value defined by

$$\langle \chi(\vec{r}, t) \rangle = \frac{1}{n(\vec{r}, t)} \int_{-\infty}^{+\infty} f(\vec{r}, \vec{v}, t) \chi(\vec{r}, \vec{v}, t) \, d\vec{v} \tag{3.41}$$

where $n$ is the mean number density of particles per unit volume as defined by Eq. (3.38).

The mean velocity of the particles in volume $d\vec{r}$ is given by

$$\langle \vec{v}(\vec{r}, t) \rangle = \frac{1}{n(\vec{r}, t)} \int_{-\infty}^{+\infty} \vec{v} f(\vec{r}, \vec{v}, t) \, d\vec{v} \tag{3.42}$$

This mean velocity, denoted below by $\vec{v}_g$, is referred to as the species fluid (or average) velocity relative to some laboratory frame of reference.

For collision-dominated gases, it is more convenient to consider particle velocities with respect to a local frame of reference that moves with the mean mass velocity (center of gravity) of the fluid $\vec{v}_g(\vec{r}, t)$ rather than with respect to a laboratory frame of reference. Therefore, the so-called peculiar velocity is introduced:

$$\vec{U} = \vec{v} - \vec{v}_g(\vec{r}, t) \tag{3.43}$$

and by definition

$$\langle \vec{U} \rangle = \langle \vec{v} - \vec{v}_g \rangle = 0 \tag{3.44}$$

which holds only if the gas consists of a single chemical species (see Chapter 7).

The concept of fluxes is very important for transport phenomena, which are determined by the calculation of fluxes of various quantities. Let us consider a surface element $dA$ (with its normal to $\vec{n}$) that divides the whole space in two regions: (+) on the side of $\vec{n}$, (−) on the other side (see Fig. 3.6). This surface may, for example, move with the mean velocity $\vec{v}_g(\vec{r}, t)$. In general, the peculiar velocities $\vec{U}$ are much higher

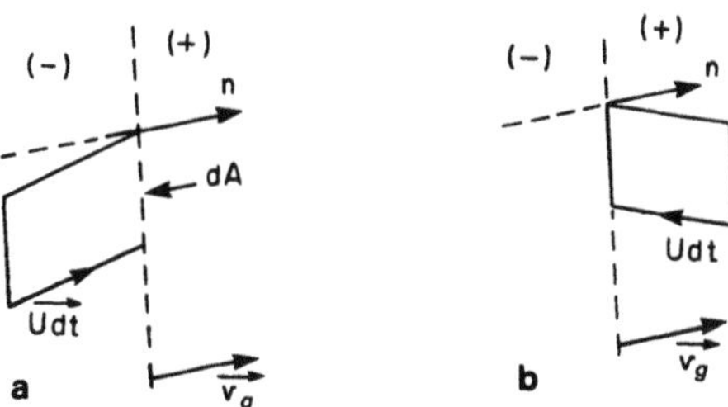


**FIG. 3.6.** Particles crossing the area element $dA$ moving with the gas velocity $v_g$ in time $dt$ from (a) the (−) to the (+) side, and (b) from the (+) to the (−) side.

than $\vec{v}_g$, and thus particles will cross $dA$ in both directions carrying different physical properties $\chi(\vec{r}, \vec{v}, t)$ with them. The normal component of the flux $F_n(\vec{r}, t)$ of $\chi$ through $dA$ is then defined as the total amount of $\chi$ carried per unit time and unit surface from the $(-)$ side to the $(+)$ side of the specified surface.

If $U_n = \vec{n} \cdot \vec{U} > 0$, particles will cross $dA$ from $(-)$ to $(+)$ and the number of particles crossing $dA$ during time $dt$ corresponds to the particles contained in a cylinder of base $dA$ and length $\vec{U}\,dt$ and a corresponding height of $|\vec{n} \cdot \vec{U}\,dt|$. This cylinder contains

$$f(\vec{r}, \vec{v}, t)\, d\vec{v}\, \left| \vec{n} \cdot \vec{U}\, dt \right| dA \tag{3.45}$$

particles, and each of these particles transports the property $\chi$. The total amount of $\chi$ transported through $dA$ from $(-)$ to $(+)$ is then obtained by integrating over all velocities for which $\vec{n} \cdot \vec{U} > 0$:

$$F_n^+ = \int_{\vec{n}\cdot\vec{U}>0} f(\vec{r}, \vec{v}, n\, |\vec{n} \cdot \vec{U}|\, \chi(\vec{r}, \vec{v}, t)\, d\vec{v} \tag{3.46}$$

Similarly, for the particles from $(+)$ to $(-)$ and corresponding to $\vec{U} \cdot \vec{n} < 0$, one finds

$$F_n^- = \int_{\vec{n}\cdot\vec{U}<0} f(\vec{r}, \vec{v}, t)\, |\vec{n} \cdot \vec{U}|\, \chi(\vec{r}, \vec{v}, t)\, d\vec{v} \tag{3.47}$$

The total flux $F_n$ per unit area and unit time is then

$$F_n = F_n^+ - F_n^- = \int_v f\vec{n} \cdot \vec{U}\chi\, d\vec{v} \tag{3.48}$$

where the integration has to be performed over all possible velocities. This expression is the key formula for determining the transport properties of any gas.

### 3.5.3. The Boltzmann Equation

This distribution function can be derived from the Boltzmann equation. The Boltzmann equation represents a particle balance equation in phase space, a six-dimensional space defined by the position $(x, y, z)$ and velocity $(v_x, v_y, v_z)$ coordinates of each particle.

The net inflow of particles into the volume element $d\vec{r}$ can be found by considering the fluxes of such particles across the six faces of the volume element $dx\,dy\,dz$. For the sake of simplicity in the following, particles will be assumed to be of the same species. Particles with velocities between $\vec{v}$ and $\vec{v} + d\vec{v}$ crossing the $y$, $z$ plane at the location $x$ are given by

$$v_x f \, d\vec{v} \, dy \, dz \tag{3.49}$$

where $v_x$ is the velocity component in the $x$ direction at location $x$.

In a similar fashion, particles crossing at $x + dx$ in the $x$ direction are given by

$$v_x f \, d\vec{v} \, dy \, dz + \frac{\partial}{\partial x}(v_x f \, d\vec{v} \, dy \, dz) \, dx \tag{3.50}$$

The net flux across the two faces of $d\vec{r}$ that are normal to $x$ is then

$$v_x f \, d\vec{v} \, dy \, dz - \left[ v_x f \, d\vec{v} \, dy \, dz + \frac{\partial}{\partial x}(v_x f \, d\vec{v} \, dy \, dz) \, dx \right]$$

$$= -\frac{\partial}{\partial x}(v_x f) \, dx \, d\vec{v} \, dy \, dz \tag{3.51}$$

Summing over all six faces of $d\vec{r}$, the net inflow is

$$-\vec{\nabla}_r(\vec{v} \cdot f) \, d\vec{v} \, d\vec{r} \tag{3.52}$$

with the divergence operator $\vec{\nabla}_r$ defined in Cartesian coordinates as

$$\vec{\nabla}_r \vec{W} = \frac{\partial W}{\partial x} + \frac{\partial W}{\partial y} + \frac{\partial W}{\partial z} \tag{3.53}$$

A similar argument can be applied to the effect of acceleration resulting from external forces $\vec{F}$ acting on the particles. Accordingly, we have to consider the particle flux in velocity space across the faces of the volume element $d\vec{v}$. In this velocity space the differential flux associated with the acceleration $\vec{F}_x/m$ in the direction $x$ is given by

$$-\frac{F_x}{m} f \, d\vec{r} \, dv_y \, dv_z \tag{3.54}$$

and the total flux is then

$$-\vec{\nabla}_v \left( \frac{\vec{F}}{m} f \right) d\vec{r} \, d\vec{v} \tag{3.55}$$

where $\vec{\nabla}_v$ denotes the divergence operator in Cartesian coordinates expressed in terms of the velocity components.

Similarly, the time variation of the particle number density in the control volume results in a net flux of

$$\frac{\partial f}{\partial t} \tag{3.56}$$

and must also be considered. In general, the plasma contains many different species, so Eqs. (3.52) and (3.55) must be written for each species, and must also account for chemical reactions among the species.

If $C_{ij}$ is the net rate of increase of particles in the control volume as a result of collisions between particles of species $i$ with particles of species $j$, the Boltzmann equation for particles of species $i$ is

$$\frac{\partial}{\partial t}f_i + \vec{\nabla}_r(\vec{v}_i f_i) + \vec{\nabla}_v\left(\frac{\vec{F}_i}{m_i}f_i\right) = \sum_j C_{ij} \tag{3.57}$$

Since the particle velocity $\vec{v}$ is independent of the spatial coordinates $\vec{r}$,

$$\vec{\nabla}_r(\vec{v}f) = \vec{v}\vec{\nabla}_r f \tag{3.58}$$

In plasmas, the forces $\vec{F}_i$ are either electric forces (independent of $\vec{v}$) or magnetic foces that act perpendicular to $\vec{v}$ and hence

$$\vec{\nabla}_v \vec{F}_i = 0 \tag{3.59}$$

Thus, the last term of the left-hand side of Eq. (3.57) can be written as

$$\frac{\vec{F}_i}{m_i}\vec{\nabla}_v f_i \tag{3.60}$$

The resulting Boltzmann equation for particles of species $i$ thus assumes the form

$$\frac{\partial}{\partial t}(f_i) + \vec{v}_i\vec{\nabla}_r f_i + \frac{\vec{F}_i}{m_i}\vec{\nabla}_v f_i = \sum_j C_{ij} \tag{3.61}$$

The determination of the collision terms on the right-hand side of the Boltzmann equation may be very complex:

For elastic collisions between neutral particles or between neutral and charged particles, the evaluation of the collision term is rather straightforward if we consider only binary collisions for which the mean free path is large compared to the diameter of the particles.

For elastic collisions between charged particles, the long-range nature of the Coulomb potential requires the introduction of a shielded Coulomb potential ("cut-off" potential). In this way it is possible to treat the charged particle interactions as two-body collisions.

For inelastic collisions (chemical reactions), the situation becomes even more complex.

The development of these calculations is beyond the scope of this book and the interested reader should consult Mitchner and Kruger (1973) or Hirschfelder (1954). These problems will also be discussed in Chapter 7 relative to transport phenomena.

### 3.5.4. The Maxwellian Distribution

If we consider a plasma in complete equilibrium, the distribution function is steady, uniform, and isotropic, and the number of particles $dN$ having velocities between $v$ and $v + dv$ with $v = |\vec{v}|$, in random directions at an absolute temperature $T$, is

$$\frac{dN(v)}{N} = f^0(v)\,dv = \frac{4\pi v^2\,dv}{(2\pi kT/m)^{3/2}} \exp\left(-\frac{mv^2}{2kT}\right) \tag{3.62}$$

where $m$ is the mass of the particles and $k$ is Boltzmann's constant ($k = 1.38\,10^{-23}$ J/K). The quantity $f^0(v)$ is called the Maxwell–Boltzmann distribution function.

In terms of the momentum ($p = mv$) this distribution can be written as

$$\frac{dN(p)}{N} = f^0(p)\,dp = \frac{4\pi p^2\,dp}{(2\pi mkT)^{3/2}} \exp\left(-\frac{p^2}{2mkT}\right) \tag{3.63}$$

There are two other useful forms of the distribution function. The energy distribution function follows from $p^2/2m = E$:

$$\frac{dN(E)}{N} = f^0(E)\,dE = \frac{2E^{1/2}\,dE}{\pi^{1/2}(kT)^{3/2}} \exp\left(-\frac{E}{kT}\right) \tag{3.64}$$

With the dimensionless quantity $u = E/kT$, the energy distribution function in reduced form is

$$\frac{dN(u)}{N} = f^0(u)\,du = \frac{2u^{1/2}}{\pi^{1/2}} \exp(-u)\,du \tag{3.65}$$

The most probable velocity $v_{\max}$ is given by

$$v_{\max} = \left(\frac{2kT}{m}\right)^{1/2} \tag{3.66}$$

and the corresponding value for the energy is

$$E_{max} = \frac{kT}{2} \tag{3.67}$$

The mean velocity $\bar{v}$ is given by

$$\bar{v} = \langle v \rangle = \frac{\int_0^\infty f(v)v\,dv}{\int_0^\infty f(v)\,dv} = \frac{2}{\pi^{1/2}} v_{max} = \left(\frac{8kT}{\pi m}\right)^{1/2} \tag{3.68}$$

and the mean square velocity or the effective velocity is

$$\overline{v^2} = \frac{\int_0^\infty f(v)v^2\,dv}{\int_0^\infty f(v)\,dv} = \frac{3kT}{m} \tag{3.69}$$

*Equation* (3.69) *is used as the definition of the temperature based on the Maxwellian distribution.*

This Maxwellian distribution is frequently expressed in terms of particle mean energies ($\bar{E} = \frac{1}{2}m\overline{v^2}$). Using Eq. (3.64) with the energy expressed in eV, one obtains

$$f(E) = 2.073\bar{E}^{-3/2}E^{1/2}\exp\left(-\frac{1.5E}{\bar{E}}\right) \tag{3.70}$$

where $\bar{E}$ is the mean value of the energy expressed in eV. Figure 3.7 shows this distribution function for three mean values $\bar{E} = 1$, 2, and 3 eV. Considering that rather high particle energies are required for

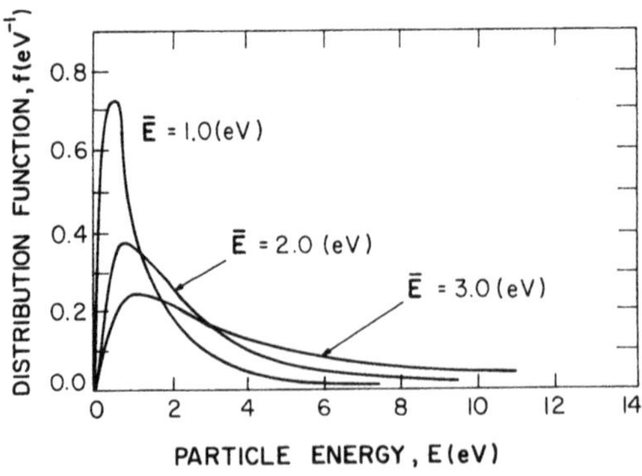


**FIG. 3.7.** Maxwell–Boltzmann distribution for various mean kinetic energies, $\bar{E}$, of the particles.

excitation and ionization (15.7 eV to ionize argon), only a fraction ($5 \times 10^{-10}$) of the particles in a Maxwellian distribution with an average energy of 1 eV will have sufficient energy to ionize argon atoms from the ground state.

It must be emphasized that all the preceding equations refer to distribution functions in the laboratory system. To express the distribution functions in the center-of-mass system, the same equations would be used but with the reduced mass $\mu$ instead of $m$. The reduced mass is obtained from

$$\frac{1}{\mu} = \sum_i \frac{1}{m_i} \tag{3.71}$$

For electrons in a plasma, the distribution function is practically the same for both the laboratory and the center-of-mass systems, since $m_e \ll m_{\text{heavy}}$. For heavy particles, however, the change from the laboratory to the center-of-mass system causes considerable changes. It must be emphasized that the distribution function in the center-of-mass system has to be used when chemical reactions between heavy particles are considered.

### 3.5.5. Collision Probabilities and Mean Free Paths in a Particle Ensemble

In a particle ensemble mean values for collision probabilities and free paths must be used. The mean collision probability was determined (see Section 3.2.3) by assuming a mean velocity for particles of type $i$, a mean relative velocity $\bar{V}_{ij}$, and a total cross section independent of $V_{ij}$. This calculation can be made more rigorous by introducing the distribution functions for averaging. The relative flux of particles $i$ with respect to a particle $j$ of velocity $\vec{v}_j$ is given by

$$f(\vec{v}_i)\, d^3v_i \vec{V}_{ij} \tag{3.72}$$

Multiplying Eq. (3.72) by the differential cross section $\sigma(V_{ij}, \theta, \varphi)$ and integrating over all solid angles $d\Omega'$ gives the total number of particles scattered by one particle $j$ in volume $d^3r$. Then it is necessary to integrate over all the scattering particles $j$ in $d^3r$ and to divide by the number of particles with velocity $v_i$ in the volume $d^3r$. The collision frequency follows as

$$\tau^{-1}(\vec{v}_i) = \frac{\int_{\vec{v}_j}\int_{\Omega'} f(\vec{v}_i)\, d\vec{v}_i \vec{V}_{ij}\sigma(V_{ij}, \theta, \varphi)\, d\Omega' f(v_j)\, d\vec{v}_j\, d\vec{r}}{f(\vec{v}_i)\, d\vec{v}_i\, d\vec{r}} \tag{3.73}$$

or

$$\tau^{-1}(\vec{v}_i) = \int_{\vec{v}_j}\int_{\Omega'} \vec{V}_{ij}\sigma(V_{ij}, \theta, \varphi)f(\vec{v}_j)\, d\Omega'\, d\vec{v}_j \tag{3.74}$$

A similar calculation gives the mean free path:

$$l = \frac{\displaystyle\int_{\vec{v}_i} \vec{v}_i f(\vec{v}_i)\, d\vec{v}_i}{\tau^{-1}(\vec{v}_i)} \tag{3.75}$$

However, a much simpler and oversimplified calculation starting from (Eq. 3.24) gives the same result as Eq. (3.75) when $\sigma_0(\bar{V}_{ij})$ can be assumed to be constant:

$$\vec{V}_{ij} = \vec{v}_i - \vec{v}_j \tag{3.76}$$

$$V_{ij}^2 = v_i^2 + v_j^2 - 2\vec{v}_i \cdot \vec{v}_j \tag{3.77}$$

$$\overline{V_{ij}^2} = \overline{v_i^2} + \overline{v_j^2} + 0 \tag{3.78}$$

because $\overline{v_i v_j} = 0$ due to the random motion of the particles. Thus Eq. (3.78) becomes

$$\overline{V_{ij}^2} = \overline{v_i^2 + v_j^2} \tag{3.79}$$

Neglecting the relatively small difference between root-mean-square and mean values, this can be written

$$\bar{V}_{ij} \sim \sqrt{\bar{v}_i^2 + \bar{v}_j^2} \tag{3.80}$$

Assuming that the particles are identical (collision between heavy species), it follows that

$$\bar{V}_{ij} = \sqrt{2}\, \bar{v}_i \tag{3.81}$$

and finally

$$l_{ij} = \frac{1}{\sqrt{2}\, n_j \sigma_{ij}} \tag{3.82}$$

It is interesting to note that this simplified calculation gives the same result as the exact equation (3.75).

## 3.6. REACTION RATES

### 3.6.1. Binary Reactions

Let us consider a simple reaction such as dissociative attachment

$$\mathrm{AB} + e \underset{\mathrm{r}}{\overset{\mathrm{d}}{\rightleftarrows}} \mathrm{A} + \mathrm{B}^-$$

which is written for simplicity as

$$1 + 2 \underset{\text{r}}{\overset{\text{d}}{\rightleftarrows}} 3 + 4$$

The arrows with the superscript d and subscript r stand, respectively, for direct reaction (d) and reverse reaction (r).

Let $\vec{v}_{12}$ be the relative velocity of the electron with respect to the molecule AB and $\sigma_{12}(\vec{v}_{12})$ the cross section for dissociative attachment. The probability $w'_{12}$ for a collision between an electron (2) and a molecule AB (1) of density $n_1$ is given (see Section 3.2.3) by

$$w'_{12}\,dt = n_1(\vec{v}_1)\sigma_{12}(\vec{v}_{12})\vec{v}_{12}\,dt \tag{3.83}$$

For an electron density $n_2(\vec{v}_2)$, the probability for collisions per unit time ($w_{12}$) becomes

$$w_{12} = n_1(\vec{v}_1)n_2(\vec{v}_2)\vec{v}_{12}\sigma_{12}(\vec{v}_{12}) = k_{12}(\vec{v}_{12}) \tag{3.84}$$

The quantity $k_{12}(\vec{v}_{12})$ is called the direct collision rate or the reaction rate. However, in general, the relative velocity $\vec{v}_{12}$ is not constant, because the electrons (2) and the molecules (1) have velocity distribution functions $f_2$ and $f_1$, respectively. Therefore, $k_{12}$ has to be calculated by averaging over the distribution function, and the total reaction rate becomes

$$k_{12} = n_1 n_2 \int_{-\infty}^{+\infty}\int_{-\infty}^{+\infty} \vec{v}_{12}\sigma_{12}(\vec{v}_{12}) f(\vec{v}_1) f(\vec{v}_2)\, d\vec{v}_1\, d\vec{v}_2 \tag{3.85}$$

where $n_1$ and $n_2$ are the total number densities of the chemical species 1 and 2 and $f_1$ and $f_2$ are the corresponding normalized distribution functions, often written as

$$n_1 \cdot n_2 \cdot \bar{k}_{12} = \langle \sigma_{12}(\vec{v}_{12})\vec{v}_{12}\rangle \cdot n_1 \cdot n_2 \tag{3.86}$$

For example, if $f_1^0$ and $f_2^0$ are Maxwellian distributions for the same temperature (or for the same mean kinetic energy), the reaction coefficient can be written as

$$\bar{k}_{12} = n_1 n_2 \frac{(m_1 m_2)^{3/2}}{(2kT)^3} \int_0^{+\infty}\int_0^{+\infty} \exp\left(-\frac{(m_1 v_1^2 + m_2 v_2^2)}{2kT}\right) v_{12}\sigma(v_{12})\, d^3v_1\, d^3v_2 \tag{3.87}$$

where $d^3v$ stands for $dv_x\, dv_y\, dv_z$.

Using the coordinates of the position vector of the center of mass $\vec{r}_i$ with the corresponding velocity

$$\vec{c} = \frac{(m_1\vec{v}_1 + m_2\vec{v}_2)}{m_1 + m_2} \tag{3.88}$$

and the relative velocity $\vec{v}_{12} = \vec{v}_1 - \vec{v}_2$, and expressing $d^3v_{12}$ in spherical coordinates:

$$d^3v_{12} = v_{12}^2\, dv_{12}\, d\Omega_v \tag{3.89}$$

where $d\Omega_v$ denotes an infinitesimal solid angle defined by the direction of $v_{12}$, integration over the whole space ($d\Omega_v = 4\pi$ if isotropic) yields

$$\bar{k}_{12} = \frac{(4\pi)^2(m_1m_2)^{3/2}}{(2\pi kT)^3}\int_0^\infty c^2 \exp\left[-(m_1 + m_2)c^2/(2kT)\right] d\vec{c}$$
$$\times \int_0^\infty v_{12}^3 \exp\left[-m_{12}v_{12}^2/(2kT)\right]\sigma_{12}(v_{12})\, d\vec{v}_{12} \tag{3.90}$$

Using the expression

$$f_2^0\int_0^\infty y^2 \exp(-y^2)\, dy = \sqrt{\pi}/4$$

one finally obtains

$$\bar{k}_{12} = \frac{4}{\sqrt{\pi}}\left(\frac{m_{12}}{2kT}\right)^{3/2}\int_0^\infty x \exp(-x)\sigma_{12}\left(\sqrt{\frac{2kT}{m_{12}}x}\right) dx \tag{3.91}$$

where $x = m_{12}v_{12}^2/(2kT)$.

This expression shows explicitly that the reaction rate is essentially a macroscopic quantity that depends on the thermodynamic state of the gas. In the expression for $\bar{k}_{12}$, the integration has been performed over all velocities (from zero to infinity). Most processes, however, have a threshold energy, and therefore integration has to be performed only from the corresponding threshold velocity to infinity.

The probability of collisions per unit time corresponds to the rate of production of species A and $B^-$. Hence the mass action law for this reaction can be written as

$$\bar{k}_{12}n_{AB}n_e = \frac{dn_A}{dt} = \frac{dn_{B^-}}{dt} \tag{3.92}$$

In general, the rate coefficient $\bar{k}_{12} = k_d$ for the forward reaction

$$AB + e \xrightarrow{d} A + B^-$$

is different from the coefficient $\bar{k}_{34} = k_r$ for the reverse reaction

$$A + B^- \xrightarrow{r} AB + e^-$$

because $\sigma_{12}$ is different from $\sigma_{34}$ and the relative velocity $\vec{v}_{12}$ of the electrons with respect to the molecules AB is higher than the relative velocity $\vec{v}_{34}$ of particles A with respect to particles $B^-$. However, if thermodynamic equilibrium prevails, a microreversibility relationship exists between $k_d$ and $k_r$.

Usually the reaction rate coefficient $k$ is expressed in $m^3\,s^{-1}$. For collisions between atoms (for example, Ar at 300 K with a Maxwellian distribution) we have

$$\sigma_{12} \sim 5.52 \times 10^{-19}\,m^2 \qquad \text{and} \qquad v_{12} \approx 4 \times 10^2\,m\,s^{-1}$$

hence

$$k \approx 2.2 \times 10^{-16}\,m^3\,s^{-1}$$

Usually, as an order of magnitude one finds $k \sim 10^{-17}\,m^3\,s^{-1} = 10^{-11}\,cm^3\,s^{-1}$ for the probability that a reaction will occur (to a first approximation, values below $10^{-17}\,m^3\,s^{-1}$ can be ignored).

### 3.6.2. Three-Body Reactions

Even at relatively low pressures, three-body reactions may become important. Schematically, a typical three-body reaction can be written as

$$A + B + M \rightarrow AB + M'$$

The third body does not participate in the reaction; it merely absorbs the excess energy ($E'_M > E_M$). If $f_1$, $f_2$, and $f_3$ are the distribution functions of the particles A, B, and M, with respective velocities of $\vec{v}_1$, $\vec{v}_2$, and $\vec{v}_3$, one finds

$$\frac{dn_{AB}}{dt} = \iiint k(\vec{v}_1, \vec{v}_2, \vec{v}_3) f_1 f_2 f_3 \, d\vec{v}_1 d\vec{v}_2 \, d\vec{v}_3 \tag{3.93}$$

where $k$ is the differential reaction rate expressed in $m^6\,s^{-1}$, i.e.,

$$k(\vec{v}_1, \vec{v}_2, \vec{v}_3) = \sigma_{12} \vec{v}_{12} V_{12,3} \tag{3.94}$$

In this expression $V_{12,3}$ is the interaction volume of the intermediate compound with the third body.

The reaction rate constant is given by

$$\bar{k} = \langle k(\vec{v}_1, \vec{v}_2, \vec{v}_3)\rangle \tag{3.95}$$

Typical mean values are in the ranges

$$\sigma_{12}\vec{v}_{12} \approx 10^{-17}\,\mathrm{m^3\,s^{-1}}, \qquad V_{12,3} \approx (10^{-9})^3\,\mathrm{m^3},$$
$$k \approx 10^{-44}\,\mathrm{m^6\,s^{-1}} = 10^{-32}\,\mathrm{cm^6\,s^{-1}}$$

### 3.6.3. Recombination

Recombination is defined as the mutual attachment of particles in the course of an encounter, for example, between a positive ion and an electron or between a positive ion and a negative ion. As previously shown, the process of recombination is generally described by a coefficient $k$, as given by the rate equation, which describes the recombination of charges during binary collisions. The three most important recombination processes for thermal plasmas will be considered here.

#### 3.6.3.1. Radiative Recombination

$$X^+ + e \rightarrow X^* + h\nu_0 \rightarrow X + h\nu_1 + h\nu_0$$

where $h\nu_0 = E_i - E^* + \frac{1}{2}m_e v_e^2$

The reverse process is photoionization, which cannot be neglected in thermal plasmas. Under thermodynamic equilibrium conditions there would be a microbalance between photoionization and photorecombination, i.e., a high probability for photoionization would also entail a high probability for photorecombination.

#### 3.6.3.2. Dissociative Recombination

In this process, part of the neutralization energy is used for the dissociation process. The electron is captured into a nonradiative repulsive state $(XY)_i^*$, which then dissociates,

$$(XY)^+ + e \rightarrow (XY)_i^* \rightarrow X^* + Y^* \rightarrow X + Y + h\nu_1 + h\nu_2$$

This process is one of the most effective recombination processes in the presence of molecular ions. The recombination coefficient increases as the density of the gas molecules increases, passes through a maximum, and then decreases for pressures exceeding one atmosphere. The maximum value is on the order of $10^{-13}\,\mathrm{m^3\,s^{-1}}$.

### 3.6.3.3. Three-Body Recombination

At higher pressures a third body, Y, which can be either a neutral particle or an electron, removes part of the neutralization energy,

$$X^+ + e + Y \rightarrow X^* + Y$$

The third particle, Y, is frequently a slow electron, because the reverse process, ionization by electron impact

$$e_{\text{fast}} + X \rightarrow X^+ + e + e_{\text{slow}}$$

has a high probability. For these recombination reactions the reaction rate coefficient $k$ is denoted by $\alpha$, which is defined by

$$\frac{dn^+}{dt} = \frac{dn_e}{dt} = -\alpha n^+ n_e \sim -\alpha n_e^2 \tag{3.96}$$

for a quasi-neutral plasma (note that $\alpha$ is defined for a two-body reaction and is expressed in $m^3\,s^{-1}$).

In the case of a Maxwellian distribution among the particles, the velocity $\vec{v}_e$ of the electrons is much higher than that of the ions, which can be considered stationary with respect to the electrons. With this assumption $\alpha$ becomes

$$\alpha = \frac{h}{\pi}\left(\frac{m_e}{2kT}\right)^{3/2} \int_0^\infty \sigma_{12}(u) f^0(u) u^{1/2}\,du \tag{3.97}$$

where $u = E/kT$ and where $f^0(u)\,du$ is given by Eq. (3.65).

Electron–ion recombination coefficients are generally lower than those for ion–ion recombination because of the higher velocities of the electrons. Radiative recombination, which results from trapping free electrons into atomic levels with emission of radiation, has a coefficient between $10^{-18}$ and $10^{-19}\,m^3\,s^{-1}$. Nevertheless, this process is sufficiently important to be partially responsible for the high luminosity of thermal plasmas. If there is a high concentration of molecular ions in a plasma, dissociative recombination becomes very important because of its large recombination coefficient, which is on the order of $10^{-13}\,m^3\,s^{-1}$.

## LIST OF SYMBOLS

| | |
|---|---|
| $\vec{c}$ | center of mass velocity of two particles (m/s) |
| $E_f$ | binding energy of two particles (eV) |
| $E_I$ | ionization energy of an atom (eV) |

| | |
|---|---|
| $E_m$ | energy of a metastable state (eV or $cm^{-1}$) |
| $E^*$ | energy of an excited state (eV or $cm^{-1}$) |
| $f^0(v)$ | Maxwellian distribution function for absolute velocities |
| $f(\vec{r}, \vec{v}, t)$ | velocity distribution function at time $t$ in the volume element $dx\,dy\,dz$ at the tip of the position vector $\vec{r}$ |
| $f_n(\vec{r}, \vec{v}, t)$ | normalized distribution function |
| $\vec{F}_i$ | external force (electric or magnetic field) acting on particle of species $i$ |
| $F_n^+$ | flux of the quantity $\chi(\vec{r}, \vec{v}, t)$ carried per unit time and unit surface from the (−) side to the (+) side of a surface defined by its normal $\vec{n}$ and moving with the gas velocity $\vec{v}_g$ |
| $F_n^-$ | flux of the quantity $\chi(\vec{r}, \vec{v}, t)$ carried per unit time and unit surface from the (+) side to the (−) side of a surface defined by its normal $\vec{n}$ and moving with the gas velocity $\vec{v}_g$ |
| $h$ | Planck's constant ($6.6 \times 10^{-34}$ W $s^2$) |
| $k$ | Boltzmann's constant ($1.38 \times 10^{-23}$ J/K particle) |
| $\bar{k}_{12}(\vec{v}_{12})$ | mean direct binary collision rate or reaction rate ($m^3\,s^{-1}$) |
| $\bar{k}(\vec{v}_1, \vec{v}_2, \vec{v}_3)$ | mean direct three-body collision rate ($m^6\,s^{-1}$) |
| $K$ | fraction of energy transferred by elastic collision from one particle of mass $m$ to another of mass $M$ averaged over all angles |
| $l_{ij}$ | average path length (or mean free path, mfp) traversed by a particle of type $i$ between two successive collisions with particles of type $j$ (m) |
| $l_i$ | mean free path of particles of type $i$ in a mixture of different chemical species (m) |
| $m$ | mass of a particle (kg) |
| $M$ | mass of a particle (kg) |
| $n$ | principal quantum number |
| $\vec{n}$ | normal to a surface $dA$ |
| $n_i$ | number density of particles of chemical species $i$ ($m^{-3}$) |
| $p$ | absolute value of momentum |
| $\vec{p}$ | momentum vector |
| $p_{ij}(V_{ij}, \theta, \varphi)$ | proportionality factor defining the differential cross section |
| $P(t)$ | probability that a particle with relative velocity $\vec{V}$ ($\vec{V} = \vec{v}_1 - \vec{v}_2$) survives a time $t$ without suffering a collision |
| $P^*(t)\,dt$ | probability that a particle, after surviving without collisions for a time $t$, suffers a collision in the time interval between $t$ and $t + dt$ |

| | |
|---|---|
| $Q_i$ | total effective cross section for collision processes ($cm^{-1}$) |
| $\vec{U}$ | peculiar velocity of a particle ($\vec{U} = \vec{v} - \vec{v}_g$) where $\vec{v}$ is the particle velocity and $\vec{v}_g$ the gas velocity |
| $U_n$ | normal component of the peculiar velocity ($U_n = \vec{U} \cdot \vec{n}$) |
| $v$ | absolute value of the velocity |
| $v_m$ | most probable velocity |
| $\bar{v}$ | mean velocity |
| $\vec{v}_i$ | velocity of a particle of chemical species $i$ (m/s) |
| $\vec{v}_g$ | mean velocity referred to as the species fluid (or average) velocity relative to some laboratory frame of reference |
| $\vec{V}_{ij}$ | relative velocity of particles of chemical species $i$ and $j$, respectively ($\vec{V}_{ij} = \vec{v}_i - \vec{v}_j$) |
| $\vec{V}$ | relative velocity of particles of type 1 and 2, respectively, before collision |
| $\vec{V}'$ | relative velocity of particles of type 1 and 2, respectively, after collision |
| $w \cdot dt$ | probability that a particle suffers a collision between $t$ and $t + dt$, also called collision rate |
| X | designation for the chemical species in its ground state |
| $X^*$ | designation for the chemical species in an excited state |
| $X^m$ | designation for the chemical species in a metastable state |
| $X^+$ | designation for the chemical species in its first ionized state |
| $X^{++}$ | designation for the chemical species in its second ionized state |
| $X_2$ | designation for the molecule made of two X atoms in its ground state |
| $X_2^+$ | designation for the ionized molecule |
| $Z'$ | number of protons in the nucleus |

## Greek Symbols

| | |
|---|---|
| $\alpha$ | recombination reaction rate coefficient ($m^3 s^{-1}$) |
| $\Delta E$ | energy difference between two excited states |
| $\theta$ | angle in spherical coordinates |
| $\lambda_i$ | maximum incident wavelength for ionization by photons (nm) |
| $\nu_{ij}$ | collision frequency between two particles of types $i$ and $j$, respectively ($s^{-1}$) |
| $\sigma_0(V_{ij})$ | total scattering cross section ($m^2$) |
| $\sigma_{ij}(\vec{V}_{ij}, \theta, \varphi)$ | differential scattering cross section |

| | |
|---|---|
| $\tau_{ij}$ | mean time between two collisions (between a particle of type $i$ and a particle of type $j$) or collision or relaxation time (s) |
| $\varphi$ | azimuthal angle in spherical coordinates |
| $\chi(v, \vec{v}, t)$ | function of position $\vec{r}$, velocity $\vec{v}$, and time $t$ |
| $\Omega$ | solid angle (ster) |

## GENERAL BIBLIOGRAPHY


Delcroix, J. L., *Physique des Plasmas, Vols. 1 and 2*, Paris: Dunod, 1966.

Fowler, G., *Statistical Thermodynamics*, Cambridge: Cambridge University Press, 1956.

Hirschfelder, J. D., *Molecular Theory of Gases and Liquids*, New York: Wiley, 1954.

Landau, L. and E. Lifshitz, *Physique Statistique*, Moscow: Mir, 1967.

Mayer, J. E. and G. M. Mayer, *Statistical Mechanics*, New York: Wiley, 1940.

Massey, H. S. W. and E. H. S. Burhop, *Electronic and Ionic Impact Phenomena. Collisions of Electrons with Atoms*, Oxford: Clarendon Press, 1969.

Massey, H. S. W., *Electronic and Ionic Impact Phenomena. Electron Collisions with Molecules—Photoionization*, Oxford: Clarendon Press, 1969.

Massey, H. S. W., *Electronic and Ionic Impact Phenomena. Slow Collisions of Heavy Particles*, Oxford: Clarendon Press, 1971.

Massey, H. S. W. and H. B. Gilbody, *Electronic and Ionic Impact Phenomena. Recombination and Fast Collisions of Heavy Particles*, Oxford: Clarendon Press, 1974.

Mitchner, M. and C. H. Kruger Jr., *Partially Ionized Gases*, New York: Wiley, 1973.

Munster, A., *Statistical Thermodynamics, Vol. 1*, Berlin: Springer-Verlag, New York: Academic Press, 1969.

Reif, F., *Fundamentals of Statistical and Thermal Physics*, New York: McGraw-Hill, 1988.


## REFERENCES


1. A. von Engel, *Ionized Gases* (Oxford: Clarendon Press, 1965).
2. S. C. Brown, *Basic Data of Plasma Physics* (Cambridge, Mass.: MIT Press, 1959).
3. R. D. Present, *Kinetic Theory of Gases* (New York: McGraw-Hill, 1958).
4. G. Francis, *Ionization Phenomena in Gases* (London: Butterworth, 1960).
5. E. W. McDaniel, *Collision Phenomena in Ionized Gases* (New York: Wiley, 1964).

Chapter 4

# Fundamental Concepts in Gaseous Electronics

It is beyond the scope of this book to even attempt a thorough treatment of this subject, which experienced immense growth during and after the Second World War. For a comprehensive treatment of gaseous electronics, the reader is referred to a number of books (see General Bibliography) that may be considered classics in this field.

## 4.1. GENERATION OF CHARGE CARRIERS

This section will be concerned with a microscopic description of the various ionization processes in a plasma. A collective treatment of such processes using the distribution functions described in Section 3.5 will be reserved for Section 4.4.

As described in Chapter 1, electrical discharges are the most common method of producing gaseous plasmas. In order to maintain a steady-state electrical discharge, charge carriers must be produced at the same rate as they vanish.

In principle, charge carriers are comprised of negative and positive particles. The former includes electrons and negative ions, and the latter, positive ions only. In the context of this book, negative ions play a minor role, and they will therefore be neglected in this section.

Charge carriers can be produced by ionization processes in a gas volume or by liberation of charged particles from confining walls and, in particular, from electrodes. Liberation of electrons from the cathode, for example, is vital for maintaining DC or AC discharges.

The following discussion will be restricted to ionization in the gas volume. Interactions among the plasma constituents (electrons, ions, neutrals, and photons) can lead to an array of different ionization processes. Among all conceivable interaction processes in a plasma, binary interactions have a much higher probability, and therefore they are particularly important.

In general, ionization is defined as the removal of one or more electrons from a neutral particle or from an ion (single or multiple ionization, respectively). Such ionization processes always produce free electrons and positive ions. These processes can occur through a simple encounter (direct ionization) or through a sequence of encounters and energy exchange processes (indirect ionization). The latter processes are characterized by intermediate energy states.

### 4.1.1. Direct Ionization

Direct ionization by photons, by electron impact, and by collisions among heavy particles was discussed in Section 3.4.2. It should be emphasized that these ionization processes refer to particles initially in the ground state. Regardless of the nature of an ionization process or any other encounter between plasma constituents, such encounters are governed by the conservation laws for mass, charge, momentum, and energy. By applying the conservation laws, we can single out the most important ionization processes in a given plasma. As we saw in Section 3.4.2, ionization by collisions among heavy particles, for example, is for practical purposes negligible because the required particle energies are beyond those encountered in the plasmas covered in this book.

### 4.1.2. Indirect Ionization

As mentioned previously, indirect ionization implies that intermediate energy states precede the actual ionization process. Such energy states (for example, excited states) can be produced by the processes described in Section 3.4.1. Typical examples of indirect ionization requiring at least two steps for producing charge carriers (associative ionization, ionization of already excited atoms by electron impact, by photons, and by excited atoms, and ionization by the Penning effect) have already been discussed in Section 3.4.3. Again, all of these various processes are governed by the conservation laws.

## 4.2. LOSS OF CHARGE CARRIERS

There are a large number of conceivable processes by which charge carriers vanish from a plasma. Only those processes that are important for the plasmas considered in this book will be discussed here.

In general, charge carrier losses can result from

(a) drift to the electrodes resulting from electric fields in the case of DC or AC discharges;
(b) diffusion to the surrounding walls with subsequent recombination;
(c) volume recombination.

Electron attachment (formation of negative ions) will not be considered here.

Drift of positive ions to the cathode results in neutralization of the ions upon impact on the cathode. Electrons drifting to the anode are absorbed into the metal lattice. Drift of charge carriers as well as diffusion of charge carriers due to particle density gradients will be discussed in more detail in Sections 4.3.1 and 4.3.2, respectively.

In the volume recombination process, an electron is captured by a positive ion; in this process both the ionization energy ($E_I$) and the kinetic energy of the colliding electron ($\frac{1}{2}m_e v_e^2$) are released:

$$e + X^+ \rightarrow X + (E_I + \tfrac{1}{2}m_e v_e^2)$$

The possible recombination mechanisms depend on the disposal of this energy within the restraints of the conservation laws. This energy, for example, cannot be assumed by the neutral particle after recombination, because this would violate momentum conservation. There are essentially three recombination mechanisms that are of interest in the context of this book: radiative recombination, three-body recombination, and dissociative recombination. These processes were discussed in Section 3.6.3.

## 4.3. MOTION OF CHARGE CARRIERS

### 4.3.1. Drift in Electric Fields

This section begins with a microscopic description of the motion of a charged particle in a uniform electric field in the absence of any other forces. By first establishing the average behavior of a charged particle in an ensemble of particles, macroscopic relations (for example, mobilities and electrical conductivity, which will be discussed in the second part of this section) can then be derived.

Charge carriers exposed to an electric field will be accelerated by the electric force acting on them according to Newton's second law of motion which, in the case of electrons, assumes the form

$$m_e \frac{d\vec{u}_e}{dt} = -e\vec{E} \tag{4.1}$$

where $\vec{u}_e$ is the electron drift velocity, $m_e$ is the electron mass, and $\vec{E}$ is the electric field strength. A similar equation holds for positive ions in an electric field:

$$M\frac{d\vec{u}_i}{dt} = e\vec{E} \tag{4.2}$$

where $\vec{u}_i$ is the ion drift velocity and $M$ is the ion mass. For simplicity it is assumed that the ion is singly charged. Because of the similarity of Eqs. (4.1) and (4.2), only electrons will be considered in the following derivations. The derivation of the final expressions for ions is left to the reader.

Integration of Eq. (4.1) assuming a constant electric field strength results in

$$\vec{u}_e = \vec{u}_{e0} - \frac{e\vec{E}}{m_e}t \tag{4.3}$$

where $\vec{u}_{e0}$ is the initial velocity of the electron at the moment when the electric field is applied. Equation (4.3) indicates that the drift velocity of the electron increases linearly with time. This increase is limited, however, by collisions with other particles in the plasma. (Only elastic collisions will be considered in the context of this analysis. This assumption is justified by the fact that elastic collisions dominate in a plasma.)

If we assume that the free flight time of an arbitrarily chosen electron is $\tau_e$, then integration of Eq. (4.3) determines the distance $\vec{s}_e$ that this electron will travel during the time interval $\tau_e$:

$$\vec{s}_e = \vec{u}_{e0} \cdot \tau_e - \frac{e\vec{E}}{2m_e}\tau_e^2 \tag{4.4}$$

In the following discussion, we will consider the average behavior of an electron rather than the behavior of an arbitrarily chosen electron. The laws of statistics may be applied to determine this average behavior, because of the large number of electrons per unit volume in a typical plasma ($n_e > 10^{20}\,\mathrm{m}^{-3}$). Since the initial velocities of the electrons are randomly distributed, the first term on the right-hand side of Eq. (4.4) will vanish, i.e., $\overline{\vec{u}_{e0} \cdot \tau_e} = 0$. As previously shown (see Section 3.2.4), the mean free path length or the mean free flight time follows a statistical distribution according to

$$\bar{\tau}_e = \int_0^\infty \frac{\tau_e}{\bar{\tau}_e}\exp(-\tau_e/\bar{\tau}_e)\,d\tau_e \tag{4.5}$$

By applying Eq. (4.5) to Eq. (4.4) for averaging, one finds

$$\bar{\vec{s}}_e = -\frac{e\vec{E}}{2m_e}\int_0^\infty \frac{\tau_e^2}{\bar{\tau}_e}\exp(-\tau_e/\bar{\tau}_e)\,d\tau_e = -\frac{e\vec{E}}{m_e}\bar{\tau}_e^2 \tag{4.6}$$

The corresponding mean drift velocity (often denoted $\vec{v}_d^e$) becomes

$$\vec{v}_d^e = \bar{\vec{u}}_e = -\frac{e\vec{E}}{m_e}\bar{\tau}_e \tag{4.7}$$

By substituting

$$\bar{\tau}_e = l_e/\bar{v}_e \tag{4.8}$$

(where $l_e$ is the mean free path length of the electrons and $\bar{v}_e$ their mean thermal velocity) into Eq. (4.7), one finds

$$\bar{\vec{u}}_e = -\frac{el_e}{m_e\bar{v}_e}\vec{E} = -\mu_e\vec{E} \tag{4.9}$$

where $\mu_e$ is the electron mobility given by

$$\mu_e = \frac{el_e}{m_e\bar{v}_e} \tag{4.10}$$

An analogous expression can be derived for the average ion drift velocity:

$$\bar{\vec{u}}_i = \mu_i\vec{E} \tag{4.11}$$

with

$$\mu_i = \frac{el_i}{m\bar{v}_i} \tag{4.12}$$

where $l_i$ is the mean free path of the ions and $\bar{v}_i$ is the mean thermal velocity of the ions. At this point it is interesting to make an order-of-magnitude comparison between the drift velocity of charged particles and their random (thermal) velocity. For this comparison, we will consider the region close to the anode of an atmospheric-pressure, high-intensity argon arc (welding arc) and specify the following parameters (order of magnitude): $T \approx 10^4$ K, $l_e \simeq l_i \simeq 10^{-6}$ m, $E = 5 \times 10^2$ V/m, $m_e = 9.1 \times 10^{-31}$ kg, and $M = 6.8 \times 10^{-26}$ kg. The drift and thermal velocities for ions and electrons are:

*Electrons*:

$$\bar{v}_e = \left(\frac{8kT}{\pi m_e}\right)^{1/2} \simeq 10^6\ \text{m/s} \qquad \text{and} \qquad |\bar{\vec{u}}_e| = \frac{el_e}{m_e\bar{v}_e}E \simeq 10^2\ \text{m/s}$$

*Ions*:

$$\bar{v}_i = \left(\frac{8kT}{\pi M}\right)^{1/2} \simeq 10^3 \text{ m/s} \qquad \text{and} \qquad |\bar{\vec{u}}_i| = \frac{el_1}{M\bar{v}_i} E \simeq 1 \text{ m/s}$$

The drift velocities are several orders of magnitude smaller than the corresponding thermal velocities. Based on these findings, we can see that the motion of the charged particles is analagous to that of a swarm of flies on a hot summer day drifting in a slight wind.

In a plasma, electrons and ions drift in opposite directions under the influence of an applied electric field, and this drift gives rise to an electric current of density

$$\begin{aligned} \vec{j} = \vec{j}_i + \vec{j}_e &= e(n_i\bar{\vec{u}}_i - n_e\bar{\vec{u}}_e) \\ &= e(n_i\mu_i + n_e\mu_e)\vec{E} \end{aligned} \tag{4.13}$$

where $n_i$ and $n_e$ are the ion and electron densities, respectively.

Since the electric field is the only driving force for electron and ion currents, Ohm's law can be written in simple form as

$$\vec{j} = \sigma_e\vec{E} \tag{4.14}$$

A comparison of Eqs (4.13) and (4.14) provides a simple expression for the electrical conductivity:

$$\sigma_e = e(n_i\mu_i + n_e\mu_e) \tag{4.15}$$

This expression can be further simplified because $n_i = n_e$ (for singly charged ions) and $\mu_i \ll \mu_e$. With this simplification, Eq. (4.15) becomes

$$\sigma_e = en_e\mu_e \tag{4.16}$$

The justification for neglecting $\mu_i$ compared to $\mu_e$ follows from Eqs. (4.10) and (4.12). The ratio

$$\frac{\mu_i}{\mu_e} = \frac{l_i}{l_e}\frac{m_e}{M}\frac{\bar{v}_e}{\bar{v}_i} \tag{4.17}$$

transforms with

$$\frac{\bar{v}_e}{\bar{v}_i} = \left(\frac{M}{m_e}\right)^{1/2} \tag{4.18}$$

into

$$\frac{\mu_i}{\mu_e} = \frac{l_i}{l_e}\left(\frac{m_e}{M}\right)^{1/2} \tag{4.19}$$

Since $l_i < l_e$ in a plasma and the ratio $(M/m_e)^{1/2}$ already exceeds 40 for hydrogen, neglecting $\mu_i$ compared to $\mu_e$ is indeed justified.

Equation (4.18) in this simple form is valid only when kinetic equilibrium ($T_e = T_h$) prevails in the plasma (see Section 4.4.3).

In deriving the expression for the electron mobility, Eq. (4.10), a number of implicit, simplifying assumptions have been made:

(a) The degree of ionization in the plasma is assumed to be small ($\xi \ll 1$), which implies that Coulomb interactions are negligible, i.e., only collisions between electrons and neutral particles have been taken into account.

(b) Averaging over the velocity distribution functions has been omitted.

(c) Local imbalances of charge neutrality have been neglected.

By averaging over the velocity distribution function, Eq. (4.10) becomes

$$\mu_e = \frac{2}{\pi}\frac{el_e}{m_e\bar{v}_e} \tag{4.20}$$

Removing all the simplifying assumptions leads to the more complex expression derived by Gvosdover:[1]

$$\mu_e = \frac{e}{\sqrt{\pi k m_e T}\left(\dfrac{1}{l_k} + \dfrac{n_e e^4}{\gamma_e (kT)^2}\right)} \tag{4.21}$$

where

$$\gamma_e = \frac{2}{\dfrac{\pi}{2}\ln\left(\dfrac{3kT}{2e^2 n_e^{1/3}}\right)}$$

and $l_k$ is a mean free path length associated with the Ramsauer cross section.

For plasmas with a low degree of ionization ($\xi \ll 1$), Eq. (4.21) reduces to

$$\mu_e = \frac{el_e}{(\pi k T m_e)^{1/2}} = \frac{2\sqrt{2}}{\pi}\frac{el_e}{m_e\bar{v}_e} \tag{4.22}$$

Except for the $\sqrt{2}$ factor, Eq. (4.22) is identical to Eq. (4.20).

In general, the mobility of a charged particle is a function of its kinetic energy. The kinetic energy that a particle can acquire is, in turn, determined by the electric field and the mean free path. The maximum kinetic energy that an electron, for example, can acquire between two collisions is given by

$$T_k = el_e E \tag{4.23}$$

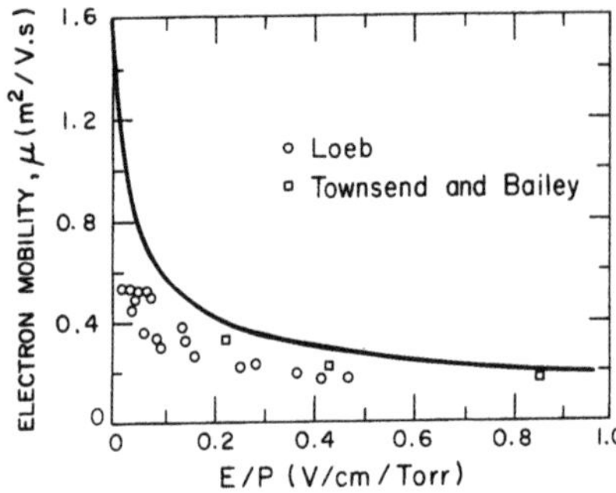


**FIG. 4.1.** Electron mobility in hydrogen versus $E/p$.

where $l_e$ and $E$ are parallel to each other. Since $l_e \sim 1/p$ [see Eq. (3.28)] the electron mobility becomes a function of $E/p$:

$$\mu_e = f(E/p)$$

Figure 4.1 shows the electron mobility in different gases as a function of $E/p$.

According to Eq. (4.12) the mobility of ions depends, in addition, on the ion mass, and therefore

$$\mu_i = f(E/p, M)$$

Figure 4.2 shows the mobility of different ions.

### 4.3.2. Diffusion of Charge Carriers

In this section we will consider diffusion of plasma charge carriers due to concentration gradients as well as ambipolar diffusion associated with the large difference between electron and ion mobilities.

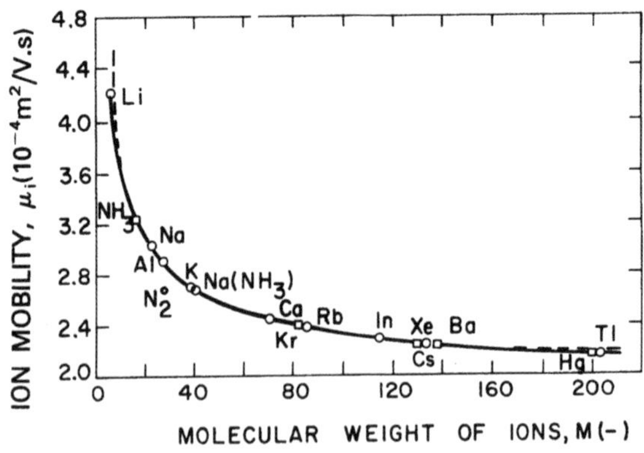


**FIG. 4.2.** Ion mobility versus molecular weight.

Ordinary diffusion can be described by Fick's law

$$\vec{j}_k = -D_k \operatorname{grad} n_k \tag{4.24}$$

where $\vec{j}_k$ represents the flux of charged particles ($m^{-2}\,s^{-1}$) of type $k$, $D_k$ is their diffusion coefficient, and $n_k$ is their number density.

The motion of charge carriers due to diffusion gives rise to an electric current of density

$$\vec{j}_e = eD_e \operatorname{grad} n_e \tag{4.25}$$

for the electrons, and the corresponding ion current current density can be written as

$$\vec{j}_i = -eD_i \operatorname{grad} n_i \tag{4.26}$$

where $D_e$ and $D_i$ are the electron and ion self-diffusion coefficients, respectively. These diffusion coefficients are related to gas kinetic parameters by

$$D_k = \frac{l_k \bar{v}_k}{3} \tag{4.27}$$

This expression follows from kinetic gas theory (see Section 7.2.1). At kinetic equilibrium (Section 4.4.3), the ratio of the electron and ion diffusion coefficients is given by

$$\frac{D_e}{D_i} = \frac{l_e \bar{v}_e}{l_i \bar{v}_i} = \frac{l_e}{l_i}\sqrt{\frac{M}{m}} = \frac{\mu_e}{\mu_i} \tag{4.28}$$

This relationship will be used later on in this section.

Charged particle density gradients are particularly steep close to plasma-confining walls. The one-dimensional situation in which a plasma borders a wall is shown in Fig. 4.3. The gradients of electron and ion density in the vicinity of the wall drive electron and ion fluxes toward the wall, but the electron flux initially exceeds the ion flux because of the higher electron mobility. Since the wall is assumed to be isolated (no net current flow), it will acquire a negative potential, producing an electric field, $E_x$, that points toward the wall (see Fig. 4.2). This field subsequently balances electron and ion fluxes (electrons are retarded and ions

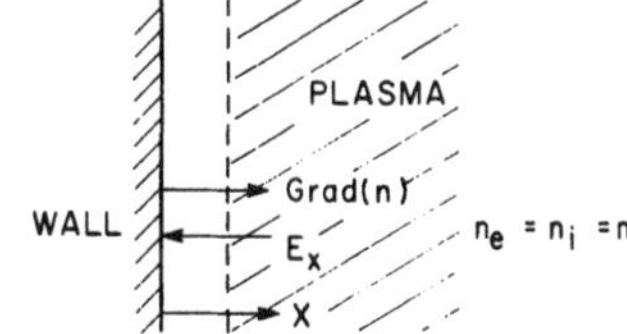


FIG. 4.3. One-dimensional situation in which a plasma borders a wall.

are accelerated), thus in a steady-state situation electrons and ions reach the wall at the same rate and recombine on impact with the wall. This process is known as *ambipolar diffusion*. In this case the wall serves as the third collision partner for three-body recombination.

The combined electron and ion current in this situation can be written as

$$j = j_e + j_i = e(n_i\mu_i + n_e\mu_e)E_x + e(D_e - D_i)\frac{dn}{dx} = 0 \tag{4.29}$$

Since there is no net current flow to an electrically isolated wall, Eq. (4.29) allows us to determine the electric field responsible for ambipolar diffusion:

$$E_x = -\frac{D_e - D_i}{n_i\mu_i + n_e\mu_e}\frac{dn}{dx} \tag{4.30}$$

For simplicity, it will be assumed that the plasma contains singly ionized species only ($n_e = n_i = n$), so Eq. (4.30) reduces to

$$E_x = -\frac{D_e - D_i}{\mu_i + \mu_e}\frac{1}{n}\frac{dn}{dx} \tag{4.31}$$

Using Eq. (4.31) and applying Eqs. (4.25) and (4.26) to the specified one-dimensional situation, we obtain the following expression for the number of charge carriers arriving at the wall per unit area and unit time:

$$\frac{j_e}{e} = -\frac{j_i}{e} = \frac{D_e\mu_i + D_i\mu_e}{\mu_i + \mu_e}\frac{dn}{dx}$$

$$= D_a\frac{dn}{dx} \tag{4.32}$$

where $D_a$ is the ambipolar diffusion coefficient. With $\mu_i \ll \mu_e$ (Section 4.3.1), the ambipolar diffusion coefficient reduces to

$$D_a = D_i + \frac{\mu_i}{\mu_e}D_e \tag{4.33}$$

and with Eq. (4.28), which is valid for kinetic equilibrium, one finds

$$D_a = 2D_i \tag{4.34}$$

This relation indicates that in a plasma in which kinetic equilibrium prevails, the ions diffuse at twice the rate they would in the absence of electrons. This finding has important consequences for situations in which heat transfer by diffusion of charged particles becomes significant.

As we have just pointed out, ambipolar diffusion plays an important role in regimes of steep particle density gradients as, for example, in the vicinity of confining walls. In such regimes, the density distribution of charged particles may be governed by the ambipolar diffusion process. In low-pressure discharges (glow discharges), where the ambipolar diffusion coefficient assumes large values, charge-carrier losses are frequently dominated by ambipolar diffusion to the walls of the plasma container.

In general, the charge-carrier continuity equation can be written as

$$\frac{\partial n_k}{\partial t} + \operatorname{div} \vec{I}_k = S_k \tag{4.35}$$

where $S_k$ represents the source term describing the net volume production of charge carriers of type $k$. By writing this equation for electrons and assuming that the source term for electron production is proportional to the electron density (i.e., $S_e = \nu_i n_e$, where $\nu_i$ is the net ionization coefficient, one finds

$$\frac{\partial n_e}{\partial t} + \operatorname{div} \vec{I}_e = \nu_i n_e \tag{4.36}$$

For steady-state conditions and considering only electron flux resulting from ambipolar diffusion ($\vec{I}_e = -D_a \operatorname{grad} n_e$), Eq. (4.36) reduces to

$$\Delta n_e + \frac{\nu_i}{D_a} n_e = 0 \tag{4.37}$$

where $\Delta$ is the Laplace operator.

As an example Eq. (4.37) will be solved for a discharge vessel consisting of a pair of infinitely long, parallel plates separated by a distance $L$. For this one-dimensional case Eq. (4.37) becomes

$$\frac{d^2 n_e}{dx^2} + \frac{\nu_i}{D_a} n_e = 0 \tag{4.38}$$

With the boundary conditions

$$n_e = 0 \qquad \text{for } x = \pm \frac{L}{2} \tag{4.39}$$

$$\frac{dn_e}{dx} = 0 \qquad \text{for } x = 0 \tag{4.40}$$

Eq. (4.38) has the solution

$$n_e = n_e(0) \cos \sqrt{\frac{\nu_i}{D_a}} \cdot x \tag{4.41}$$

where $n_e(0)$ is the electron density for $x = 0$. The first boundary condition requires that

$$n_e(0) \cos \sqrt{\frac{\nu_i}{D_a}} \cdot \frac{L}{2} = 0$$

or

$$\sqrt{\frac{\nu_i}{D_a}} \cdot \frac{L}{2} = (2k + 1)\frac{\pi}{2}, \qquad k = 0, 1, 2, \ldots \tag{4.42}$$

Equation (4.38) has solutions only in terms of eigenvalues characterized by the integer $k$. We will consider only the fundamental mode or fundamental eigenvalue for $k = 0$:

$$\frac{\nu_i}{D_a}\left(\frac{L}{2}\right)^2 = \left(\frac{\pi}{2}\right)^2$$

or

$$\frac{\nu_i}{D_a} = \left(\frac{\pi}{L}\right)^2 = \frac{1}{\Lambda^2} \tag{4.43}$$

where $\Lambda$ is defined as the characteristic diffusion length, which is a function only of the geometry.

For an infinitely long cylinder of radius $R$, the corresponding characteristic diffusion length becomes

$$\frac{1}{\Lambda^2} = \left(\frac{2.405}{R}\right)^2 \tag{4.44}$$

and the corresponding solutions of Eq. (4.37) are Bessel functions.

For a cylinder of radius $R$ and finite length $L$, which can be produced by the intersection of a pair of infinitely long parallel plates with an infinitely long cylinder, the corresponding characteristic diffusion length becomes

$$\frac{1}{\Lambda^2} = \left(\frac{2.405}{R}\right)^2 + \left(\frac{\pi}{L}\right)^2 \tag{4.45}$$

The characheristic diffusion length of more complex configurations can be similarly determined by this intersection method.

### 4.3.3. Motion of Charge Carriers in Magnetic Fields

This section will briefly discuss the effects that uniform, constant magnetic fields exert on charged particle motion. For the more complex situations involving nonuniform and/or time-varying magnetic fields, the reader should consult the pertinent textbooks on this subject.[2–6]

### 4.3.3.1. In the Absence of Collisions

If a charged particle of mass $m$, charge $q$, and velocity $\vec{v}$ is exposed to a uniform and constant magnetic field of induction $\vec{B}$, then the particle will experience a force

$$\vec{F} = m\frac{d\vec{v}}{dt} = q(\vec{v} \times \vec{B}) \qquad (4.46)$$

Multiplying Eq. (4.46) by $\vec{v}$ (dot product), one finds

$$\begin{aligned} \frac{d}{dt}(\tfrac{1}{2}mv^2) &= q\vec{v} \cdot (\vec{v} \times \vec{B}) \\ &= q(\vec{v} \times \vec{v}) \cdot \vec{B} = 0 \end{aligned} \qquad (4.47)$$

that is, the kinetic energy of the particle remains constant. This result is not surprising, because the accelerating force $\vec{F}$ is perpendicular to $\vec{v}$ according to Eq. (4.46).

In general, the velocity of a particle can have components parallel and perpendicular to the direction of the magnetic induction $\vec{B}$, i.e.,

$$\vec{v} = \vec{v}_\perp + \vec{v}_\parallel \qquad (4.48)$$

Therefore

$$m\frac{d}{dt}(\vec{v}_\perp + \vec{v}_\parallel) = q(\vec{v}_\perp \times \vec{B}) + q(\vec{v}_\parallel \times \vec{B}) \qquad (4.49)$$

or

$$\left.\begin{aligned} m\frac{d\vec{v}_\perp}{dt} &= q(\vec{v}_\perp \times \vec{B}) \\ m\frac{d\vec{v}_\parallel}{dt} &= q(\vec{v}_\parallel \times \vec{B}) = 0 \end{aligned}\right\} \qquad (4.50)$$

According to the second part of Eq. (4.50), $\vec{v}_\parallel$ = constant.

From

$$\tfrac{1}{2}mv^2 = \tfrac{1}{2}m(v_\perp^2 + v_\parallel^2) = \text{constant} \qquad (4.51)$$

it follows that $v_\perp^2$ must also be constant (or $v_\perp = |\vec{v}_\perp|$ = constant).

One can easily show that the first part of Eq. (4.50) describes a circular motion of the particle around a guiding center. Superimposed on this rotation is a linear motion with constant velocity ($\vec{v}_\parallel$) in the direction

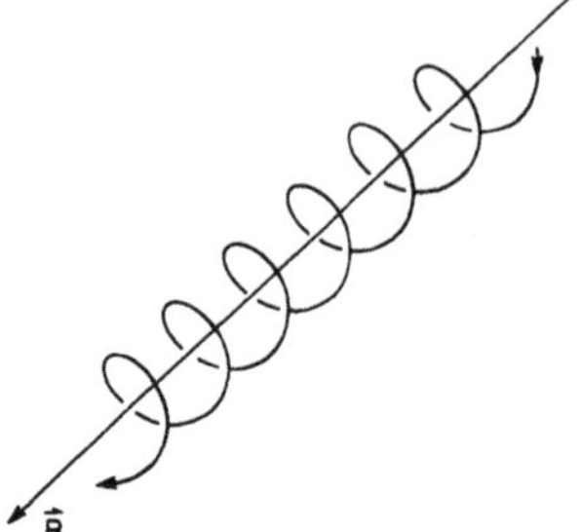


FIG. 4.4. Motion of a charge particle in a magnetic induction field.

of the magnetic induction; this motion is not affected by the magnetic induction. The superposition of $\vec{v}_\perp$ and $\vec{v}_\parallel$ results in the helical motion shown schematically in Fig. 4.4.

For the circular motion, the centrifugal force acting on the particle is balanced by the magnetic force, i.e.,

$$\frac{mv_\perp^2}{r_L} = q(v_\perp \cdot B) \tag{4.52}$$

where $r_L$ is the Larmor radius for the circular motion. It follows from Eq. (4.52) that

$$r_L = \frac{mv_\perp}{qB} \tag{4.53}$$

The frequency of rotation, known as the Larmor or cyclotron frequency, is given by

$$\omega = \frac{v_\perp}{r_L} = \frac{qB}{m} \tag{4.54}$$

Since the electric charge $q$ can be positive or negative, the direction of the circular motion will change accordingly.

As an example, let us consider an electron and a positive argon ion (singly charged), both with thermal velocity components $(v_\perp)$ corresponding to a temperature of 1 eV, moving in a magnetic field of induction $B = 1\,\text{T}$. The resulting values for the electron and ion Larmor frequencies and Larmor radii follow from Eqs. (4.54) and (4.53), respectively, as

$$\omega_e = -\frac{eB}{m_e} = -1.75 \times 10^{11}\,[\text{s}^{-1}]; \qquad \omega_i = \frac{eB}{m_i} = 2.4 \times 10^{6}\,[\text{s}^{-1}]$$

and

$$r_{L,e} = \frac{v_{\perp,e}}{\omega_e} = 3.1 \times 10^{-6}\,[\text{m}]; \qquad r_{L,i} = \frac{v_{\perp,i}}{\omega_i} = 8.4 \times 10^{-4}\,[\text{m}]$$

Since the mean free path length, $l_e$, in an atmospheric-pressure thermal plasma having a temperature of approximately 1 eV is of the same order of magnitude as the electron Larmor radius, and since the mean free path of the ions, $l_i$, is substantially smaller than their Larmor radius, the helical path of the charged particles (especially of the ions) will be interrupted by collisions.

In the presence of an electric field as well as a magnetic field, a charged particle will be exposed to the so-called Lorentz force

$$\vec{F} = m\frac{d\vec{v}}{dt} = q(\vec{E} + \vec{v} \times \vec{B}) \tag{4.55}$$

This equation will be applied to uniform and constant fields. Separating this equation into components parallel and perpendicular to the magnetic field leads to

$$m\frac{d\vec{v}_{\parallel}}{dt} = q\vec{E}_{\parallel} \tag{4.56}$$

and

$$m\frac{d\vec{v}_{\perp}}{dt} = q(\vec{E}_{\perp} + \vec{v}_{\perp} \times \vec{B}) \tag{4.57}$$

Equation (4.56) describes a constant acceleration parallel to $\vec{B}$. The electric field component in Eq. (4.57) induces a drift velocity of the particle perpendicular to both the magnetic field and $\vec{E}_{\perp}$. Figure 4.5 shows this situation for both positively and negatively charged particles with the magnetic field pointing in the $x$ direction. The positively charged

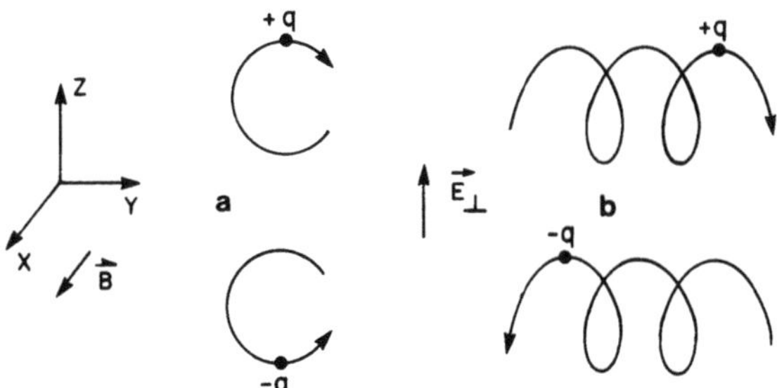


FIG. 4.5. Motion of positively and negatively charged particles (a) in a magnetic field pointing in the $x$ direction and (b) in an electric field.

particle is accelerated in the $z$ direction, and the resulting increasing velocity also increases the radius of curvature of the trajectory according to Eq. (4.53). Beyond the culmination point of the trajectory, the particle moves downward following a symmetrical path in a decelerating field, decreasing the radius of curvature of the trajectory toward the bottom turning point. This cycle repeats itself periodically as shown in Fig. 4.5.

As we have mentioned, a charged particle in this situation experiences a drift perpendicular to $\vec{E}$ and $\vec{B}$ with a certain velocity $\vec{v}_D$, where $\vec{v}_D \sim \vec{E}_\perp \times \vec{B}$. By choosing $\vec{v}_D$ as

$$\vec{v}_D = \frac{\vec{E}_\perp \times \vec{B}}{B^2} \tag{4.58}$$

it can be shown that Eq. (4.57) reduces to the same form as Eq. (4.46), which describes the gyration of a charged particle around a magnetic induction line. With

$$\vec{v}_\perp = \vec{v}_D + \vec{u} \tag{4.59}$$

Eq. (4.57) reduces to

$$m\frac{d\vec{u}}{dt} = q(\vec{u} \times \vec{B}) \tag{4.60}$$

because

$$(\vec{v}_D \times \vec{B}) = \frac{\vec{E}_\perp \times \vec{B}}{B^2} \times \vec{B} = \frac{\vec{B} \cdot (\vec{E}_\perp \cdot \vec{B} - \vec{E}_\perp \cdot (\vec{B} \cdot \vec{B}))}{B^2} = -\vec{E}_\perp \tag{4.61}$$

($\vec{E}_\perp$ is by definition perpendicular to $\vec{B}$) and

$$m\frac{d}{dt}(\vec{v}_D + \vec{u}) = m\frac{d\vec{u}}{dt} \tag{4.62}$$

This result implies that the particle motion in a coordinate system moving with velocity $\vec{v}_D$ is affected only by the magnetic field and, as previously shown, the result is a circular motion around a guiding center in the $yz$ plane (Fig. 4.5). The superimposed constant drift velocity $\vec{v}_D$ results in a cycloid as shown in Fig. 4.5.

If the electric and magnetic fields are perpendicular to each other (i.e., $\vec{E}_\parallel = 0$), then the motion of charged particles in these crossed fields will remain in a plane perpendicular to the direction of $\vec{B}$. If $\vec{E}_\parallel \neq 0$, there will be an additional drift parallel to the magnetic field as indicated by Eq. (4.56). Finally, if $\vec{E}$ and $\vec{B}$ are parallel to each other ($\vec{E}_\perp = 0$), there will be no drift velocity perpendicular to $\vec{B}$ ($\vec{v}_D = 0$), but the particles will experience a constant acceleration parallel to the direction

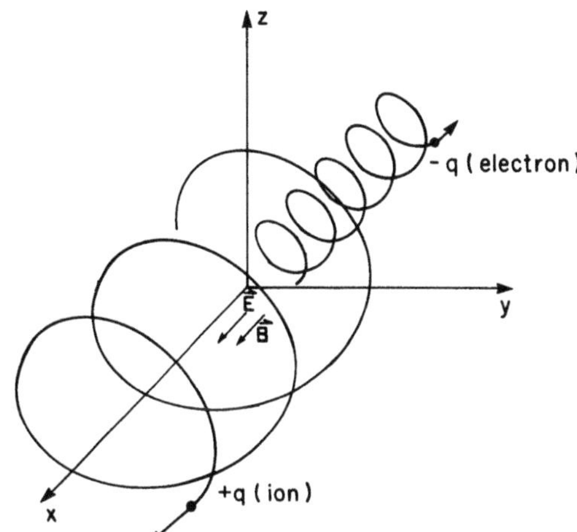


FIG. 4.6. Motion of a positive ion and an electron in parallel fields pointing in the $x$ direction.

of $\vec{B}$. Figure 4.6 shows the motion of a positive ion and an electron in parallel fields.

### 4.3.3.2. With Collisions

The situation discussed in the previous section does not generally apply to thermal plasmas that are collision dominated. Collisions among charge carriers and with other plasma constituents cause severe disturbances in the previously discussed trajectories of such particles. Between collisions, however, the individual charge carriers will move according to the drift imposed on them by the $\vec{E}$ and $\vec{B}$ fields.

This microscopic picture of the behavior of an individual charge carrier is, however, not the entire story. As soon as $\vec{E}$ and $\vec{B}$ fields are applied to a plasma, electric currents will be induced, and these currents, in turn, interact with the applied fields.

As an example, let's consider the situation in a rotationally symmetric steady arc as shown schematically in Fig. 4.7. The applied electric

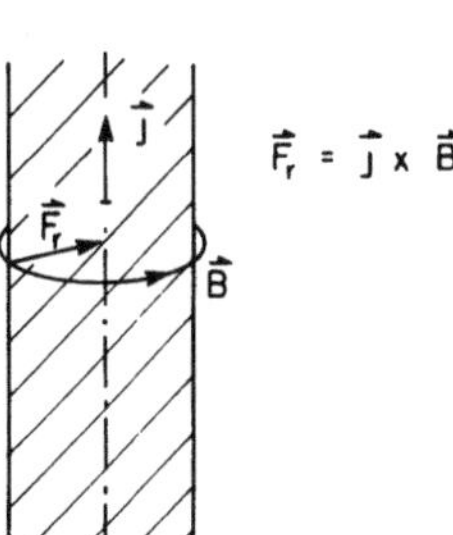


FIG. 4.7. Creation of a self-magnetic field of induction $\vec{B}$ in a rotationally symmetric arc of current density $\vec{j}$ and resulting magnetic body force.

field gives rise to a current of density $\vec{j}$, which induces a self-magnetic field of induction $\vec{B}$. The interaction of this field with the current produces a magnetic body force

$$\vec{F} = \vec{j} \times \vec{B} \tag{4.63}$$

For an analytical description of the effects of this force on the arc column, the momentum and continuity equations are required. Neglecting viscous effects, these equations can be written as

$$\rho \frac{d\vec{v}}{dt} + \text{grad}\, p = \vec{j} \times \vec{B} \tag{4.64}$$

$$\text{div}\,(\rho\vec{v}) = 0 \tag{4.65}$$

where $\rho$ is the plasma density, $\vec{v}$ the plasma velocity vector, and $p$ the pressure. The $\vec{j} \times \vec{B}$ force, which, in general, is responsible for the *pinch effect* in current-carrying plasma columns, may build up a pressure gradient and/or accelerate the plasma. Equation (4.65) determines which fraction of the magnetic body force is used for plasma acceleration. For a rotationally symmetric arc, the radial pressure gradient and the resulting overpressure in the arc can be expressed by

$$\Delta p(r) = \int_r^R j(r)B(r)\,dr \tag{4.66}$$

where $R$ is the radius of the arc periphery.

With

$$\text{rot}\,\vec{B} = \mu_0\vec{j} \tag{4.67}$$

one obtains

$$B(r) = \frac{\mu_0}{r}\int_0^R jr\,dr \tag{4.68}$$

where $\mu_0 = 1.26 \times 10^{-6}$ Hy/m is the permeability constant.

If the current density distribution $j(r)$ is known, $p(r)$ can be calculated. Assuming a uniform current density distribution (one-step model) over the cross section,

$$j = \frac{I}{\pi R^2} \tag{4.69}$$

where $I$ represents the total arc current; $\Delta p(r)$ can then be determined by combining Eqs. (4.66), (4.68), and (4.69):

$$\Delta p(r) = \frac{\mu_0 I j}{4\pi}\left(1 - \frac{r^2}{R^2}\right) \tag{4.70}$$

i.e., the overpressure in the arc is proportional to the product of total arc current and current density.

This overpressure, which is typically on the order of 1% of the total pressure in a high current arc ($I < 10^3$ A), is of little significance for a constant-diameter arc. It becomes important, however, as soon as there is a constriction of the arc channel. The resulting induced plasma flows were mentioned in Chapter 1.

## 4.4. THERMAL EXCITATION AND IONIZATION

In this section we will assume that only the *thermal energy* of the particles is responsible for excitation and ionization or, in general, for any type of energy exchange among the particles.

Assigning a temperature to an ensemble of particles (a large number of particles) implies that the particle speeds or particle energies follow a particular distribution, namely, the *Maxwell–Boltzmann* distribution. Among all conceivable distributions, the Maxwell–Boltzmann distribution represents the most probable distribution.

An important feature of the Maxwell–Boltzmann distribution appears in the speed as well as in the kinetic-energy distribution function. The kinetic energy of the particles appears in the exponential term, i.e.,

$$\exp\left(-\frac{\text{kinetic energy}}{kT}\right)$$

Boltzmann demonstrated that other forms of energy (potential energy, chemical energy, etc.) can replace the kinetic energy in this expression (this form is known as the general Boltzmann distribution).

The following section will show how to arrive at the Boltzmann distribution.

### 4.4.1. Boltzmann Distribution

Only the basics of this derivation will be described in this section. For more details the reader is referred to Lee *et al.*[7]

Let us assume a system has $N$ indistinguishable particles (as, for example, in a uniform gas) and total energy $E$. The energy can be distributed among the particles in the following way:

$$N_1, N_2, N_3, N_4, \ldots, N_i; \qquad N = \sum_k N_k$$

$$E_1, E_2, E_3, E_4, \ldots, E_i; \qquad E = \sum_k N_k E_k$$

There are, of course, many other ways in which the given energy can be distributed among the $N$ particles, and the question arises: in how many different ways is this possible?

We know that $N$ objects (particles) can be arranged in $N!$ different ways, but since $N_1$ particles have the same energy $E_1$, these repetitions have to be excluded by dividing by $N_1!$. The same argument holds for the $N_2$ particles with the same energy $E_2$, etc. The final expression can be written as

$$W = \frac{N!}{N_1!\,N_2! \cdots N_i!} = \frac{N!}{\prod_i N_i!} \tag{4.71}$$

where $W$ is known as the *thermodynamic probability,* which is usually a very large number.

In determining the thermodynamic probability, it has been assumed that every energy $E_k$ corresponds to a single energy state, i.e., the possibility of degenerate energy states has been excluded in this simple analysis. As shown in Chapter 2 for the case of the H atom, more than one eigenfunction belongs to a single energy state. This degeneracy resulting from quantum mechanics has not been included in the simple statistics that lead to Eq. (4.71). By using the principles of quantum statistics (Bose–Einstein statistics), this degeneracy can be taken into account, resulting in a modified thermodynamic probability

$$W = \prod_k \frac{(g_k + N_k - 1)!}{(g_k - 1)!\,N_k!} \tag{4.72}$$

where $g_k$ represents the multiplicity (degeneracy) of energy state $E_k$. At this point, it is useful to consider a statistical interpretation of Eq. (4.72).

For simplicity, a monatomic gas will be considered. The position as well as the momentum of every atom in a system of $N$ atoms can be described by a set of six coordinates (Cartesian coordinates) $x$, $y$, $z$, $p_x$, $p_y$, $p_z$, where the first three describe the position and the last three the momentum of an atom ($p = mv$, where $m$ is the mass of an atom).

By specifying a six-dimensional space $(x, y, z, p_x, p_y, p_z)$ known as the phase space, the state of every atom in terms of its location and momentum can be represented by a point (a phase point) in this phase space.

The phase space can be subdivided into small volume elements

$$H = dx\,dy\,dz\,dp_x\,dp_y\,dp_z \tag{4.73}$$

denoted as cells and numbered from 1 to $k$. These cells, however, must remain sufficiently large that the number of phase points per cell remains large (this is necessary for the application of statistical laws).

The distribution of the atoms with respect to location and momentum at a given instant can be translated into a distribution of phase points in phase space such that $N_1$ phase points fall into cell 1, $N_2$ phase points into cell 2, . . . , and $N_k$ phase points into cell $k$, where

$$\sum N_k = N \tag{4.74}$$

The distribution just described is called a *macrostate* of the system. Only the number of phase points per cell are specified, not their individual coordinates within a cell. In contrast, a *microstate* is defined by a complete specification of all six coordinates of the phase points in a cell. It is obvious that a given macrostate (number of phase points per cell) can correspond to a large number of different microstates.

So far there appears to be no limit on the accuracy with which the location of individual phase points in phase space can be determined. Quantum theory, however, imposes such a limit. According to Heisenberg's uncertainty principle, any quantity in physics that has the dimension of an action (energy × time) can change only in multiples of $h$ (Planck's constant), i.e.,

$$dp_x\, dx \approx h, \qquad dp_y\, dy \approx h, \qquad dz\, dp_z \approx h$$

or

$$dx \cdot dy \cdot dz \cdot dp_x \cdot dp \cdot dp_z \approx h^3 \tag{4.75}$$

As a consequence, a volume element in phase space can never be smaller than $h^3$, i.e., $H$ will have a minimum size

$$H_{\min} = h^3 \tag{4.76}$$

This minimum size is called a "compartment" ($g$). Since each cell must contain a large number of phase points, a cell will, in general, consist of many compartments,

$$g = \frac{H}{h^3} \gg 1 \tag{4.77}$$

Although the location of individual compartments within a cell can still be specified by a set of six coordinates, it is not possible to specify the coordinates of individual phase points within a compartment. This is a direct consequence of Heisenberg's uncertainty principle.

A microstate in quantum statistics is therefore defined by a complete specification of the coordinates of the compartments of the cells, and within the compartments, only the number of phase points (but not their coordinates) can be specified.

Since the atoms in a gas are moving continuously and colliding with each other, a similar behavior will be imposed on the motion of the phase points in phase space, reflecting this motion and change of momentum in actual space. Thus the microstates will continuously change, and the question arises whether there are any preferred microstates. A fundamental hypothesis of statistical mechanics states that all microstates are equally probable, i.e., over sufficiently long periods of time, any one microstate occurs as often as any other.

At a first glance, this hypothesis does not appear to be reasonable. Consider, for example, a possible microstate in which all the particles dwelling in a small volume element in actual space have the same momentum. This situation is highly improbable. Another microstate could be defined by a random distribution of the atoms in ordinary space with randomly distributed momentum vectors. This microstate would appear to have a much higher probability than the previous one. Although the specifications of the two microstates are very different, as long as both microstates are completely specified in the terms of the coordinates of the compartments in which the phase points are located, the two microstates are equally probable. The complete specification of the microstates makes them equally probable.

As an example, let us consider microstates represented by the face values of three dice (white, black, red). One possible microstate is a cast resulting in 3 sixes. The probability for this event is $p_1 = 1/6^3$. Another possible microstate has the face values 1, 3, and 6. With these dice there are six possible ways to obtain the specified face values. The corresponding probability is $p_2' = 6/6^3$, a substantially higher probability than in the first case. The second case, however, is not a microstate, because specification is not complete. A complete specification must include the color of each dice and its specified face value. For example, white-1, black-3, red-6 is a complete specification, and the probability of obtaining these face values with one cast is $p_2 = 1/6^3 = p_1$. Thus the complete specification of the two microstates shows that they are equally probable.

As previously mentioned, many microstates can represent a given macrostate. In addition, macrostates can also change due to the motion of phase points in phase space. Macrostates that appear more frequently than others will have a larger number of associated microstates.

The number of microstates associated with any given macrostate is denoted as the *thermodynamic probability* ($W$), which is generally a large number. A state for which the thermodynamic probability reaches a maximum (maximum number of microstates for this particular macrostate) is also a state of maximum entropy ($S = k \ln W$) and thus represents the equilibrium state.

To determine the Boltzman distribution (the most probable distribution), we have to find the conditions under which $W$ or $\ln W$ reaches a maximum [see Eq. (4.72)]. In the statistical interpretation, $g_k$ in Eq. (4.72) represents the number of compartments in cell $k$.

Since $N$ and the $N_k$ are large numbers, Eq. (4.72) can be simplified by introducing an approximation for factorials of large numbers,

$$\ln (x!) = \sum_{i=1}^{x} \ln i \tag{4.78}$$

If $x$ is a very large number, the sum can be approximated by an integral,

$$\ln (x!) = \sum_{i=1}^{x} \ln i \approx \int_1^x \ln i \cdot di = x \cdot \ln x - x + 1 \tag{4.79}$$

which is known as the Stirling approximation.

Since $x$ is assumed to be a large number, 1 can be neglected compared to the other terms in the last part of Eq. (4.79). Therefore,

$$\ln (x!) = x(\ln x - 1) \tag{4.80}$$

With

$$\ln W = \sum_k [\ln (g_k + N_k - 1)! - \ln (g_k - 1)! - \ln N_k!] \tag{4.81}$$

we apply the Stirling approximation and neglect 1 compared to $N_k$ and $g_k$ and thus obtain

$$\ln W = \sum_k [(g_k + N_k) \ln (g_k + N_k) - g_k \ln g_k - N_k \ln N_k] \tag{4.82}$$

As the phase points shift around, $N_k$ will change continuously and $W$ or $\ln W$ will also change. The thermodynamic probability will reach a maximum if the variation of $\ln W$ vanishes with the variation of the quantities $N_k$:

$$\delta(\ln W) = \sum_k [\ln (g_k + N_k^0) - \ln N_k^0]\, \delta N_k = 0 \tag{4.83}$$

$$\sum_k \ln \left( \frac{N_k^0 + g_k}{N_k^0} \right) \delta N_k = 0 \tag{4.84}$$

where $N_k^0$ represents the number of phase points in cell $k$ for thermodynamic equilibrium.

The quantities $N_k$ in Eq. (4.84) are not independent of each other because

$$N = \sum_k N_k = \text{constant}$$

and

$$\delta N = \sum_k \delta N_k = 0 \tag{4.85}$$

In addition, the total energy of the system remains constant, regardless of the distribution of the phase points in phase space, i.e.,

$$E_{\text{total}} = \sum_k N_k E_k = \text{constant} \tag{4.86}$$

and

$$\delta E_{\text{total}} = \sum_k E_k \, \delta N_k = 0 \tag{4.87}$$

From a mathematical point of view, Eqs. (4.84), (4.85), and (4.87) represent a variational problem with side conditions, where the side conditions refer to the constant total number of particles and the total constant energy of the system.

This problem can be solved by using Lagrange's method of undetermined multipliers. Multiplying Eq. (4.85) by $-\ln B$ and Eq. (4.87) by $-\beta$ and adding these products to Eq. (4.84) results in

$$\sum_k \left[ \ln\left(\frac{N_k^0 + g_k}{N_k^0}\right) - \ln B - \beta E_k \right] \delta N_k = 0 \tag{4.88}$$

In this equation the quantities $\delta N_k$ are independent of each other, i.e., they may be chosen arbitrarily. Therefore,

$$\ln\left(\frac{N_k^0 + g_k}{N_k^0}\right) = \ln B - \beta E_k$$

or

$$\frac{N_k^0}{g_k} = \frac{1}{B \exp(\beta E_k) - 1} \tag{4.89}$$

The last equation represents the Bose–Einstein distribution function.

The Maxwell–Boltzmann statistics can be considered a special case of the more general quantum statistics. The Maxwell–Boltzmann statistics is characterized by $N_k^0 \ll g_k$, i.e., with the Maxwell–Boltzmann statistics, many compartments in phase space will be empty. With this assumption, Eq. (4.89) reduces to

$$\frac{N_k^0}{g_k} = \frac{1}{B \exp(\beta E_k)} \tag{4.90}$$

where $B \gg 1$ and $N_k^0$ represents the most probable (equilibrium) distribution of phase points in phase space (a Maxwell–Boltzmann distribution). Since only equilibrium distributions will be considered here, the upper index (0) will be dropped. The constant $B$ follows from

$$N = \sum_k N_k = \frac{1}{B} \sum_k g_k \exp(-\beta E_k)$$

or

$$\frac{1}{B} = \frac{N}{\sum g_k \exp(-\beta E_k)} \tag{4.91}$$

The denominator of this equation is known as the partition function or sum over all states,

$$Q = \sum g_k \exp(-\beta E_k) \tag{4.92}$$

Introducing Eqs. (4.91) and (4.92) into Eq. (4.90) results in the final form of the Maxwell–Boltzmann distribution,

$$\frac{N_k}{N} = \frac{g_k}{Q} \exp(-E_k/kT) \tag{4.93}$$

where $\beta = 1/kT$ can be derived from the second law of thermodynamics.

The energy $E_k$ has not been specified in deriving the Maxwell–Boltzmann distribution, i.e., $E_k$ can represent any type of energy (kinetic, potential chemical, etc., including combinations of such energies). The special case in which $E_k$ represents the sum of chemical and kinetic energy will be considered in the next section. If $E_k$ stands for only the kinetic energy of the particles, Eq. (4.93) transforms into the Maxwell distribution of molecular energies or velocities.

Equation (4.93) can be written in terms of particle densities by dividing by the volume of the system:

$$\frac{n_k}{n} = \frac{g_k}{Q} \exp(-E_k/kT) \tag{4.94}$$

This equation represents the Boltzmann distribution of excited states if

$n_k$ is interpreted as the number density of excited atoms in the quantum state $k$, $g_k$ as their statistical weight, $E_k$ as the energy of this excited states (with $E_1 = 0$ as the energy of the ground state), $Q$ as the partition function of the atoms, and $n = \sum_k n_k$ as the total number density of atoms regardless of their excitation state.

Equation (4.94) can be generalized to $r$-times ionized atoms by adding an additional index $r$:

$$\frac{n_{r,k}}{n_r} = \frac{g_{r,k}}{Q_r} \exp(-E_{r,k}/kT) \tag{4.95}$$

where $n_{r,k}$ is the number density of $r$-times ionized atoms in the quantum state $k$, $g_{r,k}$ is their corresponding statistical weight, and $E_{r,k}$ their energy; $Q_r = \sum g_{r,k} \exp(-E_{r,k}/kT)$ represents the partition function of $r$-times ionized atoms, and $n_r = \sum_k n_{r,k}$ is the number density of $r$-times ionized atoms regardless of their excitation state.

In a state of thermodynamic equilibrium, the population of excited states as described by Eq. (4.95) follows a Boltzmann distribution. This fact is one of the cornerstones for the treatment of thermal plasmas in which the particles approach a state of thermodynamic equilibrium or of local thermodynamic equilibrium. Such Boltzmann distributions will therefore be used extensively throughout this book.

### 4.4.2. Saha Equilibrium

The equation that describes *thermal ionization*, known today as the *Saha equation*, has also been derived independently by Eggert based on thermodynamic principles.

Since ionization by collision with high-energy electrons is the most important ionization process in a thermal plasma, we will consider such a collision with a neutral atom A:

$$e_{\text{fast}} + \text{A} = \text{A}^+ + e_{\text{slow}} + e$$

The energy of the electron liberated in this process can be written as

$$E_I + \frac{1}{2m_e}(p_x^2 + p_y^2 + p_z^2) \tag{4.96}$$

The first term ($E_I$) represents the potential energy (ionization energy) of the electron with respect to the positive ion, and the second term represents the kinetic energy of the liberated electron. If $d^6n_e$ is the number density of electrons liberated by ionization in cell

$H = dx\,dy\,dz\,dp_x\,dp_y\,dp_z$, then the number of electrons per compartment will be

$$\frac{d^6 n_e}{g_k} = \frac{d^6 n_e \cdot h^3}{H}. \tag{4.97}$$

With $n_0$ as the number density of neutral atoms in the ground state we find according to the Boltzmann principle,

$$\left(\frac{d^6 n_e \cdot h^3}{H}\right) \Big/ n_0 \sim \exp\left\{-\frac{1}{kT}\left[E_I + \frac{1}{2m_e}(p_x^2 + p_y^2 + p_z^2)\right]\right\} \tag{4.98}$$

Expression (4.98) does not consider the effects of degeneracy. Integrating over the volume and all possible momenta results in

$$\frac{n_e}{n_0} \sim \frac{\Delta V}{h^3} \exp\left(\frac{-E_I}{kT}\right) \iiint_{-\infty}^{+\infty} \exp\left[\frac{-1}{2m_e kT}(p_x^2 + p_y^2 + p_z^2)\right] dp_x\,dp_y\,dp_z \tag{4.99}$$

Since

$$\int_{-\infty}^{+\infty} \exp\left(\frac{-1}{2m_e kT} p_x^2\right) dp_x = (2\pi m_e kT)^{1/2} \tag{4.100}$$

and the other two integrations over $p_y$ and $p_z$ produce the same result as Eq. (4.100), we obtain

$$\frac{n_e}{n_0} \sim \frac{\Delta V (2\pi m_e kT)^{3/2}}{h^3} \exp(-E_I/kT) \tag{4.101}$$

The electrons are derived from ionization processes that produce an equal number of positive ions. Therefore, $\Delta V$ will be chosen such that the ion from which an electron has been liberated falls into this volume element, i.e., $\Delta V \cdot n_i = 1$, where $n_i$ is the ion density. Without considering degeneracy, the expression becomes

$$\frac{n_e n_i}{n_0} \sim \left(\frac{2\pi m_e kT}{h^2}\right)^{3/2} \exp(-E_I/kT) \tag{4.102}$$

Generalizing this expression for $r$-times ionized atoms and considering degeneracy results in

$$\frac{n_e n_{r+1,0}}{n_{r,0}} = \frac{g_e g_{r+1,0}}{g_{r,0}} \left(\frac{2\pi m_e kT}{h^2}\right)^{3/2} \exp(-E_{r+1}/kT) \tag{4.103}$$

where $E_{r+1}$ represents the energy required for transforming an $r$-times ionized atom into an $(r + 1)$-times ionized atom. The statistical weight $g_e$ of the electrons is 2 (due to 2 possible spin orientations). Equation (4.103) is known as the Saha equation. Considering the total number of

$r$-times and $(r + 1)$-times ionized atoms (not only those in the ground state), one finds

$$n_r = \sum_k n_{r,k} \qquad \text{and} \qquad n_{r+1} = \sum_k n_{r+1,k}$$

resulting in an alternative form of the Saha equation:

$$\frac{n_e n_{r+1}}{n_r} = \frac{2Q_{r+1}}{Q_r}\left(\frac{2\pi m_e kT}{h^2}\right)^{3/2} \exp(-E_{r+1}/kT) \tag{4.104}$$

where $Q_r = \sum g_{r,k} \exp(-E_{r,k}/kT)$. A similar expression holds for $Q_{r+1}$. Rewriting the Saha equation for the first ionization, one finds that

$$\frac{n_e n_i}{n} = \frac{2Q_i}{Q_0}\left(\frac{2\pi m_e kT}{h^2}\right)^{3/2} \exp(-E_I/kT) \tag{4.105}$$

The term $n$ denotes the density of all neutral particles regardless of their level of excitation. With $p_e = n_e kT$ as the partial pressure of the electron gas, Eqs. (4.103) and (4.104) can be expressed by

$$\frac{n_{r+1,0}}{n_{r,0}} p_e = \frac{2g_{r+1,0}}{g_{r,0}} \frac{(2\pi m_e)^{3/2}(kT)^{5/2}}{h^3} \exp(-E_{r+1}/kT) \tag{4.106}$$

$$\frac{n_{r+1}}{n_r} p_e = \frac{2Q_{r+1}}{Q_r} \frac{(2\pi m_e)^{3/2}(kT)^{5/2}}{h^3} \exp(-E_{r+1}/kT) \tag{4.107}$$

Equations (4.106) and (4.107) apply regardless of other ionized species present in the plasma.

Let us consider the first ionization, i.e.,

$$\frac{n_e n_i}{n} = \frac{2Q_i}{Q_0}\left(\frac{2\pi m_e kT}{h^2}\right)^{3/2} \exp(-E_I/kT) \tag{4.108}$$

and let $\bar{n} = n_e + n = n_i + n$. If $\xi$ is the fraction of ionized atoms in the gas, one obtains

$$\xi = \frac{n_e}{n_e + n} = \frac{n_i}{n_i + n} \qquad \text{since } n_i = n_e \tag{4.109}$$

where

$$n_e = \xi\bar{n}, \qquad n_i = \xi\bar{n}, \qquad n = (1 - \xi)\bar{n}$$

Applying Dalton's law $p = (n_e + n + n_i)kT$, the Saha equation becomes

$$\frac{\xi^2}{1 - \xi^2} = \frac{2Q_i}{Q_0} \frac{(2\pi m_e)^{3/2}(kT)^{5/2}}{h^3 p} \exp(-E_I/kT) \tag{4.110}$$

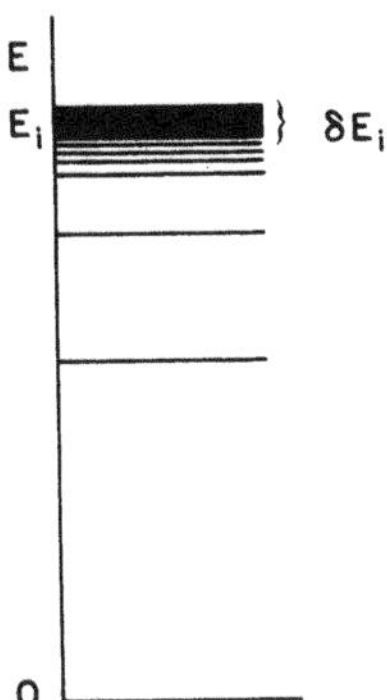


**FIG. 4.8.** Lowering of the ionization potential $\delta E_i$.

where

$$\xi = \xi(T, p, E_I) \qquad \text{and} \qquad Q_r = \sum_s g_{r,s} \exp(-E_{r,s}/kT)$$

The Saha equations as just described require a correction, because the ionization level in a plasma is slightly lowered by the effects of electric or magnetic fields. Here we will consider only electric-field effects. The charged particles in the plasma establish electric microfields which broaden the energy levels, particularly the upper levels, thus removing the degeneracy. Therefore, the upper levels overlap so that the ionization appears to be lowered by $\delta E_i$, as sketched in Fig. 4.8. The summation for the partition function is also affected by the lowering of the ionization energy, because only discrete levels are included in this summation. For moderate temperature levels, it is usually sufficient to consider the first two or three terms in the sum

$$Q_r = \sum_k g_{r,k} \exp(-E_{r,k}/kT) \tag{4.11}$$

For hydrogen and elements similar to hydrogen, Unsöld found that the lowering of the ionization potential can be expressed by

$$\delta E_I = 7 \times 10^{-9} n_e^{1/3} \tag{4.112}$$

where $\delta E_i$ has units of electron volts for $n$ in [$\text{m}^{-3}$].

As an example let us consider hydrogen under the following conditions:

$$T = 20{,}000\,[\text{K}], \qquad n_e = 2 \times 10^{17}\,[\text{cm}^{-3}], \qquad p = 1\,[(\text{atm}]$$

The lowering of the ionization potential in this case is $\delta E_I \approx 0.4$ [eV].

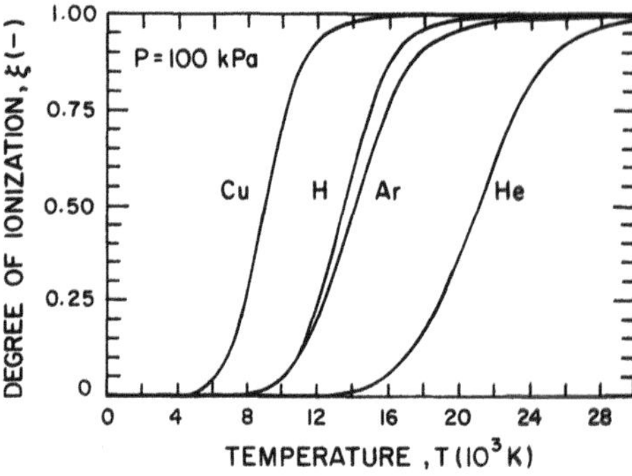


**FIG. 4.9.** Degree of ionization as a function of temperature for selected gases and vapors.

The degree of ionization as a function of temperature is shown in Fig. 4.9 for a number of selected gases and vapors. Equation (4.104) can be rewritten with the reduced ionization energy $E^*_{r+1}$ as

$$\frac{n_{r+1}}{n_r} p_e = \frac{2Q_{r+1}}{Q_r}\left(\frac{2\pi m_e}{h^3}\right)^{3/2} (kT)^{5/2} \exp\left(-E^*_{r+1}/kT\right) \qquad (4.113)$$

Substituting the appropriate values for the constants yields the working form of the Saha equation:

$$\log_{10}\left(\frac{n_{r+1}}{n_r} p_e\right) = -E^*_{r+1}\frac{5040}{T} + \frac{5}{2}\log T + \log_{10}\left(\frac{Q_{r+1}}{Q_r}\right) - 0.48 \qquad (4.114)$$

for $E^*_{r+1}$ in eV, $p_e$ in J/cm$^3$, and $n$ in cm$^{-3}$.

### 4.4.3. Complete Thermal Equilibrium (CTE)

It is useful to consider first a plasma in a state of *complete thermodynamic equilibrium* (CTE) even though this state cannot be realized under laboratory conditions. For the sake of simplicity, it will be assumed that the plasma is generated from a monatomic gas or a mixture of monatomic gases.

CTE prevails in a uniform, homogeneous plasma volume if kinetic and chemical equilibria as well as every conceivable plasma property are unambiguous functions of the temperatue. The temperature, in turn, has to be the same for all plasma constituents and their possible reactions. More specifically, the following conditions must be met:

(a) The velocity distribution functions for particles of every species $r$ that exists in the plasma, including the electrons, must follow a Maxwell–Boltzmann distribution

$$f(v_r) = \frac{4v_r^2}{\sqrt{\pi}\left(\frac{2kT}{m_r}\right)^{3/2}} \exp\left(-\frac{m_r v_r^2}{kT}\right) \tag{4.115}$$

where $v_r$ is the velocity of particles of species $r$, $m_r$ is their mass, and $T$ is their temperature, which is the same for every species $r$ and, in particular, is identical to the plasma temperature.

(b) The population density of the excited states of every species $r$ must follow a Boltzmann distribution

$$n_{r,k} = n_r \frac{g_{r,k}}{Q_r} \exp\left(-E_{r,k}/kT\right) \tag{4.116}$$

where $n_r$ is the total number density of ions of species $r$, $Q_r$ is their partition function, $E_{r,k}$ is the energy of the $k$th quantum state, and $g_{r,k}$ is the statistical weight of this state. The excitation temperature $T$, which appears explicitly in the exponential term and implicitly in the partition function $Q_r$, is identical to the plasma temperature.

(c) The particle densities (for neutrals, electrons, and ions) are described by the Saha equations which can be considered a mass action law:

$$\frac{n_{r+1}n_e}{n_r} = \frac{2Q_{r+1}}{Q_r} \frac{(2\pi m_e kT)^{3/2}}{h^3} \exp\left(-E_{r+1}/kT\right) \tag{4.117}$$

where $E_{r+1}$ represents the energy required to produce an $(r + 1)$-times ionized atom from an $r$-times ionized atom (ionization energy). The ionization temperature $T$ in this equation is identical to the plasma temperature. Lowering of the ionization potential has been disregarded in the Saha equation.

(d) The electromagnetic radiation field is that of blackbody radiation of intensity $B_\nu^0$ as described by the Planck function

$$B_\nu^0 = \frac{2h\nu^3}{c^2} \frac{1}{e^{h\nu/kT} - 1} \tag{4.118}$$

the symbol $\nu$ stands for the frequency, $h$ represents Planck's constant, and $c$ is the velocity of light. The temperature of this blackbody radiation is again identical to the plasma temperature.

A plasma that follows the ideal model described by Eqs. (4.115) to (4.118) would have to dwell in a hypothetical cavity whose walls were maintained at the plasma temperature, or else the plasma volume would have to be so large that the central part of this volume, in which CTE prevails, would not sense the plasma boundaries. In this way the plasma would be penetrated by blackbody radiation of its own temperature. An actual plasma, of course, deviates from these ideal conditions. The observed plasma radiation, for example, is much less than the blackbody radiation, because most plasmas are optically thin over a wide wavelength range. Therefore, the radiation temperature of a gaseous radiator deviates appreciably from the kinetic temperature of the plasma constituents or the already mentioned excitation and ionization temperatures. In addition to radiation losses, plasmas suffer irreversible energy losses by conduction, convection, and diffusion, which also disturb the thermodynamic equilibrium. Thus, laboratory plasmas as well as some of the natural plasmas cannot be in a CTE state. The following sections will discuss deviations from CTE and the associated concept of LTE.

### 4.4.4. Concept of Local Thermodynamic Equilibrium (LTE)

The following discussion will be restricted to optically thin plasmas, a situation that is frequently approached by laboratory arc or RF plasmas. In contrast to the CTE situation, LTE in optically thin plasmas does not require a radiation field that corresponds to the blackbody radiation intensity of the respective LTE temperature. It does require, however, *that collision processes* (*not radiative processes*) *govern transitions and reactions in the plasma* and that there be a microreversibility among the collision processes. In other words, a detailed equilibrium between each collision process and its reverse process is necessary. Steady-state solutions of the respective collision rate equations will then yield the same energy distribution as that of a system in complete thermal equilibrium, with the exception of the rarefied radiation field. LTE further requires that local gradients of the plasma properties (temperature, density, heat conductivity, etc.) be sufficiently small that a given particle that diffuses from one location to another in the plasma finds sufficient time to equilibrate, i.e., the diffusion time should be of the same order of magnitude or larger than the equilibration time. From the equilibration time and the particle velocities, an equilibration length that is smaller in regions of small plasma property gradients (for example, in the center of an electric arc) can be derived. Therefore, with regard to spatial variations LTE is more probable in such regions. Heavy-particle diffusion and resonance radiation from the center of a nonuniform plasma source help to reduce the effective equilibrium distance on the outskirts of the source.

The following sections will systematically discuss the important assumptions for LTE, based on plasmas that approach LTE.

### 4.4.4.1. Kinetic Equilibrium

It can be safely assumed that each species (electron gas, ion gas, or neutral gas) in a dense, collision-dominated, high-temperature plasma will assume a Maxwellian distribution (excluding regions close to walls and electrodes). However, the temperatures defined by these Maxwellian distributions may be different from species to species. Such a situation, which leads to a two-temperature description, will be discussed in the following paragraphs.

The electric energy fed into an arc, for example, is dissipated in the following way: The electrons because of their high mobility, pick up energy from the electric field and partially transfer it to the heavy plasma constituents through collisions. Because of this continuous energy flux from the electrons to the heavy particles, there must be a "temperature gradient" between these two species, that is, $T_e > T_h$, where $T_e$ is the electron temperature and $T_h$ the temperature of the heavy species, assuming that ion and neutral gas temperatures are the same.

In the two-fluid plasma model defined in this manner, two distinct temperatures $T_e$ and $T_h$ may exist. The degree to which $T_e$ and $T_h$ deviate from each other will depend on the thermal coupling between the two species. The difference between these two temperatures can be derived from an energy balance, assuming that the kinetic energy exchanges by elastic collisions are equal to the energy gained by electrons from the electric field $E$, neglecting inelastic collisions:

$$\tfrac{3}{2}k(T_e - T_h)(2m_e/m_h) = eEv_d^e\tau_e \tag{4.119}$$

the left-hand side represents the fraction of energy transferred from one electron of mass $m_e$ to a heavy particle of mass $m_h$ [see Eq. (3.1)]. On the right-hand side, $v_d^e$ is the drift velocity of the electrons, defined as

$$\overrightarrow{v_d^e} = \mu_e\vec{E} \tag{4.120}$$

and $\tau_e$ is the mean time between two collisions of electrons with heavy species. Using Eq. (4.22) and writing

$$\tau_e = \frac{l_e}{\bar{v}_e} \tag{4.121}$$

where $\bar{v}_e$ is the mean thermal velocity of the electrons, it follows that

$$\frac{T_e - T_h}{T_e} = \frac{\pi m_h}{24 m_e} \frac{(l_e e E)^2}{(k T_e)^2} \tag{4.122}$$

Since the term $\pi m_h/24m_e$ is already 243 for hydrogen, the amount of (directed) energy ($l_e eE$) that the electrons pick up along one mean free path length has to be very small compared with the average thermal energy ($kT_e$) of the electrons. Low field strengths, high pressures ($l_e \sim 1/p$), and high temperature levels are favorable for kinetic equilibrium among the plasma constituents. The field strength and pressure requirements are usually summarized in the parameter $E/p$. In glow discharges, which are characterized by high electron temperatures and low temperatures for the heavy species, $E/p$ assumes values on the order of $10^5$ V/m Pa, while in typical thermal arcs $E/p$ is on the order of $10^{-2}$ V/m Pa. At low pressures, for example, appreciable deviations from kinetic equilibrium may occur. The semischematic diagram in Fig. 1.3 shows how electron and gas temperatures separate in an electric arc with decreasing pressure. For an atmospheric argon high-intensity arc with $E = 1300$ V/m, $l_e = 3 \times 10^{-6}$ m, $m_{\mathrm{Ar}}/m_e = 7 \times 10^4$, and $T_e = 30 \times 10^3$ K, the deviation between $T_e$ and $T_h$ is on the order of 1% (Finkelnburg and Maecker, 1956).

#### 4.4.4.2. Excitation Equilibrium

In order to establish the criteria for excitation equilibrium, every conceivable process that may lead to excitation or de-excitation has to be taken into account. For simplicity, only the most prominent mechanisms (collisional and radiative excitation and de-excitation) will be considered (see Sections 3.4 and 3.6):

| *Excitation* | *De-excitation* |
|---|---|
| (1) electron collisions | (1) collisions of the second kind |
| (2) photoabsorptions | (2) photoemissions |

In CTE, microreversibilities would have to exist for all processes, i.e., in the above scheme, excitation by electron collisions would have to be balanced by the reverse process, namely, collisions of the second kind, and excitation by the photoabsorption process would have to be balanced by photoemission processes, which include spontaneous and induced emissions. Furthermore, the populations of excited states would have to follow a Boltzmann distribution.

Microreversibility for the radiative processes holds only if the radiation field in the plasma reaches the intensity $B_\nu^0$ of blackbody radiation. However, actual plasmas are frequently optically thin over most of the spectral range, so that the situation for excitation equilibrium seems to be hopeless. Fortunately, if collisional processes dominate, photoabsorption and photoemission processes do not have to balance: only the sum on the left-hand side and the right-hand side of the scheme above have to be equal. Since the contribution of the photon processes to the number of excited atoms is almost negligible when collisional processes dominate, the excitation process is still close to LTE.

#### 4.4.4.3. Ionization Equilibrium

For ionization equilibrium, again only the most prominent mechanisms leading to ionization and recombination will be considered.

| *Ionization* | *Recombination* |
|---|---|
| (1) electron collisions | (1) three-body recombinations |
| (2) photoabsorptions | (2) photorecombinations |

In a perfect thermodynamic equilibrium state (CTE) with cavity radiation, a microreversibility would exist among the collisional and radiative processes, and the particle densities would be described by the Saha equation. Without cavity radiation, the number of photoionizations is almost negligible, requiring a total balance of all processes involved (instead of microreversibility). Photorecombinations, especially at lower electron densities, are not negligible. The frequency of the three remaining elementary processes is a function of the electron density only, leading (for a certain electron density) to the same order-of-magnitude frequency for these elementary processes. The result is an appreciable deviation between actual and predicted (from Eq. (4.117)] electron densities. Only for sufficiently large electron densities does the Saha equation predict correct values. For smaller electron densities, the corona formula, which considers ionization by electron impact and photorecombination only, must be used. The particle concentrations in low-intensity arcs at atmospheric pressure, for example, must be calculated using this formula. Significant deviations of the electron density predicted by the Saha equation from the true electron density may also occur in the fringes of high-intensity arcs, RF discharges, and plasma jets.

In summary, LTE exists in a steady-state, optically thin plasma when the following conditions are simultaneously fulfilled:

(1) The different species that form the plasma have a Maxwellian distribution.
(2) $E/p$ is small enough and the temperature is sufficiently high that $T_e = T_h$.
(3) Collisions are the dominating mechanism for excitation (Boltzmann distribution) and ionization (Saha equilibrium).
(4) Spatial variations of the plasma properties are sufficiently small.

### 4.4.5. Deviations from LTE

In addition to the two extreme cases, namely, LTE (based on Saha ionization equilibrium) and corona equilibrium, conditions between these two limiting cases are also of interest. In this range three-body recombinations as well as radiative recombination and de-excitation are significant. Over the past 25 years a large number of investigations have been reported on the subject of radiative-collisional processes, LTE, and deviations from LTE. The results of these studies up to 1966 are summarized in two books on plasma diagnostics [Huddlestone and Leonard (1965) and Lochte-Holtgreven (1968)]. Drawin (1970) presents a comprehensive review of the validity conditions for LTE, including a discussion of *complete local thermodynamic equilibrium* (*CLTE*) and *partial local thermodynamic equilibrium* (*PLTE*). This distinction is associated with the population of excited levels that may deviate from ideal Boltzmann distributions. In the case of optically thin plasmas, the lower-lying excited energy levels tend to be underpopulated with respect to the ground state. This situation is referred to as PLTE provided that all the other conditions for LTE are met. The electron densities, $n_e$, required for CLTE in an optically thin plasma are substantially higher than those needed for the less stringent requirement of PLTE.

Griem (1964) established the following criterion for the existence of CLTE in an optically thin homogeneous plasma:

$$n_e \geq 9 \times 10^{23}\left(\frac{E_{21}}{E_{\mathrm{H}^+}}\right)^3\left(\frac{kT}{E_{\mathrm{H}^+}}\right)(\mathrm{m}^{-3}) \tag{4.123}$$

where $E_{21}$ represents the energy gap between the ground state and the first excited level, $E_{\mathrm{H}^+} = 13.58\,\mathrm{eV}$ is the ionization energy of the hydrogen atom, and $T$ is the plasma temperature. This criterion shows the sensitivity of the required electron density for CLTE to the energy of the most critical, first excited state.

It is obvious that deviations from CLTE or even from PLTE will occur in regions of low electron densities as, for example, in plasma regions adjacent to walls or in arc fringes and in all types of low-density plasmas of laboratory dimensions.

For many years the existence of CLTE in atmospheric-pressure, high-current arcs has not been questioned. Only recently have deviations from CLTE been found in such arcs. It has been shown that a pressure of approximately 300 kPa is necessary to reach a state of CLTE in the central portion of a free-burning argon arc at currents of 300 to 400 A. These conditions correspond to an electron density of approximately $10^{24}\,m^{-3}$ in the center. Deviations from CLTE still persist in the outer regions of such arcs, where the electron density drops substantially below $10^{24}\,m^{-3}$.

Numerous analytical as well as experimental studies over the past years demonstrate that LTE (CLTE or PLTE) in high-intensity arcs is the exception rather than the rule. Several studies have shown that, besides the underpopulation of lower-lying energy levels, deviations from LTE can frequently be attributed to strong gradients in the plasma and the associated diffusion effects. More details on LTE and deviations from LTE are discussed in the books by Huddlestone and Leonard (1965), Griem (1964), Lochte-Holtgreven (1968), and Mitchner and Kruger (1973).

## 4.5. RIGOROUS DEFINITION OF THE PLASMA STATE

A plasma consisting of a mixture of electrons, ions, and neutral particles in the gaseous state is overall electrically neutral, as pointed out in the first chapter. This electrical neutrality of a plasma, however, applies only for sufficiently large plasma volumes, i.e., $V > \lambda_D^3$, where $\lambda_D$, the *Debye length*, is a characteristic length in a plasma; it will be discussed in the following section.

Although a plasma can be treated in a first approximation as electrically neutral, deviations from neutrality must be considered in a second approximation; such deviations, however, are restricted to distances on the order of a Debye length.

### 4.5.1. Debye Length in a Plasma

The concept of the electric shielding of a positive ion by negative ions (and vice versa) was first developed by Debye and Hückel in 1923 for strong electrolytes. This concept, however, also applies to gaseous plasmas.

Because of Coulomb forces acting between charged particles, a positive ion is, on the average, surrounded by more than one electron; this surrounding electron cloud provides an effective shielding of the positive ion charge. The accumulation of negative charges in the vicinity of a positive ion represents a net negative space charge, i.e., a deviation from charge neutrality occurs over the dimension of the electron cloud, which is known as a Debye sphere. These Debye spheres are dynamic in nature and overlap each other.

In the following paragraph, an expression will be derived for the dimension of a Debye sphere (or Debye length) in a uniform plasma of electron density $n_e = n_i$ (singly ionized species only). In this derivation, it will be assumed that the equilibrium ion density distribution $n_i = n_{i,0}$ is not affected by the electron clouds forming around positive ions; the electrons establishing a dynamic equilibrium in the cloud retain their Maxwell–Boltzmann distribution, and no recombination with the positive ions occurs (the potential energy $eV \ll kT_e$).

Poisson's equation describes the electric field established by the negative space charge, i.e.,

$$\operatorname{div} \vec{E} = \frac{1}{\varepsilon_0} \rho_{\text{el}} \tag{4.124}$$

where $\vec{E} = -\operatorname{grad} V$ is the electric field strength, $\varepsilon_0$ is the dielectric constant, and $\rho_{\text{el}} = e(n_i - n_e)$ is the electric space charge. With the previously specified assumptions, the space charge can be expressed by

$$\rho_{\text{el}} = e[n_{i,0} - n_{e,0} \exp(eV/kT_e)] \tag{4.125}$$

where $n_{i,0} = n_{e,0}$ is the undisturbed distribution of electron and ion densities. Since $eV \ll kT_e$, the following approximation is valid:

$$\exp(eV/kT_e) \approx 1 + eV/kT_e \tag{4.126}$$

With this approximation, Eq. (4.124) can be written as

$$\Delta V = \frac{e}{\varepsilon_0} n_{e,0} \frac{eV}{kT_e}$$

or

$$\Delta V - \frac{1}{\lambda_D^2} V = 0 \tag{4.127}$$

with $\lambda_D = (\varepsilon_0 kT_e/e^2 n_{e,0})^{1/2}$ as the Debye length and $\Delta$ as the Laplace operator. By introducing spherical coordinates, one finds, because of spherical symmetry,

$$\frac{d^2V}{dr^2} + \frac{2}{r}\frac{dV}{dr} - \frac{1}{\lambda_D^2}V = 0 \tag{4.128}$$

This differential equation has the solution

$$V = A \cdot \frac{1}{r}\exp(-r/\lambda_D) + B \cdot \frac{1}{r}\exp(r/\lambda_D) \tag{4.129}$$

and the boundary conditions can be specified as

$$V = 0 \qquad \text{for } r \to \infty$$

$$V = \frac{e}{4\pi\varepsilon_0} \qquad \text{for } r \to 0$$

With these boundary conditions the final form of the solution can be written as

$$V = \frac{e}{4\pi\varepsilon_0 r}\exp(-r/\lambda_D) \tag{4.130}$$

The first term on the right-hand side of Eq. (4.130) represents the Coulomb potential of a point charge. The second term describes the action of the electron cloud around this charge, screening the positive ion potential as shown in Fig. 4.10. It can be shown that $n_e\lambda_D^3 \gg 1$, i.e., there are a substantial number of electrons in a Debye sphere of radius $\lambda_D$ (for

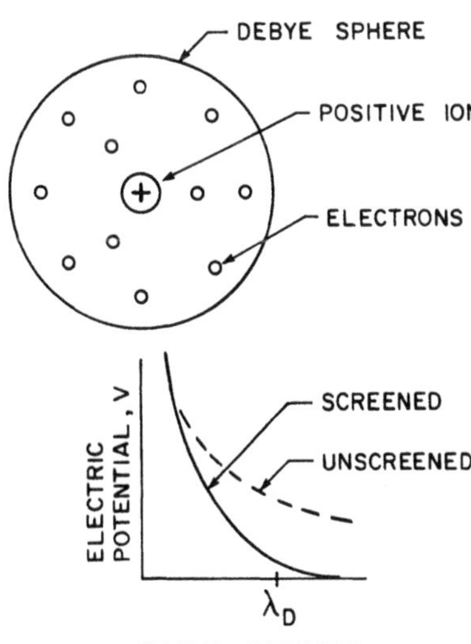


**FIG. 4.10.** The Debye sphere and evolution of the screened and unscreened Coulomb potential.

simplicity, the volume of the Debye sphere will be expressed by $\lambda_D^3$ instead of $\frac{4}{3}\pi\lambda_D^3$):

$$(n_e\lambda_D^3)^{2/3} = n_e^{2/3}\lambda_D^2 = n_e^{2/3}\frac{\varepsilon_0 kT_e}{e^2 n_e}$$

$$= \frac{kT_e}{\dfrac{e^2}{\varepsilon_0} n_e^{1/3}}$$

$$= \frac{\text{kinetic energy}}{\text{potential energy}} \gg 1 \tag{4.131}$$

Defining $\bar{d}$ as the average distance between charged particles ($\bar{d} = n_e^{-1/3}$), the term $e^2/\varepsilon_0\bar{d}$ thus represents the average potential energy between charged particles. Since the derivation of the Debye length has been based on the assumption that the kinetic energy of electrons is much larger than their potential energy, Eq. (4.131) confirms that $(n_e\lambda_D)^3 \gg 1$ is indeed valid.

### 4.5.2. Characteristic Lengths in a Plasma

At this point it is useful to establish a hierarchy of characteristic lengths in a plasma.

The *Landau parameter* is the characteristic length at which potential and kinetic energy balance each other in an electron–ion encounter:

$$\frac{e^2}{4\pi\varepsilon_0 r_{\min}} = \tfrac{3}{2}kT_e$$

or

$$r_{\min} = \frac{e^2}{6\pi\varepsilon_0 kT_e} \tag{4.132}$$

This is the smallest characteristic length in a plasma, and it depends only on the electron temperature. For an electron temperature $T_e = 10^4$ K, this length is approximately 10 Å, i.e., close to atomic dimensions. The collision cross section for Coulomb interaction between ions and electrons is directly related to the Laudau parameter:

$$\sigma_{i-e} \sim \pi r_{\min}^2 = \pi\left(\frac{e^2}{6\pi\varepsilon_0 kT_e}\right)^2 \tag{4.133}$$

**TABLE 4.1.** Typical Values of the Debye Length in Plasmas (m)

| $n_e\ (m^{-3})$ / $T_e$ (K) | $10^{16}$ | $10^{18}$ | $10^{20}$ | $10^{22}$ | $10^{24}$ |
|---|---|---|---|---|---|
| $10^4$ | $6.9 \times 10^{-5}$ | $6.9 \times 10^{-6}$ | $6.9 \times 10^{-7}$ | $6.9 \times 10^{-8}$ | $6.9 \times 10^{-9}$ |
| $10^5$ | $2.2 \times 10^{-4}$ | $2.2 \times 10^{-5}$ | $2.2 \times 10^{-6}$ | $2.2 \times 10^{-7}$ | $2.2 \times 10^{-8}$ |
| $10^6$ | $6.9 \times 10^{-4}$ | $6.9 \times 10^{-5}$ | $6.9 \times 10^{-6}$ | $6.9 \times 10^{-7}$ | $6.9 \times 10^{-8}$ |

For $T_e = 10^4$ K, this collision cross section is on the order of $10^{-18}$ m$^2$.

Another characteristic length is the previously mentioned average distance $\bar{d} = n_e^{-1/3}$ between charged particles. For an electron density of $10^{22}$ m$^{-3}$, this distance is approximately $5 \times 10^{-8}$ m. The Debye length, which is an important characteristic length in a plasma, has already been discussed and, according to Table 4.1, it assumes a value of approximately $7 \times 10^{-8}$ m for these conditions ($T_e = 10^4$ K, $n_e = 10^{22}$ m$^{-3}$).

The remaining characteristic length parameters are the mean free paths of the electrons or ions, which are on the order of $10^{-6}$ m for the previously specified conditions, and the plasma dimension itself. The evolution of these characteristic lengths with temperature is represented in Fig. 4.11 for an argon plasma at atmospheric pressure.

If the characteristic dimension $L$ of the plasma is such that $L \gg l$, the plasma is collision dominated, in contrast to a collisionless plasma, for which $L < l$. All plasmas discussed in this book are collision-dominated plasmas.

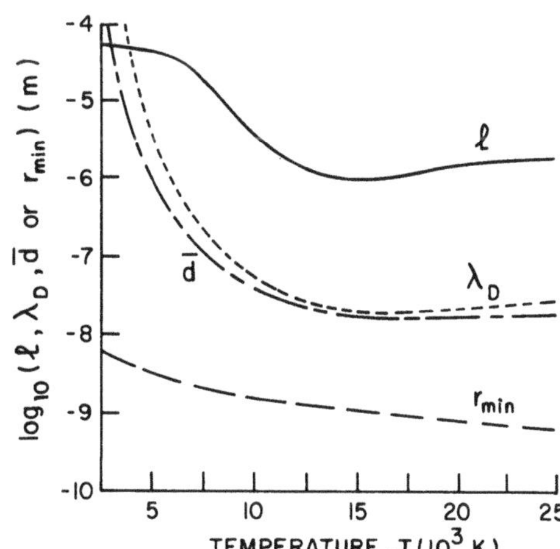


**FIG. 4.11.** Evolution with temperature of the mean free path $l$, the Debye length $\lambda_D$, the average distance between charged particles $\bar{d}$, and the Landau parameter $r_{min}$ for an argon plasma at atmospheric pressure.[8]

## 4.6. QUASI-NEUTRALITY

Quasi-neutrality prevails in a plasma if net space charges are small, for volumes large compared to the Debye length, i.e.

$$|n_i - n_e| \ll n_e, n_i(=n_c) \tag{4.134}$$

As before, it will be assumed that $n_e = n_i = n_c$ is the undisturbed plasma density.

Any deviation from quasi-neutrality will be opposed by electric fields resulting from charge separation, which tend to restore quasi-neutrality.

As an example, let us calculate the electric field resulting from a small separation ($\Delta x = 10^{-10}$ m) of charges in a plasma with an electron density of $n_e = 5 \times 10^{22}\,\text{m}^{-3}$. By using Poisson's equation in one-dimensional form, we obtain

$$\left|\frac{dE}{dx}\right| = \frac{e}{\varepsilon_0} n_e$$

or

$$\begin{aligned} |E| &= \frac{e}{\varepsilon_0} n_e \cdot \Delta x \\ &\approx 9 \times 10^4\,\text{V/m} \end{aligned} \tag{4.135}$$

This result clearly demonstrates that it is extremely difficult to separate charges in a dense plasma because of the high electric fields induced by charge separation. We will consider both diffusion due to steep gradients of the plasma properties and magnetic fields as possible mechanisms for charge separation.

### 4.6.1. Charge-Carrier Separation by Diffusion

As we have discussed, diffusion of charged particles in plasmas is governed by the ambipolar diffusion process. Charged particles joined by Coulomb forces cannot diffuse independently, unless the diffusional driving forces are able to overcome the Coulomb forces. As shown in Eq. (4.135), this process may become feasible for small electron densities (large average distances between charged particles) and correspondingly reduced Coulomb forces. It was seen in Section 4.3.2 that the electric field due to ambipolar diffusion can be expressed by

$$\vec{E} = -\frac{D_e}{\mu_e}\frac{1}{n_e}\,\text{grad}\, n_e \tag{4.136}$$

By using the Einstein relation

$$\frac{D_e}{\mu_e} = \frac{kT_e}{e} \tag{4.137}$$

Eq. (4.136) transforms into

$$\vec{E} = -\frac{kT_e}{e}\,\text{grad}\,(\ln n_e) \tag{4.138}$$

For further evaluation of Eq. (4.138), a relationship for $n_e = f(T_e)$ at atmospheric pressure is required for small values of $n_e$. This relationship is known as the Corona equation, and it will be used in the following:

$$n_e \sim \exp\,(-E_r/kT_e)$$

With this relation, Eq. (4.138) transforms into

$$\vec{E} = -\frac{E_r}{e}\frac{\text{grad}\,T_e}{T_e}$$

or

$$\text{div}\,\vec{E} = -\frac{E_r}{e}\,\text{div}\,[\text{grad}\,(\ln T_e)] \tag{4.139}$$

with

$$\frac{|n_i - n_e|}{n_c} = \frac{|\varepsilon_0\,\text{div}\,\vec{E}|}{en_c}$$

$$= \frac{\varepsilon_0 E_r}{e^2 n_c}\,\text{div}\,[\text{grad}\,(\ln T_e)] \ll 1 \tag{4.140}$$

For a relatively large value of div [grad (ln $T_e$)] $= 10^6\,\text{m}^{-2}$ in Eq. (4.140) and a typical ionization energy of $E_r = 10\,\text{eV}$, one finds that the condition for quasi-neutrality could still be met for $n_e > 10^{15}\,\text{m}^{-3}$. For charge-carrier densities $n_e < 10^{15}\,\text{m}^{-3}$, charge-carrier separation by diffusion becomes feasible, provided that large density gradients exist.

### 4.6.2. Charge-Carrier Separation by Magnetic Fields

In principle, it is possible to separate charged particles in a plasma by applied or self-induced magnetic fields, because magnetic fields that have a component perpendicular to the trajectories of charged particles give rise to the previously discussed Lorentz force (see Section 4.3.3), which acts in opposite directions for negative and positive charge carriers.

In the following derivation, the self-induced magnetic field in a rotationally symmetric arc will be considered a possible source of charge-carrier separation.

Let us assume that the undisturbed plasma is characterized by $n_e = n_i = n_c$. Then the condition for quasi-neutrality can be expressed by

$$\frac{|n_e - n_i|}{n_c} = \frac{\varepsilon_0 \operatorname{div} \vec{E}_r}{en_c} \ll 1 \tag{4.141}$$

where $\vec{E}_r$ is the induced electric field pointing in the radial direction (Fig. 4.7). This field drives electrons and positive ions in opposite directions and establishes a force balance with the magnetic body force $\vec{F}_r$,

$$\vec{F}_r = en_e\vec{E}_r = \vec{j} \times \vec{B} \tag{4.142}$$

where $\vec{j}$ is the electric current density maintaining the arc and $\vec{B}$ is the self-induced magnetic induction.

From Eq. (4.142) it follows that

$$\operatorname{div} \vec{E}_r = \frac{1}{en_e} \operatorname{div} (\vec{j} \times \vec{B}) \tag{4.143}$$

and with $\vec{j} = \vec{j}_e + \vec{j}_i \approx \vec{j}_e = en_e\vec{u}_e$ we have

$$\operatorname{div} \vec{E}_r = \operatorname{div} (\vec{u}_e \times \vec{B}) \tag{4.144}$$

where $\vec{u}_e$ is the electron drift velocity.

Since $\operatorname{div} (\vec{u}_e \times \bar{B}) = \vec{u}_e \cdot \operatorname{rot} \vec{B} - \vec{B} \cdot \operatorname{rot} \vec{u}_e$ and since $\operatorname{rot} \vec{u}_e = 0$, Eq. (4.144) transforms into

$$\operatorname{div} \vec{E}_r = \vec{u}_e \cdot \operatorname{rot} \vec{B} \tag{4.145}$$

For a steady current flow, one of Maxwell's equations states that

$$\operatorname{rot} \vec{B} = \mu_0 \vec{j} \tag{4.146}$$

and therefore

$$\operatorname{div} \vec{E}_r = \mu_0 en_e u_e^2 \tag{4.147}$$

Introducing Eq. (4.147) into Eq. (4.141) leads to

$$\frac{|n_e - n_i|}{n_e} = \varepsilon_0 \mu_0 u_e^2 \tag{4.148}$$

($\varepsilon_0 = 8.86 \times 10^{-12}$ A s/V m; $\mu_0 = 1.256 \times 10^{-6}$ V s/A m; $\varepsilon_0\mu_0 = 1/c^2$ where $c$ is the velocity of light).

Finally, one finds that the condition for quasi-neutrality is given by

$$\frac{|n_e - n_i|}{n_c} = \frac{u_e^2}{c^2} \ll 1 \tag{4.149}$$

In this case the drift velocity of the electrons would have to be close to the velocity of light to cause a substantial separation of charge carriers. For the thermal plasmas of interest here, $u_e \ll c$ (see Section 4.3.1), i.e., self-induced magnetic fields will not cause charge-carrier separation. In the case of applied magnetic fields, it can be shown that the magnitude required to separate charge carriers in a thermal plasma is extremely high.

## 4.7. PLASMA SHEATHS

As shown in the previous section, a plasma always tends to remain electrically neutral. This is especially true for thermal plasmas which, by definition, are characterized by relatively high charge-carrier densities. Charge imbalances within a plasma are restricted to regions on the order of a Debye length, as previously discussed. A region in which charge imbalances may exist in the vicinity of a solid or liquid boundary is termed a *sheath*. Sheaths are found, for example, on plasma-confining walls, on electrodes, and around probes immersed in a plasma. The plasma meets the boundary conditions at a wall by means of a sheath, which is sometimes designated as an *electrical boundary layer* accommodating the transition from electrical conduction in the plasma to that in the solid or liquid phase.

In a typical thermal boundary layer at the wall of a plasma-confining vessel, the thickness of the sheath at the bottom of the thermal boundary layer overlying the surface is several orders of magnitude smaller than the thickness of the thermal boundary layer, which encompasses many mean free path lengths of the particles. This fact becomes obvious by considering the magnitude of the Debye length,

$$\lambda_D = \left(\frac{\varepsilon_0 k T_e}{e^2 n_e}\right)^{1/2} = 69.1\left(\frac{T_e}{n_e}\right)^{1/2} \tag{4.150}$$

The numerical value gives the Debye length in m if $n_e$ is in units of $m^{-3}$ and $T_e$ in K. Typical numerical values are summarized above in Table 4.1. In the thermal plasmas of interest in this book, the Debye lengths are in the range from $10^{-8}$ to $10^{-7}$ m, while the mean free path lengths in such plasmas are in the range from $10^{-6}$ to $10^{-5}$ m. These values confirm the previous statement comparing the thickness of a sheath to that of a thermal boundary layer.

For a wall kept at floating potential (no net current flow), the electrons initially reach the wall at a higher rate than the positive ions due to their high mobility, thus charging the wall negatively. As a result,

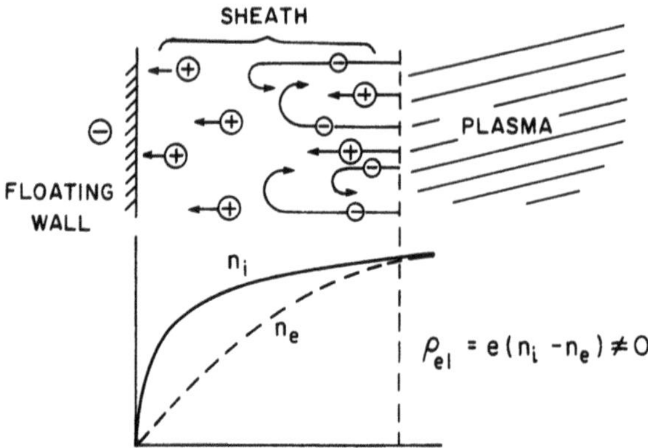


**FIG. 4.12.** Positive space charge and movements of the charged particles in the sheath.

a net positive space charge is formed in the sheath, because electrons are repelled and positive ions are attracted by the wall. This situation is sketched in Fig. 4.12, where a net positive space charge in the sheath is shown.

We will now calculate the potential distribution in the sheath adjacent to a plane wall, assuming that the wall is either floating or negatively biased (the situation would be similar in the case of a positively biased wall with a correspondingly negative space charge in front of the wall). For the sake of simplicity, a one-dimensional situation (with length of wall ≫ thickness of sheath) will be postulated.

With the Poisson equation

$$\Delta V = -\frac{\rho_{\text{el}}}{\varepsilon_0} \tag{4.151}$$

we obtain for a one-dimensional situation

$$\frac{d^2V}{dy^2} = -\frac{e}{\varepsilon_0}(n_i - n_e) \tag{4.152}$$

where $y$ is the coordinate normal to the wall. Within the sheath, $V < 0$ and the charge-carrier densities can be expressed by

$$\left.\begin{aligned} n_e &= n_{e0} \exp(eV/kT) \\ n_i &= n_{i0} \exp(-eV/kT) \end{aligned}\right\} \tag{4.153}$$

where $n_{e0} = n_{i0}$ are the undisturbed charge-carrier densities in the plasma.

The second part of Eq. (4.153) holds approximately, i.e., it is

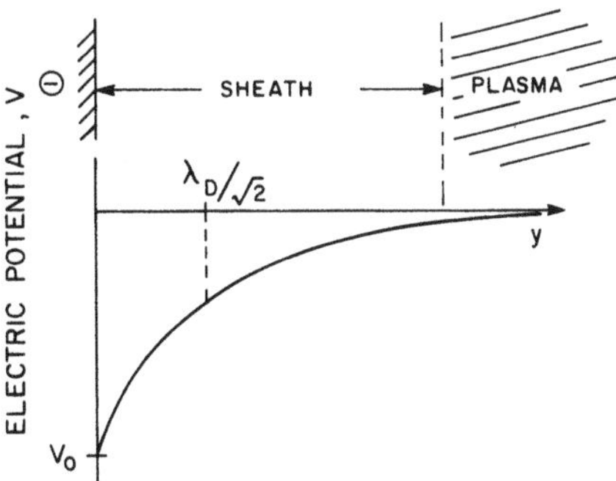


**FIG. 4.13.** Evolution of the potential in the sheath.

assumed that the positive ions also follow a Boltzmann distribution in the sheath. This approximation is only valid for $kT \gg eV$. Based on this assumption, the space charge density can be expressed by

$$\rho_{el} = en_{e0}\left[1 - \frac{eV}{kT} - \left(1 + \frac{eV}{kT}\right)\right]$$

$$= -2n_{e0}\frac{eV}{kT} \tag{4.154}$$

Introducing Eq. (4.154) into Eq. (4.152) leads to

$$\frac{d^2V}{dy^2} - \frac{1}{\lambda_D^2}V = 0 \tag{4.155}$$

with $\lambda_D = (\varepsilon_0 kT/n_{e0}e^2)^{1/2}$.

The solution of Eq. (4.155) with the boundary condition $V = V_0$ for $y = 0$ is given by

$$V = V_0 \exp(-\sqrt{2}\,y/\lambda_D) \tag{4.156}$$

where $V_0$ is either the floating potential that the wall assumes or the negative biasing potential of the wall. Figure 4.13 shows a sketch of this result.

The thickness of the sheath is on the order of a Debye length. Because of the exponential potential variation in the sheath, the definition of the sheath edge is somewhat arbitrary. Specification of the sheath edge varies in the literature from 1 to 10 Debye lengths.

Since the space charge always assumes the opposite sign of the wall potential, this potential is effectively "screened off" from the plasma, i.e., any electrical disturbance imposed on the plasma is felt only over the sheath thickness ($\sim\lambda_D$). This fact is extremely important in the application of electric probes as plasma diagnostic tools.

## LIST OF SYMBOLS

$\vec{B}$ magnetic induction (Vs/m$^2$)
$B_\nu^0$ intensity of the blackbody radiation (J/ster. m$^2$)
$c$ light velocity ($c = 3 \times 10^8$ m/s)
$D_a$ ambipolar diffusion coefficient (m$^2$/s)
$D_e$ electron diffusion coefficient (m$^2$/s)
$D_i$ ion diffusion coefficient (m$^2$/s)
$e$ electron charge ($1.6 \times 10^{-19}$ As)
$\vec{E}$ electric field (V/m)
$\vec{E}_r$ electric field in radial direction(V/m)
$E_{r+1}^*$ reduced ionization energy = $E_{r+1} - \delta E_{r+1}$ (eV)
$E_I$ ionization energy (eV)
$E_{H^+}$ hydrogen atom ionization energy (13.6 eV)
$E_{r,k}$ energy of chemical species $r$ in the excited state $k$ (cm$^{-1}$)
$f(v)$ Maxwellian distribution function (s/m)
$\vec{F}_r$ force in the radial direction (N)
$g$ number of compartments ($h^3$) in the phase-space volume $dx \cdot dy \cdot dz \cdot dp_x \cdot dp_y \cdot dp_z$
$g_k$ statistical weight of excited state $k$
$g_{r,k}$ statistical weight of chemical species $r$ in excited state $k$
$h$ Planck's constant ($6.6 \times 10^{-34}$ W s$^2$)
$H$ elementary volume in phase space: $dx \cdot dy \cdot dz \cdot dp_x \cdot dp_y \cdot dp_z$
$I$ arc current (A)
$j$ electric current density (A/m$^2$)
$j_e$ electron current density (A/m$^2$)
$j_i$ ion current density (A/m$^2$)
$\vec{j}_k$ flux of particles of type $k$ (m$^{-2}$ s$^{-1}$)
$k$ Boltzmann constant ($1.38 \times 10^{-23}$ J/K)
$l_e$ electron mean free path (m)
$l_i$ ion mean free path (m)
$m_e$ electron mass ($9.11 \times 10^{-31}$ kg)
$M$ ion mass (kg)
$n_e$ electron density (m$^{-3}$)
$n_{r,k}$ particle number density of chemical species $r$ in excited state $k$ (m$^{-3}$)
$N_k$ number of particles in excited state $k$
$p$ total pressure (Pa)
$p_x$ component of the momentum ($p_x = mv_x$)
$p_y$ component of the momentum ($p_y = mv_y$)
$p_z$ component of the momentum ($p_z = mv_z$)
$q$ electrical charge (As)

$Q_r$ partition function of chemical species $r$
$r_L$ Larmor radius for the circular motion ($r_L = mv_\perp/(qB)$) (m)
$r_{\min}$ Landau parameter (m)
$R$ arc radius (m)
$\vec{s}_e$ distance that an electron travels during time interval $\tau_e$ (m)
$S_k$ source term [see Eq. (4.35)] ($m^{-3}\,s^{-1}$)
$t$ time (s)
$T$ temperature (K)
$T_e$ electron temperature (K)
$T_h$ heavy species temperature (K)
$T_k$ maximum kinetic energy acquired by an electron between two collisions (J)
$\vec{u}_e$ electron drift velocity (m/s)
$\bar{\vec{u}}_e$ mean electron drift velocity (m/s)
$\vec{u}_i$ ion drift velocity (m/s)
$\bar{\vec{u}}_i$ mean ion drift velocity (m/s)
$\vec{v}$ particle velocity (m/s)
$\vec{v}_e$ electron velocity (m/s)
$\bar{v}_e$ mean electron velocity (m/s)
$\bar{v}_i$ mean ion velocity (m/s)
$\vec{V}_d^e$ mean electron drift velocity (m/s)
$\vec{v}_D$ electron drift velocity in magnetic and electric fields (m/s)
$v_\parallel$ charged particle velocity component parallel to the magnetic field (m/s)
$v_\perp$ charged particle velocity component perpendicular to the magnetic field (m/s)
$V$ electrical potential (V)
$W$ thermodynamic probability
$x$ position coordinate (m)
X chemical species
$X^+$ singly ionized chemical species
$y$ position coordinate (m)
$z$ position coordinate (m)

## Greek Symbols

$\beta$ Lagrange multiplier ($\beta = 1/kT$) ($J^{-1}$)
$\delta E_i$ ionization potential lowering (eV)
$\delta N$ particle number variation
$\Delta p(r)$ pressure variation along the plasma radius (Pa)
$\varepsilon_0$ dielectric constant ($\varepsilon_0 = 8.86 \times 10^{-12}$ A s/V m)
$\gamma_e$ Gvosdover parameter

| | |
|---|---|
| $\lambda_D$ | Debye length (m) |
| $\Lambda$ | diffusion length (m) |
| $\mu_e$ | electron mobility ($m^2/V\,s$) |
| $\mu_i$ | ion mobility ($m^2/V\,s$) |
| $\mu_0$ | magnetic permeability constant ($\mu_0 = 1.26 \times 10^{-6}$ Hy/m) |
| $\nu_i$ | net ionization coefficient |
| $\xi$ | fraction of ionized atoms in the gas |
| $\rho$ | specific mass ($kg/m^3$) |
| $\rho_{el}$ | electric space charge ($C/m^3$) |
| $\sigma_e$ | electrical conductivity ($ohm^{-1}\,m^{-1}$) |
| $\tau_e$ | mean free flight time for electrons (s) |
| $\omega_e$ | Larmor frequency ($s^{-1}$) |
| $\nabla$ | Laplace operator |

## GENERAL BIBLIOGRAPHY

Drawin, H. W., "Spectroscopic Measurement of High Temperatures (A Review)," *High Temp. High Pressures* 2 (1970) 359.

Finkelnburg, W. and H. Maecker, "Elektrische Bögen und thermisches Plasma," in *Encyclopedia of Physics,* S. Flügge, ed., Vol. 22, Berlin: Springer-Verlag, 1956, p. 254.

Griem, H. R., *Plasma Spectroscopy,* New York: McGraw-Hill, 1964.

Hirsh, M. N. and H. J. Oskam, eds., *Gaseous Electronics,* Vol. 1, *Electrical Discharges,* New York: Academic Press, 1978.

Huddlestone, R. H. and S. L. Leonard (eds.), *Plasma Diagnostic Techniques,* New York: Academic Press, 1965.

Lee, J. F., F. W. Sears, and D. L. Turcotte, *Statistical Thermodynamics,* 2nd ed., Reading, Mass.: Addison-Wesley, 1973.

Lochte-Holtgreven, W. (ed.), *Plasma Diagnostics,* Amsterdam: North-Holland, 1968.

Loeb, L. B., *Basic Processes of Gaseous Electronics,* Berkeley and Los Angeles: University of California Press, 1961.

Massey, H. S. and E. H. Burhop, *Electronic and Ionic Impact Phenomena,* Vol. 1, 2nd ed., Oxford: Clarendon Press, 1969.

Massey, H. S., *Electronic and Ionic Impact Phenomena,* Vol. 2, 2nd ed., Oxford: Clarendon Press, 1969.

Massey, H. S., *Electronic and Ionic Impact Phenomena,* Vol. 3, 2nd ed., Oxford: Clarendon Press, 1971.

Massey, H. S. and H. B. Gilbody, *Electronic and Ionic Impact Phenomena,* Vol. 4, 2nd ed., Oxford: Clarendon Press, 1974.

Mitchner, M. and C. H. Kruger Jr., *Partially Ionized Gases,* New York: Wiley, 1973.

## REFERENCES

1. S. D. Gvosdover, *Phys. Z. Sov.* 12 (1937): 164.
2. M. Mitchner and C. H. Kruger, Jr., *Partially Ionized Gases* (New York: Wiley, 1973).
3. A. B. Cambel, *Plasma Physics and Magnetofluid Mechanics* (New York: McGraw-Hill, 1963).

4. W. P. Allis, "Motion of Ions and Electrons," in *Encyclopedia of Physics, Vol. 21* (Springer–Verlag, Berlin, 1956): 383.
5. G. Schmidt, *Physics of High Temperature Plasmas,* 2nd ed. (New York: Academic Press, 1979).
6. M. A. Uman, *Introduction to Plasma Physics* (New York: McGraw-Hill, 1964).
7. J. F. Lee, F. W. Sears, and D. L. Turcotte, *Statistical Thermodynamics,* 2nd ed. (Reading, Mass: Addison-Wesley, 1973).
8. C. Delalondre, "Modélisation aérothermodynamique d'arcs électroniques à forte intensité avec prise en compte du déséquilibre thermodynamique local et du transfert thermique à la cathode," These (Université de Rouen, 1990).

Chapter 5

# Derivation of the Plasma Equations

In contrast to the preceding chapters, which considered the microscopic behavior of particles, this chapter will deal with the motion of the various plasma components (electrons, ions, neutrals) on *a macroscopic scale* under the influence of external as well as internal forces.*

Since this treatment will be restricted to *thermal plasmas,* which can be conveniently generated by electric arcs or high-power RF discharges, temperature is the most important parameter. Thus, the derivation of the plasma equations (for current, mass, heat flow, etc.) will be based on thermodynamics and irreversible thermodynamics. This particular approach provides a straightforward survey of the various effects that govern the behavior of such plasmas, although it leaves the absolute values of the coefficients (for example, the friction coefficient or the diffusion coefficient) open. These values will be derived from kinetic theory.

## 5.1. DEFINITIONS

A plasma may consist of a mixture of $k$ components. Chemical reactions in the plasma, particle density and temperature gradients, and macroscopic velocities for the plasma components are all permitted.

The velocity of the center of mass, $\vec{v}_g$, follows from

$$\rho\vec{v}_g = \sum_k \rho_k\vec{v}_k \tag{5.1}$$

with $\rho = \sum_k \rho_k$, where $\rho_k$ and $\vec{v}_k$ represent the mass density and velocity of an individual component, respectively.

* The material presented in this chapter has been principally derived from the comprehensive treatment in *Electric Arcs and Thermal Plasmas* by W. Finkelnburg and H. Maecker.[1]

The mass concentration is given by

$$c_k = \frac{\rho_k}{\rho} \qquad \text{where } \Sigma c_k = 1 \tag{5.2}$$

The mass flux of component $k$ relative to the center of mass is

$$\vec{I}_k = \rho_k(\vec{v}_k - \vec{v}_g) \tag{5.3}$$

and $\Sigma_k \vec{I}_k = 0$ [from Eqs. (5.1) and (5.3)].

## 5.2. CONSERVATION EQUATIONS

### 5.2.1. Conservation of Mass

There are four alternative forms of the conservation-of-mass equation; two refer to a single component and the other two refer to the entire mixture:

$$\frac{\partial \rho_k}{\partial t} + \operatorname{div}(\rho_k \vec{v}_k) = \Gamma_k \tag{5.4}$$

$$\rho \frac{dc_k}{dt} + \operatorname{div} \vec{I}_k = \Gamma_k \tag{5.5}$$

$$\frac{\partial \rho}{\partial t} + \operatorname{div}(\rho \vec{v}_g) = 0 \tag{5.6}$$

$$\frac{d\rho}{dt} + \rho(\operatorname{div} \vec{v}_g) = 0 \tag{5.7}$$

where $\Gamma_k$ represents the generation rate of particles of component $k$ (units: kg/m$^3$ s). Equation (5.6) follows from Eq. (5.4) by summation over all components, considering that $\Sigma_k \Gamma_k = 0$. Equation (5.5) reduces to Eq. (5.4) if Eqs. (5.2), (5.3), and (5.6) are applied in addition to the substantive derivative

$$\frac{d}{dt} = \frac{\partial}{\partial t} + (\vec{v}_g \cdot \operatorname{grad})$$

Finally Eq. (5.7) can be verified by introducing the substantive derivative.

### 5.2.2. Conservation of Momentum

Only external forces are able to move the center of gravity. Therefore, the momentum equations for the total mixture contain only external forces, for example:

| | force | force/unit mass |
|---|---|---|
| gravity | $m_k\vec{g}$ | $\vec{g}$ |
| electric forces | $e_k\vec{E}$ | $\dfrac{e_k\vec{E}}{m_k}$ |
| magnetic forces | $e_k(\vec{v}_k \times \vec{B})$ | $\dfrac{e_k}{m_k}(\vec{v}_k \times \vec{B})$ |
| pressure forces | $\dfrac{m_k}{\rho_k}\operatorname{grad} p_k$ | $\dfrac{1}{\rho_k}\operatorname{grad} p_k$ |

In these examples $m_k$ represents the mass, $e_k$ the charge, and $p_k$ the pressure of an individual component, $\vec{g}$ is the acceleration due to gravity, $\vec{E}$ the electric field strength, and $\vec{B}$ the magnetic field strength.

The overall momentum equation, which is identical to the basic equation of hydrodynamics, assumes the form

$$\rho\frac{d\vec{v}_g}{dt} = -\operatorname{grad} p + \sum_k \rho_k\vec{F}_k \tag{5.8}$$

The corresponding equations for the individual components cannot be derived from Eq. (5.8) because they contain, in addition to external forces, internal forces due, for example, to friction among the various components.

### 5.2.3. Conservation of Energy

We assume that the total energy of a given system is composed of kinetic and thermal energy. By multiplying Eq. (5.8) by $\vec{v}_g$, we obtain the kinetic energy:

$$\tfrac{1}{2}\rho\frac{d}{dt}v_g^2 = -\vec{v}_g\cdot\operatorname{grad} p + \vec{v}_g\cdot\sum_k \rho_k\vec{F}_k \tag{5.9}$$

From the first law of thermodynamics it follows that

$$\delta q = du_g + p\,dv_0 \tag{5.10}$$

where $q$ and $u_g$ represent heat and internal energy per unit mass, respectively, and $v_0 = 1/\rho$ is the specific volume. Equation (5.10) can be written as

$$\rho\frac{\delta q}{dt} = \rho\frac{du_g}{dt} + p\operatorname{div}\vec{v}_g \tag{5.11}$$

because

$$\rho \frac{dv_0}{dt} = -\frac{1}{\rho}\frac{d\rho}{dt} = \operatorname{div} \vec{v}_g$$

according to Eq. (5.7). The right-hand side of Eq. (5.11) represents the change in the internal energy plus the change in compression work. In general, such changes are caused by net heat fluxes and internal heat generation. Therefore, Eq. (5.11) can be written in the form of an energy balance:

$$\rho \frac{du_g}{dt} + p \operatorname{div} \vec{v}_g = -\operatorname{div} \vec{q} + \sum_k \vec{F}_k \vec{I}_k \tag{5.12}$$

The heat flow vector $\vec{q}$ contains heat conduction without mass flow as well as heat carried by the mass flows. The potential energy converted into heat by friction is represented by the last term in Eq. (5.12). By adding Eqs. (5.9) and (5.12), the total energy equation is obtained:

$$\rho \frac{d}{dt}\left(\frac{v_g^2}{2} + u_g\right) = -\operatorname{div}(p\vec{v}_g + \vec{q}) + \sum_k \vec{F}_k \rho_k \vec{v}_k \tag{5.13}$$

This equation has the typical form of a balance equation: (change of total energy) = (divergence of energy flow) + (energy source term).

### 5.2.4. Entropy Balance

A similar balance equation can also be established for the entropy. This derivation will be useful for obtaining information on mass and heat fluxes.

A combination of the first and second law of thermodynamics yields

$$\frac{\delta q}{dt} = \frac{du_g}{dt} + p\frac{dv_0}{dt} = T\frac{ds_g}{dt} + \sum_k \mu_k \frac{dc_k}{dt} \tag{5.14}$$

where $s_g$ is the entropy per unit mass and $\mu_k$ is the chemical potential of component $k$. Combining Eqs. (5.11), (5.12), and (5.14) results in

$$T\frac{ds_g}{dt} + \rho \sum_k \mu_k \frac{dc_k}{dt} = -\operatorname{div} \vec{q} + \sum_k \vec{F}_k \vec{I}_k \tag{5.15}$$

Using Eq. (5.5), the second term in Eq. (5.15) can be transformed into

$$\rho \sum_k \mu_k \frac{dc_k}{dt} = -\sum_k \operatorname{div} \vec{I}_k + \sum_k \mu_k \Gamma_k$$

or

$$\rho \frac{ds_g}{dt} = \frac{1}{T}\sum \mu_k \operatorname{div} \vec{I}_k - \frac{1}{T}\sum \mu_k \Gamma_k - \frac{1}{T}\operatorname{div} \vec{q} + \frac{1}{T}\sum \vec{F}_k \vec{I}_k \tag{5.16}$$

Equation (5.16) can be rearranged to form a typical balance equation

$$\rho \frac{ds_g}{dt} = -\mathrm{div}\left(\frac{\vec{q} - \sum \mu_k \vec{I}_k}{T}\right) + \frac{-\vec{q}\,\dfrac{\mathrm{grad}\, T}{T} + \sum \vec{I}_k\left(\vec{F}_k - T\,\mathrm{grad}\,\dfrac{\mu_k}{T}\right) - \sum \mu_k \Gamma_k}{T} \tag{5.17}$$

of the form (change of entropy) = (divergence of entropy flow) + (entropy source term). The various terms in the previous equations require further explanation. For this purpose, some simple examples will be examined.

For an interpretation of the terms in the energy and entropy balance equation, the term

$$\vec{I}_k \vec{F}_k = \vec{F}_k \rho_k (\vec{v}_k - \vec{v}_g)$$

will be considered first. Here

$\vec{F}_k \rho_k \vec{v}_k =$ work done by force $\vec{F}_k$ (per unit volume and unit time)

$\vec{F}_k \rho_k \vec{v}_g =$ work necessary for accelerating the center of mass (per unit volume and unit time)

the difference of these two terms

$$\vec{F}_k \rho_k (\vec{v}_k - \vec{v}_g) = \vec{F}_k \vec{I}_k$$

is the work that is transformed into heat (per unit volume and unit time). For further clarification, an example will be considered:

$\vec{F}_k = -\dfrac{e}{m_e}\vec{E} =$ electric field force (per unit mass)

$\vec{I}_k = -\dfrac{m_e}{e}\vec{j}_e =$ mass flux associated with the flow of electrons (per unit area and unit time)

$\vec{F}_k \vec{I}_k = \vec{j}_e \vec{E} =$ dissipated Joule heat (per unit volume and unit time)

The term $-T\,\mathrm{grad}\,(\mu_k/T)$ must be also a mechanical force that, when multiplied by the flow vector $\vec{I}_k$, describes friction work that is transformed into heat. The next term, $(\vec{q}\,\mathrm{grad}\, T)/T$, does not lend itself to a separation into mass flow and force. Nevertheless, the term $(\mathrm{grad}\, T)/T$ is, in a generalized sense, also considered a driving "force." The product $(\vec{q} \cdot \mathrm{grad}\, T)/T$ therefore also represents a heat-generation term.

All the terms that appear in the numerator of the last term of Eq. (5.17) are aptly called "energy dissipation" terms. When divided by $T$, these terms represent the entropy production (per unit volume and unit time), which must be $\geq 0$ according to the second law of thermodynamics.

The term $\mu_k \Gamma_k$ contains the chemical affinity of component $k$. At chemical equilibrium, this term vanishes. But this term will exist if deviations from chemical equilibrium occur (for example, if the chemical composition of a plasma does not follow its temperature, as in "frozen flow"). This type of deviation from chemical equilibrium may occur within the thermal boundary layer separating a plasma from an adjacent wall, as sketeched in Fig. 5.1. In this situation, electron diffusion due to the steep gradients of density and/or temperature is so fast that relaxation processes (recombination) cannot follow—the chemical processes appear to be "frozen." Since $\mu_k \Gamma_k \to 0$ as chemical equilibrium is approached, this term is useful for describing relaxation phenomena.

As stated earlier, the heat flow vector, $\vec{q}$, comprises pure heat conduction (without mass flow) and the heat flow carried by the mass flow. This heat flow is reduced by the term $\sum \mu_k \vec{I}_k$, which represents the part of the energy flow that gives rise to chemical reactions. The chemical potential $\mu_k$ is frequently interpreted as a measure of the driving force tending to cause a chemical reaction to take place.

In the last part of this section, the momentum equations for individual components will be considered. These equations will be written for a two-component mixture, assuming that there are two internal forces (friction and thermal diffusion); this leads to two additional terms in each of the two momentum equations:

$$\rho_1 \frac{d\vec{v}_1}{dt} + n_1 n_2 \varepsilon_{12}(\vec{v}_1 - \vec{v}_2) = \vec{F}_1 \rho_1 - \operatorname{grad} p_1 - \rho_1 y_1 \frac{\operatorname{grad} T}{T} \quad (5.18)$$

$$\rho_2 \frac{d\vec{v}_2}{dt} + n_1 n_2 \varepsilon_{21}(\vec{v}_2 - \vec{v}_1) = \vec{F}_2 \rho_2 - \operatorname{grad} p_2 - \rho_2 y_2 \frac{\operatorname{grad} T}{T} \quad (5.19)$$

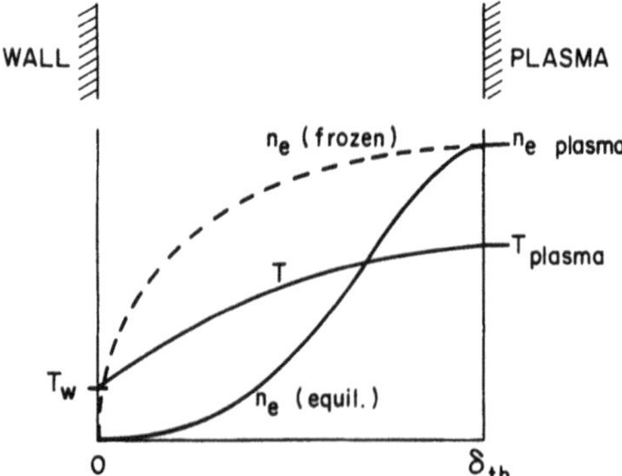


**FIG. 5.1.** Deviation from chemical equilibrium in a plasma wall boundary layer.

In these equations friction is assumed to be proportional to the relative velocity between the two components. The particle densities for components 1 and 2 are designated as $n_1$ and $n_2$, respectively, and both $\varepsilon_{12}$ and $\varepsilon_{21}$ represent friction coefficients. The last terms on the right-hand sides of Eqs. (5.18) and (5.19) are the forces responsible for thermal diffusion, with the coefficients $y_1$ and $y_2$ determining the magnitude of these forces.

The sum of Eqs. (5.18) and (5.19) should be identical to Eq. (5.8) if the last term of this equation is written for two components only. Since

$$\rho_1 \frac{d\vec{v}}{dt} + \rho_2 \frac{d\vec{v}_2}{dt} = \rho \frac{d\vec{v}_g}{dt}$$

according to Eq. (5.1), this identity is indeed fulfilled if

$$\varepsilon_{12} = \varepsilon_{21} \qquad \text{and} \qquad \rho_1 y_1 = -\rho_2 y_2$$

Friction forces act only between the two components, and therefore they are not able to move the center of mass. Thus, $\varepsilon_{12} = \varepsilon_{21}$ must be valid. A similar argument holds for the forces responsible for thermal diffusion, leading to $\rho_1 y_1 = -\rho_2 y_2$.

We will now derive an expression for individual mass flows relative to the center of mass. By combining Eqs. (5.1) and (5.3) for a two-component mixture, one finds

$$\vec{I}_1 = -\vec{I}_2 = \frac{\rho_1 \rho_2}{\rho}(\vec{v}_1 - \vec{v}_2) \tag{5.20}$$

Subtracting Eq. (5.19) from Eq. (5.18) gives

$$\vec{v}_1 - \vec{v}_2 = \frac{\rho_1 \rho_2}{\rho} \frac{1}{n_1 n_2 \varepsilon_{12}} \left[ \vec{F}_1 - \vec{F}_2 - \frac{d\vec{v}_1}{dt} + \frac{d\vec{v}_2}{dt} - \frac{1}{\rho_1} \operatorname{grad} p_1 + \frac{1}{\rho_2} \operatorname{grad} p_2 - (y_1 - y_2) \frac{\operatorname{grad} T}{T} \right] \tag{5.21}$$

Finally, introducing Eq. (5.21) into (5.20) results in

$$\vec{I}_1 = \left(\frac{\rho_1 \rho_2}{\rho}\right)^2 \frac{1}{n_1 n_2 \varepsilon_{12}} \times \left[ \vec{F}_1 - \vec{F}_2 - \frac{d\vec{v}_1}{dt} + \frac{d\vec{v}_2}{dt} - \frac{1}{\rho_1} \operatorname{grad} p_1 + \frac{1}{\rho_2} \operatorname{grad} p_2 - (y_1 - y_2) \frac{\operatorname{grad} T}{T} \right] \tag{5.22}$$

This equation will be used later.

## 5.3. ONSAGER'S RECIPROCITY RELATIONS AND SOME PHENOMENOLOGICAL THEOREMS

In the previous section we showed that the entropy production $E_s$ can be expressed in the following way:

$$E_s = \frac{\Sigma \text{ "fluxes"} \times \text{"forces"}}{T}$$

The relation between the fluxes and the "driving forces" that give rise to these fluxes can, in a first approximation, be assumed to be linear. This is a common approach that is applied to electrical and thermal conduction, for example, as well as to diffusion.

Ohm's law: $\vec{j} = -\sigma_e \operatorname{grad} V$

Fourier's law: $\vec{q} = -\kappa \operatorname{grad} T$

Fick's law: $\vec{S} = -D \operatorname{grad} n$

The justification for this approach will be discussed later. In the relations given above, the gradients of the potential $V$, the temperature $T$, and the number density $n$ are regarded as 'driving forces." The coefficients $\sigma_e$, $\kappa$ and $D$ represent electrical conductivity, thermal conductivity, and the diffusion coefficient, respectively.

The same approach can be applied to more complicated situations in which a flux depends on several driving forces.

We will now consider a two-component mixture in which each flux is assumed to depend on three different types of driving forces. This leads to the following set of phenomenological equations:

$$\vec{I}_1 = L_{11}\vec{X}_1 + L_{12}\vec{X}_2 + L_{1u}\vec{X}_u \tag{5.23}$$

$$\vec{I}_2 = L_{21}\vec{X}_1 + L_{22}\vec{X}_2 + L_{2u}\vec{X}_u \tag{5.24}$$

$$\vec{q} = L_{u1}\vec{X}_1 + L_{u2}\vec{X}_2 + L_{uu}\vec{X}_u \tag{5.25}$$

The $L_{kl}$, $L_{ku}$, and $L_{ul}$ in these equations represent coefficients that will be determined in later paragraphs, and the $\vec{X}_l$ stand for the "driving forces."

These "driving forces" will be identified with the same type of "forces" that appear in the entropy production term of Eq. (5.17), i.e.,

$$\vec{X}_1 = \vec{F}_1 - T\,\text{grad}\,\frac{\mu_1}{T}$$

$$\vec{X}_2 = \vec{F}_2 - T\,\text{grad}\,\frac{\mu_2}{T}$$

$$\vec{X}_u = -\frac{1}{T}\,\text{grad}\,T$$

These linear approximations require, as shown by the gas kinetic calculations of Enskog,[2] that the parameters of state not vary appreciably over a mean free path length. Choosing the temperature $T_i$ as the parameter of state for particles of type $i$, this requirement can be expressed by

$$\lambda_i \cdot |\text{grad}\,T_i| \ll T_i$$

where $\lambda_i$ represents the mean free path length of particles of type $i$. Similar relations hold for other parameters of state.

From $\sum_{k=1} \vec{F}_k = 0$ it follows that

$$\sum_k L_{kl} = 0 \qquad \text{and} \qquad \sum_k L_{ku} = 0 \tag{5.26}$$

because the forces can be chosen arbitrarily.

In addition, Onsager's reciprocity theorems derived from kinetic statistics[3] require that the following relations hold:

$$(L_{kl}) = (L_{lk}) \qquad \text{and} \qquad (L_{ku}) = (L_{uk}) \tag{5.27}$$

i.e., the matrices of the coefficients are symmetric. This relationship, however, is only valid if the forces $\vec{X}_k$ and the fluxes $\vec{I}_k$ are chosen in such a way that their product divided by $T$ represents an entropy-production term. Thus the forces were chosen according to the last term of Eq. (5.17).

From Eqs. (5.26) and (5.27) it follows that

$$\sum_l L_{kl} = 0 \qquad \text{and} \qquad \sum_l L_{ul} = 0$$

With this result one may write in general

$$\vec{F}_k = \sum_{l=1}^{n-1} L_{kl}(\vec{X}_l - \vec{X}_u) + L_{ku}\vec{X}_u \tag{5.28}$$

and

$$\vec{q} = \sum_{l=1}^{n-1} L_{ul}(\vec{X}_l - \vec{X}_u) + L_{uu}\vec{X}_u \tag{5.29}$$

If this result is applied to a two-component mixture, Eqs. (5.28) and (5.29) reduce to

$$\vec{I}_1 = -\vec{I}_2 = L_{11}(\vec{X}_1 - \vec{X}_2) + L_{1u}\vec{X}_u \tag{5.30}$$

and

$$\vec{q} = L_{1u}(\vec{X}_1 - \vec{X}_2) + L_{uu}\vec{X}_u \tag{5.31}$$

Returning to the previously introduced identification of the forces, we find

$$\vec{X}_1 - \vec{X}_2 = \vec{F}_1 - \vec{F}_2 - T \operatorname{grad} \frac{\mu_1 - \mu_2}{T}$$

where $\mu_k$ represents the chemical or Gibbs potential, which can be expressed by

$$\begin{aligned} \mu_k &= u_{g,k} + pv_{0,k} - Ts_{g,k} \\ &= h_{g,k} - Ts_{g,k} \end{aligned} \tag{5.32}$$

($u$, $s$, and $v$ are expressed per unit mass).

The specific volume $v_{0,k}$ of component $k$ can be expressed as

$$v_{0,k} = \frac{p_k}{p}\frac{1}{\rho_k} \quad \text{and} \quad pv_{0,k} = \frac{p_k}{\rho_k} = \frac{kT}{m_k} \tag{5.33}$$

where $k$ is the Boltzmann constant.

Statistical mechanics provides an expression for the entropy of component $k$:

$$s_{g,k} = \frac{u_{g,k}}{T} + \frac{k}{m_k} \ln Q_k \tag{5.34}$$

where $Q_k$ is the partition function of component $k$. Considering only the translational energy of particles (continuous energy distribution) we find

$$Q_k = \frac{e_n}{n_k}\left(\frac{2\pi m_k kT}{h^2}\right)^{3/2} \tag{5.35}$$

In this equation $e_n = 2.718$ is the base of natural logarithms and $h$ is Planck's constant;

$$\ln Q_k = 1 - \ln p_k + \tfrac{5}{2}\ln T + \ln C \tag{5.36}$$

In Eq. (5.36) all constants are summarized in the single term $\ln C$. By introducing Eqs. (5.33), (5.34), and (5.36) into Eq. (5.32) we obtain

$$\frac{\mu_k}{T} = \frac{k}{m_k}(\ln p_k - \tfrac{5}{2}\ln T - \ln C)$$

and

$$\operatorname{grad}\frac{\mu_k}{T} = \frac{k}{m_k}\left(\frac{1}{p_k}\operatorname{grad} p_k - \frac{5}{2}\frac{\operatorname{grad} T}{T}\right)$$

and finally

$$-T\operatorname{grad}\frac{\mu_k}{T} = h_{g,k}\frac{\operatorname{grad} T}{T} - \frac{1}{\rho_k}\operatorname{grad} p_k \tag{5.37}$$

By introducing the mole fraction $x_k = p_k/p$, we can write

$$\begin{aligned}\frac{kT}{m_k}\operatorname{grad}(\ln p_k) &= \frac{kT}{m_k}[\operatorname{grad}(\ln p) + \operatorname{grad}(\ln x_k)]\\ &= v_{0,k}\operatorname{grad} p + \frac{kT}{m_k x_k}\operatorname{grad} x_k\end{aligned} \tag{5.38}$$

For the case of binary mixture the right-hand side of Eq. (5.38) becomes

$$v_{0,1}\operatorname{grad} p + \frac{kT}{\bar{m}c_1}\operatorname{grad} c_1$$

where $x_1 = c_1 m_2/\bar{m}$ and $\bar{m} = c_1 m_2 + c_2 m_1$; $dx_1/dc_1 = m_1 m_2/\bar{m}^2$.

Thus two alternative forms of Eq. (5.37) can be given for a two-component mixture:

$$-T\operatorname{grad}\frac{\mu_1}{T} = h_{g,1}\frac{\operatorname{grad} T}{T} - \frac{1}{\rho_1}\operatorname{grad} p_1 \tag{5.39}$$

or

$$-T\operatorname{grad}\frac{\mu_1}{T} = h_{g,1}\frac{\operatorname{grad} T}{T} - v_{0,1}\operatorname{grad} p - \frac{kT}{\bar{m}c_1}\operatorname{grad} c_1 \tag{5.40}$$

By introducing Eq. (5.39) into Eqs. (5.23) and (5.24) we obtain

$$\vec{I}_1 = -\vec{I}_2 = L_{11}\left[\vec{F}_1 - \vec{F}_2 - \frac{1}{\rho_1}\operatorname{grad} p_1 + \frac{1}{\rho_2}\operatorname{grad} p_2\right] + L_{11}(h_1 - h_2)\frac{\operatorname{grad} T}{T} - L_{1u}\frac{\operatorname{grad} T}{T} \tag{5.41}$$

or, in the other form [using Eq. (5.40)], we obtain

$$\vec{I}_1 = -\vec{I}_2 = L_{11}\left[\vec{F}_1 - \vec{F}_2 - (v_{0,1} - v_{0,2})\operatorname{grad} p - \frac{kT}{\bar{m}c_1c_2}\operatorname{grad} c_1\right] + L_{11}(h_1 - h_2)\frac{\operatorname{grad} T}{T} - L_{1u}\frac{\operatorname{grad} T}{T} \tag{5.42}$$

$$\vec{q} = L_{1u}\left[\vec{F}_1 - \vec{F}_2 - \frac{1}{\rho_1}\operatorname{grad} p_1 + \frac{1}{\rho_2}\operatorname{grad} p_2\right] + L_{1u}(h_{g,1} - h_{g,2})\frac{\operatorname{grad} T}{T} - L_{uu}\frac{\operatorname{grad} T}{T} \tag{5.43}$$

An alternative expression for $\vec{q}$ can be obtained by replacing the terms in brackets in Eq. (5.43) by those in Eq. (5.42).

Equations (5.41) and (5.42) already show similarity to Eqs. (5.21) and (5.22), respectively.

## 5.4 HEAT OF TRANSITION AND ENERGY FLUXES

Assuming that there is no temperature gradient but that the other forces still exist, there will be a mass flow $\vec{I}_1 = -\vec{I}_2 \neq 0$ and also a heat flow $\vec{q} \neq 0$. Since the two mass fluxes are opposite and equal, they must carry different amounts of energy with them.

Let us assume that $\vec{I}_1$ carries the overbalance of energy defined as $q^*$ (per unit mass of the flow $\vec{I}_1$), which is called the *heat of transition.*

According to this definition

$$q^*\vec{I}_1 = \vec{q}, \qquad q^* < 0 \tag{5.44}$$

and

$$q^* = \frac{\vec{q}}{I_1} = \frac{L_{1u}}{L_{11}} \tag{5.45}$$

Equation (5.45) is a consequence of the assumption that grad $T = 0$. Equation (5.45) permits elimination of one coefficient so that Eq. (5.42) can be rewritten as

$$\vec{I}_1 = -\vec{I}_2 = L_{11}\Big[\vec{F}_1 - \vec{F}_2 - (v_{0,1} - v_{0,2})\,\mathrm{grad}\,p - \frac{kT}{\bar{m}c_1c_2}\,\mathrm{grad}\,c_1 - (q^* - h_{g,1} + h_{g,2})\frac{\mathrm{grad}\,T}{T}\Big] \tag{5.46}$$

or in the alternative form

$$\vec{I}_1 = -\vec{I}_2 = L_{11}\Big[\vec{F}_1 - \vec{F}_2 - \frac{1}{\rho_1}\,\mathrm{grad}\,p_1 + \frac{1}{\rho_2}\,\mathrm{grad}\,p_2 - (q^* - h_{g,1} + h_{g,2})\frac{\mathrm{grad}\,T}{T}\Big] \tag{5.47}$$

We will now compare Eq. (5.47) with Eq. (5.22). The two equations will be identical if

$$L_{11} = \left(\frac{\rho_1\rho_2}{\rho}\right)^2 \frac{1}{n_1n_2\varepsilon_{12}} \tag{5.48}$$

$$y_1 - y_2 = q^* - h_{g,1} + h_{g,2} \tag{5.49}$$

and

$$\frac{d\vec{v}_1}{dt} = \frac{d\vec{v}_2}{dt} \tag{5.50}$$

Elimination of $L_{1u}$ from Eq. (5.43) using Eq. (5.45) and applying Eq. (5.41) results in an interesting expression for the heat flux:

$$\vec{q} = q^*\vec{I}_1 - [L_{uu} - L_{11}q^{*2}]\frac{\mathrm{grad}\,T}{T} \tag{5.51}$$

The first term on the right-hand side represents the energy carried by the mass flow. The second term describes the energy flow due to conduction only and may be aptly termed as

$$\vec{q}_{I=0} = -[L_{uu} - L_{11}q^{*2}]\frac{\mathrm{grad}\,T}{T} \tag{5.52}$$

If we assume that we have a homogeneous mixture of two gases exposed to a temperature gradient (all other "forces" shall vanish), Eq. (5.43) reduces to

$$\vec{q}_T = -[L_{uu} - q^* L_{11}(h_{g,1} - h_{g,2})] \frac{\text{grad } T}{T} \tag{5.53}$$

The heat flux $\vec{q}_T$ is driven solely by the temperature gradient, which, however, also gives rise to a mass flow:

$$\vec{I}_{\text{TD}} = -L_{11}(q^* - h_{g,1} + h_{g,2}) \frac{\text{grad } T}{T} \tag{5.54}$$

Equation (5.54) follows from either Eq. (5.41) or (5.42) with the previously stated assumption. The mass flux $\vec{I}_{\text{TD}}$ is due to *thermal diffusion.* According to previous arguments, this mass flux carries a heat flux:

$$\vec{q}_{\text{TD}} = \vec{I}_{\text{TD}} q^* = -L_{11} q^* (q^* - h_{g,1} + h_{g,2}) \frac{\text{grad } T}{T} \tag{5.55}$$

$\vec{q}_{\text{TD}}$ represents the heat flux due to *thermal diffusion.* Since the heat flux $\vec{q}_T$ contains the contribution due to thermal diffusion as well as that caused by pure conduction, the following relation must hold:

$$\begin{aligned} \vec{q}_T - \vec{q}_{\text{TD}} &= \vec{q}_{I=0} \\ &= -[L_{uu} - L_{11} q^{*2}] \frac{\text{grad } T}{T} \end{aligned} \tag{5.56}$$

which is identical to Eq. (5.52).

Under the specified conditions, the temperature gradient initially gives rise to diffusional mass fluxes, which, however, build up concentration gradients that in turn lead to ordinary diffusion fluxes opposite to those caused by thermal diffusion. After steady-state conditions are established, both $\vec{I}_{\text{TD}}$ and $\vec{q}_{\text{TD}}$ are substantially reduced and heat transfer is essentially by conduction only. This situation is sketched in Fig. 5.2, where one-dimensional conditions are assumed. For this case we find

$$\vec{q}_T \approx \vec{q}_{I=0} = \kappa_{I=0} \, \text{grad } T$$

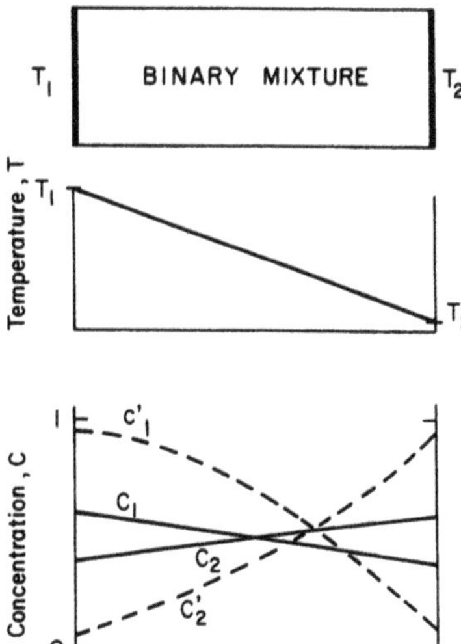


**FIG. 5.2.** Thermal diffusion and ordinary diffusion in a binary mixture. Concentration $C_i'$ is due to thermal diffusion and concentration $C_i$ is due to ordinary diffusion.

Measurements of the heat conductivity under steady-state conditions yield essentially the value of $\kappa_{I=0}$, while calculations lead to a value $\kappa$ defined by

$$\vec{q}_T = -\kappa \operatorname{grad} T$$

The difference between these two values is usually rather small except for very light particles, particularly for electrons.

## 5.5. DIFFUSION AND ENERGY FLUXES IN CHEMICALLY REACTING GASES

The results from the previous section will now be compared with the equations derived by Enskog[2] and Waldmann[4] from kinetic theory.

We will consider a binary mixture and assume that there are no external forces and no pressure gradients. For this particular case the equations for mass flux and heat flux have the following form:

$$\vec{I}_1 = -\rho D \operatorname{grad} c_1 - \alpha\rho D c_1 c_2 \frac{\operatorname{grad} T}{T} \tag{5.57}$$

and

$$\vec{q} = -\rho D\left(\frac{\alpha k T}{\bar{m}} + h_{g,1} - h_{g,2}\right) \operatorname{grad} c_1 - \kappa \operatorname{grad} T \tag{5.58}$$

or in alternate form

$$\vec{q} = \left(\frac{\alpha kT}{\bar{m}} + h_{g,1} - h_{g,2}\right)\vec{I}_1 - \kappa_{I=0}\,\text{grad}\,T \tag{5.59}$$

In these equations $D$ represents the binary diffusion coefficient, $\alpha$ the thermal diffusion factor, $\kappa$ the heat conductivity (which includes the contribution of thermal diffusion), and $\kappa_{I=0}$ the ordinary heat conductivity without mass fluxes.

Comparison of the first terms in Eqs. (5.46) and (5.57) result in the following correlation between the coefficients

$$L_{11}\frac{kT}{\bar{m}c_1c_2} = \rho D \tag{5.60}$$

or

$$L_{11} = \frac{\rho D\bar{m}c_1c_2}{kT}$$

Using the previously derived expression for $L_{11}$ in Eq. (5.48), one finds an expression for the friction coefficient:

$$\varepsilon_{12} = \frac{kT}{D(n_1 + n_2)} \tag{5.61}$$

Comparing the second terms in Eqs. (5.46) and (5.57) results in

$$L_{11}(q^* - h_{g,1} + h_{g,2}) = \alpha\rho Dc_1c_2$$

or

$$q^* - h_{g,1} + h_{g,2} = \frac{\alpha kT}{\bar{m}} = y_1 - y_2 \tag{5.62}$$

The last equality is identical to Eq. (5.49).

By introducing Eq. (5.57) into Eq. (5.59) we obtain

$$\vec{q} = -\rho Dq^*\,\text{grad}\,c_1 - q^*\alpha\rho Dc_1c_2\frac{\text{grad}\,T}{T} - \kappa_{I=0}\,\text{grad}\,T \tag{5.63}$$

The last two terms may be combined to form the term $(-\kappa\,\text{grad}\,T)$ where

$$\kappa = \kappa_{I=0} + q^*\alpha\rho Dc_1c_2\frac{1}{T} \tag{5.64}$$

Introducing the previously evaluated coefficients into Eq. (5.46) and into the alternative form of Eq. (5.43) provides the final equations

$$\vec{I}_1 = \vec{I}_2 = \frac{\rho D \bar{m} c_1 c_2}{kT}\left[\vec{F}_1 - \vec{F}_2 - (v_{0,1} - v_{0,2})\,\mathrm{grad}\,p - \frac{kT}{\bar{m} c_1 c_2}\mathrm{grad}\,c_1 - \frac{\alpha k T}{\bar{m}}\frac{\mathrm{grad}\,T}{T}\right] \tag{5.65}$$

and

$$\vec{q} = \rho D \frac{\bar{m}}{kT} c_1 c_2 \left(\frac{\alpha k T}{\bar{m}} + h_{g,1} - h_{g,2}\right) \times \left[\vec{F}_1 - \vec{F}_2 - (v_{0,1} - v_{0,2})\,\mathrm{grad}\,p - \frac{kT}{\bar{m} c_1 c_2}\mathrm{grad}\,c_1\right] - \kappa\,\mathrm{grad}\,T \tag{5.66}$$

The last equation can be written in alternative form as

$$\vec{q} = \left(\frac{\alpha k T}{\bar{m}} + h_{g,1} - h_{g,2}\right)\vec{I}_1 - \kappa_{I=0}\,\mathrm{grad}\,T \tag{5.67}$$

In Eqs. (5.65) and (5.66) the following "force" terms can be exchanged:

$$(v_{0,1} - v_{0,2})\,\mathrm{grad}\,p + \frac{kT}{\bar{m} c_1 c_2}\mathrm{grad}\,c_1 = \frac{1}{\rho_1}\mathrm{grad}\,p_1 - \frac{1}{\rho_2}\mathrm{grad}\,p_2 \tag{5.68}$$

Finally, the coefficients $y_1$ and $y_2$ can be calculated from Eqs. (5.49) and (5.62):

$$y_1 = \frac{\alpha k T}{\bar{m}} c_2 \quad \text{and} \quad y_2 = -\frac{\alpha k T}{\bar{m}} c_1 \tag{5.69}$$

Now that the various coefficients have been determined, the resulting equations will be applied to a chemically reacting gas in the following section.

## 5.6. AN EXAMPLE OF MASS AND ENERGY FLUXES IN A CHEMICALLY REACTING GAS

We will assume that a steady state prevails ($\partial/\partial t = 0$) in a binary mixture and that there are no external forces (so that $\vec{v}_g = 0$). The assumption of a binary mixture requires that

$$\mathrm{div}\,\vec{I}_1 = \Gamma_1 \quad \text{and} \quad \mathrm{div}\,\vec{I}_2 = \Gamma_2$$

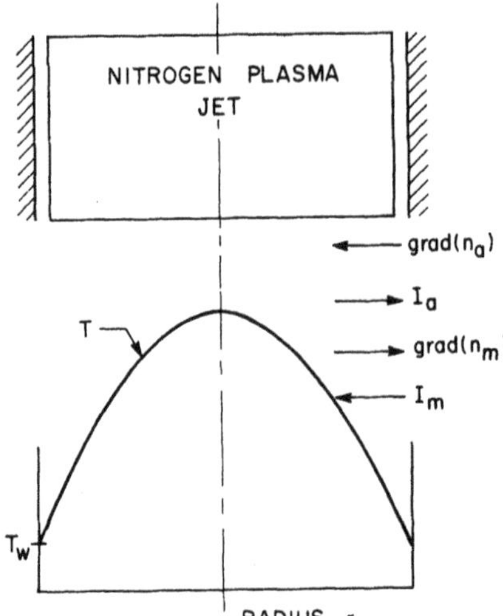


**FIG. 5.3.** Temperature distribution and gradients in a nitrogen cylindrical plasma jet (a = atoms, m = molecules.)

which means that the sources for the mass fluxes are the corresponding mass-generation terms. Since $\vec{I}_1 = -\vec{I}_2$, it follows that $\Gamma_1 + \Gamma_2 = 0$.

We will assume that the binary mixture consists of nitrogen molecules and nitrogen atoms. Such a mixture can be maintained if the temperature is sufficiently high; this can be accomplished by heating nitrogen in a plasma torch. We also assume that the plasma jet emanating from the torch is confined in a circular tube so that a rotationally symmetric temperature profile will exist with $(\text{grad}\, T)_r \neq 0$, with the exception of the axis (Fig. 5.3). Since the concentration of atoms in the plasma jet is a function of the temperature, there will be a continuous diffusional flux of $N$ atoms to colder regions (on the periphery) where the atoms will recombine into molecules, and because of the higher concentration of $N_2$ molecules in the colder regions, there is a molecular flux in the opposite direction so that

$$\text{div}\, \vec{I}_a = -\text{div}\, \vec{I}_m \qquad \text{and} \qquad \Gamma_a = -\Gamma_m$$

Here the index $a$ refers to the atoms and $m$ to the molecules. The atoms carry an enthalpy

$$h_a = \frac{5}{2}\frac{kT}{m_a} + E_a + \frac{1}{2}\frac{D_0}{m_a} \tag{5.70}$$

with them, where $m_a$ is the mass of an $N$ atom, $E_a$ the electronic excitation energy per unit mass, and $D_0$ represents the dissociation energy. It is assumed that each atom carries half of the dissociation energy as potential energy.

The corresponding enthalpy carried by the molecules is

$$h_m = \frac{5}{2}\frac{kT}{m_m} + E_m \tag{5.71}$$

where $m_m = 2m_a$ represents the mass of an $N_2$ molecule and $E_m$ is the excitation energy per unit mass; $E_m$ includes electronic excitation as well as excitation of vibrational and rotational energy states.

Combining Eqs. (5.70) and (5.71) gives

$$h_a - h_m = \frac{1}{2m_a}(\tfrac{5}{2}kT + D_0) + E_a - E_m$$

The corresponding mass flux according to Eq. (5.57) is then

$$\vec{I}_a = -\vec{I}_m = -\rho D \operatorname{grad} c_a - \alpha\rho D c_a c_m \frac{\operatorname{grad} T}{T} \tag{5.72}$$

or (assuming chemical equilibrium), with

$$\operatorname{grad} c_a = \frac{\partial c_a}{\partial T}\operatorname{grad} T \tag{5.73}$$

we have

$$\vec{I}_a = -\left(\rho D\frac{\partial c_a}{\partial T} + \alpha\rho D c_a c_m \frac{1}{T}\right)\operatorname{grad} T$$

The heat flux is given by

$$\vec{q} = \left[\frac{\alpha kT}{\bar{m}} + \frac{1}{2m_a}(\tfrac{5}{2}kT + D_0) + E_a - E_m\right]\vec{I}_a - \kappa_{I=0}\operatorname{grad} T \tag{5.74}$$

or in alternative form

$$\vec{q} = -\left[\left(\frac{\alpha kT}{\bar{m}} + \frac{5kT}{4m_a} + \frac{D_0}{2m_a} + E_a - E_m\right)\rho D \times \left(\frac{\partial c_a}{\partial T} + \alpha c_a c_m/T\right) + \kappa_{I=0}\right]\operatorname{grad} T \tag{5.75}$$

The expression in brackets in the last equation can be referred to as $\tilde{\kappa}$, so that

$$\vec{q} = -\tilde{\kappa}\operatorname{grad} T \tag{5.76}$$

The value of $\tilde{\kappa}$ contains contributions due to ordinary diffusion, thermal diffusion, enthalpy interdiffusion (or heat of transition), and pure heat conduction. It is obvious from Eq. (5.75) that diffusion and enthalpy interdiffusion are interrelated, as are thermal diffusion and enthalpy interdiffusion, i.e., the corresponding terms in Eq. (5.75) cannot be separated.

In the next section, the equations that have been derived for a binary mixture will be applied to a fully ionized plasma representing a typical case of a two-component mixture.

## 5.7. TRANSPORT EQUATIONS FOR A FULLY IONIZED PLASMA

We will assume that the plasma consists of electrons and positive ions only. This assumption excludes recombinations, at least for the time being. For the sake of simplicity it will be assumed that the ions of mass $m_i$ are singly charged ($+e$). The electron mass and charge will be denoted by $m_e$ and $-e$, respectively.

The mass fluxes of ions and electrons follow from Eq. (5.3):

$$\vec{I}_i = n_i m_i (\vec{v}_i - \vec{v}_g)$$

or

$$\frac{e}{m_i}\vec{I}_i = e n_i (\vec{v}_i - \vec{v}_g) \tag{5.77}$$

and correspondingly

$$\frac{e}{m_e}\vec{I}_e = e n_e (\vec{v}_e - \vec{v}_g) \tag{5.78}$$

The total current density can be written as

$$\vec{j} = \vec{j}_i + \vec{j}_e$$

or

$$\vec{j} = e n_i \vec{v}_i - e n_e \vec{v}_e \tag{5.79}$$

By introducing Eqs. (5.77) and (5.78) into Eq. (5.79) and using $\vec{I}_i = -\vec{I}_e$, one obtains

$$\vec{j} = -e\left(\frac{1}{m_i} + \frac{1}{m_e}\right)\vec{I}_e + e(n_i - n_e)\vec{v}_g \tag{5.80}$$

The last term of this equation contains the net electrical space charge

$$e(n_i - n_e) = \rho_{el} \tag{5.81}$$

Applying conservation of mass in the form of Eq. (5.4) to ions and electrons yields

$$\frac{\partial n_i}{\partial t} = -\operatorname{div}(n_i \vec{v}_i) \tag{5.82}$$

and

$$\frac{\partial n_e}{\partial t} = -\text{div}\,(n_e \vec{v}_e) \tag{5.83}$$

Combination of these two equations after multiplying Eq. (5.82) by $e$ and Eq. (5.83) by $-e$ leads to the current conservation equation:

$$\frac{\partial \rho_{el}}{\partial t} + \text{div}\,\vec{j} = 0 \tag{5.84}$$

In order to obtain an explicit expression for the current density, a previously derived expression for the mass flux, namely Eq. (5.47), will be applied to the flow of electrons. Introduction of this expression into Eq. (5.80) yields

$$\vec{j} = -e\left(\frac{1}{m_i} + \frac{1}{m_e}\right)L_{11}$$

$$\times \left[\vec{F}_e - \vec{F}_i - \frac{1}{\rho_e}\,\text{grad}\,p_e + \frac{1}{\rho_i}\,\text{grad}\,p_i - (q^* - h_e + h_i)\frac{\text{grad}\,T}{T}\right] \tag{5.85}$$

For the plasma densities of interests in this section, net space charges cannot exist and plasma sheaths will be excluded ($\rho_{el} = 0$). It will be shown later that net space charges may become important for tenuous plasmas for which the density of charge carriers falls below $10^{14}\,\text{cm}^{-3}$.

Equation (5.85) will be further modified by introducing, step by step, external forces that play an important role in the behavior of a plasma.

### 5.7.1. Plasma Exposed to an Electric Field

Here the electric field is the only external force acting on the charge carriers, i.e., $\text{grad}\,p_e = \text{grad}\,p_i = \text{grad}\,T = 0$. Then

$$\vec{F}_e = -\frac{e}{m_e}\vec{E} \qquad \text{and} \qquad \vec{F}_i = \frac{e}{m_i}\vec{E} \tag{5.86}$$

When Eq. (5.86) is substituted into Eq. (5.85), we obtain

$$\vec{j} = e\left(\frac{1}{m_i} + \frac{1}{m_e}\right)L_{11}\left[e\left(\frac{1}{m_i} + \frac{1}{m_e}\right)\vec{E}\right]$$

or

$$\vec{j} = e^2\left(\frac{m_i + m_e}{m_i m_e}\right)^2 L_{11}\vec{E} \tag{5.87}$$

If a plasma is exposed to an electric field only, Ohm's law can be written as

$$\vec{j} = \sigma_e \vec{E}$$

Therefore

$$\sigma_e = e^2\left(\frac{m_i + m_e}{m_i m_e}\right)^2 L_{11} \tag{5.88}$$

### 5.7.2. Plasma Exposed to an Electric Field and an Arbitrarily Directed Magnetic Field of Induction *B*

In this situation the corresponding forces acting on electrons and ions can be written as

$$\vec{F}_e = -\frac{e}{m_e}\vec{E} - \frac{e}{m_e}(\vec{v}_e \times \vec{B}) \qquad \text{and} \qquad \vec{F}_i = \frac{e}{m_i}\vec{E} + \frac{e}{m_i}(\vec{v}_i \times \vec{B}) \tag{5.89}$$

Substitution of Eq. (5.89) into Eq. (5.85) yields

$$\vec{j} = \sigma_e\left[\vec{E} + \frac{m_i m_e}{m_e(m_i + m_e)}\vec{v}_e \times \vec{B} + \frac{m_i m_e}{m_i(m_i + m_e)}\vec{v}_i \times \vec{B}\right]$$

This expression for the current density can be simplified because $m_e \ll m_i$, i.e.,

$$\vec{j} = \sigma_e\left[\vec{E} + \vec{v}_e \times \vec{B} + \frac{m_e}{m_i}(\vec{v}_i \times \vec{B})\right] \tag{5.90}$$

The drift velocities of electrons and ions can be eliminated from Eq. (5.90) by using Eq. (5.79) and the definition of the center-of-mass velocity as given in Eq. (5.1),

$$\rho_i\vec{v}_i + \rho_e\vec{v}_e = \rho_L\vec{v}_g \tag{5.91}$$

where $\rho_L = \rho_i + \rho_e$.

Multiplying Eq. (5.79) by $m_i/e$ and adding it to Eq. (5.91) yields

$$\vec{v}_e(n_e m_e + n_e m_i) = \rho_L \vec{v}_g - \frac{m_i}{e}\vec{j} \tag{5.92}$$

or

$$\vec{v}_e = \vec{v}_g - \frac{m_i}{e\rho_L}\vec{j} \quad \text{and} \quad \vec{v}_i = \vec{v}_g + \frac{m_e}{e\rho_L}\vec{j} \tag{5.93}$$

where we have used the fact that $n_e = n_i$.

Elimination of electron and ion velocities from Eq. (5.90) by using Eqs. (5.91) and (5.93) yields

$$\vec{j} = \sigma_e\left[\vec{E} + \left(\vec{v}_g \times \vec{B} - \frac{m_i}{\rho_L e}\vec{j} \times \vec{B}\right) + \frac{m_e}{m_i}\left(\vec{v}_g \times \vec{B} + \frac{m_e}{\rho_L e}\vec{j} \times \vec{B}\right)\right] \tag{5.94}$$

The last term in this equation can be neglected compared with the previous term, so we have

$$\vec{j} = \sigma_e\left[\vec{E} + \vec{v}_g \times \vec{B} - \frac{1}{en_L}\vec{j} \times \vec{B}\right] \tag{5.95}$$

where $n_L = n_e = n_i$.

Due to the presence of a magnetic field in addition to the electric field, two additional terms appear in Ohm's law. The term $\vec{v}_g \times \vec{B}$ represents the *induced field* caused by the movement of the center of mass. Only for the special case in which $\vec{v}_g$ and $\vec{B}$ are parallel will this term vanish. The last term represents a counteracting field due to the interaction of the current with the applied magnetic field.

A bulk motion of the plasma can be caused by an applied magnetic field, as can readily be seen from the momentum equation

$$\rho \frac{d\vec{v}_g}{dt} = -\operatorname{grad} p + \vec{j} \times \vec{B} \tag{5.96}$$

The interaction of the magnetic field with the current gives rise to an acceleration of the plasma and/or establishes a pressure gradient. Two extreme cases are conceivable: a free plasma flow and a completely obstructed flow. In the first case no pressure gradient can develop, i.e., the first term on the right-hand side of Eq. (5.96) vanishes. In the second case, the initial acceleration of the plasma comes to a complete stop; in this situation the Lorentz force is balanced by a pressure gradient.

An acceleration of the plasma can be initiated with the application of a magnetic field. There is, however, a counteracting force represented by the last term in Eq. (5.95). The plasma reaches its final velocity as soon as this term, which tends to reduce the current density, stops the current flow entirely. In this situation the equation

$$\vec{E} = -\vec{v}_g \times \vec{B} \tag{5.97}$$

determines the magnitude of this final velocity. This case is analogous to a running induction motor without load (zero current).

If the acceleration of the plasma is impeded, a pressure gradient will form, requiring a modification of Eq. (5.95):

$$\vec{j} = \sigma_e\left[\vec{E} + \vec{v}_g \times \vec{B} - \frac{1}{en_L}\vec{j} \times \vec{B} + \frac{m_e m_i}{e(m_i + m_e)}\left(\frac{1}{\rho_e}\operatorname{grad} p_e - \frac{1}{\rho_i}\operatorname{grad} p_i\right)\right] \tag{5.98}$$

Since $m_i \gg m_e$, the last term in this equation reduces to $(1/en_L)\operatorname{grad} p_e$. With $\operatorname{grad} p = \operatorname{grad} p_e + \operatorname{grad} p_i = 2\operatorname{grad} p_e$ and the momentum equation (5.96), the last two terms in Eq. (5.98) can be replaced to give

$$\vec{j} = \sigma_e\left[\vec{E} + \vec{v}_g \times \vec{B} - \frac{1}{en_L}\left(\tfrac{1}{2}\operatorname{grad} p + \rho\frac{d\vec{v}_g}{dt}\right)\right] \tag{5.99}$$

Besides the "forces" included in this equation, there may be a temperature gradient that is also able to drive an electric current. This current will be denoted by $j_T$ [see Eq. (5.85)]:

$$\vec{j}_T = e\left(\frac{1}{m_i} + \frac{1}{m_e}\right)L_{11}(q^* - h_e + h_i)\frac{\operatorname{grad} T}{T} \tag{5.100}$$

An alternative form is

$$\vec{j}_T = \sigma_e\frac{\alpha k}{2e}\operatorname{grad} T \tag{5.101}$$

where $q^* - h_e + h_i = \alpha kT/\bar{m}$ and $\bar{m} = 2m_e$.

The electric current driven by thermal diffusion is frequently small compared with the other contributions. In the neighborhood of electrodes, however, grad $T$ may assume such large values that the corresponding "driving force" reaches a value of several volts.

In addition to the previously discussed forces other forces (such as gravity, centrifugal forces, or inertia) may be responsible for the current flow. For laboratory plasmas, gravity does not play a significant role, the same is true for inertia.

It should be pointed out that the electrical conductivity, $\sigma_e$, which appears in the current equation, is assumed to be a scalar. This is, however, no longer true if strong magnetic fields are present. In this situation the electrical conductivity changes to a tensor, which assumes different values parallel and normal to the orientation of the magnetic field.

Finally, the energy flux in a fully ionized plasma will be considered. Applying Eq. (5.67) to a fully ionized plasma yields

$$\vec{q} = -\left(\frac{\alpha kT}{\bar{m}} + h_e - h_i\right)\frac{m_e m_i}{e(m_e + m_i)}\vec{j} - \kappa_{I=0}\,\text{grad}\,T \qquad (5.102)$$

Alternatively, if $m_e \ll m_i$ and

$$h_e = \frac{5}{2}\frac{kT}{m_e} \qquad \text{and} \qquad h_i = \frac{5}{2}\frac{kT}{m_i} + \frac{E_i}{m_i}$$

where $E_i$ is the ionization energy, we have

$$\vec{q} = -\frac{kT}{2e}\left[\alpha + 5 - \frac{m_e}{m_i}\left(5 + \frac{2E_i}{kT}\right)\right]\vec{j} - \kappa_{I=0}\,\text{grad}\,T \qquad (5.103)$$

The first term of this equation represents the energy carried by the electric current, and the second term describes pure heat conduction. The term in parentheses can be neglected compared with $(\alpha + 5)$. The current density $\vec{j}$ can be taken from Eq. (5.99).

An interesting conclusion can be drawn from Eq. (5.103) by forming $\text{div}\,\vec{q}$:

$$\text{div}\,\vec{q} = -\vec{j}\,\frac{\partial f(T)}{\partial T}\,\text{grad}\,T - \text{div}\,(\kappa_{I=0}\,\text{grad}\,T) \qquad (5.104)$$

where

$$f(T) = \frac{kT}{2e}\left[\alpha + 5 - \frac{m_e}{m_i}\left(\frac{2E_i}{kT} + 5\right)\right]$$

The scalar product $\vec{j}\cdot\text{grad}\,T \neq 0$ if $\text{grad}\,T$ is not normal to $\vec{j}$, i.e., in this case $\vec{j}\cdot\text{grad}\,T$ represents an energy-source term. This situation prevails, for example, in the boundary layers close to electrodes in an electric discharge.[5–7]

In the previously derived equations for current and energy fluxes there are coefficients ($\sigma_e$, $\kappa_{I=0}$, and $\alpha$) that must be determined. The following section contains expressions for these coefficients.

## 5.8. DETERMINATION OF TRANSPORT COEFFICIENTS

From solutions of the Boltzmann transport equation (see Chapter 3), Spitzer derived expressions for the transport properties of a fully ionized plasma:

$$\sigma_e = \frac{2m_e v_e^3}{e^2 Z \ln q}\left(\frac{2}{3\pi}\right)^{3/2} \gamma_E \tag{5.105}$$

$$\alpha = 3\frac{\gamma_T}{\gamma_E} \tag{5.106}$$

$$\kappa_{I=0} = \frac{20 m_e^2 k v_e^5}{3e^4 Z \ln q}\left(\frac{2}{3\pi}\right)^{3/2} \delta_T\left(1 - \frac{3}{5}\frac{\delta_E \gamma_T}{\delta_T \gamma_E}\right) \tag{5.107}$$

where

$$v_e^2 = \frac{3kT}{m_e}, \qquad q = \frac{kT}{e^2 Z n_i^{1/3}}, \qquad \text{and} \qquad Z = \sum \frac{n_z z^2}{n_e}$$

represents the average ion charge.

In addition, there are four transport coefficients ($\delta_T$, $\delta_E$, $\gamma_T$, and $\gamma_E$) that are functions of the average ion charge. Values for these coefficients can be found in the following table:

| | $Z = 1$ | $Z = 2$ |
|---|---|---|
| $\gamma_E$ | 0.5816 | 0.6833 |
| $\gamma_T$ | 0.2727 | 0.4131 |
| $\delta_E$ | 0.4652 | 0.5787 |
| $\delta_T$ | 0.2252 | 0.3563 |

Disregarding the minor influence of the logarithmic term in Eqs. (5.105) and (5.106), it is important to emphasize that

$$\sigma_e \sim v_e^3 \sim T^{3/2} \qquad \text{and} \qquad \kappa_{I=0} \sim v_e^5 \sim T^{5/2}$$

Both electrical and thermal conductivity are governed by the free electrons, a result that is not surprising considering the high mobility of the free electrons in a plasma.

For $Z = 1$ one finds that

$$\frac{\kappa_{I=0}}{\sigma_e} = 2\frac{k^2}{e^2}T$$

a relation that has been derived for pure metals. There is obviously a close relationship between a metal (a "solid state plasma") and a fully ionized gaseous plasma.

After considering a typical two-component mixture (a fully ionized plasma) the treatise should be extended to a three-component mixture, which can be applied in general to any plasma consisting of a mixture of neutral particles, ions, and electrons. The resulting equations are more complicated but similar to those derived for a two-component mixture. The derivation of these equations can be found elsewhere.[1]

## LIST OF SYMBOLS

| | |
|---|---|
| $\vec{B}$ | magnetic induction ($Vs/m^2$) |
| $C$ | constant, Eq. (5.36) |
| $c_k$ | mass fraction of component $k$ |
| $D$ | diffusion coefficient ($m^2/s$) |
| $D_0$ | dissociation energy (J) |
| $\vec{E}$ | electrical field strength (V/m) |
| $E_a$ | electronic excitation energy (J) |
| $E_s$ | entropy production ($W/m^3/K$) |
| $e_k$ | electric charge of component $k$ (As) |
| $e_n$ | base of natural logarithm ($e_n = 2.718$) |
| $\vec{F}_k$ | body force acting on component $k$ (N/kg) |
| $\vec{g}$ | acceleration due to gravity ($m/s^2$) |
| $h$ | Planck's constant ($6.6 \times 10^{-34}$ W $s^2$) |
| $h_g$ | enthalpy (J/kg) |
| $\vec{I}_k$ | mass flux of component $k$, relative to the center of mass ($kg/m^2s$) |
| $\vec{j}_e$ | electron current density ($A/m^2$) |
| $k$ | Boltzmann constant ($1.38 \times 10^{-23}$ J/K) |
| $L_{kl}$, $L_{ku}$, $L_{ul}$ | Onsager coefficients in Eqs. (5.23), (5.24), and (5.25) |
| $m_e$ | electron mass ($9.11 \times 10^{-31}$ kg) |
| $m_k$ | mass of component $k$ (kg) |
| $n$ | particle number density ($m^{-3}$) |
| $p_k$ | partial pressure of component $k$ (Pa) |
| $Q_k$ | partition of function of component $k$ |
| $q^*$ | heat of transition (J/kg) |
| $q$ | heat exchange with surroundings (J/kg) |
| $\vec{q}$ | heat flux ($W/m^2$) |
| $\vec{S}$ | diffusion flux ($m^{-2} s^{-1}$) |
| $s_g$ | entropy (J/kg K) |
| $t$ | time (s) |
| $T$ | absolute temperature (K) |
| $u_g$ | internal energy (J/kg) |
| $V$ | electrical potential (V) |

$\vec{v}_g$ velocity of the center of mass (m/s)
$\vec{v}_k$ velocity of component $k$ (m/s)
$v_0$ specific volume ($m^3$/kg)
$\vec{X}_l$ driving forces (N)
$x_k$ mole fraction of component $k$
$y_k$ thermal diffusion coefficient of component $k$ (J/kg)
$Z$ mean ion charge

## Greek Symbols

$\alpha$ thermal diffusion factor
$\delta_E$, $\delta_T$, $\gamma_\varepsilon$, $\gamma_T$ Spitzer coefficients in Eqs. (5.105) to (5.107)
$\varepsilon_{ij}$ friction coefficient between components $i$ and $j$
$\Gamma_k$ generation rate of particles of component $k$ ($kg/m^3s$)
$\kappa$ thermal conductivity (W/mK)
$\lambda_i$ mean free path of particle of type $i$ (m)
$\mu_k$ chemical potential of component $k$ (J/kg)
$\rho$ mass density of the mixture ($kg/m^3$)
$\rho_k$ mass density of component $k$ ($kg/m^3$)
$\sigma_e$ electrical conductivity ($ohm^{-1}\,m^{-1}$)
$\tau$ time (s)

## REFERENCES

1. W. Finkelnburg and H. Maecker, "Electric Arcs and Thermal Plasmas," in *Handbook of Physics, Vol. 22* (S. Flügge, ed.), (Berlin: Springer-Verlag, 1950): 254.
2. D. Enskog, Dissertation (University of Uppsala, Sweden, 1917).
3. L. Onsager, *Phys. Rev.* **37** (1931): 405; *Phys. Rev.* **38** (1993): 2265.
4. L. Waldmann, *Z. Naturforschung,* **4a** (1949): 105; *Z. Naturforschung,* **5a** (1950): 322.
5. K. C. Hsu, K. Etemadi, and E. Pfender, *J. Appl. Phys.* **54** (1983): 1293.
6. H. A. Dinulescu, and E. Pfender, *J. Appl. Phys.* **51** (1980): 3149.
7. N. A. Sanders, and E. Pfender, *J. Appl. Phys.* **55** (1984): 714.

## Chapter 6

# Thermodynamic Properties

## 6.1. INTRODUCTION

When a gas such as nitrogen [molecular nitrogen ($N_2$) at room temperature and atmospheric pressure] is progressively heated, the molecules first dissociate following the endothermic reaction $N_2 \rightleftarrows 2N - E_N^D$, where $E_N^D$ is the energy of dissociation. At higher temperatures ionization begins through the endothermic reaction $N \rightleftarrows N^+ + e - E_{N^+}^I$ where $E_{N^+}^I$ is the first ionization energy. Further temperature increases cause the $N^+$ ions to loose one more electron, according to $N^+ \rightleftarrows N^{++} + e - E_{N^{++}}^I$ (second ionization) and so on.

The composition of the gas (in our example, the molar fractions of $N_2$, N, $N^+$, and $e$) depends strongly on its temperature, which is a result of the energy balance between the electrical energy dissipated and the heat losses that occur mainly at the fringes of the plasma. In such a complex mixture, where a dynamic equilibrium exists between dissociation, ionization, and recombination, the total energy content depends on the energy of the various particles (frozen energy) and on the chemical reactions among them (reactive energy). Thus the thermodynamic properties (enthalpy, entropy, specific heat, and so forth) depend strongly on the composition of the plasma. The plasma composition is calculated through the minimization of the Gibbs free energy (for a given temperature and pressure), which in turn depends on the chemical potentials of the different chemical species present in the gas. At high temperatures ($T > 6000$ K) these chemical potentials can be calculated only from statistical thermodynamic considerations (see Section 4.4) through quantities called partition functions, which are related to the internal energy levels of the different chemical species of the plasma (energies can be determined spectroscopically). This chapter will first be devoted to the relationships between thermodynamic functions and partition functions for plasmas in complete thermodynamic equilibrium (CTE) and to the calculation of these partition functions. In the frame of this formulation the different ways to calculate the composition will then be presented, followed by the calculation of thermodynamic properties. However, as has been shown in Section 4.4, even if the plasma core can be assumed to be in CTE, this may not be the case in the fringes or in the plume of the thermal plasmas, where two different temperatures can be defined (one for the electrons and the other for the heavy species). The last part of the chapter will be devoted to the calculation of the composition and thermodynamic properties of a two-temperature plasma.

## 6.2. THERMODYNAMIC FUNCTIONS FOR CTE

### 6.2.1. Notation

To denote the thermodynamic function of any chemical species, the simplest notation is to use the chemical symbol as the subscript of the considered thermodymamic function. For example, for nitrogen, which contains $K$ ($K = 6$) species, we use the subscripts $N_2$ (molecules), N (atoms), $N^+$ and $N^{++}$ (atomic ions), $N_2^+$ (molecular ions), and $e$ (electrons).

However, it is quite evident that if we have a mixture of three or four elementary chemical species at room temperature, the number of chemical species in the plasma will be very large (approximately 20 to 50). The composition must then be obtained computationally and the notation defined above for pure nitrogen is no longer useful. Thus a chemical species classification is used to obtain numerical subscripts.

To obtain a linear relation between indexes and chemical species, a single index $i$, varying from 1 to $K$ (the total number of species), can be used. For example,

$$1 = e, \quad 2 = \mathrm{N}, \quad 3 = \mathrm{N}^+, \quad 4 = \mathrm{N}^{++}, \quad 5 = \mathrm{N}_2, \quad 6 = \mathrm{N}_2^+$$

The electrons, according to their particular behavior, are assigned either the index 1 or the index $K$, but most often are assigned the value 1.

Tables of energy levels for the excited states [for example, Moore (1949, 1952, 1958)] are usually arranged by increasing values (Fig. 6.1a), thus a second index for the energy of the excited state is generally

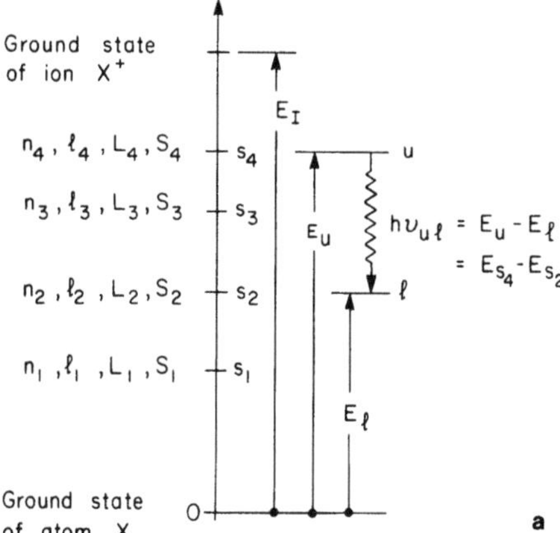


**FIG. 6.1a.** Energy diagram of an atom.

sufficient. In the following we will use $E_{i,s}$ to correspond to the energy of the $i$th nitrogen species in the excited state $s$; $E_{i,0}$ corresponds to the ground state. The total number of the species considered is thus split among the different energy levels,

$$N_i = \sum_s N_{i,s} \tag{6.1}$$

where the summation is performed over all possible excited states, including the ground state. Similar considerations hold for the diatomic molecular energy levels (see Fig. 6.1b).

With these indexes, we can write the chemical reactions in a matrix

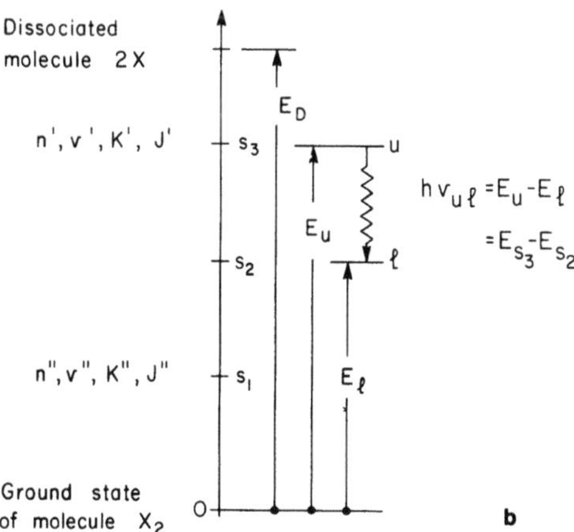


**FIG. 6.1b.** Energy diagram of a molecule.

notation suitable for computer use. However, to acquaint the reader who is not familiar with plasmas, most of the plasma properties will be presented using nitrogen as an example; the pertinent chemical species will be classified by the chemical index, followed, if necessary, by the additonal index $s$ for the excited states. When only the chemical index is given, it means that a property of the entire chemical species is considered, regardless of excitation state.

### 6.2.2. Partition Functions

The number of particles of chemical species $i$ that will be found in a quantum state $s$ of energy $E_{i,s}$ is given by Boltzmann's relationship (see Section 4.4.1)

$$\frac{N_{i,s}}{N_i} = \frac{g_{i,s} \exp\left[-E_{i,s}/(kT)\right]}{\sum_s g_{i,s} \exp\left[-E_{i,s}/(kT)\right]} \tag{6.2}$$

where $g_{i,s}$ is the statistical weight of the state $s$ (i.e., the number of wave functions related to the energy level $s$).

The denominator of Eq. (6.2) is called the atomic or molecular *internal* partition function $Q_i^{\text{int}}(T)$ for particles of species $i$:

$$Q_i^{\text{int}}(T) = \sum_s^{s^*} g_{i,s} \exp\left[-E_{i,s}/(kT)\right] \tag{6.3}$$

This quantity is dimensionless, and the summation is over all possible states of $s$, limited to a value $s^*$, which will be specified for the cases of atoms and diatomic molecules.

In general [Mayer and Mayer (1966)], for a given particle of species $i$, two types of energies have to be considered: the energy of translation (kinetic energy) and the energy related to the internal degrees of freedom. For an atom, these degrees of freedom include the excitation energy of the bound electrons while for a molecule they include the rotational energy of the molecule and the vibrational energy of the nuclei (see Section 2.4), in addition to the energy related to the excitation of the bound shared electrons.

If the translational and internal energies of the particles are supposed to be independent ($E_{i,s} = E_{i,k}^{\text{trans}} + E_{i,j}^{\text{int}}$), the partition function $Q_i$, which is then the product of the translational and internal partition functions, can be written

$$Q_i = Q_i^{\text{tr}} Q_i^{\text{int}} = \left(\frac{2\pi m_i kT}{h^2}\right)^{3/2} V Q_i^{\text{int}} = \frac{V}{\Lambda_i^3} Q_i^{\text{int}} \tag{6.4}$$

where $m_i$ is the mass of the chemical species $i$, $V$ is the plasma volume, and $\Lambda_i$ is the thermal de Broglie length:

$$\Lambda_i = \left(\frac{h^2}{2\pi m_i k T_i}\right)^{1/2} \tag{6.5}$$

All these equations hold provided the hypothesis of Boltzmann is fulfilled [for more details, see Section 4.4.1 and Landau and Lifschitz (1967)]. This means the number of accessible quantum states must be much higher than the number of particles, that is, there should be a very low probability of finding two particles simultaneously in the same quantum state. In other words, the condition

$$\left(\frac{2\pi m_i k T}{h^2}\right)^{3/2} V/N \gg 1 \tag{6.6}$$

must be fulfilled.

Notice that the Maxwell–Boltzmann distribution is a limiting case of the quantum distributions of Bose–Einstein and Fermi–Dirac.

### 6.2.3. Thermodynamic Functions

#### 6.2.3.1. Perfect Gases

Boltzmann's statistical treatment allows one to express the thermodynamic functions through the partition functions of the system under consideration. For example, according to Mayer and Mayer (1966), the Helmholtz free energy, $F$, related to a reference energy $F_0$ is given by

$$F - F_0 = -kT \ln Q_{\text{tot}} \tag{6.7}$$

where $Q_{\text{tot}}$ is the partition function of all the particles of the thermodynamic system under consideration.

The energy of the system is then assumed to be the sum of the energies of the different particles (noninteracting particles), and one finds

$$Q_{\text{tot}} = \frac{\Pi_i Q_i^{N_i}}{\Pi_i N_i!} \tag{6.8}$$

where $Q_i$ is the partition function of a single particle of the chemical component $i$, $N_i$ is the total number of particles of species $i$ in the system, and $\Pi_i = \Pi_{i=1}^{i=K}$. Using Eqs. (6.7), (6.8), and the Stirling formula ($\ln N! = N \ln N - N$), $F$ can be calculated from

$$F - F_0 = -\sum_i N_i kT \ln (e_n Q_i / N_i) \tag{6.9}$$

where $e_n$ is the base of natural logarithms (2.7182818) and $\Sigma_i = \Sigma_{i=1}^{K}$.

However, for formula (6.8) to be valid it is essential that all interconnected energy levels be referred to the same reference, for example, to free atoms in their lowest quantum states (ground states).

The right-hand side of Eq. (6.9) [valid when Eq. (6.6) is fulfilled, i.e., no quantum effects] is a function of the temperature $T$ and either the volume or the pressure. The right-hand side can be determined easily provided the energies and degeneracies of the excited states (as well as the plasma composition) are known.

For example, in the case of nitrogen at temperature $T$ and pressure $p$ we obtain the following equation, which considers only the species $N_2$, N, $N^+$, and $e$. The initial mixture is one mole of $N_2$ at $T = 0\,K$, and $p = 1$ atm (with free energy $F_0$):

$$F - F_0 = kT\left[N_{N_2} \ln\left(\frac{e_n Q_{N_2}}{N_{N_2}}\right) + N_N \ln\left(\frac{e_n Q_N}{N_N}\right) + (N_{N^+}) \ln\left(\frac{e_n Q_{N^+}}{N_{N^+}}\right) + N_e \ln\left(\frac{e_n Q_e}{N_e}\right)\right] \tag{6.10}$$

It is thus possible to calculate all other thermodynamic functions through the partition functions for the various types of particles present in the plasma. For example, the pressure $p$ is given by

$$p = -\left(\frac{\partial F}{\partial V}\right)_{T,N_i} = \sum_i N_i kT\left(\frac{\partial \ln Q_i}{\partial V}\right)_{T,N_i} \tag{6.11}$$

and the internal energy by

$$E - E_0 = T^2\left(\frac{\partial\left(\frac{F}{T}\right)}{\partial T}\right)_{V,N_i} = \sum_i N_i\left[3kT/2 + kT^2\left(\frac{\ln Q_i^{\text{int}}}{\partial V}\right)_{T,N_i}\right] \tag{6.12}$$

### 6.2.3.2. Debye Correction

As soon as the temperature increases, the densities of the charged species also increase, and the long-range Coulomb interactions result in an interaction energy that must be added to the thermodynamic functions calculated under the perfect-gas assumption.

With the Debye model[1] [see also Mayer and Mayer (1966)], $F$ becomes

$$F - F_0 = -\sum_i N_i kT \ln(e_n Q_i/N_i) - \frac{kTV}{12\pi\lambda_D^3} \tag{6.13}$$

with

$$\lambda_D = \left( \frac{\varepsilon_0 kTV}{e^2 \sum_{i=1}^{K} Z_i^2 N_i} \right)^{1/2} \tag{6.14}$$

where $\varepsilon_0$ is the vacuum permittivity and $Z_i$ is the number of ionic charges of species $i$ (for example, $Z_i = 1$ for $N^+$, $Z_i = -1$ for $e$, $Z_i = 1$ for $N_2^+$, $Z_i = 0$ for N and $N_2$).

In a nitrogen plasma containing the species $N_2$, $N_2^+$, N, $N^+$, and $e$,

$$\sum_{i=1}^{K} Z_i^2 N_i = N_{N^+} + N_e + N_{N_2^+}$$

For higher temperatures with a gas containing mainly $N^+$, $N^{++}$, and $e$,

$$\sum_{i=1}^{K} Z_i^2 N_i = N_{N^+} + 4N_{N^{++}} + N_e$$

This Coulomb field also modifies the pressure:

$$p = \frac{N_i kT}{V} - \frac{kT}{24\pi\lambda_D^3} \tag{6.15}$$

while the Gibbs free energy becomes

$$G - G_0 = pV + F - F_0 = -\sum N_i kT \ln (Q_i/N_i) - \frac{kTV}{8\pi\lambda_D^3} \tag{6.16}$$

The Debye correction will be greatest for the highest values of charged-particle densities. For $N_2$, Ar, $H_2$, and $O_2$, which all have similar ionization potentials, the maximum electron density at atmospheric pressure occurs around 14,000 to 15,000 K (see, for example, Figs. 6.2 to 6.6 in Section 6.3.3.4) and remains almost constant for higher values of $T$. Thus the Debye correction is the highest at these temperatures.

In most cases these corrections are rather small (less than 2 or 3%). Thus it is generally sufficient to calculate the composition with no Debye correction and then to evaluate the corresponding corrections ($\lambda_D$ depends on $N_i$) to the thermodynamic functions. The Debye correction is illustrated in Table 6.1, where corrections to the enthalpy of air plasmas are given at 0.2, 1, and 5 atm.

### 6.2.3.3. Virial Correction

When two atoms approach each other, an interaction potential (for example, of the Morse type, see Section 2.4.1) governs their behavior because the electronic shells surrounding the nuclei interact with each

**TABLE 6.1.** Debye and Virial Corrections for the Enthalpy of Air

| Pressure (kPa) | $T$ (K) | Perfect gas values (J/kg) | Debye corrections | Virial corrections | Real gas values (J/kg) |
|---|---|---|---|---|---|
| 20 | 1,000 | 0.7540E + 06 | −0.00% | +0.00% | 0.7540E + 06 |
| | 2,000 | 0.1980E + 07 | +0.00% | +0.00% | 0.1980E + 07 |
| | 3,000 | 0.4212E + 07 | +0.00% | +0.00% | 0.4212E + 07 |
| | 4,000 | 0.7995E + 07 | −0.00% | +0.00% | 0.7995E + 07 |
| | 5,000 | 0.1073E + 08 | +0.00% | +0.00% | 0.1073E + 08 |
| | 6,000 | 0.1892E + 08 | +0.00% | −0.00% | 0.1892E + 08 |
| | 7,000 | 0.3355E + 08 | +0.00% | +0.00% | 0.3355E + 08 |
| | 8,000 | 0.4089E + 08 | +0.01% | +0.00% | 0.4090E + 08 |
| | 9,000 | 0.4555E + 08 | +0.06% | +0.00% | 0.4558E + 08 |
| | 10,000 | 0.5333E + 08 | +0.24% | +0.00% | 0.5346E + 08 |
| | 11,000 | 0.6904E + 08 | +0.60% | +0.00% | 0.6945E + 08 |
| | 12,000 | 0.9705E + 08 | +0.85% | −0.00% | 0.9788E + 08 |
| | 13,000 | 0.1310E + 09 | +0.54% | +0.00% | 0.1317E + 09 |
| | 14,000 | 0.1542E + 09 | +0.09% | +0.00% | 0.1544E + 09 |
| | 15,000 | 0.1653E + 09 | +0.12% | −0.00% | 0.1651E + 09 |
| 100 | 1,000 | 0.7540E + 06 | +0.00% | +0.01% | 0.7541E + 06 |
| | 2,000 | 0.1977E + 07 | +0.00% | +0.00% | 0.1977E + 07 |
| | 3,000 | 0.3772E + 07 | +0.00% | +0.00% | 0.3772E + 07 |
| | 4,000 | 0.7386E + 07 | −0.00% | +0.00% | 0.7386E + 07 |
| | 5,000 | 0.9965E + 07 | +0.00% | +0.00% | 0.9965E + 07 |
| | 6,000 | 0.1466E + 08 | +0.00% | +0.00% | 0.1466E + 08 |
| | 7,000 | 0.2575E + 08 | +0.00% | −0.00% | 0.2575E + 08 |
| | 8,000 | 0.3751E + 08 | +0.01% | −0.00% | 0.3751E + 08 |
| | 9,000 | 0.4342E + 08 | +0.05% | +0.00% | 0.4344E + 08 |
| | 10,000 | 0.4861E + 08 | +0.19% | +0.00% | 0.4870E + 08 |
| | 11,000 | 0.5705E + 08 | +0.55% | +0.00% | 0.5736E + 08 |
| | 12,000 | 0.7246E + 08 | +1.17% | −0.00% | 0.7331E + 08 |
| | 13,000 | 0.9790E + 08 | +1.58% | +0.00% | 0.9944E + 08 |
| | 14,000 | 0.1291E + 09 | +1.11% | +0.00% | 0.1305E + 09 |
| | 15,000 | 0.1532E + 09 | +0.28% | +0.00% | 0.1537E + 09 |
| 500 | 1,000 | 0.7540E + 06 | −0.00% | +0.03% | 0.7542E + 06 |
| | 2,000 | 0.1975E + 07 | +0.00% | +0.02% | 0.1976E + 07 |
| | 3,000 | 0.3555E + 07 | −0.00% | +0.01% | 0.3556E + 07 |
| | 4,000 | 0.6492E + 07 | −0.00% | +0.00% | 0.6492E + 07 |
| | 5,000 | 0.9386E + 07 | −0.00% | +0.00% | 0.9386E + 07 |
| | 6,000 | 0.1249E + 08 | +0.00% | +0.00% | 0.1249E + 08 |
| | 7,000 | 0.1908E + 08 | +0.00% | +0.00% | 0.1909E + 08 |
| | 8,000 | 0.3028E + 08 | +0.01% | −0.00% | 0.3029E + 08 |
| | 9,000 | 0.3992E + 08 | +0.04% | +0.00% | 0.3993E + 08 |
| | 10,000 | 0.4568E + 08 | +0.14% | +0.00% | 0.4574E + 08 |
| | 11,000 | 0.5125E + 08 | +0.44% | +0.00% | 0.5148E + 08 |
| | 12,000 | 0.5966E + 08 | +1.11% | +0.00% | 0.6032E + 08 |
| | 13,000 | 0.7375E + 08 | +2.14% | −0.00% | 0.7533E + 08 |
| | 14,000 | 0.9592E + 08 | +2.90% | −0.00% | 0.9870E + 08 |
| | 15,000 | 0.1242E + 09 | +2.37% | +0.00% | 0.1271E + 09 |

other (see, for example, Fig. 3.1.). This interaction potential becomes repulsive only when the two atoms come very close together (angström). If the mean free path is large compared to the distance at which the interaction starts, the deflection of the particles will result in mean trajectories quite similar to those predicted by the classical model. If $l_{ij}$ is the mean free path of particle $i$ approaching a particle $j$, then an interaction potential $V_{ij}$ is active for distances smaller than $d_{ij}$. No viral corrections will be necessary as long as

$$l_{ij} \gg d_{ij} \tag{6.17}$$

where $d_{ij}$ is the distance at which the particles start to repulse each other.

If Eq. (6.17) does not hold for all colliding species, the perfect gas law has to be modified to

$$\frac{pV}{NkT} = 1 + B(T)\frac{N}{V} \tag{6.18}$$

where $N = \sum_{i=1}^{K} N_i$ and $B(T)$ is the second virial coefficient, which can be calculated using the interaction potentials $V_{ij}$ as follows:

$$B(T) = \sum_{i,j}^{K,K} \frac{N_i N_j}{N^2} B_{ij}(T) \tag{6.19}$$

where

$$B_{ij}(T) = -2\pi \int_0^\infty \left[\exp\left(\frac{V_{ij}(r)}{kT}\right) - 1\right] r^2\, dr \tag{6.20}$$

Note that the calculation is somewhat similar to that of the collision integrals (see Section 7.3.4). In most cases, the virial corrections are negligible for the pressures generally used in thermal plasmas (20 to 500 kPa), as illustrated for air in Table 6.1. They are, of course, highest for the highest densities, i.e., for the lowest temperatures.

### 6.2.4. Calculation of Partition Functions

#### 6.2.4.1. Translational Partition Functions

To calculate a thermodynamic function, the ratio $Q_i/N_i$ is needed [see, for example, Eq. (6.9)]. For the translational degree of freedom, the desired quantity is

$$\frac{Q_i^{\mathrm{tr}}}{N_i} = \left(\frac{2\pi m_j kT}{h^2}\right)^{3/2} \cdot \frac{V}{N_i} = \Lambda_i^{-3} \cdot \frac{V}{N_i} \tag{6.21}$$

where $\Lambda_i$ is the thermal de Broglie length; $Q_i^{tr}/N_i$ depends on the result of the calculation of the composition $N_i$ but, as will be shown (see Section 6.3), the composition depends only on the quantity

$$\frac{Q_i^{tr}}{V} = \Lambda_i^{-3} = 2.77721 \cdot 10^{66}(m_i T)^{3/2}$$

$$\text{(where } m_i \text{ is in kg and } T \text{ in K)} \quad (6.22)$$

which is a straightforward calculation.

#### 6.2.4.2. Limitations of the Internal Partition Functions

The second term involved in calculating the composition is the internal partition function.

*6.2.4.2.a. Internal Partition Function for Atoms.* For atoms, this function includes only the electronically excited states and is given by Eq, (6.3). Though this formula seems simple, its evaluation poses considerable difficulties:[1] (1) The statistical weights and the energies must be obtained from solutions of the Schrödinger equation. For systems of a complicated level structure, it is very difficult to obtain exact solutions. (2) The summation in Eq. (6.3) has to be performed over all possible discrete energy states which exist only for energies $E_{i,s}$, lower than $E_I - \delta E_I$ where $\delta E_I$ is the lowering of the ionization potential $E_I$.

The bound electron is no longer bound when its energy reaches $E_I - \delta E_I$ (see Section 4.4.2 and Fig. 4.8). From a practical point of view, one of the main problems is that $\delta E_I$ depends on the composition of the plasma and it becomes necessary to use an iterative procedure for rigorous calculations.

However, as demonstrated by Fauchais *et al.*,[2–4] the various theories about the breakoff $s^*$ of the partition functions give about the same values for the thermodynamic properties, and it is thus generally sufficient to use the theory of Gurvich[5] as a first approximation to calculate $\delta E_I$. Although this theory is not rigorous, it does provide $s^*$ which is correlated with the principal quantum number $n$, as a function of $T$ and $p$:

$$\begin{aligned} n^{*2} r_1 &= Z_{\text{eff}}^{-1}\left(\frac{3V}{4\pi N}\right)^{1/3} \\ &= Z_{\text{eff}}^{-1}\left(\frac{3kT}{4\pi p}\right)^{1/3} \\ &= 1.48848 \times 10^{-8} Z_{\text{eff}}^{-1}\left(\frac{T}{p}\right)^{1/3} \end{aligned} \quad (6.23)$$

where $p$ is in Pa, $T$ is in K, and $Z_{\text{eff}}$ is the effective charge, taken as the stage of ionization plus one for the particle being considered: 1 for a neutral atom, 2 for an ion such as $N^+$, 3 for $N^{++}$ and so on.

In this case the calculation of $n^*$ is straightforward for a given $T$ and $p$. Note, however,[6] that even for differences up to 120% in the value of $n^*$ for the various theories, the composition of the main species remains the same within 5%. This is true only if the *same limitation* theory is used for the calculation of the partition functions for all the species of the plasma. The differences for the minor species (for which molar fractions are less than $10^{-3}$, for example) may be much higher (up to 40%) depending on the limitation theory used, but such differences do not change the general behavior of the plasma.

*6.2.4.2.b. Internal Partition Function for Molecules.* For molecules, the internal partition function can be written as

$$Q_{\text{int}} = \left[\sum_e g_e \exp\left(\frac{-E_e}{kT}\right)\right]\left[\sum_v g_v \exp\left(\frac{-E_v(e)}{kT}\right)\right] \times \left[\sum_r g_r \exp\left(\frac{-E_r(e, v)}{kT}\right)\right] \tag{6.24}$$

where $g_e$, $g_v$, and $g_r$ denote the electronic, vibrational, and rotational degeneracies, respectively.

For an electronic state designated by $2S + 1$ ([see Section 2.4.2 and Herzberg (1950)]

$$g_e = 2S + 1 \qquad \text{for } \Sigma \text{ states } (\Lambda = 0) \tag{6.25}$$

$$g_e = 2(2S + 1) \qquad \text{for } \Pi, \Delta, \Phi \text{ states } (\Lambda \neq 0) \tag{6.26}$$

However,

$$g_v = 1 \tag{6.27}$$

for vibrational states of quantum number $v$ and

$$g_r = 2J + 1 \tag{6.28}$$

for rotational states described by the quantum number $J$.

In Eq. (6.24) the summations have to be limited (for example, when the dissociation limit for $v$ is reached[7]).

The different methods all yield similar results (within 3% for nitrogen, for example) except when the fundamental state is a multiplet (for example, $O_2^+$ or NO), in which case a method with a limitation must be used.

The main problem in calculating the molecular partition function often arises when the interactions between the various excited states complicate the calculation [see Mayer and Mayer (1966)].

### 6.2.4.3. Data Base

Data for atoms are available in spectroscopic tables such as those of C. E. Moore (1949, 1952, 1958). These tables first give the denomination of the spectral level according to the $L$–$S$ or the $j$–$j$ coupling, (see Section 2.3.3), followed by its energy expressed in $cm^{-1}$ [in Eq. (6.3) Boltzmann's constant must be modified to $k/hc = 0.695$].

The degeneracy $g$ is given by $(2J + 1)$, where $J$ is the angular momentum of the spectral level under consideration.

The total number of levels to be accounted for depends on both the temperature and energy of each level. For example, when $T < 10{,}000$ K, $Q^{+}_{\mathrm{Ar}} = 4 + 2\exp(-2060.4/T)$. However, as soon as $T$ increases, the number of levels to be accounted for rises sharply, and the limitation theory becomes important in determining this number. Although tables give almost all the levels for low values of the principal quantum number $n$, more and more levels are missing as $n$ increases.

It is then necessary to extrapolate for these missing levels using the selection rules for the orbital and spin momenta.[6,8] Isoelectronic series (for example, CuI and OIII*) are also helpful. For example it is possible to use the energy of the OIII configuration to deduce the missing levels in one of the CuI configurations by making comparisons of the curves.

In general, these methods are still insufficient and missing levels must be completed by a different empirical method (see, for example, Drellishak[8] or Capitelli *et al.*[9]).

Another problem arises because the number of states that must be taken into account in calculating partition functions is very large (for atomic nitrogen with $n = 12$, there are 1032 states). In practice, it is thus necessary (even when using high-speed computers) to regroup the states having similar excitation energies whenever $n$ is higher than 5 to 8, depending on the atom considered. For example, Veits *et al.*[10] have proposed combining the states into a single energy level $E_m$ with a statistical weight equal to the sum of the individual statistical weights and an excitation energy equal to the weighted mean of the individual energies. However, the error is reduced if

(i) $E_m/kT$ is small. In this case only terms that are very close

*In spectroscopic notation, CuI corresponds to the copper atom, CuII to its first ion, CuIII to its second ion, and so on.

together can be regrouped (in general, the low-level terms whose energies are very different cannot be regrouped).

(ii) $E_m/kT$ is large compared with 1. In this case levels having wide-ranging $kT$ energies may be regrouped.

## 6.3. COMPOSITION OF A PLASMA IN CTE

From a thermodynamic point of view, a chemical system can be characterized either by its temperature and pressure or by its temperature and volume. In the first case the equilibrium state is reached when the Gibbs free energy assumes a minimum, while the second case refers to the Helmholtz free energy. For plasmas it is customary to use temperature and pressure as state parameters, and therefore the Gibbs free energy $G$ will be considered in this section.

Considering the system at the minimum Gibbs free energy, i.e., writing $(dG)_{p,T} = 0$, allows us to obtain mass action laws from which the equilibrium composition can be calculated. We will do this first to illustrate the behavior of a plasma gas in CTE. However, when plasmas are used to heat complex mixtures, especially those including condensed phases, computational problems for which elegant solutions have been found with minimization methods may arise.

### 6.3.1. Equilibrium Relationships

Assuming that the plasma contains $K$ species but no condensed phases, let $N_i$ equal the number of particles of species $i$ at temperature $T$ and pressure $p$. Then the Gibbs free energy $G$ can be written as

$$G(T, p) = \sum_{i=1}^{K} \mu_i N_i + G^0(T, p) \tag{6.29}$$

where $G^0(T, p)$ depends only on $T$ and $p$ and not on composition, and the chemical potential $\mu_i$ is defined by

$$\mu_i = \left(\frac{\partial G}{\partial N_i}\right)_{T,p,N_{j\neq i}} = -kT \ln (Q_i/N_i) + E_{0i}^0 \tag{6.30}$$

where $E_{0i}^0$ is an absolute reference energy that has been introduced to reference the interconnected energy levels of the different chemical species $i$ to the same level.

For an isolated system with given $T$ and $p$, the trend toward an equilibrium state is described by $dG < 0$, and equilibrium is reached when $(dG)_{p,T} = 0$. The derivation of $G$ results in

$$\sum_{i=1}^{K} (\mu_i \, dN_i + N_i \, d\mu_i) = 0 \tag{6.31}$$

and, according to the Gibbs–Duhem relationship [Fowler and Guggenheim (1956)],

$$\sum_i N_i \, d\mu_i = 0 \text{ and therefore } dG = \sum_i \mu_i \, dN_i = 0 \tag{6.32}$$

The problem then is to find the values of $N_i$ that satisfy Eq. (6.32) as well as the constraints imposed by conservation of chemical elements and Dalton's law.

The large number of species present in a high-temperature plasma greatly increases the importance of a notation system. In the following discussion, calculations using nitrogen will be performed to illustrate the problems of a simple notation system (see Section 6.2.1).

When heating $n'^0_{N_2}$ moles of diatomic nitrogen to $T_0$ at $p_0$, a mixture of $N_2$, $N_2^+$, N, $N^+$, $N^{++}$, and electrons is obtained, as expressed by the following chemical balance:

$$n'^0_{N_2} N_2(T_0, p_0) \rightarrow n'_{N_2} N_2 + n'_{N_2^+} N_2^+ + n'_N N + n'_{N^+} N^+ + n'_{N^{++}} N^{++} + n'_e e \quad \text{at } T \text{ and } p \tag{6.33}$$

where $n'_x$ is the mole number $N_x/N_{av}$ of the chemical species x at $T$ and $p$ ($N_{av}$ = Avogadro's number). All the chemical species are combinations of atoms (N) and electrons ($e$), which are the base or the elementary species and are related to the other species through the following reactions:

1. A molecule is made up of two atoms according to the chemical reaction $N_2 \rightleftarrows 2N$.
2. A molecular ion is created according to the reaction $N_2 \rightleftarrows N_2^+ + e$.
3. Atomic ions depend on the reactions $N \rightleftarrows N^+ + e$ and $N^+ \rightleftarrows N^{++} + e$.

#### 6.3.1.1. Conservation of the Elements

Regardless of the plasma's composition at high temperature, the number of moles of elementary species [here identical to the base $(N, e)$] will be conserved.

Starting from one mole of diatomic nitrogen at room temperature (i.e., two moles of nitrogen atoms $n'^{0}_{N_2} = 1$ and $n'^{0}_{N} = 2$) one has

$$n'_N + n'_{N^+} + n'_{N^{++}} + 2n'_{N_2} + 2n'_{N_2^+} = 2 \tag{6.34}$$

At high temperatures, the two moles of N (corresponding to one mole of $N_2$ at room temperature) are shared among $N_2$, $N_2^+$, N, $N^+$, and $N^{++}$.

Since the electrons come from the ions, the plasma is overall electrically neutral:

$$n'_e = n'_{N^+} + 2n'_{N^{++}} + n'_{N_2^+} \tag{6.35}$$

#### 6.3.1.2. Dalton's Law

Dalton's law is expressed by

$$p = \sum_i p_i \tag{6.36}$$

where $p_i$ are the partial pressures of individual species, or, in terms of the total number density $n_T$,

$$n_T = p/(kT) \tag{6.37}$$

### 6.3.2. Law of Mass Action

According to thermodynamics, equilibrium corresponds to $(dG)_{p,T} = 0$. Using our previous example and eliminating $N_2^+$ and $N^{++}$ ($n'_{N_2^+}$ is below $10^{-4}$ and $n'_{N^{++}}$ is negligible for $T < 18{,}000$ K), we have

$$dG = \mu_e \cdot dN_e + \mu_N \cdot dN_N + \mu_{N^+} \cdot dN_{N^+} + \mu_{N_2} \cdot dN_{N_2} \tag{6.38}$$

The N atoms are derived from the dissociation of $N_2$ and are partially transformed into $N^+$ by ionization ($N^+ = N - e$). Thus $dN_N$ can be split into two terms: $dN_{N,1}$ for the generation of N by dissociation of $N_2$ and $dN_{N,2}$ for its destruction by ionization; $dG$ can then be written as

$$dG = \mu_e \cdot dN_e + \mu_N \cdot dN_{N,2} + \mu_{N^+} \cdot dN_{N^+} + \mu_{N_2} \cdot dN_{N_2} + \mu_N \cdot dN_{N,1} \tag{6.39}$$

For the dissociation reaction van't Hoff's law gives

$$-\frac{dN_{N_2}}{1} = \frac{dN_{N,1}}{2} = d\eta_N \tag{6.40}$$

The rate of production of N ($d\eta_N$) through the reaction $N_2 \leftrightarrow 2N$ is proportional to the stoichiometric coefficients of the reaction and has a positive sign for the species produced (N) and a negative sign for the species dissociated ($N_2$).

Similarly, for ionization we have

$$-\frac{dN_{N,2}}{1} = \frac{dN_{N^+}}{1} = \frac{dN_e}{1} = d\eta_{N^+} \tag{6.41}$$

Thus Eq. (6.39) becomes

$$dG = (\mu_e + \mu_{N^+} - \mu_N) \cdot d\eta_{N^+} + (2\mu_N - \mu_{N_2}) \cdot d\eta_N \tag{6.42}$$

Equation (6.42) yields the following two expressions for equilibrium conditions:

$$\mu_e + \mu_{N^+} - \mu_N = 0 \tag{6.43}$$

and

$$2\mu_N - \mu_{N_2} = 0 \tag{6.44}$$

To derive the mass action law from Eqs. (6.43) and (6.44) one starts by replacing $V/N_i$ by $kT/p_i$ and by writing $Q_i/N_i$ as

$$Q_i/N_i = kTQ_i^{\text{int}}(T)\Lambda_i^{-3}p_i^{-1} \tag{6.45}$$

where $\Lambda_i$ was introduced earlier [see Eq. (6.5)] and depends only on $T \cdot Q_i^{\text{int}}$, which depends on $T$ and $p$ when the limitation theory of Gurvich (or any other one related to ionized species density) is used [see Eq. (6.23)]. Note that the partial pressure $p_i$ is given by

$$p_i = \frac{N_i}{N_T} \cdot p \tag{6.46}$$

with

$$N_T = \sum_{i=1}^{K} N_i \qquad \left(\sum_{i=1}^{K} p_i = p\right) \tag{6.47}$$

Equation (6.30) can be written as

$$\mu_i = \mu_i^0(T) + kT \ln p_i \tag{6.48}$$

with

$$\mu_i^0 = -kT \ln [\Lambda_i^{-3} kTQ_i^{\text{int}}(T)] + E_{0i}^0 \tag{6.49}$$

The superscript 0 indicates that $\mu_i^0$ depends only on temperature.

From Eqs. (6.43), (6.44), and (6.48) it follows that

$$K_p(N) = \frac{p_N^2}{p_{N_2}} = \exp\left[-\frac{1}{kT}(2\mu_N^0 - \mu_{N_2}^0)\right] \tag{6.50}$$

and

$$K_p(\mathrm{N}^+) = \frac{p_{\mathrm{N}^+} \cdot p_{e^-}}{p_{\mathrm{N}}} = \exp\left[-\frac{1}{kT}(\mu^0_{\mathrm{N}^+} + \mu^0_e - \mu^0_{\mathrm{N}})\right] \quad (6.51)$$

where $K_p(\mathrm{N})$ and $K_p(\mathrm{N}^+)$ are respectively the partial pressure equilibrium constants for dissociation and ionization.

These equilibrium constants can also be calculated for the particle densities, $n_i = N_i/V$. By using Eq. (6.46) we obtain

$$K_p(\mathrm{N}) = \frac{(N_{\mathrm{N}}/N_T)^2}{N_{\mathrm{N}_2}/N_T} \cdot p = \frac{(N_{\mathrm{N}})^2}{N_{\mathrm{N}_2}} \frac{p}{N_T} = \frac{(N_{\mathrm{N}})^2}{N_{\mathrm{N}_2}} \cdot \frac{kT}{V}$$

or

$$K_p(\mathrm{N}) = \frac{n^2_{\mathrm{N}}}{n_{\mathrm{N}_2}} \cdot kT = K_c(N) \cdot kT \quad (6.52)$$

If $\mu^0_i$ is replaced in both Eq. (6.50) and Eq. (6.51) by the expression given in Eq. (6.49), the equilibrium constant for dissociation becomes

$$K_p(\mathrm{N}) = \frac{(Q^{\mathrm{int}}_{\mathrm{N}} \cdot \Lambda^{-3}_{\mathrm{N}} \cdot kT)^2}{(Q^{\mathrm{int}} \cdot \Lambda^{-3}_{\mathrm{N}_2} \cdot kT)} \exp\left(-\frac{2E^0_{0\mathrm{N}} - \mathrm{E}^0_{0\mathrm{N}_2}}{kT}\right) \quad (6.53)$$

In the exponential term $2E^0_{0\mathrm{N}} - E^0_{0\mathrm{N}_2}$ represents the formation energy of N (in other words, the dissociation energy of $N_2$). By considering a typical Morse potential such as the one shown in Fig. 2.11 this energy can be given by

$$2E^0_{0\mathrm{N}} - E^0_{0\mathrm{N}_2} = D_e - hv/2 \quad (6.54)$$

where $D_e$ is the energy difference between the equilibrium position of the nuclei ($r = r_e$) and the free atoms ($r = \infty$), while $\frac{1}{2}hv = \frac{1}{2}\omega_e$ is the vibrational energy of the molecule at rest (see Section 2.4.1).

For ionization, with $m_{\mathrm{N}} \approx m_{\mathrm{N}^+}$,

$$K_p(\mathrm{N}^+) = 2kT(Q^{\mathrm{int}}_{\mathrm{N}^+}/Q^{\mathrm{int}}_{\mathrm{N}})\Lambda^{-3}_e \exp\left(-\frac{E^0_{0\mathrm{N}^+} + E^0_{0e} - E^0_{0\mathrm{N}}}{kT}\right) \quad (6.55)$$

The number 2 in Eq. (6.55) corresponds to the internal partition function of the electrons, and $E^0_{0\mathrm{N}^+} + E^0_{0e} - E^0_{0\mathrm{N}}$ is the ionization energy for nitrogen atoms. To account for the interactions between charged particles in the plasma, the ionization energy must be corrected by its lowering, $\delta E_{N^+}$ (see Sections 6.2.4.2 and 4.4.2). Thus, $E^0_{0\mathrm{N}^+} + E^0_{0e} - E^0_{0\mathrm{N}} = E_{\mathrm{N}^+} - \delta E_{\mathrm{N}^+}$.

The ionization energy $E_I$ for different gases is given in tables (see, for example, Table 2.3). Its lowering, $\delta E_{N^+}$ is calculated on the basis of Coulomb interactions between charged particles in the Debye–Hückel model [see Mayer and Mayer (1966)]. This model gives

$$\delta E_{N^+} = \frac{e^2}{4\pi\varepsilon_0} \cdot \frac{e}{(\varepsilon_0 k)^{1/2}} \cdot [(n_e + n_{N^+})/T]^{1/2} \tag{6.56}$$

and

$$\delta E_{N^+} = 3.34324 \times 10^{-28}[(n_e + n_{N^+})/T]^{1/2} \tag{6.57}$$

If multiply charged ions such as $N^{++}$ and $N^{+++}$ are involved, their densities are not merely added to $n_e$ and $n_{N^+}$ but are instead multiplied by the square of their effective charge: $n_e + n_{N^+} + 4n_{N^{++}} + 9n_{N^{+++}}$.

Equation (6.55) is the Saha–Eggert equation (see Section 4.4.4; here it is deduced from thermodynamic principles).

Notice that Eq. (6.55) is the same as Eq. (4.108) when $K_p(N^+)$ is divided by $kT$ to obtain

$$K_c = \frac{n_{N^+} \cdot n_e}{n_N}$$

### 6.3.3. Calculation of the Plasma Composition

#### 6.3.3.1. NASA Method Based on Equilibrium Constants

The composition of a plasma at a particular $T$ and $p$ can be found from the solution of Eq. (6.32) [i.e., $(dG)_{p,T} = 0$] together with the conservation of the chemical elements [Eq. (6.34)] and the principle of electrical neutrality [Eq. (6.35)]. Since plasmas can contain up to 40 species, it is not possible to solve such a system of nonlinear equations in closed form, so a computer-based numerical method must be used. A review of these numerical methods can be found in the monograph of Storey and Van Zeggeren[11] as well as in the thesis of Bourdin.[12]

Brinckley's method[13,14] is probably the oldest one and is the origin of the rationalization treatment of complex chemical equilibria. It consists of solving the system of Eqs. (6.34), (6.35), (6.36), (6.50), and (6.51) with methods that lead to equations similar to those obtained by a more applicable method. An example of such a method is the NASA method developed by Huff *et al.*[15] which will be briefly described here.

It is important to note that derivatives are calculated in these methods and, in the process, problems may arise from the nonlinear

Debye corrections and the Virial corrections. This difficulty is overcome by calculating the corrections at step $k + 1$ from the densities obtained at step $k$, and the corrections are assumed to be constant for all derivatives to simplify the formalism.[16,17]

The first step of the NASA method of Huff *et al.*[15] is to provide, for a given temperature and pressure, an initial guess of the number of each type of particle (in our example, electrons, nitrogen atoms, and ions, and nitrogen molecules) and the total number of particles $N_T$ (where $N_T = \sum_{i=1}^{K} N_i$). Naturally, it is highly unlikely that these guessed values will satisfy the mass action laws, namely Eqs. (6.50) and (6.51) for the equilibrium constants, the conservation equations (6.34) and (6.35), and Dalton's law expressed by Eq. (6.36), so the method next provides the means for improving on the initial guess.

Let $\Delta g$ represent the differences in the mass action laws. These differences are expressed in logarithmic coordinates to account for the steep variations in composition. When a new species is appearing or disappearing, the composition can change by orders of magnitude for a change in temperature of just a few hundred degrees. If we start by expressing the equilibrium constants given by Eqs. (6.50 and 6.51) as functions of the particle numbers, then for dissociation

$$\Delta g(\mathrm{N}) = 2 \ln \frac{N_{\mathrm{N}}}{N} - \ln \frac{N_{\mathrm{N}_2}}{N_T} + \ln p - \ln K_p(\mathrm{N}) \tag{6.58}$$

and for ionization

$$\Delta g(\mathrm{N}^+) = \ln \frac{N_{\mathrm{N}}^+}{N_T} + \ln \frac{N_e}{N_T} - \ln \frac{N_{\mathrm{N}}}{N_T} - \ln p - \ln K_p(\mathrm{N}^+) \tag{6.59}$$

where $N_{\mathrm{N}}$, $N_{\mathrm{N}^+}$, $N_e$, $N_{\mathrm{N}_2}$, and $N_T$ are the guessed values and $K_p$ the theoretical values for a given $T$ and $p$.

Let $\Delta a$ represent the differences for the conservation laws [Eqs. (6.34) and (6.35)] for the elements. Here the laws are expressed using the particle number divided by $N_T$.

For nitrogen conservation we have

$$\Delta a(\mathrm{N}) = \frac{N_{\mathrm{N}}}{N_T} + \frac{2N_{\mathrm{N}_2}}{N_T} + \frac{N_{\mathrm{N}^+}}{N_T} - \frac{2N_{\mathrm{av}}}{N_T} \tag{6.60}$$

and for electrical neutrality

$$\Delta a(e) = \frac{N_e}{N_T} - \frac{N_{\mathrm{N}^+}}{N_T} \tag{6.61}$$

Finally, $\Delta p$ represents the difference between the real pressure $p$ and the pressure obtained from the summation of partial pressures that were deduced from the guessed values:

$$\Delta p = \left(\frac{N_e}{N_T} + \frac{N_{\mathrm{N}}}{N_T} + \frac{N_{\mathrm{N}^+}}{N_T} + \frac{N_{\mathrm{N}_2}}{N_T}\right) \cdot p - p \tag{6.62}$$

Equilibrium will be reached when $\Delta g = \Delta a = \Delta p = 0$. The most classical solution of Eqs. (6.58) through (6.62) consists of first-order linearization of the equations to find an initial set of particle-number corrections. These corrections then give new values of $\Delta g$, $\Delta a$, and $\Delta p$. This iterative procedure is stopped when the desired accuracy is reached. Finally, the partial derivatives are calculated and the system $F(X) = 0$ is assumed to be equivalent in the first order to

$$-F(X) = J(X) \cdot dX \tag{6.63}$$

where $J(X)$ is the Jacobian of $F$ and $dX$ the corrections.

However, instead of using linear variations of the particle numbers for the species $i$, logarithmic corrections may be used. For example, for $\Delta g$, these corrections would be

$$-\Delta g(\mathrm{N}) = 2d(\ln N_{\mathrm{N}}) - d(\ln N_{\mathrm{N}_2}) - d(\ln N_T) \tag{6.64}$$

and

$$-\Delta g(\mathrm{N}^+) = d(\ln N_{\mathrm{N}}^+) + d(\ln N_e) - d(\ln N_{\mathrm{N}}) - d(\ln N_T) \tag{6.65}$$

It is then possible to express $d(\ln N_{\mathrm{N}_2})$ and $d(\ln N_{\mathrm{N}^+})$ as functions of $d(\ln N_{\mathrm{N}})$ and $d(\ln N_e)$ by using Eqs. (6.64) and (6.65). Quantities $N_{\mathrm{N}}$ and $N_e$, as already stated, are the fundamental free species through which the other species (bound species) can be expressed. By introducing the expressions for $d(\ln N_{\mathrm{N}_2})$ and $d(\ln N_{\mathrm{N}^+})$ into similar expressions obtained using $\Delta a$ and $\Delta p$, a system of linear equations is obtained that allows the calculation of $d(\ln N_{\mathrm{N}})$ and $d(\ln N_{\mathrm{e}})$ as functions of the numerical coefficients and of $\Delta g$, $\Delta a$, and $\Delta p$. The values of $d(\ln N_{\mathrm{N}})$ and $d(\ln N_{\mathrm{e}})$ also allow one to calculate new particle-number values $[\ln N_{\mathrm{x}} + d(\ln N_{\mathrm{x}})]$, which are used to make corrections $\Delta g$, $\Delta a$, and $\Delta p$. This iterative process is repeated until convergence. Once the particle numbers are known, it is very easy to calculate the plasma volume (using, for example, the perfect gas relationship: $pV = N_T kT$) and finally the density of the various species.

#### 6.3.3.2. Optimization Methods

The main advantage of optimization methods [minimization of $G$ taking into account the constraints given by Eqs. (6.34) to (6.36)], such as the one of White, Johnson, and Dantzig,[16] is that they no longer make it necessary to define the free and bound species. Unfortunately, they have the disadvantage of requiring the calculation of all the unknowns, including the Lagrange multipliers. However, the fact that no assumptions about the reaction routes are required and that the variance rule is automatically satisfied is really a big advantage in multiphase plasmas, where mass action laws sometimes imply artificial elimination of certain phases to verify the variance rule.[17,18] Moreover, in spite of the introduction of Lagrange multipliers, the convergence formalism is more efficient than that of the nonlinear system obtained with mass action laws. For a detailed description of the methods applied to plasmas see, for example, Refs. 19–21.

#### 6.3.3.3. Data Base

The Gibbs free energies and/or the chemical potentials are calculated either by using standard thermodynamic tables[22–25] or through partition functions. However, when using either the tables or partition functions with complex mixtures, the volume of data required becomes very extensive and it is better to have the data in the form of polynomials.

Some tables, such as those by NASA[24] or Barin and Knacke,[25] give the data in polynomial form; otherwise it is necessary to determine the corresponding polynomial by fitting the data using a least-squares method[26] to functions of the following type:

$$h(x) = r(x)\left[b + p(x)\int g(x)h_0(x)\,dx\right] \tag{6.66}$$

For example, the relationship between the entropy $S = h(T)$ and the specific heat $c_p = h_0(T)$ corresponds to

$$g(x) = 1/T, \qquad p(x) = 1, \qquad b = 0, \qquad r(x) = 1$$

resulting in

$$S(T) - S(T_0) = \int_{T_0}^{T} \frac{c_p(T)}{T}\,dT \tag{6.67}$$

It has also been shown that expressions of the type

$$\frac{c_p}{R} = \frac{a_1}{T^2} + \frac{a_2}{T} + a_3 + a_4T + a_5T^2 + a_6T^3 + a_7T^4 \tag{6.68}$$

lead to quite satisfactory results.[12]

Unfortunately, all these expressions are valid *only for a given temperature range.* In most tables, data for thermodynamic functions are limited to temperatures up to 5000 or 6000 K, and it is therefore necessary to carry out the calculations through the partition functions (see Section 6.2.4). When the partition functions have been calculated, it is possible to fit them with polynomial functions of temperature such as[12,24]

$$Q_{\text{int}} = a_1 + a_2T + a_3T^2 + a_4T^3 + a_5T^4 + a_6T^5 + a_7T^6 \quad (6.69)$$

The differentiation of such polynomials is much more expedient for the calculation of $\partial \ln Q_i/\partial T$ and $\partial^2 \ln Q_i/\partial T^2$ than any direct calculation would be (such values are necessary to determine enthalpy and specific heat, respectively). Results obtained from the derivatives of these polynomials are generally in good agreement with the values obtained from direct calculation (within 3%).

Another important problem is the accuracy of the partition function values and their derivatives, both of which are very sensitive to the limitation theories used for their calculation (see Section 6.2.4.2).[4] However, when the same limitation theory is used for all components of a plasma, its composition and thermodynamic properties are not very different (less than 4% for the major species) from those calculated using another limitation theory.[3,4] This is due to the fact that when limitation theories play an important role (for example, at $T > 13{,}000$ K for an argon atom), the density of the considered species is already low (see Fig. 6.2 below). It should be pointed out, however, that the use of different limitation theories for the different plasma components could introduce rather significant errors (up to 30%) in the composition and thermodynamic properties, It is therefore important when using tables at high temperatures ($T > 8000$ K) to employ data from only one source, if possible.

### 6.3.3.4. Composition of Simple Plasma Gases

First we will discuss the gases most commonly used in plasma generators: argon, nitrogen, hydrogen, helium, and oxygen (see Fauchais *et al.*,[2–4] Drellishak,[8] Capitelli *et al.*,[9] and Pateyron *et al.*[17,18,27]).

Argon and helium are the simplest. For argon at atmospheric pressure and temperatures below 35,000 K, the following species must be taken into account:

$$1 \text{ mole of Ar } (T_0, p_0) \rightarrow n'_{\text{Ar}}\text{Ar} + n'_{\text{Ar}^+}\text{Ar}^+ + n'_{\text{Ar}^{++}}\text{Ar}^{++} + n'_{\text{Ar}^{+++}}\text{Ar}^{+++} + n'_e e^- \qquad (T, p)$$

Three equilibrium constants corresponding to the three successive

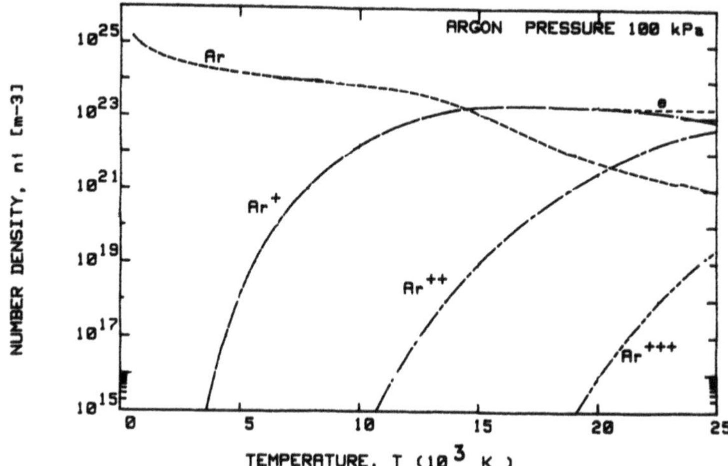


**FIG. 6.2.** Temperature dependence of the composition (species number densities) of an argon plasma at atmospheric pressure (starting from one mole of Ar at room temperature).[8]

ionization steps, $Ar \rightleftarrows Ar^+ + e$, $Ar^+ \rightleftarrows Ar^{++} + e$, and $Ar^{++} \rightleftarrows Ar^{+++} + e$, are necessary.

Figure 6.2 shows the temperature dependence of the argon plasma equilibrium composition at atmospheric pressure, starting from room temperature. As the temperature increases, the particle density ($N_{Ar}/V$) of argon atoms decreases monotonically due to progressive ionization, which is completed at about 15,000 K. At still higher temperatures, the density of $Ar^+$ decreases steadily with the appearance of $Ar^{++}$, while the electron density remains almost constant. The total particle density $n_T$ decreases with temperature for a given pressure ($n_T = p/kT$ in the ideal case). If the calculations are performed at higher temperatures, they show that the density of $Ar^{+++}$ starts to rise over $10^{15}$ at 19,000 K.

Figure 6.2 illustrates a very important point: the steep variation in the particle densities with temperature. For example, $n_e$ varies over five orders of magnitude between 4000 and 10,000 K while remaining almost constant for $T > 15{,}000$ K.

It is also important to note that for $T < 15{,}000$ K, Ar atoms and $Ar^+$ ions are the only heavy species present. $Ar^{++}$ and $Ar^{+++}$ are almost negligible. Furthermore, between 20,000 K and 30,000 K only $Ar^+$ and $Ar^{++}$ need to be considered. Such steep variations will, as discussed previously, lead to considerable calculational problems related to the choice of the species through which the other species are expressed. For example, the choice of Ar and $e$ as free species is quite adequate for

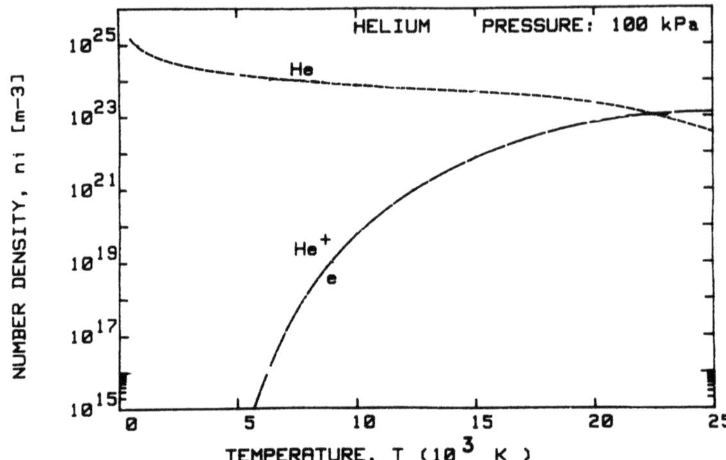


**FIG. 6.3.** Temperature dependence of the composition (species number densities) of a helium plasma at atmospheric pressure (starting from one mole of He at room temperature).[27]

$T < 15{,}000$ K. However, for $T > 15{,}000$ K both the precision and the iterative convergence will be poor because species such as $Ar^+$ and $Ar^{++}$ will be calculated from the argon species, which has a negligible concentration. The most recent computer codes include an automatic change of the base species at different temperatures and pressures.

The results obtained for helium (see Fig. 6.3) are similar to those for argon except that the maximum for $He^+$ is attained at about 25,000 K instead of the 15,000 K found for argon due to the higher ionization energy of helium (24.6 eV, compared to 15.7 eV for argon; see Table 2.3). $He^+$ density becomes higher than that of atomic He at a temperature above 22,500 K.

The results for nitrogen are also similar to those for argon (see Fig. 6.4) except that nitrogen must first dissociate. N atoms reach a maximum concentration at about 7500 K followed by a decrease in concentration at higher temperatures due to the formation of $N^+$ ions. Note that the ionization energy of nitrogen (15.5 eV) is comparable to that of argon and the maximum for $N^+$ is observed at about the same temperature as the maximum for $Ar^+$. The $N^{++}$ molar fraction is only 0.1 at 25,000 K. Similar results are also obtained for oxygen (Fig. 6.5) and hydrogen (Fig. 6.6). The maxima for $O^+$ and $H^+$ are likewise close to 15,000 K, indicating similar ionization energies. The dissociation of hydrogen, however, starts at a lower temperature than that for oxygen, while both are lower than that of nitrogen. The maximum particle densities for H, O, and N at about 3800 K, 4300 K, and 7500 K, respectively, are in direct

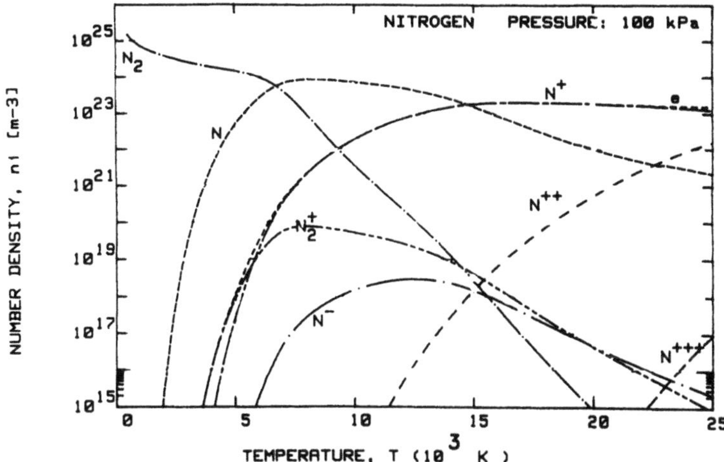


**FIG. 6.4.** Temperature dependence of the composition (species number densities) of a nitrogen plasma at atmospheric pressure (starting from one mole of $N_2$ at room temperature).[8]

correlation with the corresponding dissociation energies of their molecules (4.48 eV, 5.08 eV, and 9.786 eV). The $N^-$ and $O^-$ densities are practically negligible.

Figure 6.7, based on data gathered by Lesinski and Boulos[28] and calculated by Pateyron *et al.*,[27] shows the temperature dependence of the

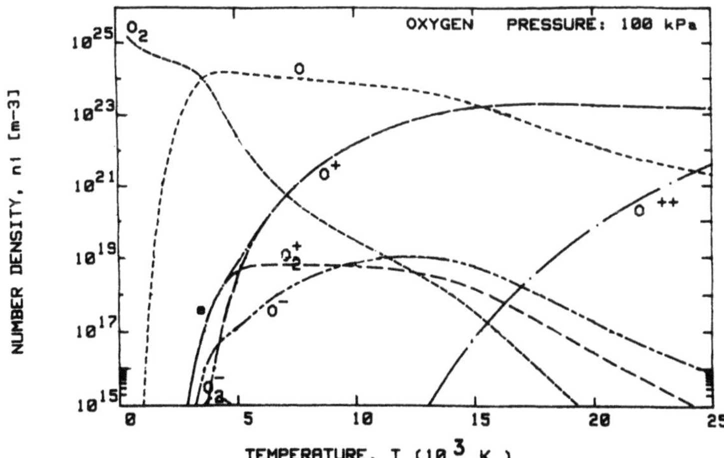


**FIG. 6.5.** Temperature dependence of the composition (species number densities) of an oxygen plasma at atmospheric pressure (starting from one mole of $O_2$ at room temperature).[27]

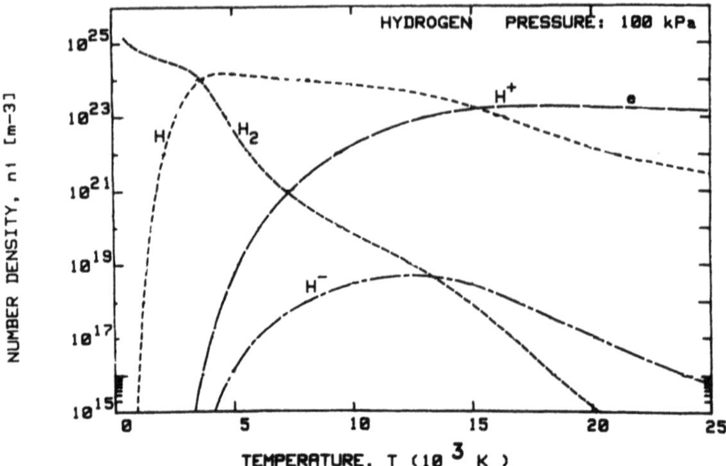


FIG. 6.6. Temperature dependence of the composition (species number densities) of a hydrogen plasma at atmospheric pressure (starting from one mole of $H_2$ at room temperature).[27]

specific volume ($1/\rho$) of various plasma gases. This figure illustrates a very important point about hot plasmas: a gas has a lower specific mass at 8000–10,000 K than at 1000–2000 K. For example, $\rho_{2000}/\rho_{10000}$ is 5.1 for argon, 10.2 for nitrogen, and 10.4 for oxygen. These temperature variations also mean that the momentum of the cold gas flowing around the hot plasma core in the nozzle of a spray jet is not negligible, even though the plasma velocity is high (about 1500 m/s for nitrogen). The data used for Fig. 6.7 also show very good agreement among different authors.

### 6.3.3.5. Composition of Complex Mixtures

*6.3.3.5.a. Air Plasma.* An air plasma is more complex than the plasma of any single gas, and if the calculation is limited to 15,000 K close to atmospheric pressure, the following species are pertinent:

$$e, N, O, Ar, N^+, O^+, Ar^+, N_2, N_2^+, O_2, O_2^+, NO, NO^+, NO_2, N_2O$$

Here, air is considered to be a mixture of nitrogen, oxygen, and argon at room temperature, with all other species neglected. Figure 6.8 shows the species densities versus temperature at atmospheric pressure. For an air plasma (dry air with no water vapor), the main ions are $NO^+$ below 6000 K, while $N^+$ and $O^+$ are dominant above 9000 K.

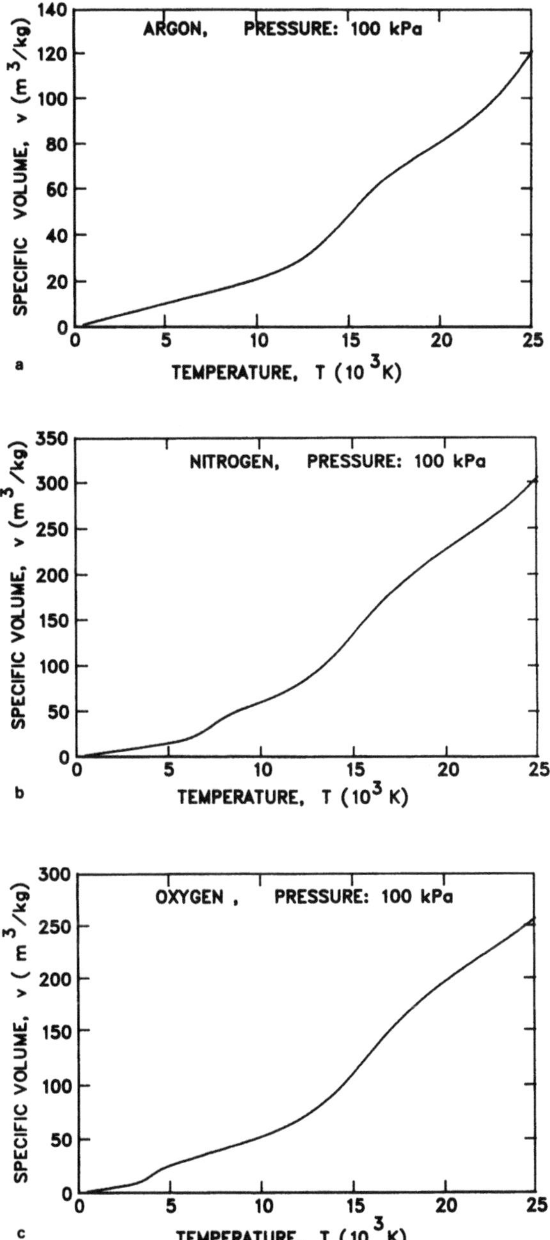


**FIG. 6.7.** Temperature dependence at atmospheric pressure of the specific volume of (a) argon,[28] (b) nitrogen,[28] (c) oxygen,[28] (d) hydrogen,[27] and (e) helium.[27]

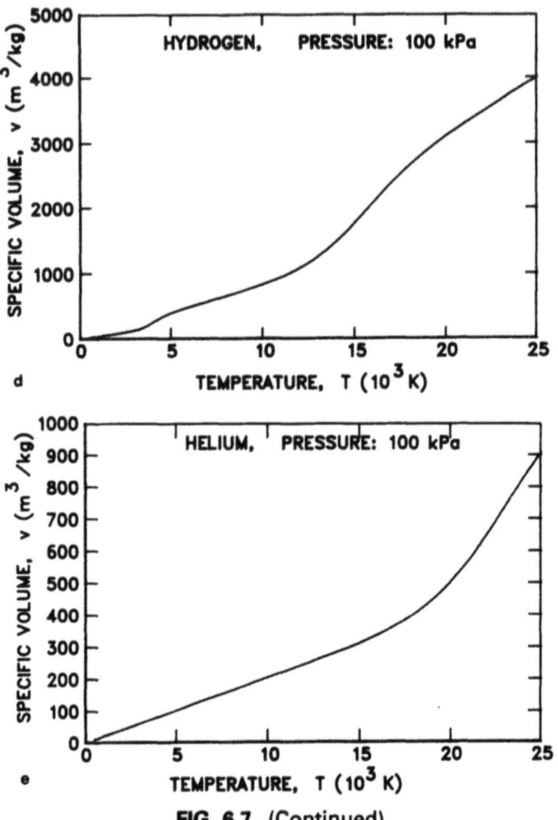


FIG. 6.7. (Continued).

Figure 6.9 also shows the species densities versus temperature, but in this figure the pressure is 500 kPa. It can be seen that dissociation of $N_2$ and $O_2$ occurs at higher temperatures (approximately 2000 K higher). The obtained NO has a higher density, but this occurs at a temperature higher by a few hundred degrees, and naturally all densities are higher. At 20 kPa, dissociation occurs at lower temperatures and all densities are lower (see Fig. 6.10).

*6.3.3.5.b. Ar–$H_2$ Mixture.* This mixture is very common in plasma spraying, especially when good heat transfer is required, as in ceramic spraying, for example. In most cases the volumetric hydrogen percent is between 15% and 30%. Figure 6.11 shows the temperature evolution (at atmospheric pressure) of the composition of an Ar–$H_2$ mixture with a hydrogen volume percent of 20%. Because no chemical reactions occur between Ar and $H_2$, the general tendencies observed with pure gases (see Fig. 6.2 for Ar and Fig. 6.6 for $H_2$) are preserved: $H_2$ dissociation occurs

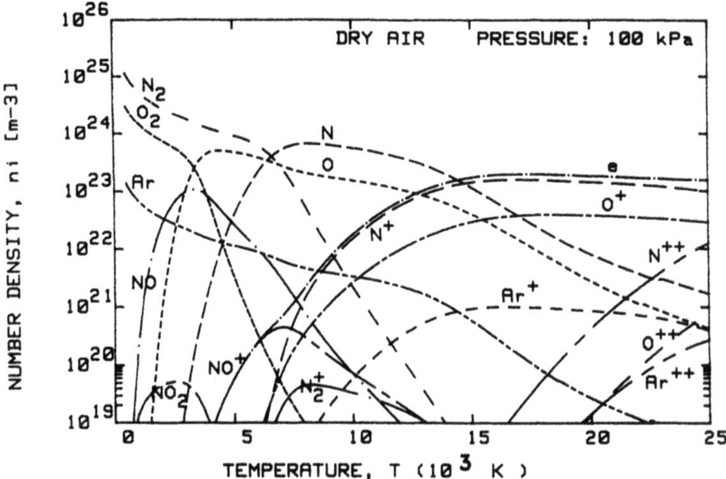


**FIG. 6.8.** Temperature dependence of the composition (species number densities) of an air plasma at atmospheric pressure (starting from one mole of air at room temperature).[27]

at the same temperature as in pure $H_2$, and ions and electrons come from Ar and H species ionizations. Thus the plasma's electrical behavior, which is linked to the density of charged species, will be about the same for pure Ar, pure $H_2$, or any mixture of these gases (up to 60% by volume $H_2$; see Section 7.6.1). However, as will be shown later (see

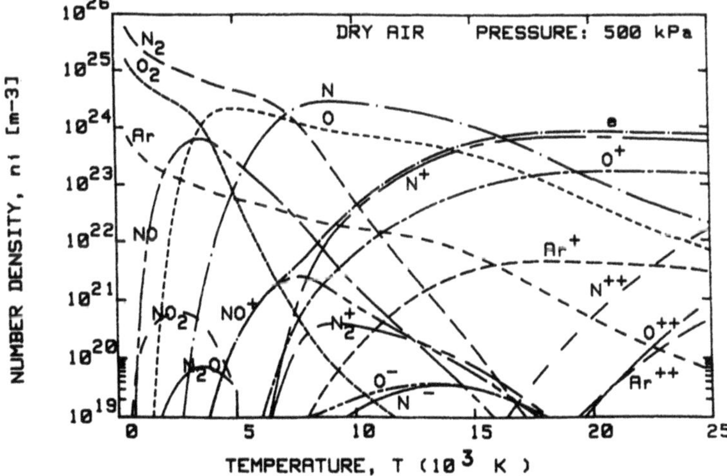


**FIG. 6.9.** Temperature dependence of the composition (species number densities) of an air plasma at five atmospheres (starting from one mole of air at room temperature).[27]

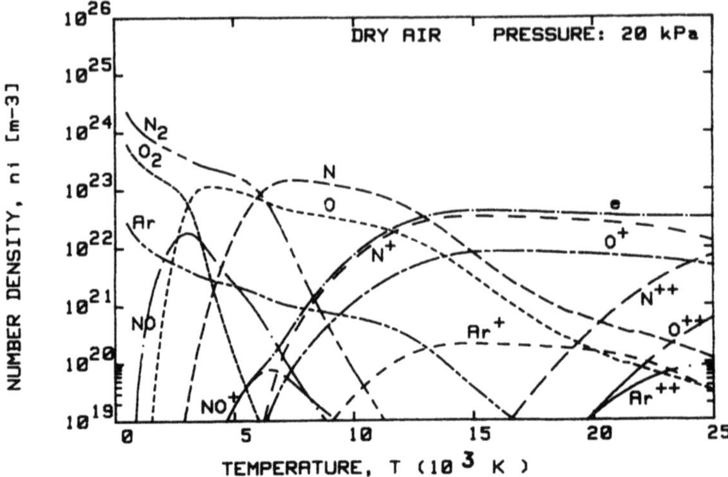


FIG. 6.10. Temperature dependence of the composition (species number densities) of an air plasma at 20 kPa (starting from one mole of air at room temperature).[27]

Section 7.5.2), the thermal conductivity of the mixture will be drastically increased by hydrogen dissociation.

*6.3.3.5.c. Ar–He Mixture.* As in the Ar–$H_2$ mixture, there is no chemical reaction between the two gases (Ar and He). Figure 6.12a shows the temperature evolution of the composition of the mixture

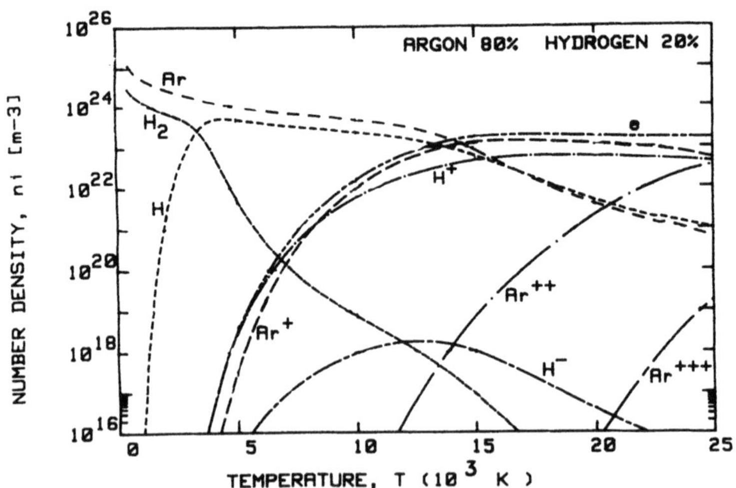


FIG. 6.11. Temperature dependence of the composition (species number densities) of an Ar–$H_2$ (20 vol%) plasma at atmospheric pressure.[29]

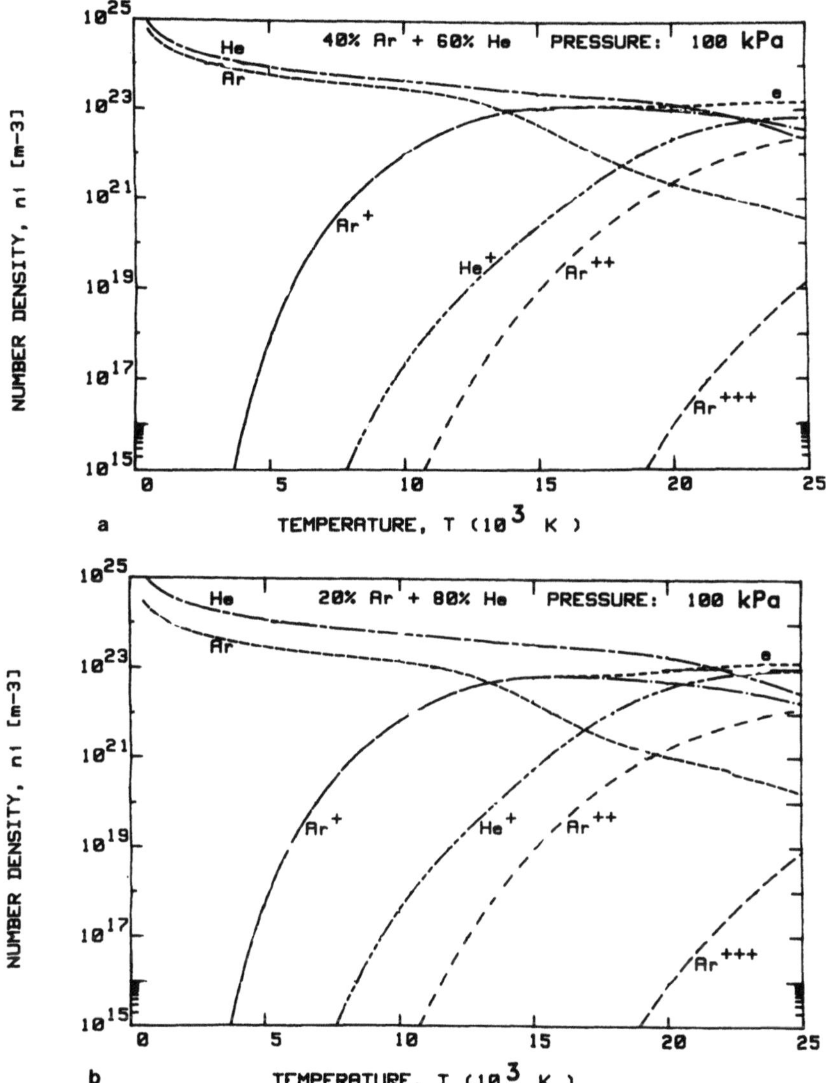


**FIG. 6.12.** Temperature dependence of the composition (species number densities) of atmospheric pressure of an Ar–He plasma: (a) Ar + He (60 vol%), (b) Ar + He (80 vol%).[29]

containing 40% Ar + 60% He by volume and Fig. 6.12b shows that of 20% Ar + 80% He mixture. The general tendencies observed with the two pure gases (see Fig. 6.2 for Ar and Fig. 6.3 for He) are the same with the mixtures. Note that even when the mixture contains 80% He, ions and electrons below 14,000 K come mainly from the Ar species. Thus the electrical behavior of the plasma will be controlled by Ar up to these temperatures. Only for a He volume percent higher than 90% will the plasma behave like a He plasma, which requires temperatures higher than 12,000 K to be self-sustaining (compared with 7000 K for Ar). At high temperature ($T > 20{,}000$ K) $He^+$ density is only 3 to 4 times higher than $Ar^{++}$ density.

## 6.4. THERMODYNAMIC PROPERTIES OF PLASMAS IN CTE

### 6.4.1. Specific Heat at Constant Pressure

By definition,

$$c_p = \left(\frac{\partial h_g}{\partial T}\right)_p \tag{6.70}$$

where $h_g$ is the specific enthalpy defined by

$$h_g = \frac{\sum_{i=1}^{K} x_i H_i}{\sum_{i=1}^{K} x_i M_i} \tag{6.71}$$

and expressed in kJ/kg, with $c_p$ in kJ/kg K.

In Eq. (6.71) $M_i$, $H_i$, and $x_i$ are, respectively, the mass of one mole, the enthalpy of one mole, and the molar fraction $x_i = N_i/N_T$ of the chemical species $i$. Note that this formula assumes that the plasma behaves as a perfect gas with a zero mixing enthalpy.

For example, using a nitrogen mixture containing $e$, N, $N^+$, and $N_2$,

$$h_g = (x_e H_e + x_N H_N + x_{N^+} H_{N^+} + x_{N_2} H_{N_2}) \cdot \frac{1}{m_g} \tag{6.72}$$

with

$$m_g = \sum_{i=1}^{K} x_i M_i \tag{6.73}$$

The total mass of the mixture is given by

$$m_g = x_e M_e + x_N M_N + x_{N^+} M_{N^+} + x_{N_2} M_{N_2} \tag{6.74}$$

Taking the partial derivative of Eq. (6.71) with respect to $T$ gives

$$c_p = \frac{1}{m_g}\left(\sum_{i=1}^{K} x_i c_{pi} + \sum_{i=1}^{K} x_i (H_i - M_i \cdot h_g)\left(\frac{\partial \ln x_i}{\partial T}\right)_p\right) \tag{6.75}$$

which is generally written as

$$c_p = c_p^f + c_p^r \tag{6.76}$$

For nitrogen, $c_p^f$ and $c_p^r$ are respectively given by

$$c_p^f = \frac{1}{m_g}(x_e c_{pe} + x_{\mathrm{N}} c_{p_{\mathrm{N}}} + x_{\mathrm{N}^+} c_{p_{\mathrm{N}_+}} + x_{\mathrm{N}_2} c_{p_{\mathrm{N}_2}}) \tag{6.77}$$

and

$$\begin{aligned} c_p^r = {} & x_e(H_e - M_e h_g)\left(\frac{\partial \ln x_e}{\partial T}\right)_p + x_{\mathrm{N}}(H_{\mathrm{N}} - M_{\mathrm{N}} h_g)\left(\frac{\partial \ln x_{\mathrm{N}}}{\partial T}\right)_p \\ & + x_{\mathrm{N}^+}(H_{\mathrm{N}^+} - M_{\mathrm{N}^+} h_g)\left(\frac{\partial \ln x_{\mathrm{N}^+}}{\partial T}\right)_p \\ & + x_{\mathrm{N}_2}(H_{\mathrm{N}_2} - M_{\mathrm{N}_2} h_g)\left(\frac{\partial \ln x_{\mathrm{N}_2}}{\partial T}\right)_p \end{aligned} \tag{6.78}$$

The first term, $c_{\mathrm{p}}^{\mathrm{f}}$, represents the sum of contributions of all different species $i$ in the system; it is called the frozen specific heat, and it corresponds to a mixture with no chemical reactions. The second term, $c_p^r$, is the reactional specific heat and is associated with the chemical reactions at a given temperature. Its influence will often dominate in a plasma, as will be shown.

Obtaining the frozen term is relatively easy once the composition of the plasma has been calculated, because the $c_{pi}$ are usually given in tables or can be calculated through the second derivative of the partition functions.

Calculation of the reaction term is more complex, since it is necessary to use the laws of conservation in addition to the equilibrium constants when calculating the term $(\partial \ln x_i / \partial T)_p$.

Applying van't Hoff's law to a nitrogen plasma gives

$$\left(\frac{\partial \ln K_p(\mathrm{N})}{\partial T}\right)_p = \frac{2H_{\mathrm{N}} - H_{\mathrm{N}_2}}{RT^2} = \frac{H_{\mathrm{N}}^{\mathrm{D}}}{RT^2} \tag{6.79}$$

and

$$\left(\frac{\partial \ln K_p(\mathrm{N}^+)}{\partial T}\right)_p = \frac{H_{\mathrm{N}^+} + H_e - H_{\mathrm{N}}}{RT^2} = \frac{H_{\mathrm{N}^+}^{\mathrm{I}}}{RT^2} \tag{6.80}$$

where $H_{\mathrm{N}}^{\mathrm{D}}$ and $H_{\mathrm{N}^+}^{\mathrm{I}}$ are, respectively, the molar enthalpy variations (in kJ/mol) for the dissociation and first ionization reactions. In Eq. (6.80), the ionization potential lowering has been neglected.

The mass action laws, Eqs. (6.53) and (6.55), can be written as functions of molar fractions as follows:

$$\ln K_p(\mathrm{N}) = 2 \ln x_{\mathrm{N}} - \ln x_{\mathrm{N}_2} + \ln p \tag{6.81}$$

and

$$\ln K_p(\mathrm{N}^+) = \ln x_{\mathrm{N}^+} + \ln x_e - \ln x_{\mathrm{N}} + \ln p \tag{6.82}$$

Taking the derivative of these equations and substituting van't Hoff's laws [see Eqs. (6.40) and (6.41)] results in

$$\frac{H_{\mathrm{N}}^{\mathrm{D}}}{RT^2} = 2\left(\frac{\partial \ln x_{\mathrm{N}}}{\partial T}\right)_p - \left(\frac{\partial \ln x_{\mathrm{N}_2}}{\partial T}\right)_p \tag{6.83}$$

and

$$\frac{H_{\mathrm{N}^+}^{\mathrm{I}}}{RT^2} = \left(\frac{\partial \ln x_{\mathrm{N}^+}}{\partial T}\right)_p + \left(\frac{\partial \ln x_e}{\partial T}\right)_p - \left(\frac{\partial \ln x_{\mathrm{N}}}{\partial T}\right)_p \tag{6.84}$$

From these equations it follows that $(\partial \ln x_{\mathrm{N}_2}/\partial T)_p$ and $(\partial \ln x_{\mathrm{N}^+}/\partial T)_p$ can be expressed as functions of $(\partial \ln x_{\mathrm{N}}/\partial T)_p$ and $(\partial \ln x_e/\partial T)_p$.

Finally, by using the conservation equation (6.34) for nitrogen and the electrical neutrality equation (6.35), it becomes possible to calculate the derivatives of $x_{\mathrm{N}}$ and $x_e$ as functions of $H_{N^+}^{I}$, $H_{\mathrm{N}}^{\mathrm{D}}$, and the molar fractions, thus obtaining the reactional specific heat.

One of the important characteristics of thermal plasmas is the nonlinearity of the thermodynamic properties with temperature, as is clearly shown by the variation of specific heat with temperature at constant pressure. Figure 6.13 shows a plot of frozen and total specific heat versus temperature for a nitrogen plasma at atmospheric pressure. While the frozen term varies approximately linearly, the total $c_p$ exhibits three peaks corresponding, respectively, to the dissociation at 7600 K, the first ionization at 14,500 K, and the second ionization close to 30,000 K. These peaks show clearly the position of the maxima for dissociation and ionization.

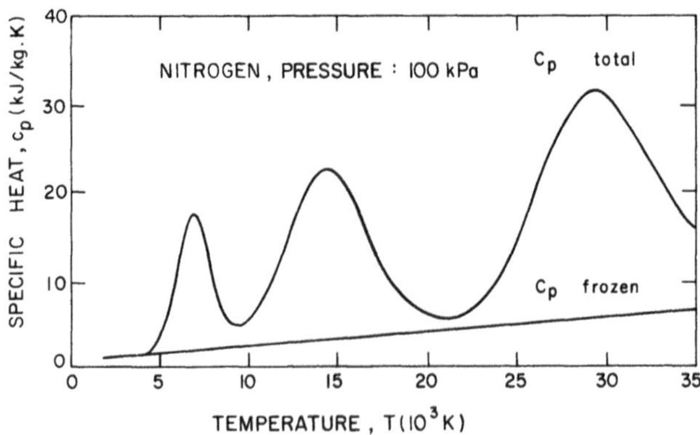


**FIG. 6.13.** Temperature dependence of the specific heat at constant pressure (100 kPa) of a nitrogen plasma.[8]

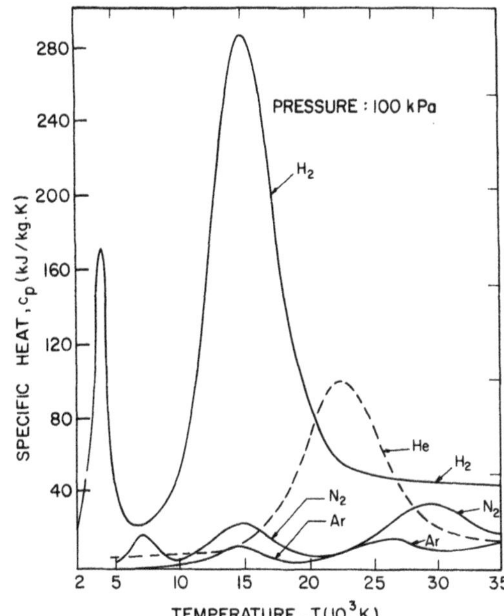


**FIG. 6.14.** Temperature dependence of the specific heat at constant pressure (100 kPa) of various plasmas ($H_2$, $N_2$, Ar, He).[8]

At the maxima, the reactive specific heat is almost an order of magnitude higher than the frozen one. Figure 6.14 shows the total specific heat for Ar, He, $N_2$, and $H_2$. It is clear that the maximum heat capacity is obtained for hydrogen.

Figure 6.15 shows the temperature dependence of the specific heat

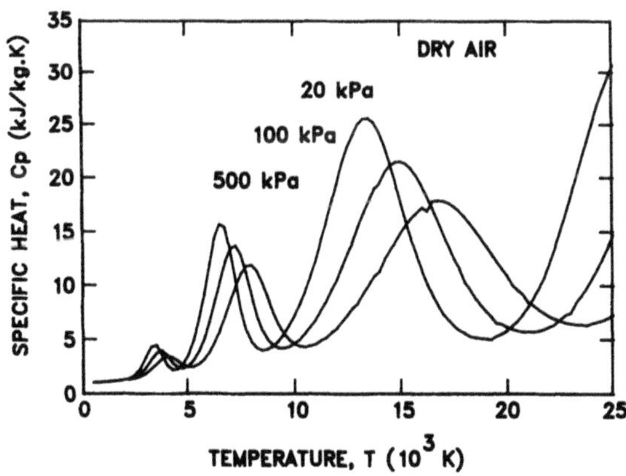


**FIG. 6.15.** Temperature dependence of the specific heat of air at 20, 100, and 500 kPa.[27]

of air at pressures of 500, 100, and 20 kPa. It can be seen that when the pressure increases, the successive peaks corresponding to oxygen dissociation, nitrogen dissociation, and then nitrogen and oxygen ionizations are shifted to higher temperatures and their maximum values reflect the composition of the plasma (see Figs 6.8 to 6.10). The maximum values of the peaks decrease with the increase of the pressures.

### 6.4.2. Enthalpy and Entropy

These functions can be calculated either directly [see Eq. (6.71)] or by integrating over $c_p$,

$$h_g - h_g^0 = \int_0^T c_p(T)\, dT \tag{6.85}$$

where $h_g$ is the total enthalpy of the mixture at $T$ and $p$, while $h_g^0$ is the total enthalpy at the reference state $T = 0\,\text{K}$, $p = 1$ atm. The entropy, $S$, is found from

$$S = \int_0^T \frac{c_p(T)}{T}\, dT \tag{6.86}$$

Integration of $c_p$ is used when $c_p$ data are available from tables. When calculated from partition functions, the enthalpy is determined through Eq. (6.71) and its numerical derivation gives $c_p$ more simply than Eqs. (6.77) and (6.78).

These functions can also be calculated directly by introducing the frozen and reactive enthalpies and entropies, as shown in the following example:

One mole of $N_2$ at $T_0$, $p_0$ gives the following composition at $T$ and $p$:

$$N_2 \rightarrow n'_{N_2}N_2 + n'_N N + n'_{N^+}N^+ + n'_e e^-$$

where $n'$ is the number of moles ($n' = N/N_{av}$).

The species conversation of nitrogen gives

$$2 = 2n'_{N_2} + n'_N + n'_{N^+}$$

and electrical neutrality gives

$$n'_{N^+} = n'_e$$

The only reactions are

- the dissociation $N_2 \rightarrow 2N$ for which the enthalpy change is $2H_N - H_{N_2} = H_N^D$
- the ionization $N \rightarrow N^+ + e^-$ with the change in enthalpy $H_{N^+} + H_{e^-} - H_N = H_{N^+}^I$

The total enthalpy at $T$ and $p$ is then

$$H = n'_{N_2} H_{N_2} + n'_N H_N + n'_{N^+} H_{N^+} + n'_e H_e$$

while at $T_0$ and $p_0$ the total enthalpy is $H_{N_2}^0$.

Thus the enthalpy change to produce the plasma is $\Delta H = H - H_{N_2}^0$.

Using conservation of nitrogen and electrical neutrality, $H$ is written as

$$\begin{aligned}\Delta H &= H_{N_2} - H_{N_2}^0 + n'_N H_N - \frac{n'_N}{2} H_{N_2} + n'_{N^+}\left(H_{N^+} + H_e - \frac{H_{N_2}}{2}\right) \\ &= \Delta H_{N_2} + \frac{n'_N}{2} H_N^D + n'_{N^+}(H_{N^+} + H_e - H_N) + \frac{n'_{N^+}}{2} H_N^D \\ &= \Delta H_{N_2} + \left(\frac{n'_N}{2} + \frac{n'_N}{2}\right) H_N^D + n'_N H_{N^+}^I \qquad (6.87)\end{aligned}$$

The first term represents the frozen enthalpy with no reaction occurring (no dissociation or ionization) while the molecule $N_2$ is heated from $T_0$ to $T$. The two remaining terms correspond to the reaction enthalpies due to dissociation ($H_N^D$) and ionization ($H_{N^+}^I$), respectively.

It is clear from this equation that, because of the existence of the reaction terms, the total enthalpy of a binary mixture is generally not the simple sum of the enthalpies of the two individual gases. Similar calculations can also be performed for the entropy.

Figure 6.16 represents the specific enthalpy (in MJ/kg) as a function of temperature for various plasma gases ($N_2$, $H_2$, $O_2$, Ar, and He) at atmospheric pressure. The steep variations in enthalpy are essentially due to the heats of reaction (dissociation and ionization). The very high enthalpy of $H_2$ is also due to its low mass. At the maximum temperature shown in Fig. 6.16 the ionization of He has not really started yet; thus, in spite of its low mass, its enthalpy is much lower than that of $H_2$ but higher than that of Ar. This figure illustrates the important economics of using plasmas, in which the energy supply is independent of the gas and the temperature is not determined by the chemical reactions (as in flames). Specifically, if an oxygen-fuel flame at 3000 K is used to heat a body to

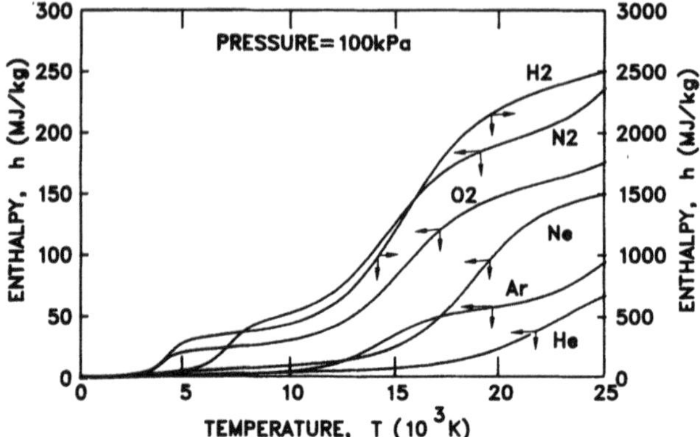


**FIG. 6.16.** Specific enthalpy (MJ/kg) of various gases at atmospheric pressure versus temperature.[27]

2500 K, only 20% of its energy is used, and the rest has to be recovered through, for example, heat exchangers. If a nitrogen plasma at 10,000 K is used for the same purpose, it is possible to recover almost 95% of the available energy in the gas.

The influence of pressure on the specific enthalpy of an air plasma is shown in Fig. 6.17 at 20, 100, and 500 kPa. At a given temperature, the enthalpy of a gas increases as the pressure decreases due to the lower dissociation and ionization temperatures; a great deal of energy is needed to dissociate or ionize a gas, and thus the energy content is higher at temperatures exceeding the dissociation of ionization levels.

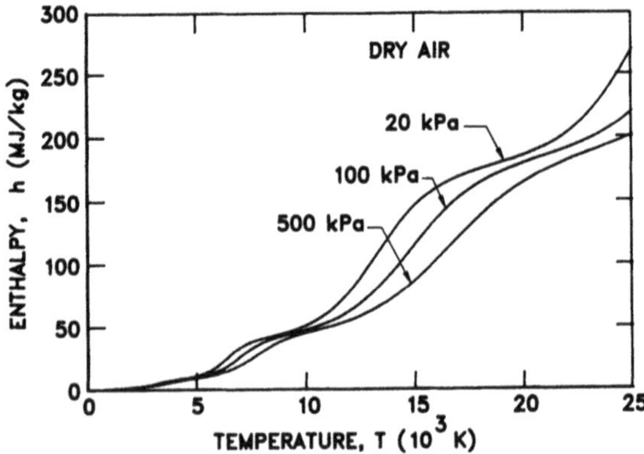


**FIG. 6.17.** Specific enthalpy (MJ/kg) of air at 20, 100, and 500 kPa versus temperature.[27].

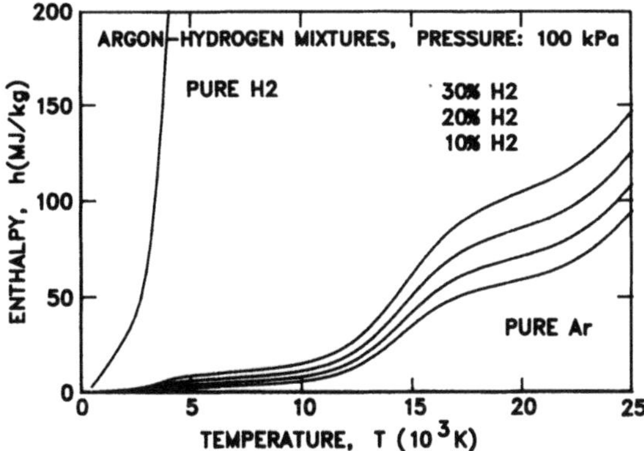


**FIG. 6.18.** Specific enthalpy (MJ/kg) of various Ar–$H_2$ mixtures (vol%) at atmospheric pressure versus temperature.[29]

Figures 6.18 and 6.19 illustrate the enthalpy evolution of various mixtures of Ar and $H_2$, and Ar and He, respectively. It can be seen that the addition of $H_2$ to Ar (see Fig. 6.18) increases the enthalpy of the mixture, especially when dissociation and first ionizations occur, but due to the high mass of argon compared with that of hydrogen, the increase in enthalpy is not as drastic for less than 30 vol% $H_2$ as that obtained with pure $H_2$. The addition of He (see Fig. 6.19) drastically modifies the argon enthalpy for temperatures higher than 17,000 K, but here again the mass of He is small compared with that of Ar, and thus the modification becomes very sensitive for He percents higher than 70%. Compared with a pure argon plasma at 12,000–14,000 K, the enthalpy of the mixtures is doubled with either 80% He or 30% $H_2$.

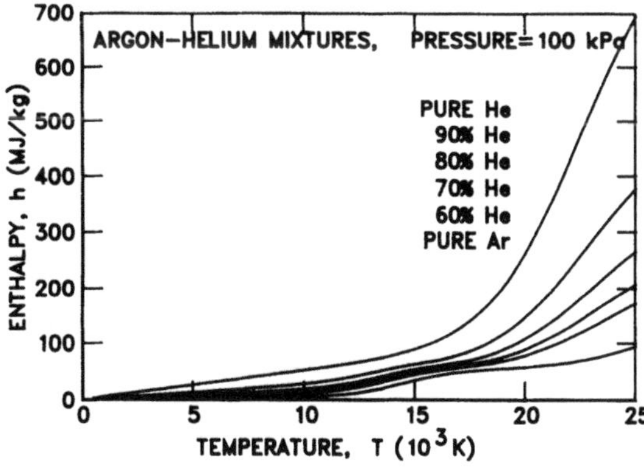


**FIG. 6.19.** Specific enthalpy (MJ/kg) of various Ar–He mixtures (vol%) at atmospheric pressure versus temperature.[29]

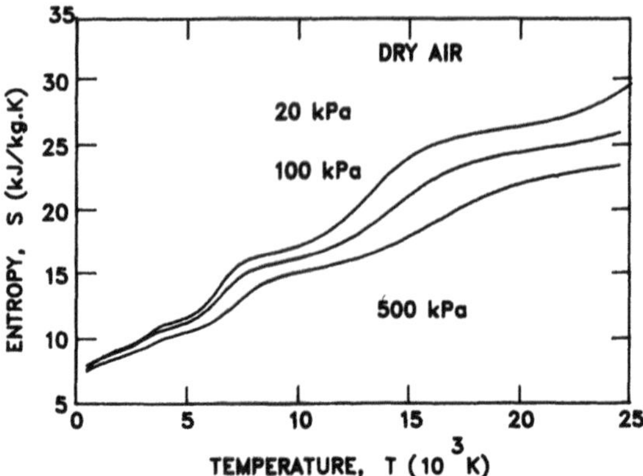


**FIG. 6.20.** Specific entropy (KJ/kgK) of air at 20, 100, and 500 kPa versus temperature.[27]

Figure 6.20 shows the same behavior for the entropy of air. However, in this case (according to the logarithmic dependence of the entropy on pressure) the entropy increase with decreasing pressure is almost independent of dissociation or ionization. The observed steep variations of the entropy are due to dissociation and ionization phenomena.

## 6.5. COMPOSITION AND THERMODYNAMIC PROPERTIES OF A TWO-TEMPERATURE PLASMA

Up to now it has been assumed that ions, neutral particles, and electrons have Maxwellian distributions and that collisions between these particles are numerous enough to allow an equalization of energies between the heavy and light particles. Such a hypothesis leads to a unique temperature (defined as the mean kinetic energy: $\frac{1}{2}m\overline{v_i^2} = \frac{3}{2}kT_i$) for both the heavy and the light particles.

However, if the pressure is lowered or the electric field increased (for example, near the electrodes of arcs at atmospheric pressure) and when large gradients exist (near cold walls or when a cold gas is injected into the plasma), equilibrium will no longer exist. In Section 4.4.4, Eq. (4.122) shows that the electron temperature $T_e$ differs from that of the heavy species $T_h$ when the energy gained by the electrons between two collisions ($l_e eE$) is no longer negligible compared with the thermal energy $kT_e$. At atmospheric pressure in the hot core of the plasma, the difference between $T_e$ and $T_h$ is less than 2% [see Eq. (4.122)] and equilibrium prevails. At lower pressures (~1 kPa), $l_e$ increases drastically, and even in the hot core of the jet $T_e$ will be up to two times higher than $T_h$.

We must also keep in mind that the regions close to walls and electrodes were excluded when Eq. (4.122) was established. In these regions the boundary layers correspond to about $10^2$ mean free paths. The heavy particles reach the wall with almost the same temperature as that of the wall, while the electrons loose only a few percent of their kinetic energy and maintain almost the same temperature as in the equilibrium regions.

Furthermore, equilibrium is not readily attained when a cold gas is injected into the plasma. The energy exchange among heavy particles is very fast, but the thermalization of the electrons with the heavy particles requires rather long distances ($10^3$ to $10^4$ times the mean free path).[33]

### 6.5.1. Composition

Since the pressure $p$ is supposed to be constant, the total pressure can be expressed by

$$p + \delta p = p_1 + \sum_{i=2}^{K} p_i \tag{6.88}$$

Here $\delta p$ is the pressure correction term resulting from the electrostatic interactions [see Section 6.2.3.2 and Eq. (6.15)] that cannot be neglected for nonequilibrium regions, and the index 1 stands for electrons. Equation (6.88) can be rewritten as

$$p + \delta p = n_1 k T_e + \sum_{i=2}^{K} n_i k T_h \tag{6.89}$$

Introducing the parameter $\theta = T_e/T_h$, the latter equation becomes

$$\frac{\theta(p + \delta p)}{kT_e} = \theta n_1 + \sum_{i=2}^{K} n_i \tag{6.90}$$

with

$$\delta p = \frac{kT_e}{24\pi\lambda_D^3} = 2.147 \times 10^{-24}\left(\frac{T_e}{\lambda_D^3}\right) \quad \text{(Pa)} \tag{6.91}$$

where $\lambda_D$ is the Debye length. Equation (6.14) for the Debye length has to be modified by introducing the electron temperature $T_e$ and the electron density along with the heavy-particle temperature for all the heavy species for a nitrogen plasma containing only $N_2$, N, $N^+$, and $e$, it follows that

$$\lambda_D^{-2} = \left(\frac{e^2}{\varepsilon_0 k}\right)\left(\frac{n_e}{T_e} + \frac{n_{N^+}}{T_h}\right) \quad \text{or} \quad \lambda_D^2 = \left(\frac{\varepsilon_0 k T_e}{e^2 n_e}\right)\frac{1}{(1 + T_e/T_h)}$$

since $n_{N^+} = n_e$.

### 6.5.1.1. Assumptions for the Partition Functions

The composition calculation for a two-temperature plasma depends (as done by Potapov[30]) on the Gibbs free enthalpy, which in turn depends on the partition functions. The partition functions are themselves functions of the electron and heavy-particle temperatures. It is commonly assumed[31] that

$$Q_{\text{int}}(T_h, T_e) \approx Q_{\text{int}}(T_e) \tag{6.92}$$

for atoms and their ions, as soon as ionization is pertinent. This assumption seems reasonable since the main contributors to the partition-function calculation at high temperatures are the highly excited levels (whose contribution is weighted by $2n^2$, $n$ being the principal quantum number) where the excitation temperature is close to the electron temperature.

For dissociation reactions (for example, $N_2 \to 2N$) or for recombination reactions ($N + O \rightleftarrows NO$, for example) one has

$$Q_{\text{int}}(T_e, T_h) = Q_{\text{int}}(T_h) \tag{6.93}$$

because such reactions are mainly caused by collisions between heavy particles. The same is true for molecular ions. This assumption can be justified for two reasons; (1) The vibration–rotation levels make the main contribution to $Q_{\text{int}}$ for molecules (with only a few electronically excited states) having temperatures close to the kinetic temperature (the heavy-particle temperature); the relaxation time between rotation and translation is on the order of $10^{-9}$ to $10^{-12}$ s. (2) The molecular dissociation is effective before the electronically excited levels are populated enough to contribute significantly to $Q_{\text{int}}$.

The chemical potentials [see Eqs. (6.48) and (6.49)] are then written for the electrons and heavy particles as

$$\mu_1 = \mu_1^0 + kT_e \ln (p_1/p) \tag{6.94}$$

and

$$\mu_i = \mu_i^0 + kT_h \ln (p_i/p) \tag{6.95}$$

Thus, replacing $\mu_1^0$ [see Eq. (6.49)] with the corresponding values of the partition functions[30–32] and taking species conservation [Eqs. (6.34) and (6.35)] into account, it follows that

$$\mu_1 = -kT_e \ln \left[ \frac{kT_e}{p} \left( \frac{2\pi m_e kT_e}{h^2} \right)^{3/2} \right] - kT_e \ln 2 + kT_e \ln \left( \frac{p_1}{p} \right) + E_{01}^0 \tag{6.96}$$

In Eq. (6.96) ln 2 stands for $\ln Q_e$ ($Q_e = 2$ for electrons). For atoms and their ions,

$$\mu_i = -kT_h \ln\left[\frac{kT_h}{p}\left(\frac{2\pi m_i kT_h}{h^2}\right)^{3/2}\right] - kT_h \ln Q^i_{\text{int}}(T_h) + kT_h \ln\left(\frac{p_i}{p}\right) + E^0_{0i} \tag{6.97}$$

### 6.5.1.2. Equations

Nitrogen will be used to illustrate the calculation. At equilibrium, the following species will be considered: $e$, N, $N^+$, $N_2$, $N_2^+$. Figure 6.4 shows that for $T < 15{,}000$ K, $N_2^+$ ions are present but with a low molar fraction (between $10^{-4}$ and $10^{-6}$), the main ion being $N^+$ (even when $N_2^+$ is at its maximum, the $N^+$ density is more than one order of magnitude greater). However, for the same electron temperature (at 15,000 K, for example, if the ratio is $\theta = 3$, then $T_h = 5000$ K), the dissociation of $N_2$, which is mainly controlled by collision between heavy particles, is low while the electron energy is high. Since the main component of the plasma is $N_2$ at $T_h = 5000$ K, the high-energy electrons will have a high probability of removing an electron from $N_2$ but a low probability of taking an electron off N, whose population is very low. Thus under these conditions, the main ion will be $N_2^+$.

The composition at equilibrium is obtained by satisfying $(dG)_{p,T} = 0$ together with the other conditions of nitrogen conservation [Eq. (6.34)] and electrical neutrality [Eq. (6.35)]. Proceeding as in Section 6.3.1 for the mass action law in CTE,

$$dG = (2\mu_N - \mu_{N_2})\, d\eta_N + (\mu_{N^+} + \mu_e - \mu_N)\, d\eta_{N^+} + (\mu_{N_2^+} + \mu_e - \mu_{N_2})\, d\eta_{N_2^+} \tag{6.98}$$

where $d\eta_N$, $d\eta_{N^+}$, and $d\eta_{N_2^+}$ are respectively the production rates of N ($N_2 \rightleftarrows 2N$), $N^+$ ($N^+ + e \rightleftarrows N$), and $N_2^+$ ($N_2^+ + e \rightleftarrows N_2$).

According to Eqs. (6.96) and (6.97), it follows for dissociation that

$$K_p(N) = \frac{p_N^2}{p_{N_2}} = \frac{\left[Q^N_{\text{int}}(T_e)\left(\frac{2\pi m_N kT}{h^2}\right)^{3/2} kT_h\right]^2}{Q^{N_2}_{\text{int}}(T_h)\left(\frac{2\pi m_{N_2} kT}{h^2}\right)^{3/2} kT_h} \exp\left(-\frac{2E^0_{0N} - E^0_{0N_2}}{kT}\right) \tag{6.99}$$

This is the same as for the CTE case. This was a predictable result because, as already emphasized in Section 6.5.1.1, dissociation is due mainly to collisions between heavy particles.

For the nitrogen atomic ion formation, substitution of Eqs. (6.96) and (6.97) into $\mu_{N^+} + \mu_e - \mu_N = 0$ gives

$$- kT_h \ln\left(\frac{kT_h}{p} Q^{tr}_{N^+}(T_h)\right) - kT_e \ln Q^{int}_{N^+}(T_e) + kT_h \ln\left(\frac{p_{N^+}}{p}\right) + E^0_{0N^+}$$

$$-kT_e \ln\left(\frac{kT_e}{p} Q^{tr}_e(T_e)\right) - kT_e \ln 2 + kT_e \ln\left(\frac{p_e}{p}\right) + E^0_{0e}$$

$$+kT_h \ln\left(\frac{kT_h}{p} Q^{tr}_N(T_h)\right) + kT_h \ln Q^{int}_N(T_e)$$

$$-kT_h \ln\left(\frac{p_N}{p}\right) - E^0_{0N} - \delta E_{N^+} = 0$$

Thus, finally,

$$K_p(N^+) = p_e\left(\frac{p_{N^+}}{p_N}\right)^{1/\theta} = kT_e\left(\frac{2\pi m_e kT_e}{h^2}\right)^{3/2} 2\left(\frac{Q^{int}_{N^+}(T_e)}{Q^{int}_N(T_e)}\right) \times \exp\left(-\frac{E^I_{N^+} - \delta E_{N^+}}{kT_e}\right) \tag{6.100}$$

where $E^I_{N^+}$ is the ionization potential for the reaction ($N \rightleftarrows N^+ + e$) and $\delta E_{N^+}$ its lowering.

Similarly, for a nitrogen molecular ion

$$K_p(N_2^+) = p_e\left(\frac{p_{N_2^+}}{p_{N_2}}\right)^{1/\theta} = kT_e\left(\frac{2\pi m_e kT_e}{h^2}\right)^{3/2} 2\left(\frac{Q^{int}_{N_2^+}(T_h)}{Q^{int}_{N_2}(T_h)}\right)^{1/\theta} \times \exp\left(-\frac{E^I_{N_2^+} - \delta E_{N_2^+}}{kT_h}\right) \tag{6.101}$$

where $E^I_{N_2^+}$ is the ionization potential for the reaction ($N_2 \rightleftarrows N_2^+ + e$) and $\delta E_{N_2^+}$ its lowering. In Eq. (6.101) it has been assumed that $m_{N_2} \approx m_{N_2^+}$ and the main difference for the molecular ions compared to dissociation is the exponent $1/\theta$ for the ratio of partition-functions.

Once the ratio $\theta = T_e/T_h$ is given, the composition calculation can be developed in exactly the same way as in equilibrium using the NASA method (see Section 6.3.3.1) with the new equilibrium constants given by Eqs. (6.99) to (6.101).

To keep the computation time reasonable, it is best to calculate the partition functions in such a way that they are independent of the charged particles (which depend on the final result) and the mole ratio in the starting cold mixture. This is possible using Gurvich's limitation, which depends only on pressure and temperature. This procedure allows one to store the partition functions as polynomials of $T_e$, $T_h$, and $p$.

### 6.5.1.3. Example: Composition of Ar–$H_2$ Plasmas

The following species will be taken into account: $e$, H, Ar, $H^+$, $Ar^+$, $H_2^+$.[31] The composition is calculated at atmospheric pressure for $T_e$ varying from 5000 to 15,000 K in steps of 500 K. Calculations are made for five values of the ratio $n^0_{H_2}/n^0_{Ar}$ corresponding to the transition from pure $H_2$ to pure Ar in steps of 25%. For pure $H_2$ at equilibrium ($\theta = 1$), the results of Capitelli *et al.*[9] and Fauchais *et al.*[3] are found to agree within 2%. Figure 6.21, for pure argon, shows the variation in the numerical density of $n_{Ar}$ and $n_{Ar^+} = n_e$ with electron temperature $T_e$ for $\theta = 1, 2, 3$, and 6. This figure is strictly valid only for a $T_e/T_h$ ratio lower than 3, but in order to get an idea of what happens in a strong nonequilibrium case the results have been extended to $\theta = 6$. Note first, that the electron density curves show a crossover point at about 14,000 K (Hsu and Pfender[33]) and, second, that at $T_e = 5000$ K, $n_{Ar}(\theta) = \theta \cdot n_{Ar}(1)$ where $n_{ar}(1)$ is the argon atom number density at equilibrium.

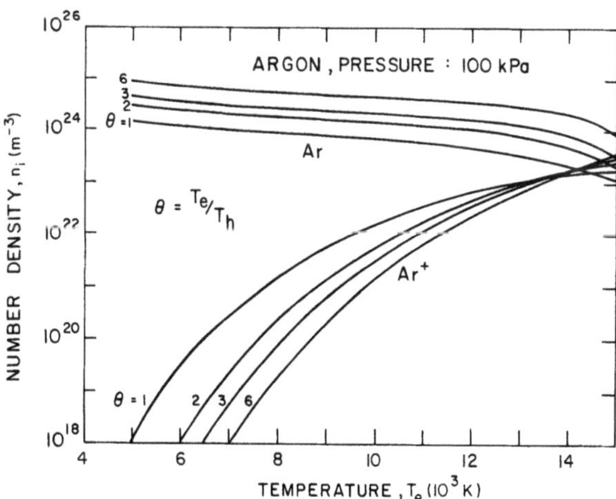


**FIG. 6.21.** Dependence of the number densities of Ar and $Ar^+$ on electron temperature at atmospheric pressure for a two-temperature plasma with $\theta = T_e/T_h = 1, 2, 3, 6$.[31]

This second point reflects the fact that the low electron density decreases with $\theta$. Figure 6.22a, b, and c emphasize the influence of $\theta$ (which has values 1, 2, 3, respectively) on the composition of a mixture consisting of 3 moles of $H_2$ for 1 mole of Ar. It can be seen that when $\theta$ increases from 1 to 3, $H^+$ and H are progressively reduced. For example, when $T_e$ = 15,000 K with $\theta$ = 1, the main species are Ar, $Ar^+$, H, $H^+$, and $e$, while for $\theta$ = 3 they are Ar, $H_2$, $H_2^+$, H, and $e$. For $\theta$ = 3, $T_h$ is only 5000 K, and the dissociation of $H_2$, which at this temperature depends only on the heavy-particle temperature, is just completed. For $\theta$ = 1, the

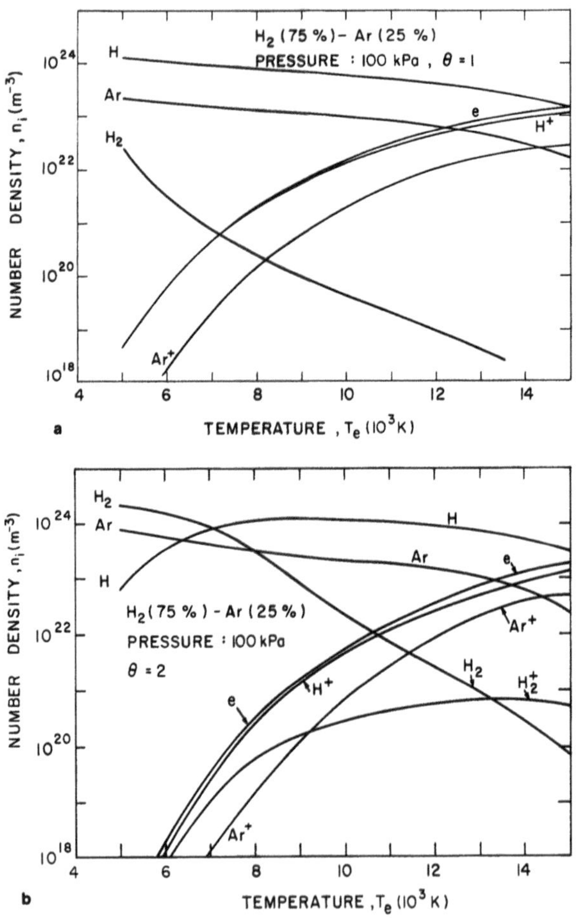


FIG. 6.22. Dependence of the number densities of a mixture of argon and hydrogen (75% in mole fractions) on electron temperature at atmospheric pressure: for ratios $\theta = T_e/T_h$ = 1, 2, and 3.[31]

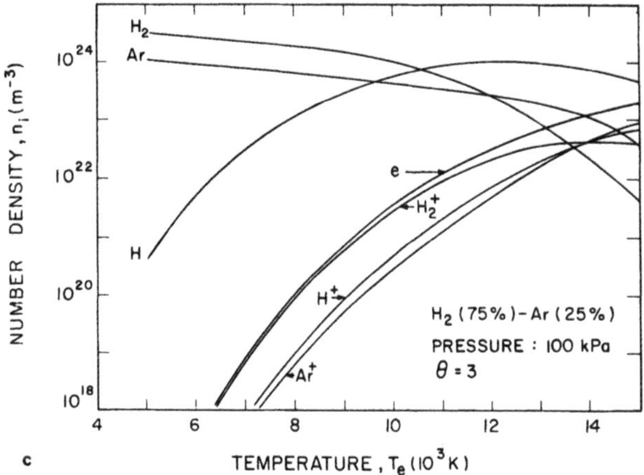


**FIG. 6.22.** (Continued).

main ion is $H^+$ (H has a somewhat lower ionization potential than Ar), while for $\theta = 3$, the main ion is $H_2^+$.

*Limitations of this model.* The preceding approach is valid only for a small disequilibrium. To be rigorous, such calculations should be performed using a kinetic model such as the one developed by Richley and Tuma.[34] This model involves finding the steady state by noting that the reaction rates $d\eta_i$ are related to the reaction rate coefficients $k_i^d$ and $k_i^r$ for the $i$th chemical reaction (the upper indexes refer to direct and reverse, respectively). The main problem with this method is finding data for $k_i^d$ and $k_i^r$ and making sure that no reaction is neglected. When one coefficient, $k^d$ or $k^r$, is found in the literature, the other one can be calculated from the equilibrium constant $K_i$ ($k_i^d/k_i^r = K_i$). These reaction rates can also be calculated from the reaction cross sections, if they are known. Aubreton[32] used this method to calculate the composition of an Ar–$H_2$ plasma taking into account 24 different reactions, and he then compared the results to those of the preceding method. He found that up to $\theta = 3$ the results of the two methods are similar. For a greater deviation from equilibrium the main discrepancies come from the following reactions:

$e + H_2^+ \rightarrow 2H$ for low densities ($n_e < 10^{18}$ elec/m$^3$)

$2e + H \rightarrow H^- + e$ and $H^+ + H^- \rightarrow 2H$

for high densities ($n_e > 10^{20}$ elec/m$^3$)

These reactions are not taken into account in the method derived from equilibrium.

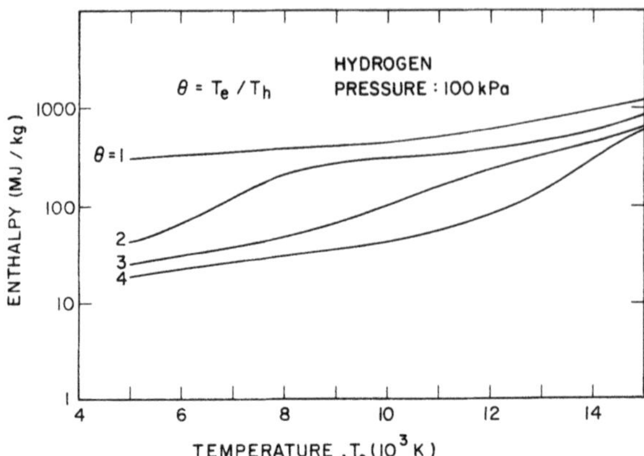


**FIG. 6.23.** Dependence of the enthalpy of pure hydrogen in a two-temperature plasma on electron temperature at atmospheric pressure for $\theta = T_e/T_h = 1, 2, 3, 4$.[31]

### 6.5.2. Thermodynamic Properties

Once the composition is determined, the thermodynamic properties are calculated using the partition functions. Their derivatives are then calculated from the proper temperature ($T_e$ or $T_h$) using the calculated numerical densities for a given ratio of $\theta$.

Figure 6.23 shows the logarithm of the enthalpy for one mole of pure hydrogen as a function of $T_e$ for various values of $\theta$. For $\theta = 1$, the change in enthalpy due to dissociation is not represented because it takes place between 3000 and 4500 K. However, for $\theta = 2$, this change takes place at the same heavy-particle temperatures, i.e., between 6000 and 8000 K for $T_e$, while it begins at $T_e = 12{,}000$ K for $\theta = 3$. For $\theta = 4$ the increase in enthalpy that starts at about $T_e = 12{,}000$ K corresponds to the formation of $H_2^+$ by ionization of $H_2$.

## LIST OF SYMBOLS

| | |
|---|---|
| $B_{ij}(T)$ | second virial coefficient related to the interaction between particles of types $i$ and $j$ |
| $B(T)$ | second virial coefficient |
| $c_p$ | specific heat at constant pressure (kJ/kg K) |
| $c_p^r$ | reactional specific heat at constant pressure (kJ/kg K) |
| $c_p^f$ | frozen specific heat at constant pressure (kJ/kg K) |
| $d_{ij}$ | distance at which interaction between particles of types $i$ and $j$ has to be taken into account (m) |

$d\eta_i$ rate of production of species $i$

$D_e$ energy difference between the equilibrium position of the nuclei and the free atoms for a diatomic molecule (eV)

$e_n$ base of natural logarithm: 2.7182818

$E_{i,s}$ energy of the chemical species $i$ in the excited state $s$ ($cm^{-1}$)

$E^0_{0i}$ energy of the chemical species $i$ in its ground state related to an absolute reference state

$E_e$ electronic energy of a given particle (eV or $cm^{-1}$)

$E_v(e)$ vibrational energy of a given particle in the electronic state $e$ (eV or $cm^{-1}$)

$E_r(e, v)$ rotational energy of a given particle in the electronic state $e$ and the vibrational state $v$ (eV or $cm^{-1}$)

$E^D_X$ dissociation energy of the diatomic molecule $X_2$ (eV)

$E^I_{X^+}$ ionization energy of the atom X (eV)

$F$ Helmholtz free energy (J)

$g_{i,s}$ statistical weight of the chemical species $i$ in the excited state $s$

$g_v$ vibrational statistical weight

$g_r$ rotational statistical weight

$g_e$ electronic statistical weight

$G$ Gibbs free enthalpy (J)

$h$ Planck's constant ($6.6 \times 10^{-34}$ W $s^2$)

$h_g$ specific enthalpy of the gas (kJ/kg)

$H_i$ molar enthalpy of the chemical species $i$ (J/mole)

$H^D_X$ dissociation enthalpy of the molecule $X_2$ (J)

$H^I_X$ ionization enthalpy of the molecule $X_2$ (J)

$i$ index of the chemical species

$k$ Boltzmann's constant ($1.38 \times 10^{-23}$ J/K part)

$K_c(X)$ equilibrium constant to produce the species X

$K_p(X^+)$ partial pressure equilibrium constant for ionization of X

$K_p(X)$ partial pressure equilibrium constant for dissociation of $X_2$

$l_{ij}$ mean free path between collision of particles of type $i$ and $j$ (m)

$m_i$ mass of a particle of chemical species $i$ (kg)

$M_i$ atomic mass of chemical species $i$ (kg)

$n$ principal quantum number

$n^*$ maximum value of $n$ for the calculation of the partition function

$n_i$ number density of chemical species $i$ regardless of their excitation state ($m^{-3}$)

$n_{i,s}$ number density of chemical species $i$ in the excited state $s$ ($m^{-3}$)

$n'_i$ mole number of chemical species $i$ regardless of their excitation state

$n_T$ total number density ($n_T = p/kT$) ($m^{-3}$)

$N_i$ number of particles of chemical species $i$ whatever their excited state may be

$N_{i,s}$ number of particles of chemical species $i$ in the excited state $s$
$p$ pressure of the gas (Pa)
$p_i$ partial pressure of the chemical species $i$ (Pa)
$Q_i$ partition function of the chemical species $i$
$Q_i^{tr}$ translational partition function of the chemical species $i$
$Q_i^{int}$ internal partition function of the chemical species $i$
$Q_{tot}$ total partition function of the gas ($Q_{tot} = \prod_i Q_i^{N_i}/\prod_i N_i!$)
$s$ index of the excited state
$s^*$ maximum value of $s$ for the calculation of the partition function
$S$ entropy of the gas (kJ/kg K)
$T$ equilibrium temperature
$T_h$ heavy species temperature
$T_e$ electron temperature
$V$ volume of the gas or of the plasma ($m^{-3}$)
$V_{ij}$ interaction potential between particles of type $i$ and $j$
$x_i$ molar fraction of chemical species $i$
$Z_i$ effective charge of the chemical species $i$ ($Z_i = 0$ for atoms or molecules, $Z_i = 1$ for the first ion, and so on)

## Greek Symbols

$\delta E^{I}_{X^+}$ ionization potential lowering for $X^+$ ion production (eV)
$\delta p$ lowering of the pressure (Pa)
$\Delta H$ enthalpy change of a thermodynamic state at $T$ and $p$ with respect to a reference state (J)
$\varepsilon_0$ vacuum permittivity (As/Vm)
$\lambda_D$ Debye length (m)
$\Lambda_i$ de Broglie's wavelength for the chemical species $i$ (m)
$\mu_i$ chemical potential for the chemical species $i$ (J/part)
$\mu_i^0$ part of the chemical potential of chemical species $i$ that depends only on temperature (J/part)
$\nu$ vibrational frequency ($s^{-1}$)
$\omega_e$ vibrational energy of the molecule at rest expressed in units of $cm^{-1}$ ($E/hc$)
$\theta$ ratio of electron to heavy-particle temperature

# GENERAL BIBLIOGRAPHY

Fowler, F. H., and E. A. Guggenheim, *Statistical Thermodynamics,* Cambridge: University Press, 1956.

Herzberg, G., *Molecular Spectra and Molecular Structure I. Spectra of Diatomic Molecules,* New York: Van Nostrand, 1950.

Huber, K. P., and G. Herzberg, *Constants of Diatomic Molecules,* Litton Educational Publishing, 1979.
Landau, L. and E. Lifschitz, *Physique Statistique,* Moscow: Mir, 1967.
Mayer, J. E. and G. M. Mayer, *Statistical Mechanics* (11th ed.), New York: Wiley, 1966.
Moore, C. E., *Atomic Energy Levels,* NBS Circular 467, Vol. 1 (1949), Vol. 2 (1952), Vol. 3 (1958).
Münster, A., *Statistical Thermodynamics, Vol. 1,* Berlin: Springer-Verlag; New York: Academic Press, 1969.
Wilson, A. H., *Thermodynamics and Statistical Mechanics,* Cambridge: University Press, 1957.

## REFERENCES

1. H. W. Drawin, "Thermodynamic Properties of the Equilibrium and Nonequilibrium States of Plasmas," in *Reactions under Plasma Conditions,* M. Venugopalan, ed. (New York: Wiley Interscience, 1972).
2. P. Fauchais, *Rev. Int. Hautes Temp. Réfract.* 6 (1969): 77.
3. P. Fauchais, A. Vasseur, and N. Manson, *Rev. Int. Hautes Temp. Réfract.* 6 (1969): 5.
4. P. Fauchais, J. M. Baronnet, and S. Bayard, *Rev. Int. Hautes Temp. Réfract.* 12 (1975): 221.
5. L. V. Gurvich and V. A. Kvlividze, *Zh. Fiz. Khim.* 35 (1961): 1672.
6. P. Fauchais, "Etude des propriétés thermodynamiques des plasmas produits par un générateur à arc", Thèse de Doctorat d'Etat (Université de Poitiers, 1962).
7. B. J. McBride and S. Gordon, *Fortran IV Program for Calculation of Thermodynamic Data* (NASA TN-D-4097, 1967).
8. K. S. Drellishak, "Partition Functions and Thermodynamic Properties of High Temperature Gases," Ph.D. Thesis (Northwestern University, Illinois, 1963).
9. M. Capitelli, E. Ficocelli, and E. Molinari, "Equilibrium Compositions and Thermodynamic Properties of Mixed Plasmas I, He–$N_2$, Ar–$N_2$ and Xe–Ne Plasmas at One Atmosphere between 5000 K and 35,000 K," Internal Report (University of Bari, Italy 1969).
10. I. V. Veits, L. V. Gurvich, and N. P. Rtischeva, Zh. Fiz. Khim. 32 (1958): 2532.
11. S. H. Storey and F. Van Zeggeren, *The Computation of Chemical Equilibria* (Cambridge: University Press, 1970).
12. E. Bourdin, "Contribution à l'étude théorique et expérimentale des nitrures et oxynitrures par réaction de jets de plasma d'azote avec des poudres d'aluminium, de silicium et de leurs oxyde," Thèse 3ème cycle (Université de Limoges, France, March 1976).
13. S. R. Brinckley, *J. Chem. Phys.* 14 (1946): 9.
14. S. R. Brinckley, *J. Chem. Phys.* 34 (1947) 2.
15. V. N. Huff, S. Gordon, and F. V. Zeleznick, "A general method for automatic computation of equilibrium compositions and theoretical rocket performance of propellants." *NASA TN* D-132 Oct. 1959.
16. S. M. White, S. Johnson, and G. B. Dantzig, *J. Chem. Phys.* 28 (1958): 5.
17. B. Pateyron, J. Aubreton, M. F. Elchinger, and G. Delluc, Paper delivered at the Second Symposium on Critical Evaluation and Prediction of Phase Equilibrium in Multicomponent Systems (Paris, 11–12 Sept. 1985).
18. B. Pateyron, J. Aubreton, M. F. Elchinger, and G. Delluc, Paper delivered at the First

Codata Symposium on Chemical Thermodynamic and Thermophysical Properties. Data Base (Paris, 9–10 Sept. 1985).
19. J. Amouroux, *Ann. Chimie* 3 (1978): 59.
20. P. Sutre, and J. P. Malenge, *Entropie* 40 (1968): 285.
21. J. F. Coudert, "Contribution à l'étude des oxydes d'azote par chalumeau à plasma," Thèse de 3ème cycle (Université de Limoges, France, Juin 1978).
22. V. P. Glouchko, "Propriétés thermodynamiques des corps purs" (Moscow: Mir, 1962).
23. JANAF, thermochemical data compiled and calculated by the Dow Chemical Company, Thermal Laboratory, Midland, Michigan (1971).
24. S. Gordon, and B. J. McBride, *Thermodata,* NASA SP 273.
25. I. Barin, and O. Knacke, *Thermochemical Properties of Inorganic Substances* (Berlin and New York: Springer-Verlag, 1973 and 1977).
26. F. J. Zeleznick, and S. Gordon, *Simultaneous Least Squares Approximation of a Function and its First Integrals with Applications to Thermodynamic Data, Can. J. Phys.* 44, 877 (1966).
27. B. Pateyron, J. Aubreton, M. F. Elchinger, G. Delluc, and P. Fauchais, "Thermodynamic and Transport Properties of $N_2$, $O_2$, $H_2$, Ar, He and their Mixtures," Internal Report, Laboratoire Céramiques Nouvelles URA 320 CNRS, University of Limoges, France (1986).
28. J. Lesinski, and M. Boulos, "Thermodynamic and Transport Properties of Argon, Nitrogen and Oxygen at Atmospheric Pressure over the Temperature Range 300–30,000 K," Internal Report, University of Sherbrooke, Quebec, CN (May 1984).
29. B. Pateyron, M. F. Elchinger, C. Delluc, and P. Fauchais, "Thermodynamic and Transport Properties of Ar–$H_2$ and Ar–He Plasma Gases for Spraying at Atmospheric Pressure—Part 1: Properties of the Mixture," *Plasma Chemistry, Plasma Processing* (submitted).
30. A. V. Potapov, *High Temp.* 4 (1966): 48.
31. C. Bonnefoi, "Contribution à l'étude des méthodes de résolution de l'équation de Boltzmann dans un plasma à deux températures: exemple le mélange argon hydrogène," Thèse de Doctorat d'Etat (University of Limoges, France, 1983).
32. J. Aubreton, "Etude des propriétés thermodynamiques et de transport dans les plasmas thermiques à l'équilibre et hors équilibre thermodynamique. Applications aux plasmas de mélange Ar–$H_2$, Ar–$O_2$," Thèse de Doctorat d'Etat (University of Limoges, France, Feb 22, 1985).
33. K. C. Hsu, and E. Pfender, "Calculation of Thermodynamic and Transport Properties of a Two-Temperature Argon Plasma," Proceedings of the Fifth Int. Symp. on Plasma Chemistry, Vol. 1, (Heriot-Watt Univ., Edinburgh, Scotland, 1981): 144.
34. E. Richley and D. T. Tuma, *J. Appl. Phys.* 53 (1982): 8537.

Chapter 7

# Transport Properties

## 7.1. DEFINITION

Consider a gas in a nonequilibrium steady state with a constant flow in one direction (for example, between two parallel plates, one heated at $T_1$ and the second at $T_2$, with $T_1 > T_2$). Energy will flow from the plate at $T_1$ to the plate at $T_2$. The temperature gradient in the gas is the driving force and the physical quantity that is transported in this process is energy.

In most transport processes there is a linear relationship between fluxes and driving forces [Child (1974), Hirschfelder *et al.* (1964)]:

$$\text{flux} = (\text{coefficient}) \times (\text{driving force}) \tag{7.1}$$

provided, of course, that the force is not too large. Such a relation is called a phenomenological law (such as Ohm's law for electricity), and in the general case the corresponding transport coefficient is related to the flux density of a quantity $\chi$:

$$\begin{aligned} J_\chi &= \text{flux density of } \chi \\ &= \text{net quantity of } \chi \text{ transported across per unit area and unit time.} \end{aligned}$$

where $\chi$ may be a number of particles ($n$), a transverse momentum ($mv_x$), an energy ($\frac{3}{2}kT$), or a charge ($Ze$), corresponding to the following coefficients:

diffusion ($D$) expressed in $m^2/s$:

$$\vec{J}_n = -D \operatorname{grad} n \tag{7.2}$$

viscosity ($\mu$) expressed in kg/m · s:

$$\frac{\vec{F}_x}{A} = \vec{J}_{p_x} = -\mu \frac{dv_x}{dz} \tag{7.3}$$

thermal conductivity ($\kappa$) expressed in W/m · K:

$$\vec{J}_E = -\kappa \operatorname{grad} T \tag{7.4}$$

electrical conductivity $\sigma_e$ expressed in $(\Omega \cdot \mathrm{m})^{-1}$ or mho/m:

$$\vec{J}_e = -\sigma_e \operatorname{grad} V \tag{7.5}$$

where $-\operatorname{grad} V = \vec{E}$ is the electric field.

This chapter will be devoted to the transport phenomena occurring in plasmas when gradients exist.

The calculation of the transport coefficients related to the gas particles is quite complicated, even in the rather simple case of ideal gases. It is therefore quite valuable to develop very simple approximate methods that yield physical insight into basic mechanisms. Indeed, it is very often found that such simple approximations lead to the correct dependence of all significant parameters (such as pressure or temperature), even if the numerical values sometimes differ by up to 50% from the results of the rigorous calculation.

Thus, the first part of this chapter will deal with the simplest approximate method that can be used to calculate the transport properties, i.e., the elementary kinetic theory. The second part will give the guidelines for dealing with these phenomena more rigorously, and the third will present the results of the rigorous calculations.

## 7.2. SIMPLIFIED DERIVATION OF THE TRANSPORT COEFFICIENTS

In a gas, particles interact with each other through collisions. If the gas is not initially in an equilibrium situation, the distribution function is given by the solution of the Boltzmann equation (see Section 3.5.3), but these collisions are also responsible for bringing about the ultimate equilibrium situation in which a Maxwell–Boltzmann velocity distribution prevails. Section 7.3 will give the guidelines for calculating the transport properties through the Boltzmann equation, but only for the case of a tenuous gas ($p < 1$ MPa), which is relatively simple because of the following assumptions [Mitchner and Kruger (1973), Reif (1988)]:

(a) The gas is sufficiently tenuous that only two-particle collisions need to be taken into account.
(b) The mean free path $l$ is much greater than the range of intermolecular forces, and thus two particles originate, before their encounter, at a relative separation that is on the order of $l$ and thus sufficiently large that a correlation between their initial velocities is unlikely (this is known as the "molecular chaos" assumption). In this section an even more restrictive hypothesis will be used: between two collisions, the particles travel with their mean random velocity $\bar{v}$.

### 7.2.1. Self-Diffusion Coefficient

Consider a system with a particle density gradient (corresponding to a chemical potential gradient). Particles travel freely with a velocity $\bar{v}$ over distances on the order of the mean free path $l$ before they collide. Suppose that the gradient is in the $z$ direction and that, in a collision at position $z$, the particles come into a local equilibrium at the local chemical potential $\mu(z)$ and local concentration $n(z)$. Let $l_z$ be the $z$ component of the mean free path.

Across the plane at $z$, there is a particle flux density in the positive direction equal to $n(z - l_z) \cdot \bar{v}_z/2$ and a flux density in the negative $z$ direction equal to $-n(z + l_z) \cdot \bar{v}_z/2$, where $\bar{v}_z$ is the mean random velocity of the particles in the $z$ direction. In these equations, $n(z - l_z)$ stands for the particle density at $z - l_z$; the factor $\frac{1}{2}$ is inserted because, when a particle leaves a given point, its probability of going in the forward (or backward) direction is $\frac{1}{2}$. The net particle flux density is the average over all directions of a hemisphere:

$$J_{nz} = [n(z - l_z) - n(z + l_z)] \cdot \bar{v}_z/2 = -\frac{dn}{dz}\bar{v}_z \cdot l_z \tag{7.6}$$

Since $n$ and its derivative are already averaged values, this means that the average of $\bar{v}_z l_z$ has to be calculated over the surface of a hemisphere, because all forward directions are equally likely. Using spherical coordinates $\theta$, $\varphi$ (the values of $l_z$ and $\bar{v}_z$ are then $l_z = l\cos\theta$ and $\bar{v}_z = \bar{v}\cos\theta$), the element of surface area is $2\pi \sin\theta\, d\theta$ and thus

$$\langle v_z l_z \rangle = \bar{v}l \frac{2\pi \int_0^{\pi/2} \cos^2\theta \sin\theta\, d\theta}{2\pi} = \bar{v}l/3 \tag{7.7}$$

so that

$$J_{n,z} = -\bar{v}l\frac{dn}{dz}\Big/3 = -D\frac{dn}{dz} \tag{7.8}$$

and finally

$$D = \bar{v}l/3 \tag{7.9}$$

For a Maxwellian distribution [see Eq. (3.68)], $\bar{v}$ is given by

$$\bar{v} = \left(\frac{8kT}{m\pi}\right)^{1/2} \tag{7.10}$$

As a first approximation [see (Eq. 3.82)]

$$l = \frac{1}{\sqrt{2}\,n\sigma_0} \tag{7.11}$$

where $\sigma_0$ is the total collision cross section and $n$ is the particle density.

For a perfect gas $n = p/kT$, and finally, the self-diffusion coefficient

$$D = \frac{2}{3\sqrt{\pi}}\cdot\frac{1}{\sigma_0 p}\sqrt{\frac{(kT)^3}{m}} \tag{7.12}$$

is a function of pressure and temperature; under identical conditions, it will be the highest for the lightest gas. Using Eq. (7.12), the order of magnitude of $D$ at room temperature and atmospheric pressure for $N_2$ gas is $5 \times 10^{-5}\,\text{m}^2/\text{s}$, and the measured value at 273 K at 1 atm is $1.85 \times 10^{-5}\,\text{m}^2/\text{s}$. In view of the crude nature of this simple calculation, the agreement between theory and experiment is quite satisfactory.

### 7.2.2. Viscosity

When a fluid moves with a local macroscopic velocity $\vec{u}$, it is clear that the average microscopic velocity $\vec{v}$ can no longer vanish over a small element of volume $d\vec{r}$ ($d\vec{r} = dx\,dy\,dz$) but must be averaged to $\vec{u}$, so that

$$u_x = \langle v_x \rangle \tag{7.13}$$

The same is true for $y$ and $z$. Thus a transport of momentum $m\vec{v}$ is set up because molecules arriving at some point from a distance $l$ carry with them, on the average, the velocity that was acquired during the last collision. This transport of momentum, which gives rise to internal friction, is responsible for the viscosity of a gas (or a liquid). Consider a flow field in which $u_y = u_z = 0$ and where $u_x(z)$ varies with the $z$ coordinate only. In a plane at $z$, there must develop a stress (force per unit area) $\sigma_{zx}$ that tends to slow down the faster fluid (above $z$, for example) and to accelerate its slower layers. This force results from the fact that the molecules which arrive from below move with a velocity whose $x$ component is smaller than that prevailing at $z$. Similarly, the reverse flux of molecules carries with it a momentum whose $x$ component is larger than that prevailing at $z$. The change in momentum is equivalent to a force (or stress per unit area) in the $x$ direction. Following our analysis of self-diffusion and assuming a uniform particle density $n$, we have

$$\begin{aligned}\sigma_{zx} &= \tfrac{1}{2}n\bar{v}_z m[u_x(z - l_z) - u_x(z + l_z)] \\ &= -nm\frac{\partial u_x}{\partial z}\overline{v_z l_z}\end{aligned} \tag{7.14}$$

and, using the value of $\overline{v_z l_z}$ calculated previously,

$$\sigma_{zx} = -nml\bar{v}\frac{\partial u_x}{\partial z}\Big/3 \tag{7.15}$$

By definition, the viscosity $\mu$ is related to $\sigma_{zx}$ by

$$\sigma_{zx} = -\mu\frac{\partial u_x}{\partial z} \tag{7.16}$$

and finally,

$$\mu = nml\bar{v}/3 \tag{7.17}$$

Replacing $l$ and $\bar{v}$ by their values [see Eqs. (7.10) and (7.11)] yields

$$\mu = \frac{2}{3\sqrt{\pi}}\frac{\sqrt{mkT}}{\sigma_0} \tag{7.18}$$

In this oversimplified theory, $\mu$ is independent of pressure. The apparent paradox of the independence of $\mu$ with the pressure can be explained the following way: if the number of particles is doubled in the same volume, there are twice as many particles available to transport momentum from one plate to the other, but the mean free path of each

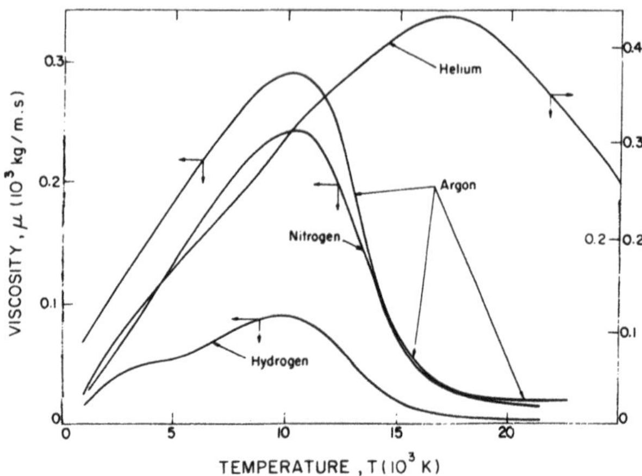


**FIG. 7.1.** Temperature dependence of the viscosity of various gases ($H_2$, $N_2$, Ar, He) at atmospheric pressure.[1,2]

molecule is also halved, so that it can transport this momentum only half as effectively, and thus the net rate of momentum transfer is unchanged. Quantity $\mu$ depends only on the square root of temperature and mass. However, one has to keep in mind that the values of $\sigma_0$ are not necessarily independent of the masses and thus $\mu$ is not strictly proportional to $\sqrt{m}$. When the temperature increases beyond room temperature, the values of $\mu$ are first controlled by the interactions between neutral molecules (for diatomic molecules), then as dissociation starts they are controlled by neutral atoms, and even at higher temperatures by the charged particles. In reality, when ionization is pertinent (above 10,000 K for Ar, $N_2$, $H_2$, and $O_2$ and above 17,000 K for He at atmospheric pressure), $\mu$ decreases with increasing temperatures (see the data in Fig. 7.1 for Ar, He, $H_2$, and $N_2$). This drop in viscosity is due to ionization of the gas, resulting in Coulomb forces of relatively long range between particles. The mobility of the charged particles decrease when charged-particle densities increase; the corresponding reduction in the transport of momentum is not accounted for by Eq. (7.18), and the kinetic theory is thus strictly limited to short-range interactions.

Comparing the self-diffusion coefficient [see Eq. (7.12)] with viscosity [see Eq. (7.18)], it follows that

$$\frac{D}{\mu} = \frac{1}{nm} = \frac{1}{\rho} \tag{7.19}$$

where $\rho$ is the mass density of the gas. Experimentally, it is found that

$(\rho D/\mu)$ lies in the range between 1.3 and 1.5, compared with the value of 1 given by the oversimplified theory.

### 7.2.3. Thermal Conductivity

If $q_z$ is the heat flux due to temperature gradients in the $z$ direction, the thermal conductivity $\kappa$ is defined by

$$q_z = -\kappa \frac{\partial T}{\partial z} \tag{7.20}$$

If $\bar{e}$ is the mean kinetic energy of the particles, $q_z$ will result from the energy flux of particles crossing the elementary surface $dA$ (orthogonal to $z$) without collisions. Following the same analysis we used for viscosity, we can develop an expression for the thermal conductivity:

$$q_z = \tfrac{1}{2} \cdot n \cdot \bar{v}_z \cdot [\bar{e}(z - l_z) - \bar{e}(z + l_z)]$$

$$= -\tfrac{1}{2} \cdot n \cdot \bar{v}_z \cdot 2 \cdot l_z \cdot \frac{\partial \bar{e}}{\partial T} \cdot \frac{\partial T}{\partial z}$$

$$q_z = -n \cdot \bar{v}_z \cdot l_z \cdot c_v \cdot \frac{\partial T}{\partial z} = -\tfrac{1}{3} \cdot n \cdot \bar{v} \cdot c_v \cdot l \cdot \frac{\partial T}{\partial z} \tag{7.21}$$

Thus,

$$k = \tfrac{1}{3} n \bar{v} c_v l \tag{7.22}$$

Or, replacing $l$ and $\bar{v}$ by the expressions given in Eqs. (7.10) and (7.11), it follows that

$$\kappa = \tfrac{2}{3} \frac{c_v}{\sigma_0} \sqrt{\frac{kT}{m\pi}} \tag{7.23}$$

In this oversimplified theory, the thermal conductivity $\kappa$ is independent of pressure and depends only on the square root of temperature. This independence of pressure can be explained using the same arguments as those for viscosity (see Section 7.2.2). However, the increase with temperature, even for $c_v$ constant (i.e., taking into account only the kinetic energy of the particles), occurs faster than Eq. (7.23) would suggest, and thus another theoretical development is necessary. Moreover, when the evolution of $c_v$ versus temperature (see Fig. 6.14) is compared to the increase in thermal conductivity with temperature (see

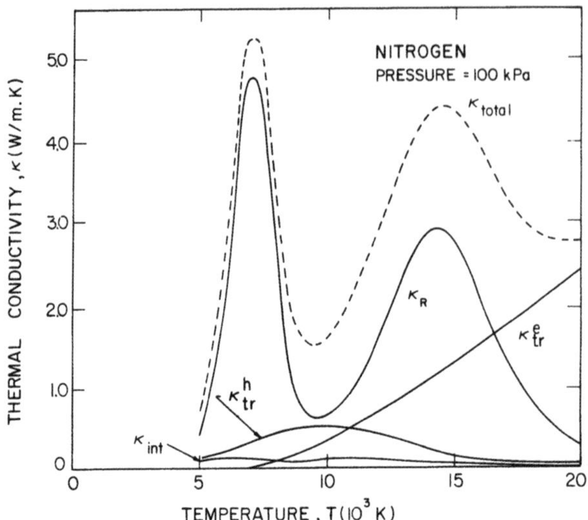


**FIG. 7.2.** Temperature dependence of the different components of the thermal conductivity of nitrogen plasmas at atmospheric pressure.[1,2]

Fig. 7.2, from Refs. 1 and 2, for a nitrogen plasma at atmospheric pressure), we see that the thermal conductivity shows peaks at temperatures similar to those obtained for $c_v$, although the peak values are not quite proportional to those obtained for $c_v$. This discrepancy occurs because, in this oversimplified calculation, it has been assumed that the system was frozen (no chemical reactions). However, in a real plasma dissociations, ionizations, and chemical reactions cannot be neglected, and they result in a supplementary heat transfer, for example, when atom recombination occurs in lower-temperature regions due to the existence of temperature (and thus composition) gradients. It is thus necessary to introduce a new coefficient $\kappa_R$, called the reactive thermal conductivity, whose calculation is quite different from that of the kinetic thermal conductivity.

Finally, it is important to emphasize that this oversimplified kinetic theory gives transport properties for monatomic gases only. While the equation for viscosity provides us with a workable first approximation even for polyatomic gases, the analysis for thermal conductivity does not because internal degrees of freedom in polyatomic molecules or even in monatomic gases in highly excited states contribute significantly to the transport of energy. In 1913, Eucken[3] advanced a very simple, intuitive argument that allowed one to extend the validity of Eq. (7.23) to polyatomic gases by introducing a constant factor. Eucken's theory was later reexamined in terms of the more exact Chapman–Enskog theory [see Hirschfelder *et al.* (1964) and Chapman and Cowling (1952)] by Manson and Monchick,[4] especially to account for the contribution of excited states. Thus the thermal conductivity $\kappa$ is the sum of three terms: $\kappa_{tr}$, which results from the translation of particles (considered to behave as hard spheres), $\kappa_R$, from the chemical reactions (dissociations, ionizations, pure chemical reactions), and finally $\kappa_{int}$, from the internal degrees of freedom:

$$\kappa = \kappa_{tr} + \kappa_R + \kappa_{int} \tag{7.24}$$

where $\kappa_{tr}$ is often split into $\kappa_{tr}^h$ for heavy particles and $\kappa_{tr}^e$ for electrons (see Fig. 7.2). These terms will be discussed later (see Sections 7.3.5.3, 7.4, and 7.5).

### 7.2.4. Electrical Conductivity

The electrical conductivity was treated in Section 4.3 in connection with the motion of charge carriers and charge mobility $\mu_e$ (neglecting the ion motion), and it was established in Chapter 4 that

$$\sigma_e = en_e\mu_e$$

with

$$\mu_e = \frac{el_e}{\sqrt{\pi kTm_e}}$$

Taking the expression for $l_e$ from Eq. (7.11), we have

$$\sigma_e = \frac{n_e e^2}{\sqrt{2\pi Tm_e}\, n_a\sigma_{en}} \tag{7.25}$$

where $n_a$ is the neutral particle number density and $\sigma_{en}$ the electron–neutral collision cross section. Here, it has been assumed that the electron and ion densities are low compared with that of the neutral species and thus that the mean free path of the electron $l_e$ [see Eq. (7.11)] is mainly due to collisions with neutral particles.

This oversimplified expression shows clearly that $\sigma_e$ depends mainly on the electron density, which (as shown in Section 6.3.3.4) varies almost exponentially with temperature. This is why $\sigma_e$ is almost negligible for the most common plasma gases (Ar, $H_2$, $N_2$, ...) below 6000 K, as illustrated

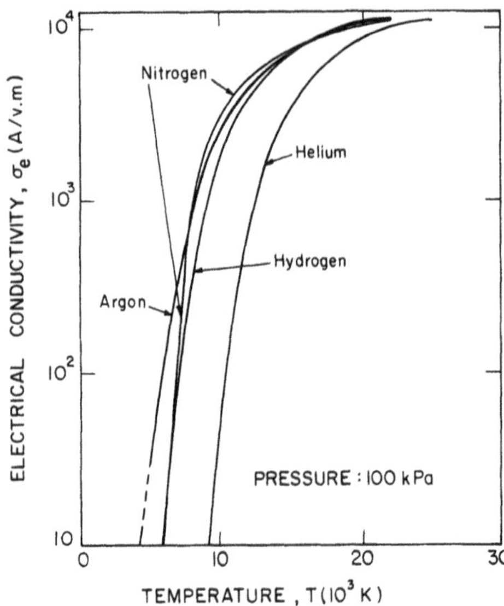


**FIG. 7.3.** Temperature dependence of the electrical conductivity of various gases ($H_2$, $N_2$, Ar, He) at atmospheric pressure.[1,2]

in Fig. 7.3; $\sigma_e$ for helium reaches significant values only for $T > 13{,}000$ K due to the higher ionization potential of He (see also Fig. 6.3, which shows the evolution of $n_e$ with $T$ for a He plasma).

Addition of metallic vapors to the plasma gases strongly affects $\sigma_e$ because the lower ionization potential of the metal results in a much higher electron density at lower temperatures, as is illustrated in Fig. 7.18 for an Ar plasma with copper vapor.

## 7.3. DERIVATION OF THE TRANSPORT COEFFICIENTS FROM THE BOLTZMANN EQUATION

The preceding simplified treatment (kinetic theory) leaves much to be desired. It assumes that the particles travel at their mean random velocity, which is much higher than the flow velocity; it does not treat the collisions in a detailed way, and it neglects correlations between particle velocities before and after a collision, i.e., the persistence of velocity effects. Nor does it account for the particle velocity distributions. Moreover, it assumes that there is a single chemical specie in the flow while, in fact, many species are present. Thus the problem must be formulated in a more rigorous way. However, since this would be a fairly complex task, far beyond the scope of this book, only the main guidelines of this treatment will be given here. The reader who is not interested in this development can proceed directly to Section 7.5, which gives some examples of the results from the calculation of transport coefficients.

A more exact formulation implies calculating (taking into account the velocity distributions) the fluxes (mass, momentum, and energy) of all species through a reference surface $dA$ moving with the mean fluid velocity.

### 7.3.1. Basic Equations

In Section 3.5.2, mean values were defined through the distribution function $f(\vec{r}, \vec{v}, t)$ for a gas containing only one chemical specie. When numerous species are considered, the mean value of any function $\chi_i(\vec{r}, \vec{v}, t)$ related to the chemical species $i$ is defined as

$$\bar{\chi}_i = \frac{\displaystyle\int \chi_i f_i \, d\vec{v}_i}{\displaystyle\int f_i \, d\vec{v}_i} = \langle \chi_i \rangle \tag{7.26}$$

As indicated in Section 3.5.2, the integrals are calculated over all possible values of the velocity components.

By definition, the number density of the species $i$ is given by

$$n_i = \int f_i \, d\vec{v}_i \tag{7.27}$$

and $\bar{\chi}_i$ can be written as

$$\bar{\chi}_i = (1/n_i) \int \chi_i f_i \, d\vec{v}_i \tag{7.28}$$

The average flow velocity or macroscopic velocity is defined by

$$\rho \vec{v}_0 = \sum_{i=1}^{K} m_i n_i \langle \vec{v}_i \rangle \tag{7.29}$$

where $\langle \vec{v}_i \rangle$ is the mean velocity of species $i$ [defined through Eq. (7.26) with $\chi_i = v_i$], while the mass density $\rho$ is given by

$$\rho = \sum_{i=1}^{K} n_i m_i \tag{7.30}$$

It is clear that $\vec{v}_0$ is a function only of position and time but not of the velocities $\vec{v}_i$.

The relation of the particle velocity $U_i$ with respect to the gas flow velocity $v_0$ is given by

$$\vec{U}_i = \vec{v}_i - \vec{v}_0 \tag{7.31}$$

where $\vec{U}_i$ is known as the peculiar velocity. Unlike the results we have established for a unique chemical specie (see Section 3.5.2), $\langle\vec{U}_i\rangle \neq 0$ and the relationship equivalent to Eq. (3.44) becomes

$$\sum_{i=1}^{K} n_i m_i \langle\vec{U}_i\rangle = 0 \tag{7.32}$$

This allows us to define the diffusion velocity of chemical specie $i$, which is the flow rate of particle of type $i$ with respect to the mass average velocity of the gas:

$$\vec{U}_i(r, t) = \langle\vec{v}_i - \vec{v}_0\rangle \tag{7.33}$$

Clearly, it is also the average of the peculiar velocity:

$$\langle\vec{U}_i\rangle = (1/n_i)\int \vec{U}_i f_i \, d\vec{v}_i \tag{7.34}$$

All these quantities can be calculated only if the distribution functions $f_i$ are known through the solution of the Boltzmann equation.

As was shown in Section 3.5.3, the Boltzmann equation for a single chemical specie $i$ can be written as

$$\mathcal{D} f_i = \mathcal{D}_c f \tag{7.35}$$

where $\mathcal{D}$ is the operator

$$\mathcal{D} = \frac{\partial}{\partial t} + \vec{v}_i\vec{\nabla}_r + \frac{\vec{F}_i}{m_i}\vec{\nabla}_v \tag{7.36}$$

$\vec{\nabla}_v$ being the gradient of the velocity components, i.e., $\partial/\partial v_x$, $\partial/\partial v_y$, $\partial/\partial v_z$, while $\vec{\nabla}_r$ corresponds to $\partial/\partial x$, $\partial/\partial y$, $\partial/\partial z$, and $\mathcal{D}_c f$ is the collision term (written as $\sum_j c_{ij}$ in Section 3.5.3).

Since collisions are the key point allowing calculation of the distribution function and thus the transport properties, deviation angles, and cross sections for collisions must first be defined [see Child (1974)].

The first assumption is that the internal degrees of freedom of the particles are unaffected by elastic collisions. The interactions between particles depend only on their relative positions and velocities.

The two colliding particles, 1 and 2, have initial velocities $\vec{v}_1$ and $\vec{v}_2$, so they move with a relative velocity $\vec{V} = \vec{v}_1 - \vec{v}_2$. After the collision, their velocities are $\vec{v}_1'$ and $\vec{v}_2'$, and they move with a relative velocity $\vec{V}_i' = \vec{v}_1' - \vec{v}_2'$.

Conservation of the total momentum allows a change of coordinates because the change implies that the velocity of the center of mass $\vec{c}$ is

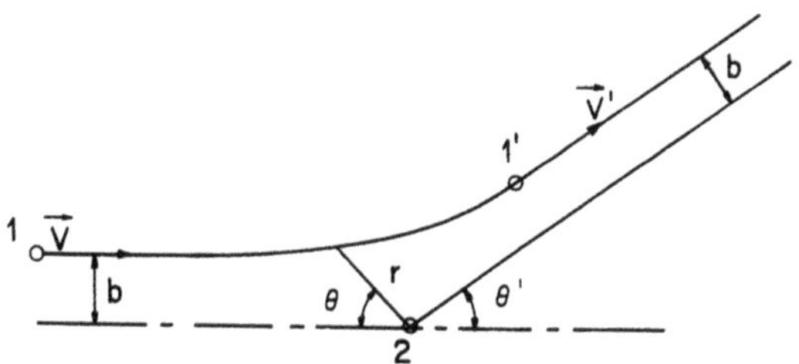


**FIG. 7.4.** Schematic of the scattering process.

time independent, and thus calculations can be developed in a frame of reference that moves with the center of mass.

Assuming elastic collisions implies that the total kinetic energy of the two colliding particles remains unchanged in a collision, so that $V^2 = V'^2$. Thus, $|\vec{V}'| = |\vec{V}|$, and the only effect of the collision is to change $\vec{V}$ in direction but not in magnitude.

In the frame relative to the center of mass, the position vectors $\vec{r}_1^*$ and $\vec{r}_2^*$ have opposite directions and their magnitude has a fixed ratio. It is then easy to demonstrate that the motion of the two particles can be reduced to the solution of a simple one-particle problem in which particle 2 is thought of as the target. The scattering process appears (as shown in Fig. 7.4) to be the motion of a particle of reduced mass $\mu_m = m_1 m_2/(m_1 + m_2)$ and velocity $\vec{V}$. In Fig. 7.4, the impact parameter $b$ is defined as the distance of closest approach of the particles if there is no interaction between them.

The collision process can thus be described by merely specifying the polar angle $\theta'$ and azimuthal angle $\varphi'$ of the final relative velocity $\vec{V}'$ with respect to the relative velocity $\vec{V}$ before the collision.

In Fig. 7.4, the motion of particle 1, whose mass is $\mu_m$ and velocity is $\vec{V}$, can be described in polar coordinates $(r, \theta)$ with the origin at the position of particle 2 ($\vec{r} = \vec{r}_1^* - \vec{r}_2^*$). From the equations of motion, it can be demonstrated that

$$\vec{r} \times \vec{V} = \vec{K} \tag{7.37}$$

which means that the plane defined by $\vec{V}$ and $\vec{r}$ moves parallel to itself (orthogonal to the fixed vector $\vec{K}$).

The force acting on the particles is a function of the distance between them. It is more convenient to use the potential energy of interaction $V(r)$ rather than the force of interaction

$$\vec{F}(r) = -\frac{dV(r)}{dr} \tag{7.38}$$

The deviation angle $\theta'$ is then defined as

$$\theta' = \pi - 2\int_{r_m}^{\infty} \frac{dr}{\sqrt{r^4/b^2 - r^2 - 2r^4V(r)/(\mu_m g_{12}^2 b^2)}} \tag{7.39}$$

where $r_m$ is the distance of closest approach, defined by $dr/d\theta = 0$, and $g_{12} = |\vec{V}| = |\vec{v}_1 - \vec{v}_2|$.

The quantity $\theta'$ is not an observable quantity because it is impossible to single out a particular impact parameter on a molecular scale. The observable quantity is $\sigma'$, the scattering cross section, which can be measured from molecular beam experiments or calculated through $\theta'$ once the interaction potential $V(r)$ is known.

In the classical mechanical approach, particles having initial velocities $\vec{v}_1$ and $\vec{v}_2$ can be scattered in their relative motion through various angles $\theta'$ and $\varphi'$. The angles are functions of $b$, and the scattering process must be described in statistical terms in order to take into account the particle distributions $f_1$ and $f_2$ before the collision and $f_1'$ and $f_2'$ after the collision. This is done in terms of the quantity $\sigma'$, which is defined so that $\sigma'(\vec{v}_1, \vec{v}_2 \rightarrow \vec{v}_1', \vec{v}_2')\, \overrightarrow{dv_1'}\, \overrightarrow{dv_2'}$ is the number of particles per unit time (per unit flux of type 1 particles incident with relative velocity $\vec{V}$ upon a type 2 particle) emerging after scattering with respective final velocities between $\vec{v}_1'$ and $\vec{v}_1' + \overrightarrow{dv_1'}$ and between $\vec{v}_2'$ and $\vec{v}_2' + \overrightarrow{dv_2'}$. According to conservation of momentum and energy, $\vec{v}_1'$ and $\vec{v}_2'$ must have a constant center-of-mass velocity and $|\vec{V}| = |\vec{V}'|$. Thus, $\sigma' = 0$ if these conditions are not fulfilled.

Some symmetry properties of $\sigma'$ are important in deriving the Boltzmann equation: $\sigma'$ is invariant

- for reversed collisions (reversed velocities),
- for space inversion ($\vec{r} \rightarrow -\vec{r}$), especially for the "inverse" collision (original collision with initial and final states interchanged).

A simpler but less symmetrical quantity, the differential scattering cross section, can also be introduced (see Section 3.2.2): $\sigma(\vec{V}')\, d\Omega'$ is the number of particles scattered per unit time and emerging with a final relative velocity $\vec{V}'$ in a direction given by the solid angle $d\Omega'$ about $\theta'$ and $\varphi'$.

From these definitions it follows that

$$\sigma(\vec{V}')\, d\Omega' = \int_{\vec{c}'} \int_{\vec{V}'} \sigma'(\vec{v}_1, \vec{v}_2 \rightarrow \vec{v}_1', \vec{v}_2')\, \overrightarrow{dv_1'}\, \overrightarrow{dv_2'} \tag{7.40}$$

where $\vec{c}'$ is the velocity of the center of mass after collision. The integration is over all values of $\vec{c}'$ and $\vec{V}'$.

Quantity $\sigma(\vec{V}')$, often defined as $\sigma(\theta', \varphi')$, can be easily related to the impact parameter $b$ when considering all the particles of the incident beam whose impact parameter is between $b$ and $b + db$. The number of collisions, if $\phi$ is the flux of incident particles with a unique relative velocity $\vec{V}$, is given by $\phi \cdot 2\pi b\, db$, while the number of scattered particles is $\phi\sigma(\vec{V}, \theta)2\pi \sin\theta'\, d\theta'$ (the scattering is supposed to be symmetric for the $\varphi'$ angles, whose integration gives $2\pi$), and thus

$$\sigma(\vec{V}, \theta') = \left|\frac{b}{\sin\theta'} \cdot \frac{db}{d\theta'}\right| \tag{7.41}$$

The absolute value is taken, because in experiments $\sigma$ is measured in crossed beams of atomic or molecular jets and, since diffusion occurs in all directions, it is not possible to distinguish between positive and negative deflection angles.

The total elastic collision cross section is defined as

$$\begin{aligned} Q(V) &= \int_0^{2\pi}\int_0^{\pi} \sigma'(\theta', \varphi', V)\sin\theta'\, d\theta'\, d\varphi' \\ &= 2\pi\int_0^{\pi} \sigma'(\theta', V)\sin\theta'\, d\theta' \end{aligned} \tag{7.42}$$

With these definitions, it is not too difficult, for a single chemical specie, to show that the collision term $\not{D}_c f$ is given by the following triple integral (integrated over all the possible velocities $\vec{v}_2$ of the target particles and the resulting velocities $\vec{v}_1'$ and $\vec{v}_2'$ of particles 1 and 2 after collision):

$$\not{D}_c f = \int_{v_2}\int_{v_2'}\int_{v_1'} (f_2'f_1' - f_1f_2)\vec{V}\sigma'(\vec{v}_1, \vec{v}_2 \rightarrow \vec{v}_1', \vec{v}_2')\, \overrightarrow{dv_1'}\, \overrightarrow{dv_2'}\, \overrightarrow{dv_2} \tag{7.43}$$

The distribution functions $f_1$ and $f_2$ are related to the colliding particles before collision when $f_1'$ and $f_2'$ are the distributions after collision.

If $\sigma(\Omega')$ [see Eq. (7.40)] is used instead of $\sigma'$, we have

$$\not{D}_c f = \int_{v_2}\int_{\Omega'} (f_2'f_1' - f_2f_1)V\sigma'\, d\Omega'\, \overrightarrow{dv_2} \tag{7.44}$$

Of course, the $\not{D}_c f$ term causes all the difficulties in solving the Boltzmann integrodifferential equation.

It is important to note that the expression of the collision term has been established (see Section 3.5.3) assuming that $f$ does not vary appreciably during the time interval on the order of the free flight time, i.e., over a spatial distance on the order of the range of intermolecular forces. Thus a shielded Coulomb's potential is used to keep this assumption not too far from reality for charged particles, with their long-range interaction potential.

When $K$ different chemical species are considered (electrons and heavy species), $K$ equations must be written for the $K$ species. For example, given a particle of chemical species $i$ it follows that

$$\mathcal{D}f_i = \sum_{j=1}^{K} \iint (f_i'f_j' - f_if_j)\vec{V}_{ij}\sigma_{ij}\, dv_j\, d\Omega = \mathcal{D}_c f_i \tag{7.45}$$

The solution of these $K$ Boltzmann equations gives $K$ distribution functions, allowing us to calculate mean values of the various properties (i.e., macroscopic properties) and fluxes.

### 7.3.2. Fluxes

As defined in Section 3.5.2, the total flux of a quantity $\chi_i$ (related to a chemical species $i$) across an elementary surface oriented by its normal $\vec{n}$ and moving in the flow direction with the mean flow velocity $\vec{v}_0$ [see Eq. (7.29)] is given by

$$\vec{J}_i = \int \chi_i f_i \vec{U}_i \vec{n}\, \overrightarrow{dU}_i \tag{7.46}$$

The different fluxes outlined in Section 7.1 are related to the transport of mass, momentum, and energy.

With $\chi_i = m_i$, we have

$$\vec{J}_i = m_i \int f_i \vec{U}_i \overrightarrow{dU}_i = n_i m_i \langle \vec{U}_i \rangle \tag{7.47}$$

where $\vec{J}_i$ is the mass flux vector and $\langle U_i \rangle$ the mean peculiar velocity [see Eq. (7.34)].

With $\chi_i = m_i U_{ix}$, where $U_{ix}$ is the component of peculiar velocity $\vec{U}_i$ in the $x$ direction, we can write

$$J_{ix} = m_i \int U_{ix} \vec{U}_i f_i \overrightarrow{dU}_i = n_i m_i \langle U_{ix} \cdot \vec{U}_i \rangle \tag{7.48}$$

This is the flux vector associated with the transport of *the $\chi$ component of momentum* (relative to $\vec{v}_0$). This vector has components proportional to $U_{ix} \cdot U_{ix}$, $U_{ix} \cdot U_{iy}$, and $U_{ix} \cdot U_{iz}$. Similar flux vectors are obtained for the $y$ and $z$ components of the momentum, making a total of three flux vectors associated with momentum transfer. The components of the three vectors form a symmetrical second-order tensor $\bar{\bar{p}}_i$ corresponding to a partial pressure and given by

$$\bar{\bar{p}}_i = n_i m_i \langle \vec{U}_i \cdot \vec{U}_i \rangle \tag{7.49}$$

representing the flux of momentum through the gas.

With $\chi_i = m_i U_i^2/2$, it follows that

$$\vec{q}_i = m_i \left( \int U_i^2 \cdot \vec{U}_i f_i \overrightarrow{dU}_i \right) \Big/ 2 = m_i n_i \langle U_i^2 \cdot \vec{U}_i \rangle / 2 \tag{7.50}$$

where $\vec{q}_i$ is the flux vector associated with the transport of kinetic energy of particles of species $i$. The sum of these vectors over all components $K$ gives the heat flux vector $\vec{q}$, whose components $q_x$, $q_y$, and $q_z$ represent the flux of kinetic energy in the $x$, $y$, and $z$ directions, respectively.

Calculation of the mass flux vector, the pressure tensor, and the heat flux vector $\vec{q}$ will allow the determination of the transport coefficients, provided $f$ is known. Thus, the main problem is to determine $f$.

The fundamental hydrodynamic equations of continuity, motion, and energy balance can be derived from the Boltzmann equation without determining the form of the distribution function $f_i$. To do this [see Hirschfelder *et al.* (1964)], the Boltzmann equation (7.45) for the $i$th component is multiplied by the quantity $\chi_i$ associated with the $i$th species and integrated over $\vec{v}_i$. This new set of equations is called the transfer equations.

It can then be demonstrated that the integrals of the right-hand side, i.e., the ones related to collisions, vanish if $\chi_i$ is $m_i$, $m_i \vec{U}_i$, or $\frac{1}{2} m_i U_i^2$, yielding the well-known Navier–Stokes equations relative to mean values.

### 7.3.3. Calculation of Distribution Functions

As in the preceding section, only the outline of the calculation procedure will be presented.

#### 7.3.3.1. Equilibrium Plasma

The theoretical study of transport coefficients of equilibrium plasmas was first developed for monatomic gases through the Boltzmann equation. The results were then modified [Hirschfelder *et al.* (1964), Chapman and Cowling (1952) and Ref. 5] to account for high temperatures and the presence of excited particles[6] and ions.[7]

A series solution of the Boltzmann equation can be obtained by introducing a perturbation parameter $\xi$ in such a manner that the frequency of collisions can be varied in an arbitrary manner without affecting the relative number of collisions of a particular kind. The Boltzmann equation is then written as

$$\mathcal{D}f_i = \frac{1}{\xi}\mathcal{D}_c f_i \tag{7.51}$$

where $1/\xi$ measures the frequency of collisions (and is proportional to $\sigma$, see Section 3.2.3). (If $\xi$ is very small, collisions would be very frequent and the gas would behave like a continuum in which local equilibrium is maintained everywhere.) The distribution function is expanded in a series as follows:

$$f_i = f_i(0) + \xi f_i(1) + \xi^2 f_i(2) + \cdots \tag{7.52}$$

where $f_i(0)$ is obtained by assuming that the right-hand side of the Boltzmann equation, which depends only on collisions, must be zero when equilibrium is reached. With the help of the H-theorem of Boltzmann, it can be shown [Hirschfelder *et al.* (1964)] that

$$f_i(0) = n_i\left(\frac{m_i}{2kT}\right)^{3/2} \exp\left(-\frac{m_i U_i^2}{2kT}\right) \tag{7.53}$$

which corresponds to Maxwellian distributions for the peculiar velocity $\vec{U}_i$ at a unique temperature $T$ [$f_i(0)$ is usually denoted $f_i^0$].

In Enskog's first-order approximation, $f_i(1)$ is written as

$$f_i(1) = f_i(0) \cdot \phi_i \tag{7.54}$$

where $\phi_i$ is a perturbation function ($\phi_i \le 1$). The perturbation function $\phi_i$ depends only on space and time via the species densities, the average flow velocity, the temperature, and their space derivatives [see Eqs. (7.56) and (7.57)]; $\phi_i$ is linear in the derivatives. The coefficients appearing in the expression of $\phi_i$ are then expanded in a finite series of Sonine's polynomials.[8–13] The final results, i.e., the transport coefficients, are expressed as functions of complex quantities called bracket integrals, which are themselves functions (through collision integrals) of the interaction potentials characterizing the different collisions.

### 7.3.3.2. Two-Temperature Plasmas

When the plasma is no longer in equilibrium, i.e., when the electron temperature $T_e$ is different from the heavy-particle temperature $T_h$ (with a given ratio $\theta = T_e/T_h$), a new kinetic model has to be developed, as has been done by Devoto[14] and Chmieleski.[15] However, Chmieleski's approach, in spite of its rigorous mathematical development, does not allow for easy integration of the interaction potentials and does not lead to numerical results. Bonnefoi[16] recently developed a complete calculation procedure of the transport properties in a two-temperature model, using the approach of Devoto. The advantage of this model is that it gives a general solution for two-temperature plasma, equilibrium being the limiting case when $\theta = 1$.

Devoto,[11] in order to emphasize the rapidity with which the distribution functions of either electrons or heavy particles relax toward Maxwellian distributions, introduces (as in the equilibrium case) a factor $1/\xi$ in the right-hand side of the Boltzmann equation for electron–electron as well as for heavy–heavy particle collisions, assuming that these two categories of particles are in a quasilocal equilibrium but with different temperatures.

Thus Eq. (7.45), where the index $j = 1$ corresponds to electrons, can be separated into two types of equations:

- For electrons, the left-hand side equals $\mathcal{D}f_1$ minus the integrals of collision related to the collisions between heavy particles, and the right-hand side is the integral of collision related to collisions between electrons multiplied by $1/\xi$.
- For each heavy species, the left-hand side equals $\mathcal{D}f_i$, with $i = 2$ to $K$, minus the integrals of collision related to the collisions between electrons and heavy particles, and the right-hand side equals the integrals of collision related to the collisions of all species of heavy particles with the heavy species $i$, multiplied by $1/\xi$.

As in the equilibrium case, the solution of these equations is written as

$$f_i = f_i^0(1 + \phi_i) \tag{7.55}$$

where $f_i^0$ are the Maxwellian distributions for species $i$. These distributions are related both to electrons ($\frac{3}{2}kT_e = \frac{1}{2}m_1\overline{U_1^2}$) and to heavy particles ($\frac{3}{2}kT_h = \frac{1}{2}m_1\overline{U_1^2}$ for $i \neq 1$). In these expressions, $T_e$ is the electron temperature and $T_h$ is a *unique* temperature for all the heavy particles (molecules, atoms, and ions).

Taking into account the low value of the ratio $m_1/m_i$ ($i \neq 1$) and provided $T_e/T_h = \theta$ is not too high ($\theta \leq 3$), Devoto[14] shows that the $\phi_i$ can be related to $\mathcal{D}f_i^0$ and to integrals depending on $\phi_i$ and $f_i^0$. By using transfer equations, Devoto[14] and also Bonnefoi[16] give complex expressions for $\phi_i$:

$$\phi_1 = \vec{A}_1 \cdot \vec{\nabla} \ln T_e - \overleftrightarrow{B}_1 : \vec{\nabla} v_0 + \vec{C}_1 \cdot \vec{d}_1' + D_1 Q_1^0 \tag{7.56}$$

and for heavy particles:

$$\phi_i = +\vec{A}_i \cdot \vec{\nabla} \ln T_h - \overleftrightarrow{B}_1 : \vec{\nabla} v_0 + \sum_{j=1}^{K} \vec{C}_i^j \cdot \vec{d}_j' + D_i \cdot Q_i^0 \tag{7.57}$$

In these equations, $\vec{A}_i$ $(i = 1, \ldots, K)$, $\vec{C}_1$, and $\vec{C}_i^j$ are vectors, $\overleftrightarrow{B}_1$ $(i = 1, \ldots, K)$ are second-order tensors, $D_i$ $(i = 1, \ldots, K)$ are scalars, and $Q_i^0$ is a complex expression given by

$$Q_i^0 = \sum_{\substack{j=1 \\ j \neq i}}^{N} \iiint \tfrac{1}{2} m_i (V_i'^2 - V_i^2) f_i^0 f_j^0 V \sigma_{ij} \, d\Omega \, d\vec{v}_i \, d\vec{v}_j \tag{7.58}$$

where $\vec{\nabla} \ln T$ and $\vec{\nabla} v_0$ are space derivatives of temperature and flow velocity [see Eq. (7.29)].

Quantities $Q_i^0$ are complex integrals depending on the differences between the squares of the relative velocities before and after collision, and $\vec{d}_i'$ are diffusion forces introduced by Bonnefoi[16] as

$$\vec{d}_i' = \frac{n_i m_i}{\rho} \sum_{j=1}^{K} n_j \vec{F}_j - n_i \vec{F}_i - n_i m_i \frac{\vec{\nabla} p}{\rho} + \vec{\nabla} p_i \tag{7.59}$$

where $\vec{F}_i / m_i$ is the acceleration of the particles of chemical species $i$ [see the Boltzmann equation (7.35) and (7.36)].

The summation $\sum_{i=1}^{K} n_i m_i = \rho$ implies that

$$\sum_{i=1}^{K} \vec{d}_i' = \sum_{i=1}^{K} \vec{\nabla} p_i - \vec{\nabla} p = \vec{\nabla} \left( \sum_{i=1}^{K} p_i - p \right) \tag{7.60}$$

and, according to Dalton's law, which pertains to partial pressures,

$$\sum_{i=1}^{K} \vec{d}_i' = 0 \tag{7.61}$$

This definition of diffusion forces is different from the one of Devoto, whose definition does not lead to (7.61) but, however, has been often applied in nonequilibrium plasmas.

Expression (7.61) shows that for the case of a two-temperature gas, the diffusion forces are dominated by the gradients of partial pressures, while at equilibrium the density gradients play a leading role[16] (in this case $p_i/p = n_i/n$).

The unknowns $D_1$ and $D_i$ are not involved in the calculation of the transport coefficients through the expression of the flux vectors (in the flux vectors, the integrals containing $D_1$ and $D_i$ vanish because their argument is odd).

### 7.3.4. Interaction Potentials and Collision Integrals

The expressions for $\phi_1$ and $\phi_i$ in eqs. (7.56) and (7.57) allow the determination of the new unknown coefficients $\vec{A}$, $\vec{B}$, and $\vec{C}$.[16] It is assumed, as Chapman and Cowling (1952) assumed for equilibrium calculations, that these coefficients can be expanded in finite series of Sonine polynomials $S_m^n(x)$. For example, when considering electrons

$$\vec{A}_1 = A_1(W_1^2)\vec{W}_1 \tag{7.62}$$

with

$$\vec{W}_1 = \left(\frac{m_1}{2kT_e}\right)^{3/2} \vec{U}_1 \tag{7.63}$$

and

$$A_1(W_1^2) = \sum_{p=0}^{l-1} a_{1p}(l) S_{3/2}^p(W_1^2) \tag{7.64}$$

where $l$ is the degree of expansion; for a first-order approximation, $l = 1, \ldots$.

Depending on the type of calculation, Eq. (7.64) is expanded to beween the second and fourth approximation.

The new unknown coefficients $a_{ip}(l)$ for $\vec{A}_i$, $b_{ip}(l)$ for $\vec{B}_i$, and $c_{ip}(l)$ for $\vec{C}_i$ are solutions of linear equations whose coefficients, written as $q_{ij}^{ls}$, are very complex expressions depending on what are called "bracket integrals." In these bracket integrals there appear quantities $\bar{Q}_{ij}^{l,s}$, which depend on:

- collision (or interaction potentials) between particles of type $i$ and those of type $j$,
- the expansion degree $l$ of the coefficients $a$, $b$, and $c$,
- Sonine type of polynomials.

Quantities $\bar{Q}_{ij}^{l,s}$ can be written in the form

$$\bar{Q}_{ij}^{l,s} = \sigma_{ij}\bar{\Omega}_{ij}^{l,s} \tag{7.65}$$

where $\sigma_{ij}$ is the cross-sectional area presented by particles considered to be hard spheres:

$$\sigma_{ij} = \pi[(d_i + d_j)/2]^2 \tag{7.66}$$

$d_i$ and $d_j$ being the diameters of the colliding particles; $\bar{\Omega}_{ij}^{ls}$ is a reduced collision integral that is calculated from the collision integral $\Omega_{ij}^{ls}$ defined by Hirschfelder *et al.* (1964) as

$$\Omega_{ij}^{ls} = \left(\frac{kT}{2\pi\mu_{ij}}\right)^{1/2} \int_0^\infty \exp(-\gamma_{ij}^2) \cdot \gamma_{ij}^{2s+1} Q_{ij}^l(g_{ij})\, d\gamma_{ij} \tag{7.67}$$

with

$$\gamma_{ij} = \frac{\mu_{ij} g_{ij}^2}{2kT} \quad \text{and} \quad \mu_{ij} = \frac{m_i m_j}{m_i + m_j} \tag{7.68}$$

Here, $g_{ij} = |\vec{v}_i - \vec{v}_j|$ is the absolute value of the relative velocity of the colliding particles, and $Q_{ij}^l$ is the quantum collision cross section.

Why a quantum collision cross section? The fundamental defect in the classical theory is that no particle trajectory can ever be precisely defined because of the inherent uncertainties associated with it. The impact parameter can be prescribed within an uncertainty $\Delta b$ only by accepting an uncertainty $\Delta p_\perp \sim \hbar/\Delta b$ in the transverse component of momentum ($\hbar = h/2\pi$, $h$ being Planck's constant). Hence an uncertainty

$$\Delta\theta = \frac{\Delta p_\perp}{mV} = \frac{\hbar}{mV\,\Delta b} \tag{7.69}$$

appears in the incident direction of the motion and results in an equal uncertainty in the final scattering angle. This uncertainty problem is solved by employing a wave function and its associated interference pattern in place of the limiting deflection for a well-defined classical trajectory (see Section 7.3.1). The precise connection between the scattering amplitude of this wave function and the differential cross section is obtained by comparing the incident flux with the rate of scattering into a given solid angle $d\Omega$. It can be shown that for the case of elastic scattering with a potential $V(r)$ the wave function can be expanded with Legendre functions. Depending on the degree $l$ of the development, the weighted corresponding cross sections are given by

$$Q^l(\vec{V}) = 2\pi \int_0^\pi \sigma(\theta', \vec{V}) \cdot (1 - \cos^l \theta') \sin\theta' \, d\theta' \tag{7.70}$$

or, in terms of the impact parameter,

$$Q^l(\vec{V}) = 2\pi \int_0^\pi (1 - \cos^l \theta') b \, db \tag{7.71}$$

As already emphasized, these collision cross sections depend on the interaction potential $V(r)$.

From a practical point of view, collision integrals are calculated as follows:

- If the transfer cross section $Q^l$ is known (from experience or theory), Eq. (7.67) is integrated numerically. This is generally the case for collisions between electrons and neutrals.
- If the interaction potential of the collision is known ($Q^l$ depends

directly on it through the scattering angle $\theta'$), the determination of $\bar{\Omega}^{l,s}$ is performed with the help of tables. Calculated or tabulated collision integrals for the most useful potentials have been gathered from the literature and are presented in the thesis of Gorse[1] and in Aubreton,[17] as well as in Aubreton and Fauchais.[18] When the interaction involves different excited states, each with its own potential, collision integrals are calculated successively for each potential and then a mean collision integral such as

$$\bar{\Omega}^{l,s} = \frac{\sum_k g_k \bar{\Omega}_k^{l,s}}{\sum_k g_k} \tag{7.72}$$

is defined, where $g_k$ is the statistical weight of the potential $k$.

### 7.3.5. Transport Properties

#### 7.3.5.1. Thermal and Ordinary Diffusion

For electrons, the diffusion velocity is given [see Eq. (7.34)] by

$$\langle \vec{U}_1 \rangle = 1/n_1 \cdot \int \vec{U}_1 \cdot f_1 \cdot \overrightarrow{dU}_1 = 1/n_1 \cdot \int \vec{U}_1 \cdot f_1^0 \cdot \phi_1 \cdot \overrightarrow{dU}_1 \tag{7.73}$$

This can be rewritten after some calculations as

$$\langle \vec{U}_1 \rangle = \frac{1}{n_1 k T_e} D_{11}(1) \vec{d}_1' - \left( \frac{D_{1T}(1)}{n_1 m_1 T_e} \right) \vec{\nabla} T_e \tag{7.74}$$

where $\vec{d}_1'$ is the diffusion force [see Eq. (7.59)], and $D_{1T}(1)$ and $D_{11}(1)$ are, respectively, the thermal and ordinary diffusion coefficients given by

$$D_{1T} = \frac{n_1 m_1}{2} \left( \frac{2kT_e}{m_1} \right)^{1/2} a_{10}(1) \tag{7.75}$$

and

$$D_{11}(1) = \frac{n_1 k T_e}{2} \left( \frac{2kT_e}{m_1} \right)^{1/2} c_{10}(1) \tag{7.76}$$

These can both be expressed as functions of the $q_{11}$ coefficients (see Section 7.3.4). All the corresponding expressions as functions of the collision integrals can be found in the thesis of Bonnefoi.[16]

Diffusion velocities for heavy particles are given by expressions similar to Eq. (7.74) for electrons, but with index $i$ substituted for index 1.

A similar calculation using Sonine's expansion of $\vec{C}_i(W_i)$ and $\vec{A}_i(W_i)$

[see Eq. (7.57)] gives expressions for $\langle \vec{U}_i \rangle$ that contain $D_{iT}$ and $D_{ij}$, the thermal and ordinary diffusion coefficients, respectively.

The expressions for these coefficients (which are very complex determinants) can be found in the thesis of Bonnefoi.[16]

Such expressions allow us to define the ordinary diffusion coefficients, which to a first-order approximation can be written as

$$D_{ij} = \frac{2.638 \times 10^{-7}}{p} \frac{T^{3/2}}{\bar{\Omega}_{ij}^{1,1}} \left( \frac{m_i + m_j}{2m_i m_j} \right)^{1/2} \tag{7.77}$$

where $D_{ij}$ is given in m$^2$/s, $p$ in atm, and $\bar{\Omega}_{ij}^{1,1}$ in Å$^2$.

This expression explains why $\bar{\Omega}_{ij}^{1,1}$ is often called the diffusion collision integral. Of course, Eq. (7.77) shows that $D_{ij}$ increases when the pressure decreases [this was also shown by the oversimplified expression in Eq. (7.12)]. Usually the $\bar{\Omega}_{ij}^{1,1}$ decrease with temperature, and thus the $D_{ij}$ values increase with temperature.

#### 7.3.5.2. Electrical Conductivity

The electronic contribution to the electric current flux is, by definition [Hirschfelder *et al.* (1964)],

$$\vec{J}_1 = -en_1 \langle \vec{U}_1 \rangle \tag{7.78}$$

Taking into account Eqs. (7.74) and (7.59) as well as the electrical neutrality, and assuming that the electrical conductivity is the coefficient of the electrical field $\vec{E}$ in the resulting expression [in Eq. (7.59) the forces are written $\vec{F}_i = eZ_i' \vec{E}$, where $eZ_i'$ is the electrical charge of the particle of species $i$], it follows that

$$\sigma_e(1) = \frac{e^2 n_1}{kT_e} D_{11}(1) \tag{7.79}$$

Since ion contributions can be neglected in a good approximation [Hirschfelder *et al.* (1964)], $\sigma_e(1)$ is therefore the total electrical conductivity of the gaseous mixture.

For the fourth term in the Sonine expansion, $\sigma_e$ is given by

$$\sigma_e(4) = \frac{e^2 n_1}{kT_e} D_{11}(4) \tag{7.80}$$

$$= 1.85926 \times 10^{-15} \frac{n_1}{T_e} D_{11}(4) \tag{7.81}$$

The expression allowing calculation of $\sigma_e(4)$ can be found in the thesis of

Bonnefoi.[16] It is noteworthy that when the calculation is limited to the first-order approximation (Devoto[10] has demonstrated that third-order approximation is a requisite), we have

$$\sigma_e(1) = \frac{3e^2\sqrt{\pi}}{8\sqrt{m_e k T_e}} \frac{n_1}{\sum_{j=2}^{K} n_j \bar{Q}_{ij}^{11}} \tag{7.82}$$

where $\bar{Q}_{ij}^{11}$ is related to the collision integrals [see Eq. (7.65)]. Also note that this expression is quite similar to the simplified one from kinetic theory [see Eq. (7.25), where $n_e = n_1$] except for the coefficient and the quantity $n\sigma_{en}$, which is replaced by $\sum_{j=2}^{K} n_j \bar{Q}_{ij}^{11}$. If Eq. (7.82) is oversimplified, it shows the general trend of the evolution. At all temperatures $\sigma_e$ (see Fig. 7.3) is essentially controlled by the electron density, which rises rapidly with temperature for $T > 6000$ K for Ar, $N_2$, $H_2$, and $O_2$ plasmas and reaches a quasi-plateau at $T > 14{,}000$ K. First the collisions with neutrals play an essential role: for example, if $\bar{Q}_{e\mathrm{Ar}}^{11}$ increases slowly with temperature[17] in an Ar plasma, $n_{\mathrm{Ar}}$ decreases much faster, resulting in a faster rise of $\sigma_e$ with $T$. For $T > 9000$–$10{,}000$ K, $\mathrm{Ar}^+$ has to be accounted for, and $n_{\mathrm{Ar}^+}$ increases with $T$ when $\bar{Q}_{e\mathrm{Ar}^+}^{(1,1)}$ decreases.

### 7.3.5.3. Translational Thermal Conductivity

The translational thermal conductivity is related to elastic collisions. By definition, the heat flux vector for electrons is written as

$$\vec{q}_1 = \frac{m_1}{e} \int \vec{U}_1 \cdot U_1^2 \cdot f_1^0 \phi_1 \, \overrightarrow{dU}_1 \tag{7.83}$$

and the thermal conductivity of the electrons $\kappa_1$ (also denoted $\kappa_{\mathrm{tr}}^e$) is defined as the coefficient of $-\vec{\nabla} T_e$. The corresponding expressions are given in the thesis of Bonnefoi.[16]

Even when limiting the development of $\kappa_1$ to the second-order approximation, its expression is complex:

$$\kappa_1(2) = \tfrac{75}{8} n_1^2 k \left(\frac{2\pi k T_e}{m_1}\right)^{1/2} \frac{1}{q_1^{11}} \tag{7.84}$$

with

$$q_1^{11} = 8\sqrt{2}\, n_1^2 \bar{Q}_{11}^{2,2} + 8 n_1 \sum_j n_j \left(\tfrac{25}{4} \bar{Q}_{1j}^{1,1} - 15 \bar{Q}_{1j}^{1,2} + 12 \bar{Q}_{1j}^{1,3}\right) \tag{7.85}$$

where again the $\bar{Q}_{1j}^{ls}$ are related to collision integrals [see Eq. (7.65)]. However, for Ar, $H_2$, $O_2$, and $N_2$ plasmas and their mixtures at atmospheric pressure, $\kappa_1(4)$ (see Fig. 7.2) is negligible for $T < 7000$ K (again illustrating the importance of $n_e$), and it rises steadily with increasing temperature (at least up to 20,000 K).

A similar analysis for heavy particles shows that the heavy-particle heat flux vector is given by the expression

$$\vec{q}' = \tfrac{5}{2} \cdot kT_h \sum_{j=2}^{K} n_j \langle \vec{U}_j \rangle - \kappa^{*\prime} \vec{\nabla} T_h - \sum_{k=2}^{K} \frac{D_{kT}(l)}{n_k m_k} \vec{d}_1' \tag{7.86}$$

where

$$\kappa^{*\prime} = -\frac{5k}{4} \sum_{j=2}^{K} n_j \left( \frac{2kT_h}{m_j} \right)^{1/2} a_{j1}(1) \tag{7.87}$$

According to the work of Muckenfus and Curtis,[19] the term proportional to the thermal diffusion coefficient in Eq. (7.86) contains a new coefficient of $-\vec{\nabla} T_h$, which when added to $\kappa^{*\prime}$ yields the true heavy-particle translational thermal conductivity $\kappa'$. It is quite remarkable that $\kappa'(1)$ can be derived[16] from Eq. (7.87), in which it is sufficient to insert the series expansion of $A_i(W_i)$.

The expression for the heavy-particle thermal conductivity coefficients [$\kappa'(4)$, up to the fourth term of the Sonine expansion] are given in the thesis of Bonnefoi.[16]. Of course, in a two-temperature plasma Eq. (7.86) has to be calculated at $T_h$. This expression is quite independent of the electrons, and Devoto[10] has demonstrated that the Sonine expansion to order 2 is sufficient. Figure 7.2 shows the evolution of $\kappa_{\mathrm{tr}}^h$ for a nitrogen plasma at atmospheric pressure: $\kappa_{\mathrm{tr}}^h$ rises slowly up to 8000 K, and after a plateau it decreases slowly at temperatures higher than 11,000 K, following the evolution of the nitrogen atoms, which are progressively replaced by nitrogen ions (see the corresponding evolution of $\kappa_{\mathrm{tr}}^e$ in Fig. 7.2).

#### 7.3.5.4. Viscosity

Starting from the stress tensor expression [Hirschfelder *et al.* (1964)] for the electrons and using a theorem due to Chapman and Cowling (1952), the viscosity $\mu_1$ of the electrons is given[16] by

$$\mu_1 = \frac{1}{15} \frac{m_1^2}{2kT_e} \int f_1^0 U_1^4 B_1(W_1) \cdot \overrightarrow{dU}_1 \tag{7.88}$$

where $B_1(W_1)$ is the coefficient of the second-order tensor: $\overleftrightarrow{B}_1 = B_1(W_1) \cdot \vec{W}_1 \cdot \vec{W}_1$. The derivation of this expression for $\mu$ for the electrons can be found in the thesis of Bonnefoi.[16]

In most cases, the electron viscosity is neglected[10] and only the heavy-species viscosity is considered. However, in the expression for the heavy-species viscosity, collisions with electrons are taken into account. For a completely ionized gas, neglecting the electron viscosity modifies the viscosity calculated from the heavy species by about 2%.[10,11]

Starting from the heavy-particle stress tensor expression and using the variational method [Hirschfelder *et al.* (1964)], the heavy-particle viscosity $\mu'(4)$ (to the fourth order of the Sonine expansion) has an expression similar to that obtained in an equilibrium plasma. Here again, the subscript 1, which is reserved for electrons, is excluded from the lower indexes $i$ and $j$. The calculations in a two-temperature plasma have to be performed at $T_h$. The corresponding expression can be found in the thesis of Bonnefoi.[16]

From such expressions the viscosity of a pure gas can be defined by

$$\mu_i(1) = \frac{5}{16\sqrt{\pi}} \frac{\sqrt{mkT_h}}{\bar{Q}_{ii}^{22}} \tag{7.89}$$

which is similar (except for the constant) to the kinetic theory formula [see Eq. (7.18)], where $\sigma_0$ has been replaced by $\bar{Q}_{ii}^{22}$. That is why the $\bar{\Omega}_{ij}^{22}$ collision integrals, related to $\bar{Q}_{ij}^{22}$ by Eq. (7.65), are often called viscosity integrals. Figure 7.1 shows the evolution of the viscosity of different plasma gases at atmospheric pressure. First the interactions between neutral molecules control the value of $\mu_i$, whose contribution is rather well represented by Eq. (7.89), then with dissociation it will be the neutral atoms, and afterward the charged particles. It can be seen for temperatures where ionization is pertinent (about 10,000 K for Ar, $N_2$, $H_2$, and $O_2$ and 17,000 K for He) that $\mu$ decreases with increasing temperature. Because of the relatively long-range Coulomb forces between charged particles, their mobility decreases with increasing charged particle densities, resulting in a reduction of the viscosity with a corresponding increase in the collision integrals for charged particles.

## 7.4. CONTRIBUTION OF OTHER TRANSPORT MECHANISMS TO THE THERMAL CONDUCTIVITY

### 7.4.1. Reactional Contribution

Until now it has been assumed that the system is in a frozen state, i.e., that chemical reactions among the components (dissociation, ionization, and so on) could be neglected.

In actuality, these reactions do take place, and here we will assume that dissociations occur at $T_h$ (mainly due to collisions between heavy particles) and ionizations at $T_e$ (primiarly due to collisions with electrons). These two types of reactions result in additional heat transfer when recombination occurs in lower-temperature regions. Note that both reactions are due to the existence of temperature gradients (and thus composition gradients) in the arc or the RF plasma at pressures between a few tenths of an atmosphere and an atmosphere. It is thus necessary to introduce a new coefficient, $\kappa_R$, the so-called reactive or reactional thermal conductivity. Dissociations and ionizations may appear simultaneously, and it is not possible, in spite of the temperature differences, to consider this new thermal conductivity as a corrective term to either the thermal conductivity of the electrons or that of the heavy particles. It has to be treated as a coefficient that depends on the whole mixture.

A study of $\kappa_R$ has been performed in a unique-temperature gas (equilibrium conditions) by Butler and Brokaw,[20] and their expressions were subsequently extended to ionized gases by Meador and Stanton.[21] Butler and Brokaw's method uses expressions that are rigorous at equilibrium but meaningless in a two-temperature plasma, where their extension is not possible.

To develop the calculation in a two-temperature plasma, the reactional heat flux vector is introduced [Hirschfelder *et al.* (1964)] in the form

$$\vec{q}_R = \sum_{i=1}^{K} n_i' H_i \langle \vec{U}_i \rangle \tag{7.90}$$

where $H_i$ is the molar enthalpy, $n_i'$ the mole number, and $\langle \vec{U}_i \rangle$ the diffusion velocity of species $i$. Using partial pressures instead of densities, $\vec{q}_R$ can be written[16] as

$$\vec{q}_R = \sum_{i=1}^{K} H_i \langle \vec{W}_i' \rangle \tag{7.91}$$

where

$$\langle \vec{W}_i' \rangle = \frac{p_i}{RT_i} \langle \vec{U}_i \rangle \tag{7.92}$$

$R$ is the perfect gas constant, $T_i = T_e$ for $i = 1$ and $T_i = T_h$ for $i \neq 1$.

Since $\theta = T_e/T_i$, temperature gradients for heavy particles and electrons are related, and $\vec{q}_R$ can be written as

$$\vec{q}_R = -\kappa_R \vec{\nabla} T_e \tag{7.93}$$

or

$$\kappa_R \vec{\nabla} T_e = -\sum_{i=1}^{K} H_i \langle \vec{W}_i' \rangle \tag{7.94}$$

and thus $\kappa_R$ can be calculated once the quantities $\langle \vec{W}'_i \rangle$ are known as functions of the temperature gradients.

Again let us consider the example in which we begin with one mole of $N_2$ at $T_0$ and $p_0$ and take into account the following species at $T$ and $p$:

| $i$ | 1 | 2 | 3 | 4 |
|---|---|---|---|---|
| species | $e$ | $N$ | $N^+$ | $N_2$ |

Two reactions must be considered:

dissociation $N_2 \rightleftarrows 2N$ with its corresponding enthalpy variation $H_N^D = 2H_N - H_{N_2}$

ionization $N \rightleftarrows N^+ + e^-$ with $H_{N^+}^I = H_{N^+} + H_e - H_N$

along with conservation of nitrogen ($2N_{N_2} + N_N + N_{N^+} = 2N_{av}$) and electrical neutrality ($N_{N^+} = N_e$).

If the gas mixture is considered to be in an equilibrium state, the total flux of any free species, either free or bounded, must be zero, i.e., for this nitrogen plasma

$$2\langle \vec{W}'_{N_2} \rangle + \langle \vec{W}'_N \rangle + \langle \vec{W}'_{N^+} \rangle = 0 \tag{7.95}$$

and

$$\langle \vec{W}'_{N^+} \rangle = \langle \vec{W}'_E \rangle \tag{7.96}$$

Using Eqs. (7.95) and (7.96) in (7.94), it can easily be shown that

$$\kappa_R \vec{\nabla} T_e = \langle \vec{W}'_{N_2} \rangle H_N^D + \langle \vec{W}'_e \rangle H_{N^+}^I \tag{7.97}$$

and it is then necessary to calculate $\langle \vec{W}'_{N_2} \rangle$ and $\langle \vec{W}'_e \rangle$ to obtain $\kappa_R$.

By using the definition of diffusion forces [see Eqs. (7.59) and (7.60)] and the expression for the binary diffusion coefficients $D_{ij}(1)$ [see Eq. (7.77)], neglecting thermal diffusion, and introducing the molar fractions $x_i$ ($x_i = n_i/n_T$), relationships between quantities $\langle \vec{W}'_i \rangle$ and pressure gradients can be established for heavy species.[17,23]

The problem is then to relate pressure gradients to the temperature gradients that characterize thermal conductivity. This can be done by differentiating the equilibrium constants [defined in the two-temperature model by Eqs. (6.50) and (6.51)]. For example, considering the nitrogen plasma, differentiation of Eq. (6.50) gives

$$\frac{d \ln K_p}{dT_h} \frac{dT_h}{dT_e} \vec{\nabla} T_e = 2\vec{\nabla} \ln p_N - \vec{\nabla} \ln p_{N_2} \tag{7.98}$$

By using van't Hoff's law, Eq. (7.98) can be rewritten as

$$2\vec{\nabla} \ln p_{\mathrm{N}} - \vec{\nabla} \ln p_{\mathrm{N}_2} = \frac{H_{\mathrm{N}}^{\mathrm{D}}}{RT_h^2} \cdot \frac{1}{\theta} \cdot \vec{\nabla} T_e = \frac{H_{\mathrm{N}}^{\mathrm{D}}}{RT_e^2} \theta \vec{\nabla} T_e \tag{7.99}$$

It is then possible to obtain relationships between $\langle \vec{W}_i' \rangle$ and $\vec{\nabla} T_e$ and thus to write[16,22]

$$\kappa_R = \frac{\theta}{RT_e^2} \frac{\Delta_1}{\Delta} \tag{7.100}$$

with

$$\Delta_1 = \begin{vmatrix} 0 & n_{\mathrm{N}^+} H_{\mathrm{N}^+}^{\mathrm{I}} & n_{\mathrm{N}_2} H_{\mathrm{N}}^{\mathrm{D}} \\ H_{\mathrm{N}^+}^{\mathrm{I}} & A_{\mathrm{N}^+,\mathrm{N}^+} & A_{\mathrm{N}^+,\mathrm{N}_2} \\ H_{\mathrm{N}}^{\mathrm{D}} & A_{\mathrm{N}_2,\mathrm{N}^+} & A_{\mathrm{N}_2,\mathrm{N}_2} \end{vmatrix} \tag{7.101}$$

and

$$\Delta = \begin{vmatrix} A_{\mathrm{N}^+,\mathrm{N}^+} & A_{\mathrm{N}^+,\mathrm{N}_2} \\ A_{\mathrm{N}_2,\mathrm{N}^+} & A_{\mathrm{N}_2,\mathrm{N}_2} \end{vmatrix} \tag{7.102}$$

It is beyond the scope of this book to give the expressions for the $A$ coefficients. In two-temperature plasmas, $\Delta$ has no simple expressions compared with the case of the equilibrium plasmas[22] and thus must be calculated for each specific mixture following the guidelines given above. For more details the reader should refer to Refs. 16 and 22.

Figure 7.2 shows the importance of the reactional term $\kappa_R$ compared with the other contribution for nitrogen when the temperature approaches dissociation or ionization temperatures.

### 7.4.2. Contribution of Internal Energy

At low temperature it can be readily seen from Eqs. (7.17) and (7.22) that

$$\frac{\mu c_v}{\kappa} = 1 \tag{7.103}$$

or, in the more exact formulation of Chapman and Cowling (1952),

$$\frac{\mu c_v}{\kappa} = \frac{1}{c} \quad \text{with } c = 2.5 \tag{7.104}$$

Experiments show that Eq. (7.104) does not work for polyatomic gases.

In 1913 Eucken[3] advanced a very simple intuitive argument to extend the validity of the equations developed for monoatomic gases to polyatomic ones. He remarked that the internal degrees of freedom contribute little to the transport of momentum but play a more important role in thermal conductivity, especially at high temperatures.

So he proposed that $c_v = c_v^{tr} + c_v^{i}$ ($c_v^{i}$ being due to internal motions) and concluded that the factor $c$ should be applied only to $c_v^{tr}$ and not to $c_v^{i}$. After a few calculations this gives

$$\frac{\mu c_v}{\kappa} = \frac{1}{f} \qquad \text{with } f = \tfrac{1}{4}(9\gamma - 5) \tag{7.105}$$

where $\gamma = c_p/c_v$.

Of course, for a monatomic gas with $\gamma = 5/3$, $f = c = 2.5$.

This Eucken theory has been reexamined by Mason *et al.*[5] and then extended [Hirschfelder *et al.* (1964)] to internal degrees of freedom.

To calculate the internal heat flux vector related to excited states, Hirschfelder *et al.* (1964) assume that heavy particles of species $i$ that can exist in an excited state $s$ are in fact different chemical species $i, s$. They assume that all the distributions of particles of type $i$ in excited states $s$ are Maxwellian at temperature $T_h$, that pressure and field gradients are zero, and that the interparticle potentials are equal (no thermal diffusion). Thinking of collisions between excited states $s$ of species $i$ as chemical reactions, we can write for species $i$, as we did in the preceding section,

$$(\kappa_{\text{int}})_i \vec{\nabla} T_h = \frac{p_i}{RT_h} \sum_{j=1}^{J} x_j H_j \langle \vec{U}_j \rangle \tag{7.106}$$

where the index $j$ stands for the different excited states of the species $i$, which have a total number $J$; $H_j$ is the molar enthalpy corresponding to the excited state $j$. After some calculations [see Ref. 16 and Hirschfelder *et al.* (1964)] it can be shown that

$$(\kappa_{\text{int}})_i = n_i k D_{ii}(1) - \left(\frac{c_{pi}}{R} - \frac{5}{2}\right) \tag{7.107}$$

where $c_{pi}$ is the molar specific heat of species $i$ at constant pressure and $D_{ii}(1)$ is the self-diffusion coefficient.

Then, following Hirschfelder *et al.* (1964), the calculation can be extended to the mixture of heavy particles by

$$\kappa_{\text{int}} = \sum_{i=2}^{K} \frac{x_i(\kappa_{\text{int}})_i}{\sum_{j=2} x_j D_{ii}(1)/D_{ij}(1)} \tag{7.108}$$

Figure 7.2 shows the contribution of $\kappa_{\text{int}}$ to the total thermal conductivity of a nitrogen plasma at atmospheric pressure; its contribution is not very important.

## 7.5. TRANSPORT COEFFICIENTS OF SIMPLE GASES AND COMPLEX GAS MIXTURES IN CTE

As mentioned earlier, when the distribution functions are known, transport properties can be calculated if either the interaction potentials or cross sections of the different elementary collisions involved are known. However, the accuracy of these calculations is far below that of the thermodynamic properties due to uncertainties in the determination of the distribution functions as well as to difficulties in the experimental work involved in the measurements of collisional cross sections. Another problem related to the accuracy of these calculations is the number of terms to take into account in the development of Sonine polynomials for the translational properties, i.e., the number of terms in the transport-property expressions. Most of the comments about this accuracy were made by Devoto,[10] and they relate to his calculations for argon:

- When calculating viscosity, the contribution of the electrons is generally negligible (less than 1%). The difference between the results obtained using the first and the second approximation for the heavy particles is less than 0.1% (see Fig. 7.5, which represents this difference as a percent). When the contribution of the

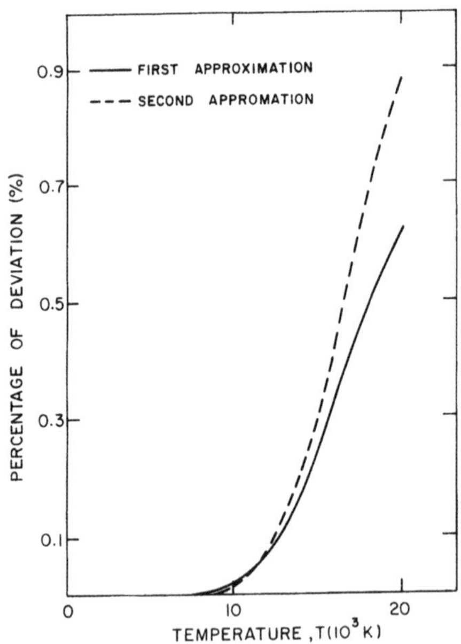


**FIG. 7.5.** Viscosity data of argon calculated using the first and second Sonine approximations neglecting the contribution of the electrons.[11]

electrons is neglected, the calculations for the heavy particles are performed to either a first or a second approximation.

- For the translational thermal conductivity, the difference between the third and the fourth approximation (used for electron and heavy particles) is less than 0.1%. The difference between using only the second approximation compared with using the third for the heavy particles while maintaining the third approximation for the electrons is less than 0.9%.
- For the electrical conductivity Devoto has also shown that the third approximation is generally sufficient.

Thus most authors have calculated their transport properties using the following approximations:

- For viscosity: taking into account only heavy particles in the first or second approximation,
- for thermal conductivity: generally the third approximation,
- for electrical conductivity: the third or fourth approximation (most often the third).

### 7.5.1. Examples for Simple Gases

The results obtained for the properties of the simple gases most commonly used in plasmas ($H_2$, $N_2$, Ar, and He) are of primary importance in understanding heat and momentum transfer between plasmas and condensed particles. In the following examples, all the results will be presented at atmospheric pressure.

#### 7.5.1.1. Viscosity

The viscosity coefficient establishes the proportionality between the friction force in the direction of the flow and the velocity gradient in the orthogonal direction. Calculation of this coefficient requires the first approximation of the pressure tensor according to Chapman–Enskog [see Chapman and Cowling (1952)].

As shown in Fig. 7.1, the viscosities of Ar, He, $H_2$, and $N_2$ are rather different. As a first approximation [see Eq. (7.89)], $\mu$ is proportional to the square root of the product of mass and temperature and inversely proportional to the $\bar{Q}_{ij}^{22}$ collision integral, at least for charged-particle molar fractions below 1%. This is illustrated by comparing Ar and He values: in spite of the fact that $m_{\mathrm{Ar}}/m_{\mathrm{He}} \simeq 10$, the viscosity of He up to 10,000 K is almost equal to that of Ar because $\bar{Q}_{\mathrm{Ar-Ar}}^{22}/\bar{Q}_{\mathrm{He-He}}^{22}$ is close to $\sqrt{10}$. Of course, as soon as $T$ reaches 10,000 K, where ionization is pertinent (see Fig. 6.2), the viscosity of Ar starts to decrease due to the reduction of the charged-particle mobility as the long-range Coulomb forces induce a lowering of the transport of momentum.[23] For He, this decrease is not observed until $T \approx 17{,}000$ K due to the higher ionization potential of He (see Fig. 6.3). Up to these two maxima the variation of $\mu$ with $\sqrt{T}$ is a good first approximation because the $\bar{Q}_{ij}^{22}$ decrease slowly with temperature. For diatomic gases, dissociation results in a change of the slope of $\mu(T)$, as is illustrated by the behavior of $H_2$ (see Figs. 6.6 and 7.1), where the change in the slope at 3500 K is due to the change in collision integrals (from $H_2$–$H_2$ to H–H) and the corresponding change from $\sqrt{2}$ for $H_2$ to 1 for H. As for Ar, $\mu$ for $H_2$ starts to decrease at 11,000 K due to the presence of charged species.

The viscosity of plasmas at 10,000 K lies typically in the range from $0.03 \times 10^{-3}$ to $0.33 \times 10^{-3}$ (kg/m s). The maximum values are approximately ten times higher than those of the same gases at room temperature. This difference partly explains the difficulty of mixing cold gases with plasmas or introducing solid particulates into a thermal plasma stream in the temperature range from 5000 to 10,000 K. Note that the choice of the interaction potentials to be used in the calculation of the collision integrals plays a very important role in calculating the viscosity, as illustrated in Fig. 7.6 (after Lesinski and Boulos[24]) for argon. The

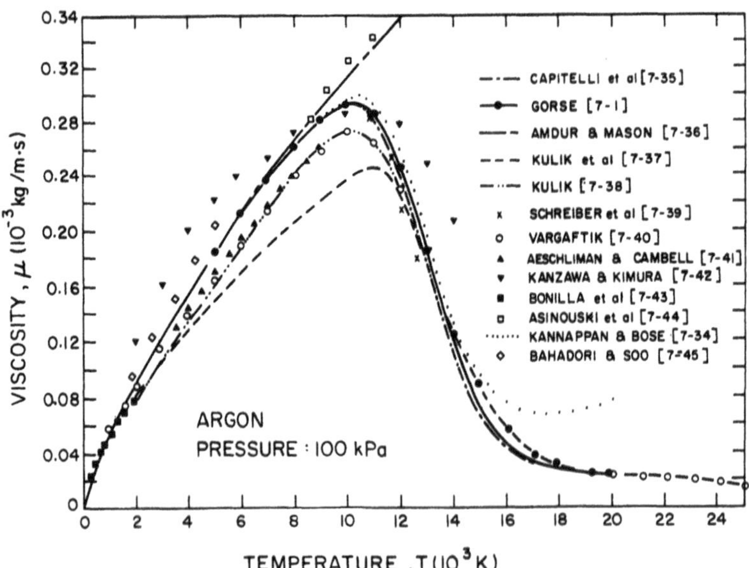


**FIG. 7.6.** Temperature dependence of the viscosity of argon calculated by various authors at atmospheric pressure.[24]

agreement is better for temperatures above 12,000 K because most authors use the same calculations for the interactions between charged particles.

#### 7.5.1.2. Electrical Conductivity

In the presence of an electric field (in arcs or RF discharges, for example) or under the action of other driving forces such as temperature, pressure, or concentration gradients, transport within the plasma occurs by electric chages that constitute an electric current.

Figure 7.3 shows the evolution of the electrical conductivity $\sigma_e$ versus temperature at atmospheric pressure for He, Ar, $N_2$, and $H_2$. The electrical conductivity increases sharply with temperature and reaches a limiting value of 10,000 $(\Omega\,\mathrm{m})^{-1}$ before second ionizations occurs. The influence of ionization on the electrical conductivity is obvious: the plasma is electrically conducting when the gas is sufficiently ionized [Eqs. (7.25) and (7.80) show that $\sigma_e$ is proportional to the electron density $n_e$, i.e., the degree of ionization of the plasma]. Ar, $N_2$, and $H_2$, which have comparable ionization energies, consequently have about the same electrical conductivity ($\sigma_e$) at the same temperature. For example, at 10,000 K, $\sigma_e$ is about 2000 $(\Omega\,\mathrm{m})^{-1}$ for these three gases, while He has a very low value of $\sigma_e$ = 500 $(\Omega\,\mathrm{m})^{-1}$ due to its higher ionization energy.

#### 7.5.1.3. Thermal Conductivity

The thermal conductivity is of primary importance for thermal plasmas. It controls the energy losses in the arc of RF discharge fringes and thus the discharge behavior as well as the heat transfer to solid materials or particulates in flight. The translational thermal conductivity (see Section 7.3.5.3) can be written as the sum of two terms, one due to electrons ($\kappa_{\mathrm{tr}}^{e}$) and the other due to heavy particles ($\kappa_{\mathrm{tr}}^{h}$). It was shown in Section 6.3.3 in the discussion on equilibrium composition that at higher temperatures the plasma gas is dissociated and ionized. The dissociation and ionization phenomena make a large contribution to energy transport, and the corresponding $\kappa_R$ term (see Section 7.4.1) has a high value at the temperatures where these phenomena occur. Since each species has an internal energy component (due to vibrational, rotational, and electronic excitation) at high temperature, energy is transferred through inelastic collisions of the second kind. This effect is taken into account by the internal thermal conductivity (see Section 7.4.2). The total thermal conductivity is considered to be the sum of these three contributions: $\kappa = \kappa_{tr} + \kappa_R + \kappa_{int}$. The contributions of these three terms will now be considered using nitrogen as an example, since it has been studied in detail from an equilibrium point of view.

Figure 7.2 shows that for nitrogen, the translational and internal conductivities of neutral particles make a small contribution to the total thermal conductivity. The main contribution is clearly the reactional one, which shows two maxima corresponding to the dissociation and ionization energies. At temperatures higher than 10,000 K, the translational thermal conductivity of electrons becomes significant. It is interesting to compare Fig. 7.2 with Fig. 7.7, which shows the thermal conductivity of argon as a function of temperature. No dissociation occurs for argon at low temperatures ($T < 10{,}000$ K), where the translational contribution of neutral species is the most significant. The reactional contribution, along with the contribution of the electrons, becomes important at 10,000 K because Ar is significantly ionized (see Fig. 6.2). At very high temperatures, when the first ionization is completed, the translational thermal conductivity of the electrons is mainly responsible for the energy transfer until second ionization occurs.

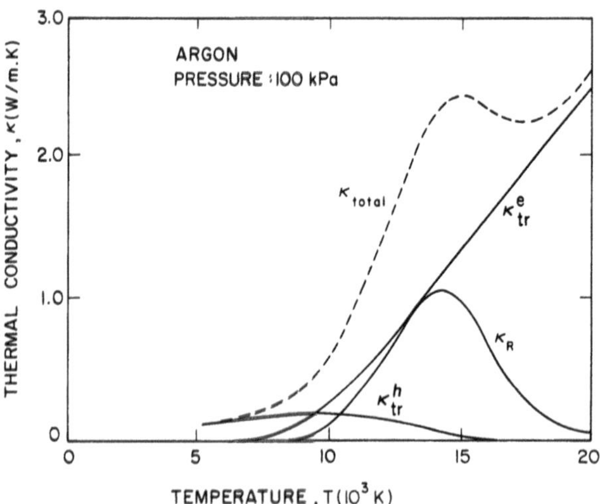


**FIG. 7.7.** Temperature dependence of the different components of the thermal conductivity of argon plasmas at atmospheric pressure.[1,2]

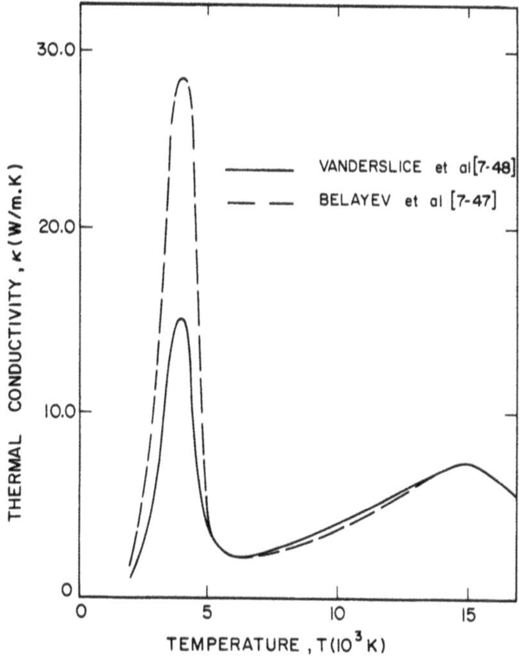


**FIG. 7.8.** Temperature dependence of the thermal conductivity of the hydrogen plasma for two different values of the H–H interaction potential at atmospheric pressure.[1]

A large volume of data has been published for the thermodynamic and transport properties of gases under thermal plasma conditions. These data have been reviewed in a progress report of the IUPAC Subcommision on Plasma Chemistry[2] as well as by Lesinski and Boulos.[24] One has to be very careful when using such data, especially transport properties, because the collision integrals on which the calculations are based are not always well known, and thus large uncertainties may result. For example, Fig. 7.8 shows the influence of such uncertainties on the H–H interaction potentials for the thermal conductivity of hydrogen. Errors of a factor of 2 can occur in the translational thermal conductivity due to uncertainties in the interaction potentials. The same uncertainties may occur in the other transport properties. The data in Figs. 7.9 and 7.10 represent the results of both experimental and theoretical studies compiled by Lesinski and Boulos[24] for argon and nitrogen at atmospheric pressure. Due to significant experimental difficulties and uncertainties in the theoretical data bases used, important differences between the calculated and measured transport properties can be observed.

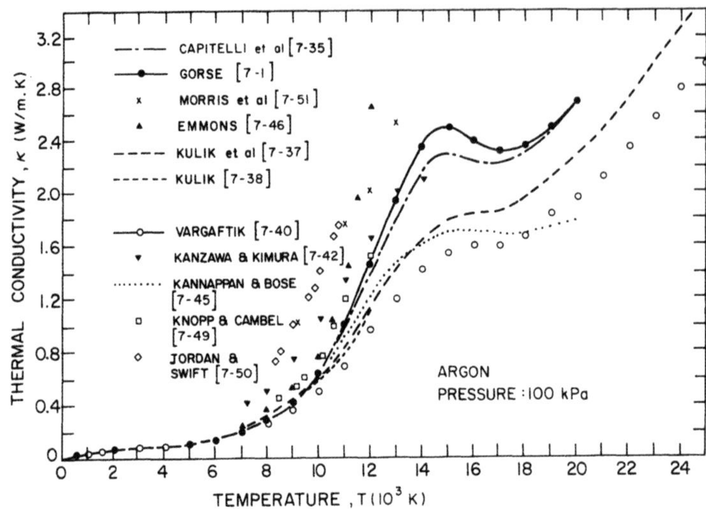


**FIG. 7.9.** Temperature dependence of the thermal conductivity of argon at atmospheric pressure (data calculated by various authors and gathered by Lesinski and Boulos[24]).

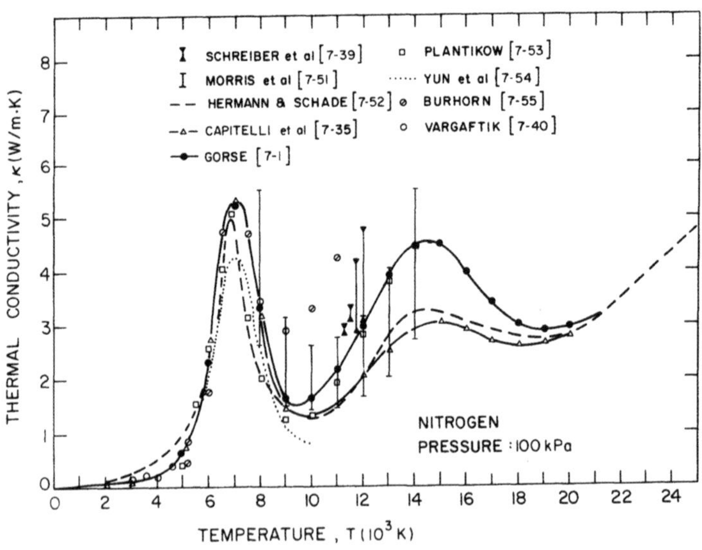


**FIG. 7.10.** Temperature dependence of the thermal conductivity of nitrogen at atmospheric pressure (data calculated by various authors and gathered by Lesinski and Boulos[24]).

### 7.5.2. Examples for Complex Gas Mixtures

For complex gas mixtures the calculations become more difficult, first because the number of interaction potentials to be taken into account increases drastically ($C_K^2$ for $K$ species, including electrons), and second, because of the lack of data about interaction potentials (when they are unknown, the species are supposed to be interacting as hard spheres). In the following paragraphs the transport properties of mixtures such as air, Ar–$H_2$, and Ar–He will be described. Because of their important influence on certain transport properties, mixtures containing metal vapors (e.g., Ar–Cu and air–Cu) will also be included. These combinations of gases have been chosen because air is commonly used in cold electrode vortex plasma torches, Ar–$H_2$ and Ar–He are mainly used in plasma spraying, and finally, Ar–Cu and air–Cu mixtures show the influence of electrode erosion on the plasma properties.

#### 7.5.2.1. Viscosity

Figure 7.11 from Pateyron *et al.*[25–27] gives the viscosity of air at different pressures (50–500 kPa). The evolution of $\mu$ with temperature is quite similar to that of nitrogen (see Fig. 7.1). The small variations in the slope are due to the dissociation of oxygen between 3000 and 5000 K and to $NO^+$ formation and destruction between 5000 and 8000 K. The rapid reduction in the slope at $T > 7000$ K corresponds to nitrogen dissociation. At temperatures higher than 10,000 K, the viscosity decreases more rapidly when the pressure is reduced, in agreement with the charged particle densities, which increase when the pressure decreases. Between

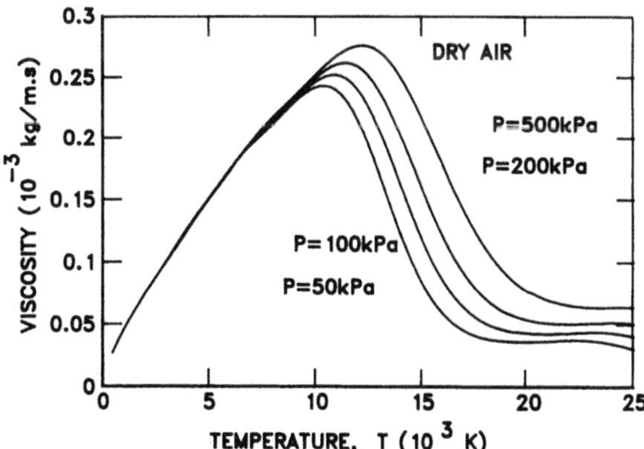


**FIG. 7.11.** Temperature dependence of the viscosity of air at four different pressures (50, 100, 200, and 500 kPa).[26]

3000 and 10,000 K, the pressure effects, which are not very significant, are linked successively to the $O_2$ dissociation, $NO^+$ formation and destruction, and $N_2$ dissociation. For $T > 20{,}000$ K, when the densities of the charged particles are almost constant the viscosity reaches a constant value which is highest for the higher densities of charged particles, i.e., for the highest pressure.

In an Ar–$H_2$ mixture at atmospheric pressure, the viscosity is mainly dominated by Ar for $H_2$ volume percents up to 40–50% (see Fig. 7.12). Intuitively this can be explained by using the first approximation for viscosity and Eq. (7.89) with the following values for the $\bar{Q}_{ij}^{22}$ at 8000 K: $\bar{Q}_{\text{Ar-Ar}}^{22} = 4.8$, $\bar{Q}_{\text{H-H}}^{22} = 2.16$, $\bar{Q}_{\text{Ar-H}}^{22} = 3.4$. Assuming that Eq. (7.89) can be used for the diffusion of H in Ar with $m = \mu$ (reduced mass), it can be seen that the contributions of H–H and H–Ar to the viscosity (proportional to $\sqrt{m}/\bar{Q}_{ij}^{22}$) are low compared with that of Ar (as long as the Ar density is high). At temperatures above 22,000 K the density of $Ar^{++}$ increases (see Fig. 6.2) and the viscosity decreases again.

In contrast, for the Ar–He mixture at 8000 K (with $\bar{Q}_{\text{He-He}}^{22} = 1.2$ and $\bar{Q}_{\text{Ar-He}}^{22} = 1$) the contributions of these interactions, especially that of Ar–He, is not important compared to that of Ar–Ar. Thus the viscosity of this mixture, up to 11,000 or 12,000 K, is very close to that of the pure initial components (see Fig. 7.13). At temperatures above $10^4$ K, the presence of neutral He atoms causes a drastic increase of the viscosity of the mixture compared to that of pure argon. At temperatures beyond $2 \times 10^4$ K, ionization of the He atoms increases rapidly and correspondingly the viscosity drops.

A few attempts have been made recently to calculate the influence of metallic vapors (electrode vapors, for example) on the transport properties.[28,29] For the case of an Ar–Cu mixture, Mostaghimi and

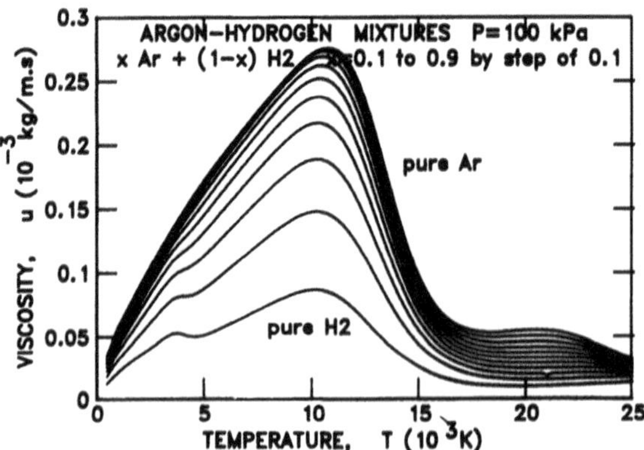


FIG. 7.12. Temperature dependence of the viscosity of an Ar–$H_2$ mixture at atmospheric pressure.[27]

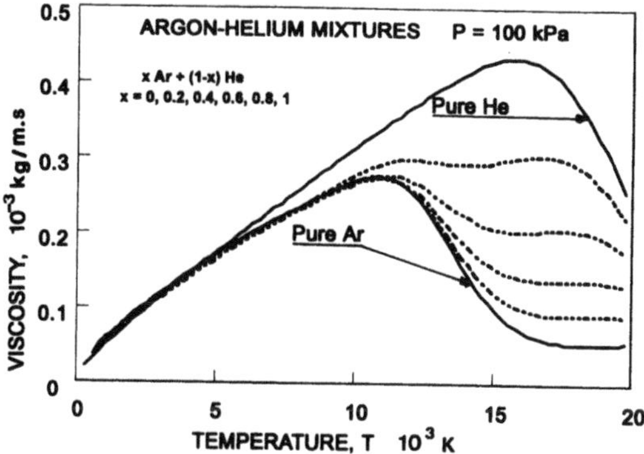


**FIG. 7.13.** Temperature dependence of the viscosity of an Ar–He mixture at atmospheric pressure.[27]

Pfender[28] used the first approximation of Sonine for the heavy particles and the sixth approximation for the electrons. The unknown potentials were assumed to be hard-sphere potentials. The plasma was at atmospheric pressure with temperatures ranging from 1000 K to 24,000 K, and the contamination level of copper ranged from 0.01% to 5% by volume. both the first and second ions of copper and argon were considered. The main conclusion was that the effect of Cu vapor on the viscosity was, for all practical purposes, negligible (see Fig. 7.14).

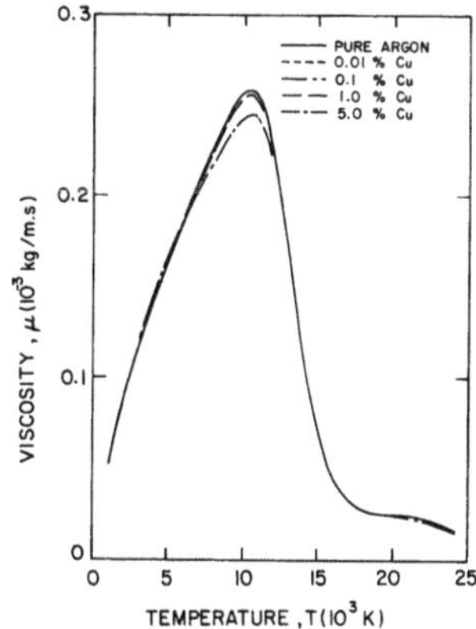


**FIG. 7.14.** Temperature dependence of the viscosity of an argon plasma containing a few mole percent of Cu vapor at atmospheric pressure.[28]

### 7.5.2.2. Electrical Conductivity

The electrical conductivity of air is almost independent of the molar ratios of nitrogen or oxygen, since the electrical conductivities of the two components are almost the same. The influence of pressure is illustrated in Fig. 7.15, which shows the variation of $\sigma_e$ versus temperature for an air plasma at four different pressures. When the pressure increases, ionization is delayed, and $\sigma_e$ decreases slightly at temperatures lower than 12,000 K. At higher temperatures the electron density increases with pressure, and thus $\sigma_e$ increases, too.

Because $\sigma_e$ is proportional to $n_e$, $\sigma_e$ is almost the same for the $Ar–H_2$ mixture (see Fig. 7.16) whatever the percent of hydrogen (ionization starts at a temperature 1000 K higher for $H_2$ than for Ar). For $T > 22{,}000$ K the increase of the $Ar^{++}$ density modifies the collision cross sections, and thus $\sigma_e$. For the Ar–He mixture (see Fig. 7.17), $\sigma_e$ is mostly controlled by the argon present in the mixture (up to 80 vol% He). Even for 90 vol% He and temperatures up to 13,000 K, the electrical conductivity is close to that of pure argon.

When copper vapor is introduced into the plasma, the electrical conductivity, unlike the viscosity, may be drastically changed, as shown by Mostaghimi and Pfender[28] for an argon plasma with contamination levels of copper ranging from 0.01 to 5% by volume. Figure 7.18 shows the corresponding results. The electrical conductivity of the argon plasma at $T = 5000$ K increases by a factor of 28 when 1% Cu is added. Similar results are obtained for air at atmospheric pressure (see Fig. 7.19 from Ref. 26). At $T > 17{,}000$ K the $Cu^{++}$ ions have only little influence on the electrical conductivity of the plasma.

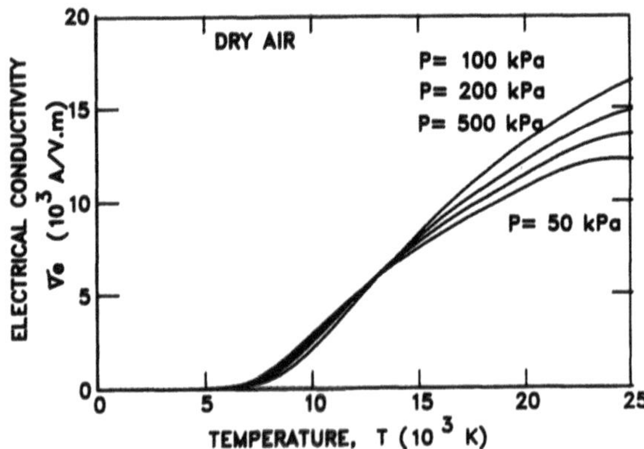


**FIG. 7.15.** Temperature dependence of the electrical conductivity of an air plasma at pressures of (50, 100, 200, and 500 kPa).[28]

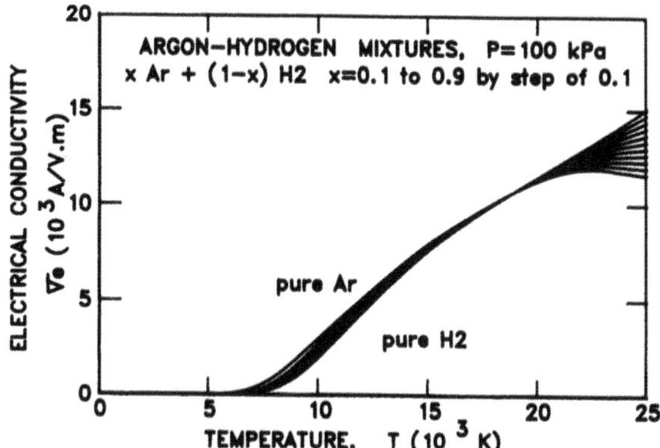


FIG. 7.16. Temperature dependence of the electrical conductivity of Ar–$H_2$ plasmas at atmospheric pressure.[27]

### 7.5.2.3. Thermal Conductivity

In air we see the peak due to oxygen dissociation at 3500 K and the peak for nitrogen dissociation at 7000 K. The formation of NO (less than 6%) occurs at 3500 K (see Fig. 7.20). The ionization peaks of O and N occur at almost the same temperature. The peaks are somewhat broadened compared to those of pure gases, but their amplitudes are in almost the same ratio as that of $N_2$ and $O_2$ in air at room temperature. At higher pressure (above atmospheric), the $O_2$ and $N_2$ dissociation peaks are shifted to higher temperatures and lower maximum values are obtained. The ionization peaks are also shifted to higher temperatures and their values increase slightly with pressure.

When $H_2$ is added to Ar (recall the high dissociation and ionization

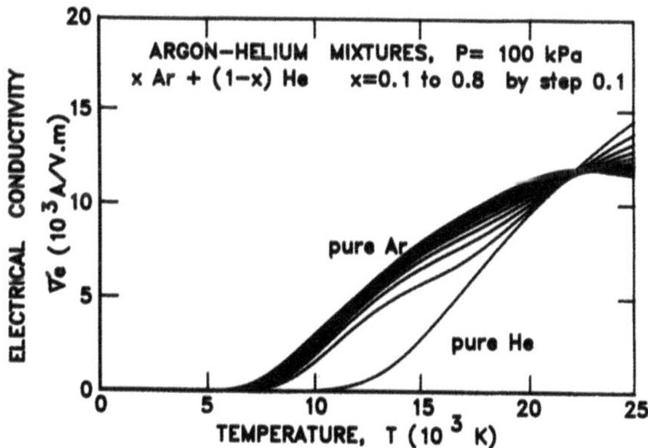


FIG. 7.17. Temperature dependence of the electrical conductivity of Ar–He plasmas at atmospheric pressure.[27]

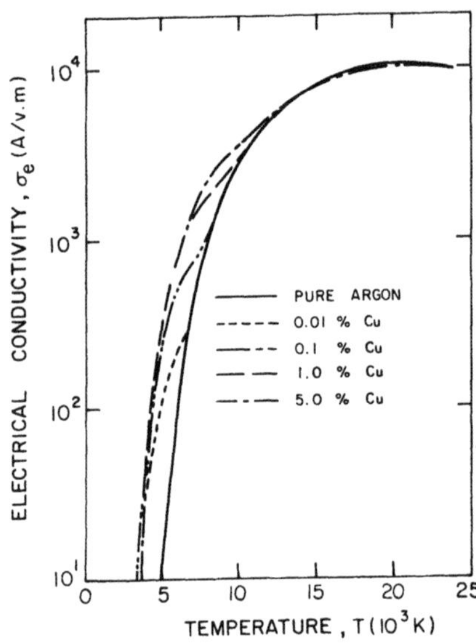


**FIG. 7.18.** Temperature dependence of the electrical conductivity of an argon plasma containing a few mole percent Cu vapor at atmospheric pressure.[28]

peaks of $H_2$, in the Ar–He mixture), thermal conductivity of the mixture increases with the $H_2$ vol%, especially, near the dissociation and ionization temperatures (see Fig. 7.21). For $T > 20{,}000$ K, the curves merge together due to the fact that the hydrogen atom/ion composition remains unchanged while the maximum ionization of $Ar^{++}$ is not yet reached.

In the Ar–He mixture it can also be seen that an increase in the He content results in a steady increase in the thermal conductivity of the mixture; the increase is almost proportional to the He vol% (see Fig. 7.22). For $T > 17{,}000$ K the ionization peak for He is noted. Argon

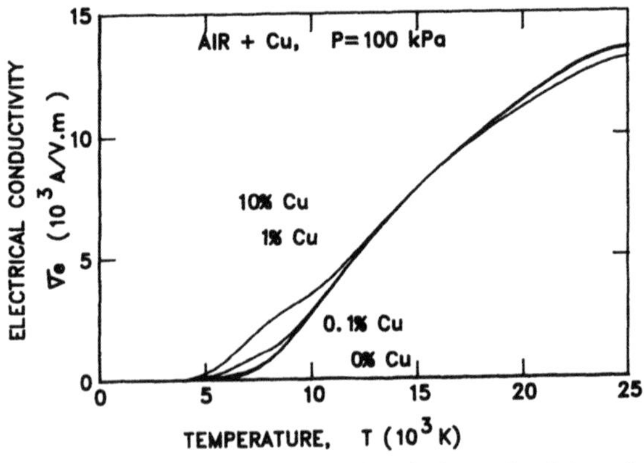


**FIG. 7.19.** Temperature dependence of the electrical conductivity of an air plasma containing a few mole percent Cu vapor at atmospheric pressure.[26]

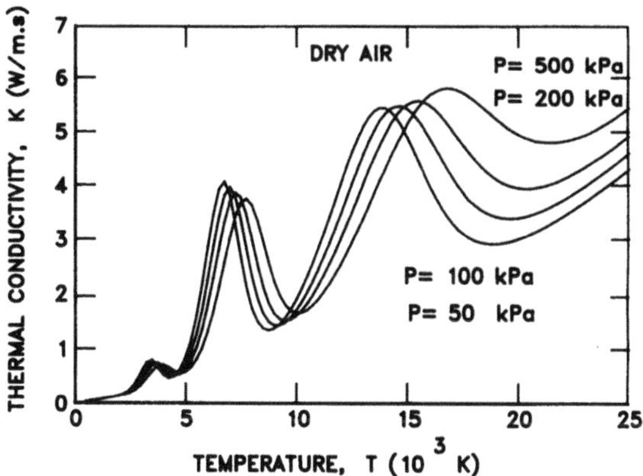


FIG. 7.20. Temperature dependence of the thermal conductivity of air at 50, 100, 200, and 500 kPa.[26]

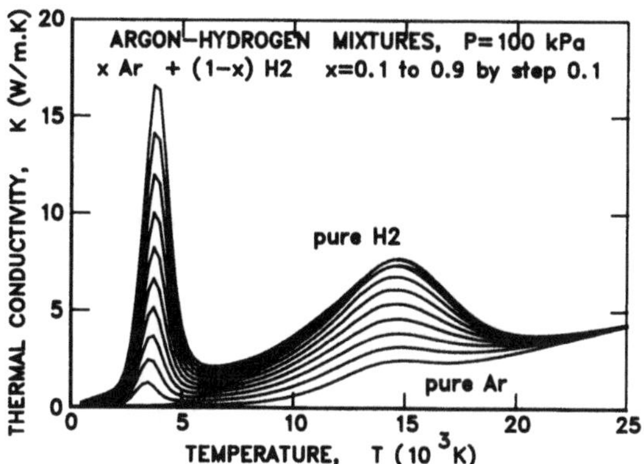


FIG. 7.21. Temperature dependence of the thermal conductivity of Ar–$H_2$ plasmas at atmospheric pressure.[27]

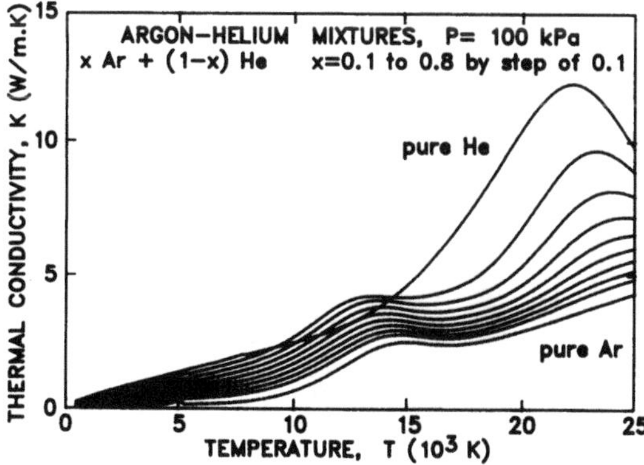


FIG. 7.22. Temperature dependence of the thermal conductivity of Ar–He plasmas at atmospheric pressure.[27]

shows a minor peak in the range from 12,000 to 16,000 K. In both mixtures ($Ar–H_2$ and Ar–He), close to the Ar ionization temperature for high volume percentages of $H_2$ or He the values $\kappa$ are above the highest value obtained for the pure component ($H_2$ for $Ar–H_2$ and He for Ar–He).

In the presence of steep temperature gradients careful attention must be given to the way in which the mean effective thermal conductivity $\bar{\kappa}$ is calculated. Bourdin *et al.*[30] have suggested using the mean integrated thermal conductivity defined as

$$\bar{\kappa} = [1/(T_p - T_s)] \int_{T_s}^{T_p} \kappa(t)\, dt \tag{7.109}$$

where $T_p$ and $T_s$ are the temperature limits across which the energy transfer is taking place.

Figure 7.23 represents $\bar{\kappa}$ for the $Ar–H_2$ mixture. Note that the important advantage of adding $H_2$ to the mixture is that the mean integrated thermal conductivity is increased as soon as the temperature exceeds 4000 K. In the Ar–He mixture (see Fig. 7.24) the increase with temperature is more regular than for $Ar–H_2$. He ionization increases $\bar{\kappa}$ slightly for $T > 17{,}000$ K. A comparison between the thermal conductivity of two gas mixtures typically used in plasma spraying is given in Fig. 7.25. These are Ar + 30% $H_2$ (vol) and Ar + 60% He (vol). For $T >$ 24,000 K, due to He ionization, $\bar{\kappa}(Ar–He) > \bar{\kappa}(Ar–H_2)$.

The contribution of the various species to the thermal conductivity of a contaminated plasma depends on temperature (see Fig. 7.26, which shows data for an Ar–Cu plasma). For $T < 6000$ K the heavy species dominate, while for $6000 < T < 10{,}000$ K the electrons derived from Cu vapor make a substantial contribution to the total thermal conductivity.

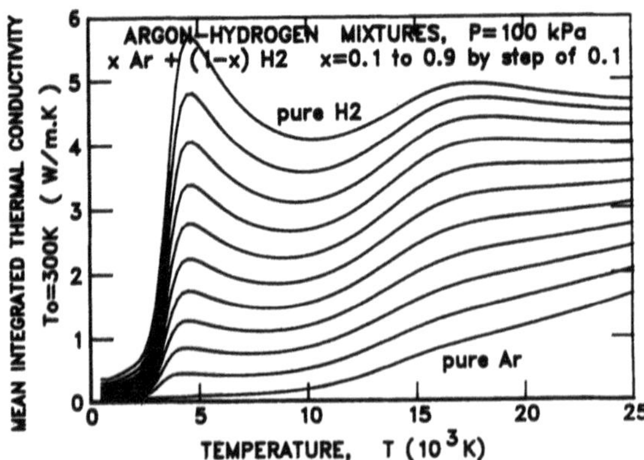


FIG. 7.23. Temperature dependence of the mean integrated thermal conductivity of $Ar–H_2$ plasmas at atmospheric pressure.[27]

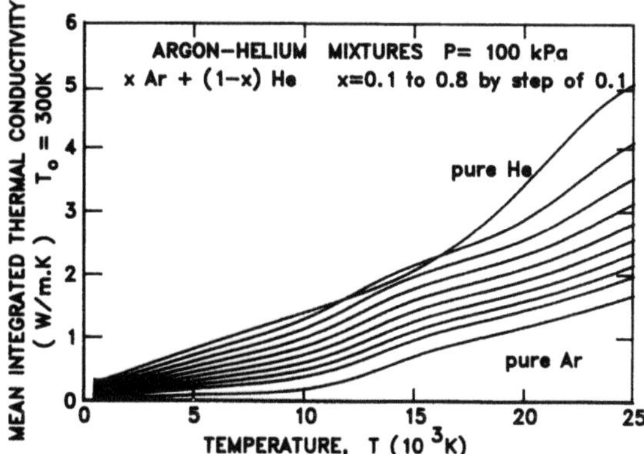


FIG. 7.24. Temperature dependence of the mean integrated thermal conductivity of Ar–He plasmas at atmospheric pressure.[27]

Finally, for $T > 10^4$ the total thermal conductivity is governed by the contributions of chemical reactions (ionization) and free electrons.

### 7.5.3. Mixing Rules and Their Limitations

Due to the complexity of the calculation procedure for the accurate determination of the transport properties of gaseous mixtures, there is a strong need for simplified mixing rules that can be used to calculate the

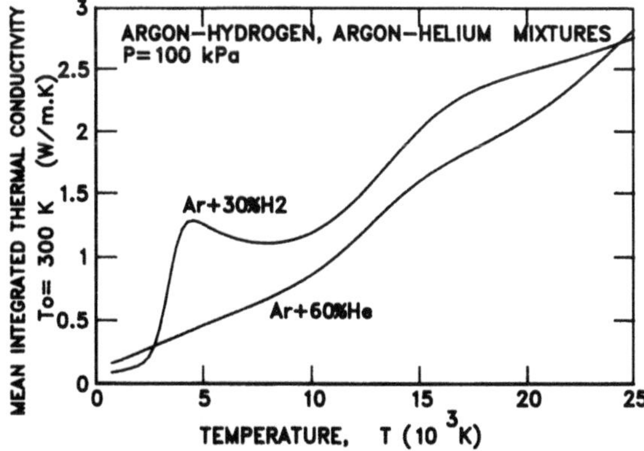


FIG. 7.25. Temperature dependence of the mean integrated thermal conductivity of a mixture of 70 vol% Ar, 30 vol% $H_2$, and 40 vol% Ar, 60 vol% He at atmospheric pressure.[27]

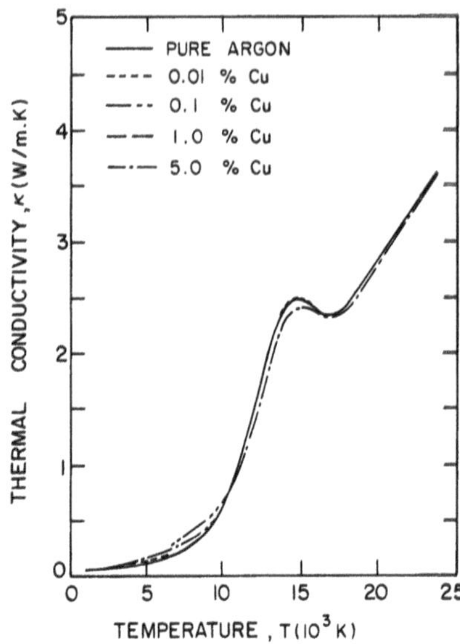


**FIG. 7.26.** Temperature dependence of the total thermal conductivity of an argon plasma containing a few mole percent Cu vapor at atmospheric pressure.[28]

properties of a given gaseous mixure as a function of its composition and the corresponding properties of the pure gases. By neglecting all interactions between the components of the two gases and using only a first-order development of Chapman–Enkog's expressions, Wilke[31] has developed a semiempirical expression giving the viscosity of the mixture:

$$\mu_{\text{mix}} = \sum_{i=1}^{K} \left( x_i \mu_i \Big/ \sum_{j=1}^{K} x_j Z_{ij} \right) \tag{7.110}$$

where

$$Z_{ij} = \frac{1}{\sqrt{8}} (1 + M_i/M_j)^{-1/2} [1 + (\mu_i/\mu_j)^{1/2} (M_j/M_i)^{1/4}]^2 \tag{7.111}$$

$K$ is the number of chemical species in the mixture, $x_i$ and $x_j$ are the mole fractions of species $i$ and $j$, $\mu_i$ and $\mu_j$ are the viscosities of species $i$ and $j$ at the mixture temperature and pressure, and $M_i$ and $M_j$ are the corresponding molecular weights ($Z_{ij}$ is dimensionless and, for $i = j$, $Z_{ij} = 1$).

Figure 7.27 shows that the agreement between the exact calculation and Wilke's is rather good for Ar–$H_2$ mixtures. For Ar–He mixtures (for example, 20 vol% Ar–80 vol% He), the exact calculation gives quite a different result from those calcuated using either Wilke's expression or a linear interpolation (see Fig. 7.28). These differences can be explained easily by considering the relative importance of the neglected $\bar{Q}_{ij}^{22}$ values

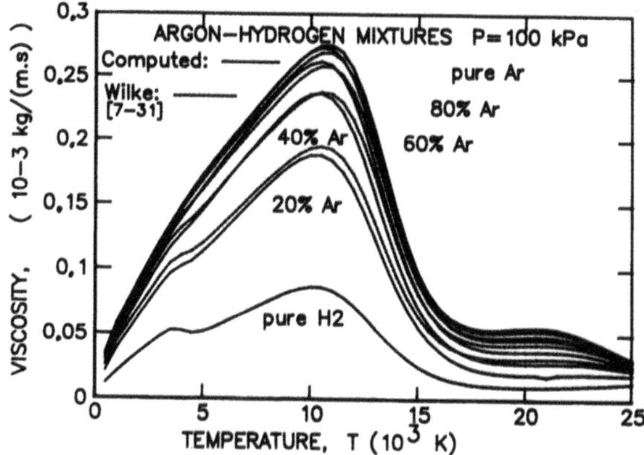


FIG. 7.27. Comparison of the viscosity of an Ar–$H_2$ mixture at atmospheric pressure calculated rigorously (solid lines) and using Wilke's expression (dashed lines).

especially for the interaction of He atoms with oxygen ions. For $T > 10^4$ K, a simple linear mixing rule is worse than Wilke's expression.

When dealing with the calculation of the thermal conductivity of mixtures, it is important to keep in mind that Wilke's formula [Eq. (7.110)] also neglects all reactive contributions to $\kappa$. It is therefore not surprising that differences of up to 60–70% were observed between the exact calculation and those obtained using the mixing rule, especially close to dissociation of ionization temperatures.

To conclude, the mixing rule has to be used very cautiously for viscosity and cannot be used for thermal conductivity, especially when dissociation and ionization take place.

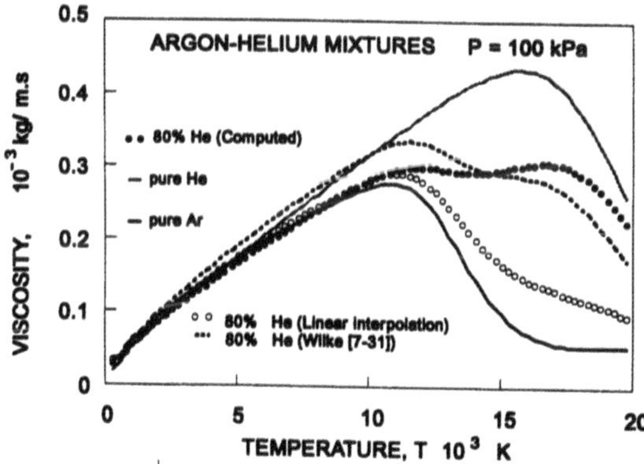


FIG. 7.28. Comparison of the viscosity of an Ar–He mixture at atmospheric pressure calculated rigorously (solid lines) and using Wilke's expression (∘ ∘ ∘) or linear interpolation (△△△).

## 7.6. TRANSPORT COEFFICIENTS FOR A TWO-TEMPERATURE PLASMA: EXAMPLE FOR AN Ar–$H_2$ PLASMA MIXTURE

As mentioned earlier, while LTE conditions may be satisfied in atmospheric-pressure plasmas, deviations from LTE can occur in the presence of steep temperature gradients in the boundary-layer region around plasma jets, especially when a cold gas is injected (see Section 4.4.4). Moreover, as soon as the pressure drops below approximately 25 kPa (200 Torr) nonequilibrium effects need to be taken into account in the plasma core itself.

Once the nonequilibrium composition is determined (see Section 6.5.1), the transport properties are calculated using the extension of Devoto's model by Bonnefoi,[16] a model that allows the use of collision integrals calculated at equilibrium. The main change introduced by Bonnefoi is related to diffusion forces that essentially modify the diffusion velocities and thus the reactional thermal conductivity, as shown by Aubreton.[17] The main parameter in these nonequilibrium calculations is the ratio $\theta = T_e/T_h$, and one question that arises is to what extent calculations using the Saha–Potapov equation for nonequilibrium composition are valid. Aubreton[17] compared an Ar–$H_2$ plasma composition calculated using the Saha–Potapov model with the one calculated from a kinetic model such as that of Richely and Tuma[32] and found that agreement between the two is acceptable up to $\theta = 2$ to 3. Thus in the following discussion Bonnefoi's[16] and Aubreton's[17] results will be presented only for Ar–$H_2$ plasmas with $\theta \leq 3$.

Figure 7.29 represents the viscosity of an argon plasma (calculated by Aubreton[17] using the compositions deduced from the kinetic model) as a function of $T_e$ for different values of $\theta$. The curves have a shape independent of $\theta$ and are shifted downward as $\theta$ increases. This shift is due to the decrease in heavy-particle density as $\theta$ increases (see Fig.

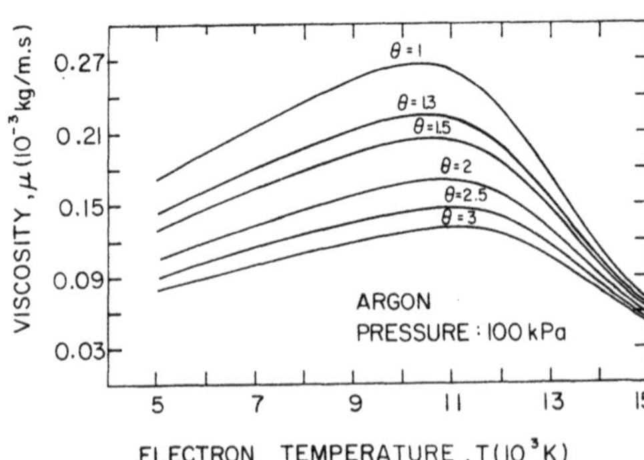


**FIG. 7.29.** Electron temperature dependence of the viscosity of an argon plasma for different values of the ratio $\theta = T_e/T_h$ at atmospheric pressure.[17]

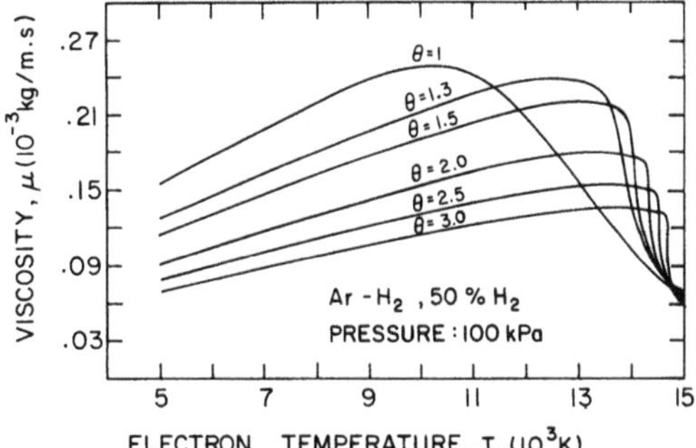


**FIG. 7.30.** Electron temperature dependence of the viscosity of an argon hydrogen plasma (50% by mole) for different values of the ratio $\theta = T_e/T_h$ at atmospheric pressure.[17]

6.21); note that density is proportional to $T_h$ and not to $T_e$. The position of the peak ($T_e = 10{,}000$ K for $\theta = 1$ and $T_e = 11{,}500$ K for $\theta = 3$) indicates the limit between the viscosity controlled by neutral–neutral collisions and the one controlled by ion–ion collisions. This is due to the fact that $\Omega^{1,s}$ (ion–ion) $\gg$ $\Omega^{1,s}$ (neutral–neutral) and $\Omega^{1,s}$ (neutral–ion). The shift of the peak toward higher temperatures when $\theta$ increases corresponds to a delay in ionization.[17]

In Ar–$H_2$ mixtures, the viscosity is dominated by argon, which has a much higher mass. However, hydrogen plays a role in the delay of ionization (dissociation of $H_2$, controlled by $T_h$, must be achieved before $Ar^+$ can be involved) and thus the peaks are shifted toward higher temperatures (see Fig. 7.30, which shows the viscosity of a mixture containing 50 vol% Ar and 50 vol% $H_2$).

The electrical conductivity $\sigma_e$ (see Fig. 7.31 from Ref. 17) exhibits the same trend to be reduced when $\theta$ increases for $T_e < 12{,}000$ K. However, all the curves intersect at about 14,000 K because $\sigma_e$ is strongly

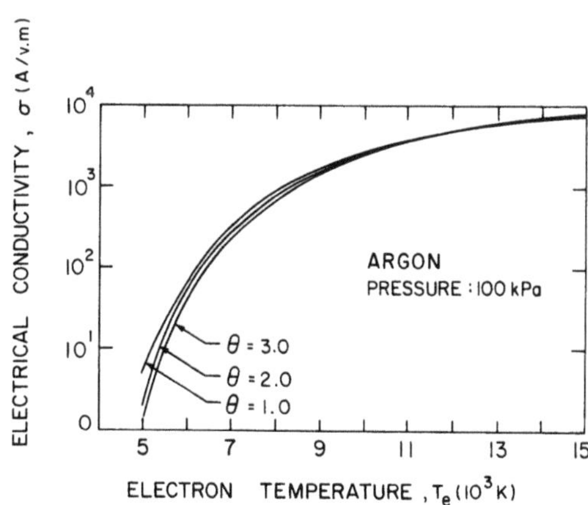


**FIG. 7.31.** Electron temperature dependence of the electrical conductivity of an argon plasma for different values of the ratio $\theta = T_e/T_h$ at atmospheric pressure.[17]

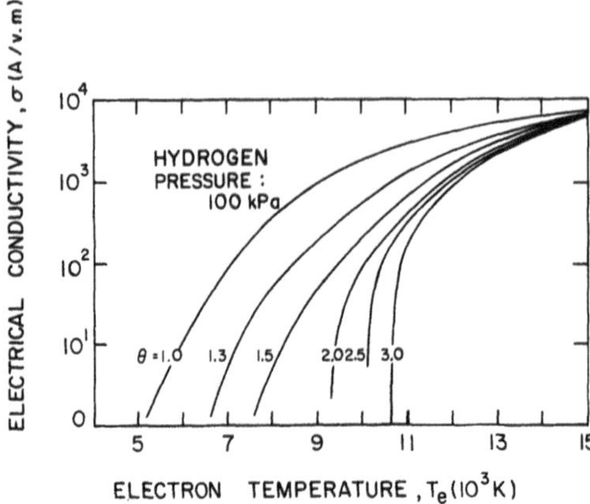


**FIG. 7.32.** Electron temperature dependence of the electrical conductivity of a hydrogen plasma for different values of the ratio $\theta = T_e/T_h$ at atmospheric pressure.[17]

dependent on the electron density, which exhibits the same behavior (see Fig. 6.21). At high values of $\theta$ and low values of $T_e$ (below 10,000 K), thermal ionization is low and $n_e$ is reduced in parallel with $T_h$, but at high values of $T_e$ direct ionization by electrons becomes possible and $n_e$ increases. In logarithmic coordinates, $\sigma_e$ is almost independent of $\theta$ for argon, while for hydrogen the delay in ionization plays a major role (see Fig. 7.32 from Ref. 17).

Figure 7.33 shows the thermal conductivity of hydrogen as a function of $T_e$ for $\theta = 2$.[17] The reaction peak occurring under equilibrium conditions at 3500 K is now obtained at $T_e = 7200$ K, corresponding

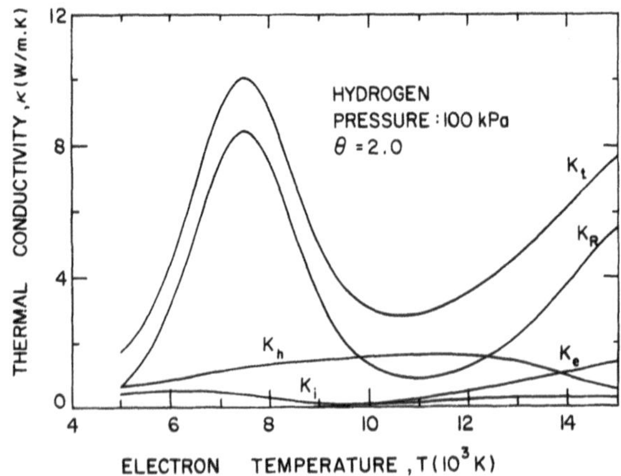


**FIG. 7.33.** Electron temperature dependence of the thermal conductivity and its different components for a hydrogen plasma for $\theta = T_e/T_h = 2$ at atmospheric pressure.[22]

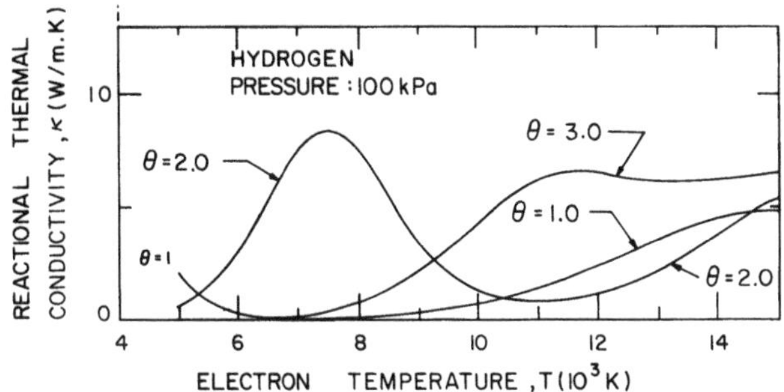


**FIG. 7.34.** Electron temperature dependence of the reactional thermal conductivity of a hydrogen plasma for different values of the ratio $\theta = T_e/T_h$ at atmospheric pressure.[22]

effectively to $T_h = 3600$ K. The value of $\kappa_h^{tr}$ for the heavy particles is lower than the same term for $\theta = 1$, but $\kappa_e^{tr}$ is not changed. At high temperatures ($T_e > 12{,}000$ K) the ionization peak is about the same as at equilibrium. Figure 7.34 shows this phenomenon clearly for the reactional thermal conductivity; the dissociation peak is shifted from $T_h = T_e = 3600$ K at equilibrium to 7200 K for $\theta = 2$ and 10,600 K for $\theta = 3$.

The results for pure argon are shown in Fig. 7.35,[17] It is very important to note that when using the model of Bonnefoi and Aubreton with the new diffusion forces, the values of the total thermal conductivity do not increase very fast with increases in $\theta$, contrary to Hsu[33] and Kannapan.[34] They used a simple extension of the model of Butler and

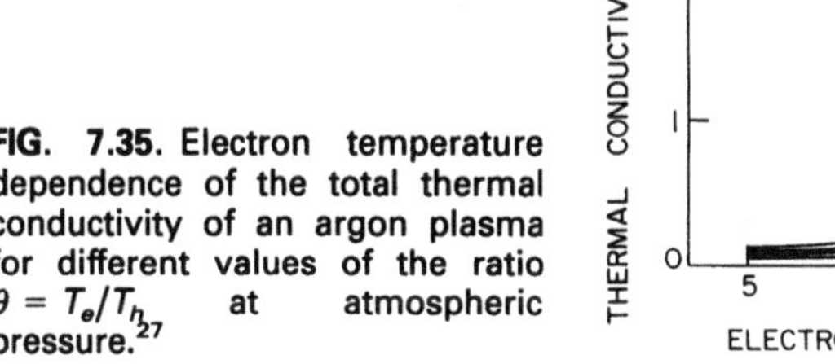
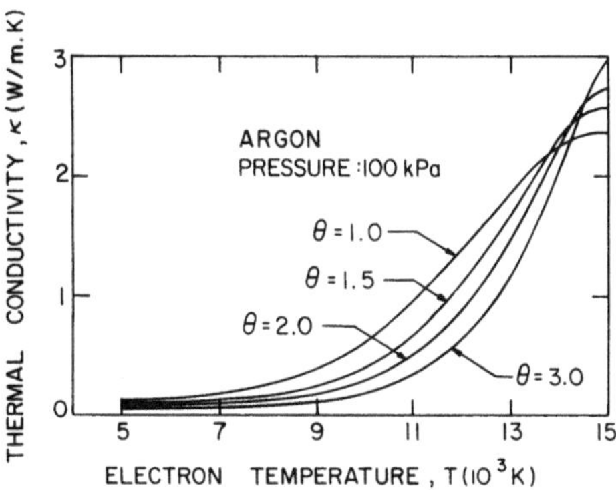


**FIG. 7.35.** Electron temperature dependence of the total thermal conductivity of an argon plasma for different values of the ratio $\theta = T_e/T_h$ at atmospheric pressure.[27]

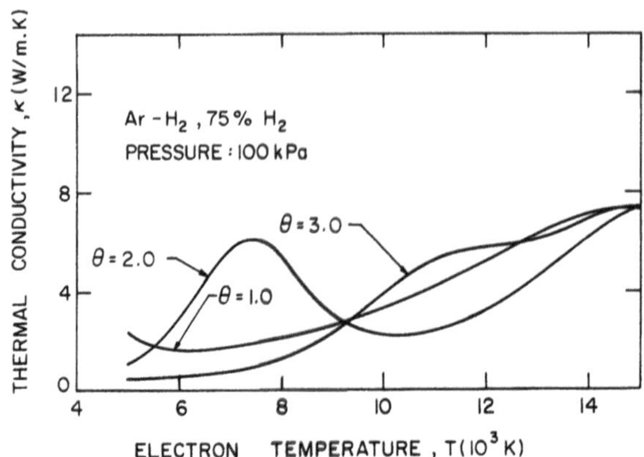


**FIG. 7.36.** Electron temperature dependence of the total thermal conductivity of a mixture of argon and hydrogen (25 mol% Ar–75 mol% $H_2$) for different values of the ratio $\theta = T_e/T_h$ at atmospheric pressure.[22]

Brockaw[20,21] to nonequilibrium. The following table gives $\kappa$ at 15,000 K (in W/m K), where $\kappa_R$ is at a maximum for argon.

| Reference | Kannapan[34] | Hsu[33] | Aubreton[17] |
|---|---|---|---|
| $\theta = 1$ | 2.4 | 2.4 | 2.38 |
| $\theta = 2$ | 15 | 5 | 2.76 |
| $\theta = 3$ | 76 | 8 | 2.99 |
| $\theta = 4$ | — | 14 | 3.11 |

The more reasonable results obtained by Aubreton are due to the term in $\theta/T_e^2$, which appears in the expression in Eq. (7.100) for the reactional thermal conductivity and sensibly reduces its value at high temperatures.[17]

When a mixture of Ar–$H_2$ (75 vol% $H_2$) is considered (see Fig. 7.36, which shows the total thermal conductivity of this mixture), a similar phenomenon is observed. The intensities of the dissociation peaks for different $\theta$ are reduced compared to those of pure hydrogen due to the reduction in the mole number of hydrogen.

## LIST OF SYMBOLS

| | |
|---|---|
| $A_{X,X}$ | coefficients for calculating the reactional thermal conductivity |
| $\vec{A}_i$ | vectorial coefficients for calculating the perturbation function $\phi_i$ |
| $b$ | impact parameter |

| | |
|---|---|
| $\overleftrightarrow{B}_i$ | second-order tensor allowing calculation of the perturbation function $\phi_i$ |
| $\vec{c}$ | center-of-mass velocity for two particles of type $i$ and $j$ (m/s) |
| $c_{pi}$ | molar specific heat of species $i$ at constant pressure (J/mole K) |
| $c_v$ | specific heat at constant volume (J/kg K) |
| $\vec{d}_i'$ | diffusion forces |
| $dA$ | elementary surface ($m^2$) |
| $d\vec{r}$ | elementary volume in ordinary space ($m^3$) |
| $d\vec{v}$ | elementary volume in velocity space: $d\vec{v} = dv_x\, dv_y\, dv_z$ ($m^3/s^3$) |
| $D$ | diffusion coefficient ($m^2/s$) |
| $D_i$ | scalar coefficients for calculating the perturbation function $\phi_i$ |
| $D_{ij}$ | ordinary diffusion coefficients between species $i$ and $j$ ($m^2/s$) |
| $D_{iT}$ | thermal diffusion coefficients |
| $\mathcal{D}_i f$ | left-hand side of the Boltzmann equation |
| $\mathcal{D}_c f$ | right-hand side of the Boltzmann equation: collision term |
| $\vec{E}$ | electric field (V/m) |
| $f_i$ | distribution function |
| $f_i^0$ | equilibrium distribution function |
| $F_x$ | component of the force in the $x$ direction (N) |
| $g_{12}$ | relative velocity ($g_{12} = \lvert\vec{v}_1 - \vec{v}_2\rvert$ m/s) |
| $h$ | Planck's constant ($6.6 \times 10^{-34}$ W $s^2$) |
| $H_i$ | molar enthalpy of species $i$ (kJ/mole) |
| $H_N^D$ | molar dissociation enthalpy of nitrogen molecule (kJ/mole) |
| $H_{N^+}^I$ | molar ionization enthalpy of nitrogen atom (kJ/mole) |
| $\vec{J}_E$ | energy flux ($W/m^2$) |
| $\vec{J}_n$ | particle flux (part/$m^2$ s) |
| $\vec{J}_{px}$ | momentum flux in the $x$ direction |
| $\vec{J}_x$ | flux of the quantity $x$ |
| $k$ | Boltzmann constant ($1.38 \times 10^{-23}$ J/K) |
| $l$ | mean free path (m) |
| $l_z$ | mean free path in the $z$ direction (m) |
| $m_i$ | mass of the particle of chemical species $i$ (kg) |
| $M_i$ | atomic mass of chemical species $i$ (kg) |
| $n$ | number density (particles/$m^3$) |
| $n_i$ | number density of chemical species $i$ (particles/$m^3$) |
| $n_i'$ | mole number of chemical species $i$ |
| $p$ | pressure (Pa) |
| $\bar{\bar{p}}$ | pressure tensor |
| $\vec{q}_i$ | flux vector associated with the transport of the kinetic energy of particles of species $i$ ($J/m^2s$) |

$\vec{q}_R$ reactional heat flux vector (J/m$^2$ s)
$\vec{q}_z$ heat flux vector in the $z$ direction (J/m$^2$ s)
$q_{ij}^{ls}$ bracket integral
$\bar{Q}_{ij}^{ls}$ product of the reduced collision integral by the cross-sectional area presented by particles considered as hard spheres (m$^2$)
$Q_{ij}^{l}$ quantum collision cross section
$Q^{l}(\vec{v})$ elastic collision cross section at the degree $l$ of the Legendre functions
$\vec{r}$ position vector (m)
$\vec{r}^*$ relative position vector ($\vec{r}^* = \vec{r}_1 - \vec{r}_2$) (m)
$r_m$ distance of closest approach (m)
$R$ perfect gas constant (8.32 J/K mol)
$T$ absolute temperature (K)
$T_e$ absolute electron temperature (K)
$T_h$ absolute heavy-species temperature (K)
$\vec{U}_i$ peculiar velocity ($\vec{v}_i - \vec{v}_0 = \vec{U}_i$) of particles of species $i$ (m/s)
$\vec{U}_{ix}$ peculiar velocity in the $x$ direction (m/s)
$\langle\vec{U}_i\rangle$ diffusion velocity of particles of species $i$
$\vec{v}_i$ velocity of particles of species $i$ (m/s)
$\bar{v}$ mean velocity of particles (m/s)
$\vec{v}_0$ mean flow velocity (m/s)
$V(r)$ interaction potential
$\vec{V}_{ij}$ relative velocity before collision ($\vec{V}_{ij} = \vec{v}_i - \vec{v}_j$) (m/s)
$\vec{V}_{ij}'$ relative velocity after collision ($\vec{V}_{ij}' = \vec{v}_i' - \vec{v}_j'$) (m/s)
$\vec{W}_i$ velocity proportional to the peculiar velocity, $\vec{W}_i = (m_i/2kT_i)^{3/2}\vec{U}_i$
$\vec{W}_i'$ velocity proportional to the diffusion velocity, $\vec{W}_i = p_i/RT_i\langle\vec{U}_i\rangle$

## Greek Symbols

$\theta$ angle in spherical coordinates
$\theta$ ratio of the electron temperature to the heavy-species temperatures
$\theta'$ deviation angle
$\kappa$ thermal conductivity (W/m K)
$\bar{\kappa}$ mean integrated thermal conductivity $\bar{\kappa} = 1/(T_p - T_s)\int_{T_s}^{T_p}\kappa(s)\,ds$ (W/m K)
$\kappa_{\text{int}}$ internal thermal conductivity (W/m K)
$\kappa_R$ reactional thermal conductivity (W/m K)
$\kappa_{\text{tr}}$ translational thermal conductivity (W/m K)

$\bar{\kappa}_{tr}^{e}$ electron translational thermal conductivity (W/m K)
$\bar{\kappa}_{tr}^{h}$ heavy-species translational thermal conductivity (W/m K)
$\mu$ molecular viscosity (kg/m s)
$\mu_e$ electron mobility ($m^2$/V s)
$\mu_m$ reduced mass, $\mu_m = m_i m_j/(m_i + m_j)$ (kg)
$\mu_{min}$ molecular viscosity of a mixture (Wilke's formula)
$\xi$ inverse of the frequency of collisions
$\rho$ specific mass (kg/$m^3$)
$\sigma_e$ electrical conductivity (mho/m)
$\sigma_0$ total collision cross section ($m^2$)
$\sigma_{ij}$ total collision cross section between particles, considered as hard spheres, of type $i$ and $j$ ($m^2$)
$\sigma_{en}$ total collision cross section between electrons and neutral particles ($m^2$)
$\sigma'(\vec{v}_1, \vec{v}_2 \rightarrow \vec{v}_1', \vec{v}_2')\, d\vec{v}_1', d\vec{v}_2'$ collision cross section for particles with initial velocities between $\vec{v}_1$ and $\vec{v}_1 + d\vec{v}_1$ and $\vec{v}_2$ and $\vec{v}_2 + d\vec{v}_2$ and final velocities (after collision) between $v_1'$ and $v_1' + dv_1'$ and $v_2'$ and $v_2' + dv_2'$
$\sigma(\vec{V})\, d\Omega$ collision cross section for particles emerging after collision with a relative velocity $\vec{V}'$ into a solid angle range $d\Omega'$ about $\theta'$ and $\varphi'$
$\sigma_{zx}$ stress in a plane at $z$ in direction $x$ (N/$m^2$)
$\phi_i$ perturbation function for calculating the distribution function
$\chi_i(\vec{r}, \vec{v}, t)$ dummy variable for the chemical species $i$
$\langle \chi_i \rangle$ mean value of $\chi_i$
$\bar{\chi}_i$ mean value of $\chi_i$
$\Omega_{ij}^{ls}$ Hirschfelder's collision integral for particles of species $i$ and $j$ ($m^2$)
$\bar{\Omega}_{ij}^{ls}$ reduced collision integral for particles of species $i$ and $j$
$\vec{\nabla}_r$ position gradient vector $(\partial/\partial x, \partial/\partial y, \partial/\partial z)$
$\vec{\nabla}_v$ velocity gradient vector $(\partial/\partial v_x, \partial/\partial v_y, \partial/\partial v_z)$

## GENERAL BIBLIOGRAPHY

Chapman, S. and T. Cowling, *The Mathematical Theory of Non-Uniform Gases,* New York: Cambridge University Press, 1952.

Child, M. S., *Molecular Collision Theory,* New York: Academic Press, 1974.

Hirschfelder, J. O., C. F. Curtiss, and R. B. Bird, *Molecular Theory of Gases and Liquids,* 2nd ed., New York: Wiley, 1964.

Mitchner, M. and C. H. Kruger Jr., *Partially Ionized Gases*, New York: Wiley, 1973.

Pauly, H., Chapter 4 in *Atom Molecule Collision Theory* (B. B. Bernstein, ed.), New York and London: Plenum Press, 1979.

Reif, F., *Fundamentals of Statistical and Thermal Physics,* New York: McGraw-Hill, 1988.

## REFERENCES


1. C. Gorse, "Contribution au calcul des propriétés de transport des plasmas des mélanges $Ar–H_2$ et $Ar–N_2$," Thèse 3e cycle (University of Limoges, France, 1975).
2. IUPAC Subcommission on Plasma Chemistry, "Thermodynamic and Transport Properties of Pure and Mixed Thermal Plasmas at LTE," Pure Appl. Chem. 6 (1982): 1221.
3. A. Eucken, *Phys. Z.* **14** (1913): 324.
4. E. A. Mason, and L. Monchick, *J. Chem. Phys.* **36** (1962): 1622.
5. E. Mason, J. T. Vanderslice, and J. M. Yos, *Phys. Fluids* **2** (1959): 688.
6. C. Nyeland and E. M. Mason, *Phys. Fluids* **10** (1967): 985.
7. H. Grad, in Proceedings of the 5th International Conference on Ionization Phenomena in Gases, Munich (1961).
8. W. F. Athye, *A Critical Evaluation of Methods for Calculating Transport Coefficients of Partially and Fully Ionized Gases,* NASA TN, ND-2611 (1965).
9. R. S. Devoto, *Phys. Fluids* **9** (1966): 1230.
10. R. S. Devoto, *Phys. Fluids* **10** (1967): 2105.
11. R. S. Devoto and C. P. Li, *J. Plasma Phys.* **2** (1968): 17.
12. R. S. Devoto, *AIAA J.* **7** (1979): 789.
13. R. S. Devoto, *Phys. Fluids* **16** (1973): 616.
14. R. S. Devoto, "The Transport Properties of a Partially Ionized Monoatomic Gas," Ph.D. Thesis (Stanford University, 1965).
15. R. M. Chmieleski, "Transport Properties of a Nonequilibrium Partially Ionized Gas," Ph.D. Thesis (Stanford University, 1967).
16. C. Bonnefoi, "Contribution à l'étude des méthodes de résolution de l'équation de Boltzmann dans un plasma à deux températures: exemple le mélange $Ar–H_2$," Thèse de doctorat d'Etat (University of Limoges, France, May 1983).
17. J. Aubreton, "Étude des propriétés thermodynamiques et de transport dans les plasmas thermiques à l'équilibre et hors équilibre thermodynamique. Applications aux plasmas de mélange $Ar–H_2$, $Ar–O_2$," Thèse de doctorat d'État (University of Limoges, France, 22 February 1985).
18. J. Aubreton and P. Fauchais, *Rev. Phys. Appl.* **18** (1983): 51.
19. C. Muckenfus and C. F. Curtiss, *J. Chem. Phys.* **29** (1958): 1273.
20. J. N. Butler and R. S. Brokaw, *J. Chem. Phys.* **26** (1957): 1636.
21. W. E. Meador and L. D. Stanton, *Phys. Fluids* **8** (1965): 1694.
22. C. Bonnefoi, J. Aubreton, and J. M. Mexmain, *Z. Naturforsch. A* **40a** (1085): 885.
23. H. W. Emmons, *Modern Developments in Heat Transfer* (W. Ibelee, ed.) (New York: Academic Press, 1963).
24. J. Lesinski and M. Boulos. "Thermodynamic and Transport Properties of Argon, Nitrogen and Oxygen at Atmospheric Pressure Over the Temperature Range 300–30,000 K," Internal Report (University of Sherbrooke, Quebec, Canada).
25. B. Pateyron, J. Aubreton, M. F. Elchinger, G. Delluc, and P. Fauchais, "Thermodynamic and Transport Properties of $N_2$, $O_2$, $H_2$, Ar, He and their mixtures," Internal Report LMCTS (University of Limoges, France, 1986).
26. B. Pateyron, M. F. Elchinger, G. Delluc, and P. Fauchais, "Thermodynamic and Transport Properties of Air and Air–Cu at Atmospheric Pressure, Internal Report, LMCTS (University of Limoges, 1990).
27. B. Pateyron, M. F. Elchinger, G. Delluc, and P. Fauchais, "Thermodynamic and Transport Properties of $Ar–H_2$ and Ar–He Plasma Gases for Spraying at Atmospheric Pressure—Part 1: Properties of the Mixtures," *Plasma Chemistry, Plasma Processing,* submitted.

28. J. Mostaghimi-Tehrani and E. Pfender, *Plasma Chemistry, Plasma Processing* **4(2)** (1984): 129.
29. H. Wilhelmi, W. Lyhs, and E. Pfender, *Plasma Chemistry, Plasma Processing* **4(4)** (1984): 315.
30. E. Bourdin, M. Boulos, and P. Fauchais, *Int. J. Heat Mass Transfer* **26** (1983): 567.
31. C. R. Wilke, *J. Chem. Phys.* **18** (1950): 517.
32. E. Richely and D. T. Tuma, *J. Appl. Phys.* **53** (1982): 8537.
33. K. C. Hsu and E. Pfender, "Calculation of Thermodynamics and Transport Properties of a Two-Temperature Argon Plasma," Proceedings of the Fifth International Symposium on Plasma Chemistry, Vol. 1 (Heriot-Watt University, Edinburgh, 1981): 144.
34. D. Kannappan, and T. K. Bose, *Phys. Fluids* **16** (1973): 491.
35. M. Capitelli, C. Gorse, and P. Fauchais, *J. Chim. Phys.* **7** (1976): 755.
36. I. Amdur and E. A. Mason, *Phys. Fluids* **1** (1958): 370.
37. P. P. Kulik, I. G. Panevin, and V. I. Khvesyuk, *Teplofizika Vysokikh Temperatur* **1** (1963): 56.
38. P. P. Kulik, *Teplofizika Vysokikh Temperatur* **9** (1971): 431.
39. P. W. Schreiber, A. M. Hunter, and K. R. Benedetto, *AIAA J.* **10** (1972): 670.
40. N. B. Vargaftik, *Tables on the Thermophysical Properties of Liquids and Gases* (Hemisphere Publishing Corporation, Washington, London, 1975).
41. D. P. Aeschliman and A. B. Cambell, *Phys. Fluids* **13** (1970): 2466.
42. A. Kanzawa and I. Kimura, *AIAA J.* **5,** no. 7 1315 (1967).
43. C. F. Bonilla, S. J. Wang, and M. Weiner, "The Viscosity of Steam, Heavy-Water Vapor, and Argon at Atmospheric Pressure up to High Temperatures," Transactions of the ASME (1956): 1285.
44. E. I. Asinovskii, E. V. Drokhanova, A. V. Kirillin, and A. N. Lagarkov, *Teplofizika Vysokikh Temperatur* **5** (1967): 739.
45. M. N. Bahadori and S. L. Soo, *Int. J. Heat Mass Transfer* **9,** 17 (1966).
46. M. W. Emmons, *Phys. Fluids* **10** (1967): 1125.
47. Y. N. Belyaev and V. B. Leonas, *Teplofizika Vysokikh Temperatur* **5** (1967): 1123.
48. J. T. Vanderslice, S. Weissmann, E. A. Mason, and R. J. Fallon, *Phys. Fluids* **5** (1962): 155.
49. C. F. Knopp and A. B. Cambell, *Phys. Fluids* **9** (1966): 989.
50. D. L. Jordan and J. D. Swift, *Int. J. Electron.* **35** (1973): 595.
51. J. C. Morris, R. P. Rudis, and J. M. Yos, *Phys. Fluids* **13** (1970): 608.
52. W. Hermann and E. Schade, *Z. Phys.* **233** (1970): 333.
53. U. Plantikow, *Z. Phys.* **237** (1970): 388.
54. K. S. Yun, S. Weissman, and E. A. Mason, *Phys. Fluids* **5** (1962): 672.
55. F. Burhorn, *Zeitschrift Für Physik* **155,** 42 (1959).

Chapter 8

# Radiation Transport

## 8.1. GENERAL CONCEPTS

### 8.1.1. Definitions

The radiation energy that passes through a cross section $dS$ within the solid angle element $d\Omega$ (measured in Steradian) in the direction $\theta$ with respect to the surface normal $\vec{n}$ (see Fig. 8.1), during a time interval $dt$ at frequencies between $\nu$ and $\nu + d\nu$, contains an amount of energy (see General Bibliography) given by

$$dE_\nu(\theta, \varphi) = I_\nu(\theta, \varphi)\, d\nu\, dS \cos\theta\, d\Omega\, dt \tag{8.1}$$

The quantity $I_\nu(\theta, \varphi)$, which refers to unit surface, unit time, and unit frequency, is called the monochromatic radiation intensity and is usually

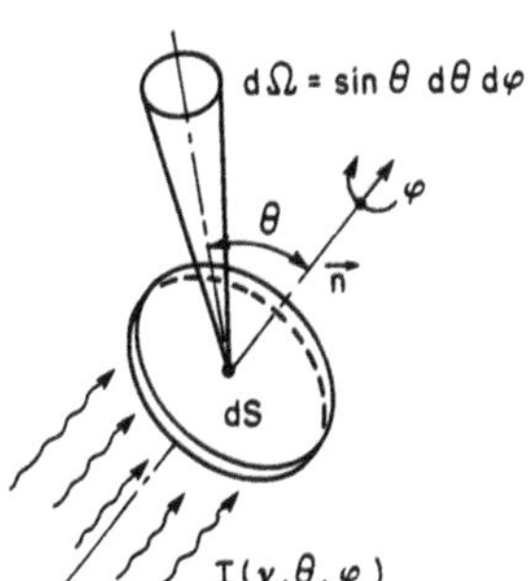


FIG. 8.1. Definition of the monochromatic radiation intensity of the radiation field.

expressed in J/ster m$^2$ (ster is the unit of the solid angle). The total intensity is obtained by integrating over all frequencies:

$$I(\theta, \varphi) = \int_0^{\infty} I_\nu(\theta, \varphi)\, d\nu \tag{8.2}$$

Here $I(\theta, \varphi)$ is expressed in W/ster m$^2$ and one should note that the dimensions of $I_\nu(\theta, \varphi)$ and $I(\theta, \varphi)$ are different.

By integrating over all possible angles $\theta$ (from 0 to $\pi/2$) and $\varphi$ (from 0 to $2\pi$) and all frequencies, we obtain the total radiation intensity

$$I = \iiint I_\nu(\theta, \varphi)\, d\nu\, d\theta\, d\varphi \tag{8.3}$$

expressed in W/m$^2$. The intensity can also be expressed as a function of the wavelength $\lambda$ instead of the frequency $\nu$. In this case, since $I_\nu\, d\nu = I_\lambda\, d\lambda$ and $d\nu = -c\, d\lambda/\lambda^2$, it follows that

$$I_\nu = \frac{\lambda^2}{c} I_\lambda \tag{8.4}$$

and (units of $I_\nu$) = (units of $I_\lambda$) m s. Thus, $I_\lambda$ is expressed in W/m$^3$ ster.

The total radiation flux passing through $dS$ from left to right, with $d\Omega = \sin\theta\, d\theta\, d\varphi$, is

$$H^+ = +\int_{\varphi=0}^{2\pi} d\varphi \int_{\theta=0}^{\pi/2} d\theta \int_{\nu=0}^{\infty} I_\nu(\theta, \varphi)\, d\nu \cos\theta \sin\theta \tag{8.5}$$

and the radiation from right to left is

$$H^- = -\int_{\varphi=0}^{2\pi} d\varphi \int_{\pi/2}^{0} d\theta \int_{\nu=0}^{\infty} I_\nu(\theta, \varphi)\, d\nu \cos\theta \sin\theta \tag{8.6}$$

The quantities $H^+$ and $H^-$ are called radiation fluxes and are expressed in W/m$^2$.

For isotropic radiation (which is independent of $\theta$ and $\varphi$), $I_\nu(\theta, \varphi) = I_\nu$, and since the radiation is the same in all directions,

$$H^+ = H^- = \pi I \tag{8.7}$$

Radiation propagates with the velocity of light $c$; in time $dt$ photons will travel a distance $l = c\, dt$. It is then possible to define a corresponding volume $dV = dS \cdot l$. For the special case of an isotropic radiation field, the photons pass through $dV$ under all solid angles from 0 to $4\pi$, and hence

$$u_\nu = \frac{4\pi}{c} I_\nu \tag{8.8}$$

where $u_\nu$ is expressed in J s/m$^3$.

The integration of $u_\nu$ over all frequencies gives the total radiation density, expressed in J/m$^3$. If the volume contains matter of refractive index $n_r$ instead of a vacuum, the radiation density must be multiplied by $n_r^3$. In plasmas, however, it is generally assumed that $n_r = 1$.

## 8.1.2. Blackbody Radiation

### 8.1.2.1. Planck's Law

The spectral dependence of the blackbody radiation field in a vacuum is a function only of $T$ and is given by Planck's radiation law. The density of the radiation field in the frequency interval $\nu$ to $\nu + d\nu$ follows from the formula

$$u_\nu^0(T)\, d\nu = \frac{8\pi h\nu^3}{c^3} \frac{d\nu}{\exp[h\nu/(kT)] - 1} \tag{8.9}$$

The superscript 0 indicates that $u^0$ is related to blackbody radiation.

In a vacuum, the radiation field is isotropic and the blackbody radiation intensity $I_\nu^0$, often written as $B_\nu$, is obtained from Eq. (8.8):

$$B_\nu\, d\nu = \frac{2h\nu^3}{c^2} \frac{d\nu}{\exp[h\nu/(kT)] - 1} \tag{8.10}$$

By using the relationship $I_\nu\, d\nu = I_\lambda\, d\lambda$, one can also express Eqs. (8.9) and (8.10) as functions of wavelength:

$$u_\lambda^0(T)\, d\lambda = \frac{8\pi hc}{\lambda^5} \frac{d\lambda}{\exp[hc/(kT\lambda)] - 1} \tag{8.11}$$

and

$$B_\lambda\, d\lambda = \frac{2hc^2}{\lambda^5} \frac{d\lambda}{\exp[hc/(kT\lambda)] - 1} \tag{8.12}$$

with $u_\lambda^0$ expressed in J/m$^4$ and $B_\lambda$ in W/m$^3$ ster.

Since $2hc^2 = 2c_1 = 1.1909 \times 10^{-16}$ W m$^2$ and $hc/k = c_2 =$

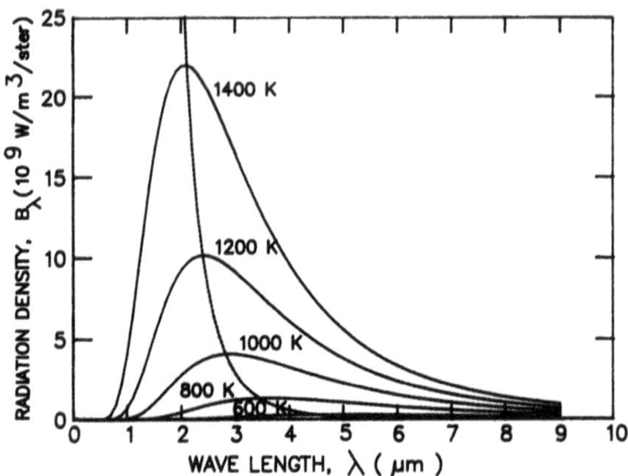


**FIG. 8.2.** Blackbody radiation density versus wavelength.

0.014386 m deg (all spectroscopic tables are given in $cm^{-1}$, and $c_2$ is given in cm deg), the numerical form of Eq. (8.12) is

$$B_\lambda(T) = 1.1909 \times 10^{-16} \times \lambda^{-5} \frac{1}{\exp[0.014386/(\lambda T)] - 1} \tag{8.13}$$

where $\lambda$ is in m and $T$ in K. Figure 8.2 represents $B_\lambda(T)$ versus $\lambda$ for various temperatures.

### 8.1.2.2. Wien's Law

For a fixed temperature $T$, Planck's formula gives an intensity maximum $I_{max}$ at a distinct wavelength, $\lambda_{max}$. The important relationship

$$\lambda_{max} T = \text{const.} = 2886\ \mu\text{m K} \tag{8.14}$$

follows from $dI_\lambda(T)/d\lambda = 0$ and is known as Wien's law. The corresponding position of maximum values of $B_\lambda(T)$ is represented by the dashed curve in Fig. 8.2.

### 8.1.2.3. The Stefan–Boltzmann Law

The total flux emitted by a black body per unit surface area and unit time into the solid angle element $d\Omega$ and in the direction $\theta$ is obtained by integrating Planck's radiation law over all frequencies $\nu$ and multiplying the result by $\cos\theta\, d\Omega$:

$$dH^0 = \int_{\nu=0}^{\infty} B_\nu(T) \cos\theta \cdot d\Omega\, d\nu = \frac{2\pi^4 k^4}{15c^2h^3} T^4 \cos\theta\, d\Omega \tag{8.15}$$

The total flux (intensity emitted per unit surface per unit time into the half-sphere) is obtained by integration. It follows that

$$\begin{aligned} H^0 &= \frac{2\pi^4 k^4 T^4}{15c^2h^3} \int_0^{\pi/2} \cos\theta \, d\Omega \\ &= \frac{2\pi^5 k^4}{15c^2h^3} T^4 = \sigma_s T^4 \end{aligned} \tag{8.16}$$

This is the Stefan–Boltzmann radiation law, and the value of $\sigma_s$ is

$$\sigma_s = 5.671 \times 10^{-8} \qquad (\mathrm{W\,m^{-2}\,K^{-4}}) \tag{8.17}$$

The radiation intensity [see Eq. (8.7)] is

$$B(T) = \frac{\sigma_s}{\pi} T^4 = 1.80513 \times 10^{-8} T^4 \tag{8.18}$$

Finally, for the case of an isotropic blackbody radiation field, the total radiation density is given by

$$u^0(T) = \frac{4\sigma_s}{c} T^4 \qquad \mathrm{J/m^3} \tag{8.19}$$

### 8.1.3. Gaseous Radiation

It will be assumed in the following calculations that the radiation field is isotropic and thus that

$$I_\nu(\theta, \varphi, T) = I_\nu(T) = I_\nu \tag{8.20}$$

#### 8.1.3.1. Volumetric Emission Coefficient

The volumetric monochromatic emission coefficient is the radiant energy $dE$ emitted by a volume element $dV$ in a frequency interval between $\nu$ and $\nu + d\nu$ into a solid angle $d\Omega$ per unit time $dt$ [see Griem (1964)]:

$$\varepsilon_\nu = \frac{dE}{d\nu \, dV \, d\Omega \, dt} \tag{8.21}$$

or

$$\varepsilon_\lambda = \frac{dE}{d\lambda \, dV \, d\Omega \, dt} \tag{8.22}$$

As already shown, the difference in units ($\varepsilon_\nu$ in J/m$^3$ ster and $\varepsilon_\lambda$ in W/m$^4$ ster) comes from Eq. (8.4). The volume element $dV$ should be so small that absorption and induced emission processes occurring within it are negligible.

The integrated volumetric emission coefficient $\varepsilon_L(\lambda_0)$ of a spectral line is

$$\varepsilon_L(\nu_0) = \int_{\text{line}} \varepsilon_\nu \, d\nu \quad \text{or} \quad \varepsilon_L(\lambda_0) = \int_{\text{line}} \varepsilon_\lambda \, d\lambda \tag{8.23}$$

where $\nu_0$ indicates that $\varepsilon_L(\nu_0)$ is related to the line centered at the frequency $\nu_0 = \nu_{ul}$.

Technically, this integral is to be taken over all frequencies since the line is theoretically infinite in extent. In practice, some judgement has to be made to decide how far out on the line the integral must be taken. The practical evaluation of the integral can be complicated by interferences from neighboring lines or when the background continuum is not constant over the range of integration. The units of $\varepsilon_L(\nu_0)$ (W/m$^3$ ster) are the same as those of $\varepsilon_L(\lambda_0)$.

### 8.1.3.2. Absorption Coefficient

If $I_\nu$ is the specific monochromatic intensity of radiation passing through an absorbing medium of thickness $dx$, then $\kappa'_\nu$, the absorption coefficient per unit length at frequency $\nu$, is given by [see Pecker-Wimel (1967)]

$$dI_\nu = -\kappa'_\nu I_\nu \, dx \tag{8.24}$$

Note that the upper index $'$ is introduced here because induced emission is not considered.

Quantity $\kappa'_\nu$ is generally expressed in cm$^{-1}$. It is in general a function of the wavelength, the gas properties at the frequency $\nu$, and the direction in which the radiation propagates. If we consider $dI_\lambda$ instead of $dI_\nu$, the relationship would be the same:

$$dI_\lambda = -\kappa'_\lambda I_\lambda \, dx \tag{8.25}$$

thus $\kappa'_\lambda = \kappa'_\nu$.

In general, after integrating over the thickness $L$ of a gaseous layer,

$$I_\nu = I_{\nu,0} \exp\left(-\int_0^L \kappa'_\nu \, dx\right) \tag{8.26}$$

$$= I_{\nu,0} \exp(-\tau_\nu) \tag{8.27}$$

where $I_{\nu,0}$ is the radiation intensity entering the gaseous layer and $\tau_\nu$ represents the optical depth of the layer (dimensionless).

### 8.1.3.3. Relationship between Emission and Absorption

For a parallel beam of radiation intensity $I_\nu$, the increase in intensity across an element of thickness $dx$ is given by the difference between emission and absorption due to the element:

$$dI_\nu = \varepsilon_\nu(x)\,dx - I_\nu(x)\kappa'_\nu(x)\,dx \tag{8.28}$$

With the boundary condition $I_\nu(0) = 0$ this differential equation has the solution

$$I_\nu(L) = \int_0^L \varepsilon_\nu(x)\exp\left(-\int_x^L \kappa'_\nu(t)\,dt\right)dx \tag{8.29}$$

If $\kappa'_\nu$ and $\varepsilon_\nu$ are constant throughout a gas at uniform temperature, Eq. (8.29) reduces to

$$I_\nu(L) = \frac{\varepsilon_\nu}{\kappa'_\nu}[1 - \exp(-\kappa'_\nu L)] \tag{8.30}$$

The ratio $\varepsilon_\nu/\kappa'_\nu$ is called the source function, $S_\nu$, and under complete equilibrium it can be shown to be equal to the Planck function $B_\nu$(T) [see Eq. (8.10)]. In the general case of a medium whose refractive index is $n_r$, the source function is given by

$$S_\nu = \varepsilon_\nu/n_r\kappa'_\nu \tag{8.31}$$

There are two limiting forms for the expression for $I_\nu(L)$. The first is the optically thin approximation in which $\kappa'_\nu L \ll 1$. This leads, after expanding the exponential as a series, to

$$I_\nu(L) = \int_0^L \varepsilon_\nu(x)\,dx \tag{8.32}$$

The second occurs when $\kappa'_\nu L$ is large compared with 1. For uniform conditions,

$$I_\nu(L) = \frac{\varepsilon_\nu}{\kappa'_\nu} = B_\nu(T) \tag{8.33}$$

where $B_\nu(T)$ is Planck's monochromatic radiation intensity. This situation corresponds to the behavior of an optically thick plasma.

## 8.2. RADIATION MECHANISMS IN PLASMAS

### 8.2.1. Spontaneous Emission

An excited atom in a higher quantum state $u$ may return to a lower energy state $l$ by emitting a photon of energy $h\nu_{ul}$ (see Fig. 8.3). In the absence of incident radiation, the number of atoms leaving $u$ in a time interval $dt$ is proportional to $N_u(t)$, the population of the state at time $t$, or

$$dN_u(t) = -A_{ul}N_u(t)\,dt \tag{8.34}$$

Integration yields

$$N_u(t) = N_u(0)\exp(-A_{ul}t) \tag{8.35}$$

where $N_u(0)$ is the population of state $u$ at initial time $t = 0$.

The constant $A_{ul}$ is called the spontaneous transition probability and is defined as the probability per second that an atom in the state $u$ will spontaneously decay to state $l$ by emitting a photon at frequency $\nu_{ul}$. In fact, Eq. (8.34) can also be written in terms of the lifetime $\tau_{ul}$, defined as the time necessary to reduce the population of state $u$ by a factor $e$. Note that

$$\tau_{ul} = 1/A_{ul} \tag{8.36}$$

and $A_{ul}$ is expressed in $s^{-1}$. Typical values of $A$ are between $10^6$ and $10^8\,s^{-1}$; the highest values are obtained for resonance states of atoms or molecules (i.e., transition from the first excited state to the ground state).

### 8.2.2. Induced Emission

Induced or forced emission plays a key role for the population inversion in lasers. When a plasma is imbedded in a radiation field of density $u_\nu^0$ then the plasma may not only exist by spontaneous emission, but also by induced emission. In the elementary process of induced

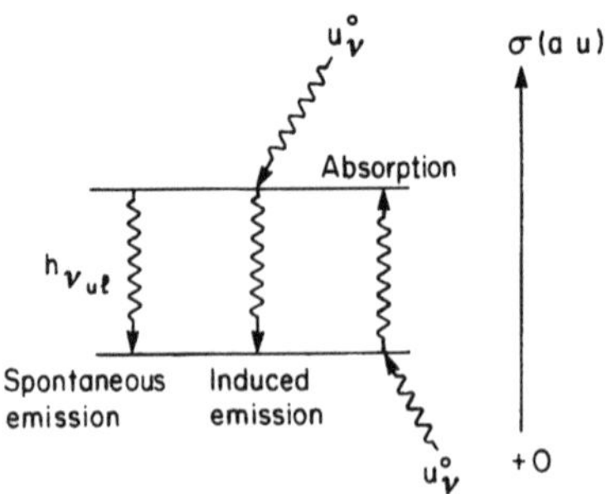


FIG. 8.3. Spontaneous emission, induced emission, and absorption.

emission, a photon of frequency $\nu$ interacts with an atom or ion in a higher quantum state and thus will force this atom or ion to emit a photon of the same frequency and in the same direction as the oncoming photon. The number of such transitions per second and $m^3$ is proportional to a transition probability $B_{ul}$, to the density of the radiation field $u_\nu$, and to the number density of atoms or ions in the appropriate excited state (see Section 8.2.4).

### 8.2.3. Absorption

Consider a volume with monochromatic radiation of density $u_\nu^0$ given by Eq. (8.9) and atoms capable of being raised from state $l$ to state $u$ by absorbing radiation of energy $h\nu_{lu}$. The probability per second that an atom in state $l$ will absorb a quantum $h\nu_{lu}$ and pass to state $u$ when exposed to isotropic radiation $u_\nu^0 d\nu$ in the frequency range between $\nu_{lu}$ and $\nu_{lu} + d\nu$ is proportional to $u_\nu^0$ and written as $B_{lu}u_\nu^0$ (see Fig. 8.3). Notice that the dimensions of the transition probability for absorption, $B_{lu}$, are $m^3/J\ s^2$.

### 8.2.4. Microreversibility Principle

In any assembly of atoms, the equilibrium must be dynamic. At any given instant, some atoms are absorbing radiation while others are emitting radiation, and it is found that equilibrium is maintained only if each individual interaction process is balanced by its own reverse process (microbalance).

Based on the principle of dynamic balance, it is then possible to consider the details of an individual process and to derive relations that connect absorption and emission processes. If absorption and emission are the only processes effective between states $l$ and $u$, then

$$u_\nu^0 B_{lu} N_l = A_{ul} N_u \tag{8.37}$$

and if the atomic populations are governed by Boltzmann's law [see Eq. (4.94)], then

$$u_\nu^0 = \frac{A_{ul}}{B_{lu}} \frac{g_u}{g_l} \exp\left(\frac{h\nu_{lu}}{kT}\right) \tag{8.38}$$

where $g_u$ and $g_l$ are the statistical weights of the states $u$ and $l$. Equation (8.38), however, is not the final balance equation, because induced emission must be included in this balance. As previously mentioned, these induced emissions are proportional to the density of the radiation field $u_\nu^0$, to the transition probability $B_{u,l}$, and to the number of atoms in the upper quantum state $u$ ($N_u$). Thus, applying the principle of detailed balance for an equilibrium situation it follows that

$$u_\nu^0 B_{lu} N_l = A_{ul} N_u + B_{ul} u_\nu^0 N_u \tag{8.39}$$

which leads to

$$u_\nu^0 = \frac{A_{ul}/B_{ul}}{\dfrac{g_l}{g_u}\dfrac{B_{lu}}{B_{ul}}\exp\dfrac{h\nu}{kT} - 1} \tag{8.40}$$

By comparing this expression with Planck's law Eq. (8.9), it follows that

$$g_l B_{lu} = g_u B_{ul} \tag{8.41}$$

and

$$A_{ul} = \frac{8\pi h \nu^3}{c^3} B_{ul} \tag{8.42}$$

or

$$A_{ul} = \frac{1.66543 \times 10^{-32}}{\lambda^3} B_{ul}$$

with $\lambda$ in meters.

These transition probabilities ($A_{ul}$, $B_{ul}$, $B_{lu}$) are called Einstein coefficients. Once one is known (calculated or measured), the two others can be deduced from Eqs. (8.41) and (8.42). They are related to the volumetric emission and absorption coefficients [see Eqs. (8.21) and (8.24)].

If $n_{i,u}$ is the number density of atoms of species $i$ excited to state $u$, the number of quantum transitions becomes $A_{ul}^i n_{i,u}$ and thus in an isotropic plasma the line emission coefficient (W/m$^3$ ster.) due to the finite width of a spectral line [Eq. (8.23)] is given by

$$\varepsilon_L(\nu) = \frac{1}{4\pi} A_{ul}^i n_{i,u} h \nu_{ul} \tag{8.43}$$

where $1/4\pi$ refers to the unit solid angle.

Similarly, an integrated absorption coefficient $\kappa_L(\nu)$ can be introduced into Eq. (8.28), integrated over the line profile. The integrated equation can then be written taking into account the relationship between $B_\nu$ and $u_\nu^0$ [see Eq. (8.8)]:

$$I_\nu^t = \frac{A_{ul}}{4\pi} h\nu_{u,l} n_u\, dx + \frac{I_\nu^0}{c} h\nu_{lu}\, dx (B_{ul} n_u - B_{lu} n_l) \tag{8.44}$$

In this equation, the intensity $I_\nu^0$ has been assumed as constant over the width of the line. Of course, $\nu_{ul} = \nu_{lu}$. Assuming a Boltzmann distribution corresponding to a temperature $T$ and using Eq. (8.41), it follows that

$$\frac{dI_\nu^t}{dx} = \frac{A_{ul}}{4\pi} n_u h\nu_{ul} - \frac{B_{lu}}{c} n_l I_\nu^0 h\nu_{ul}\left[1 - \exp\left(-\frac{h\nu_{ul}}{kT}\right)\right] \tag{8.45}$$

The second term on the right-hand side of Eq. (8.45) is the effective integrated absorption term. Without the spontaneous emission term, Eq. (8.45) becomes equivalent to the differential equation that defines the integrated absorption coefficient [see Eq. (8.24)], provided that the adsorption profile of the line is the same as that of the induced emission profile, an assumption which holds for CTE and LTE plasmas (Griem 1964).

$$\kappa_L(\nu) = \frac{h\nu_{ul}}{c} B_{lu} n_l \left[1 - \exp\left(-\frac{h\nu_{ul}}{kT}\right)\right] \tag{8.46}$$

In the case where $\exp(h\nu_{ul}/kT) \gg 1$ (induced emission is negligible), Eq. (8.46) reduces to the familiar relation

$$\kappa_L'(\nu) = \frac{B_{lu} n_l h\nu_{ul}}{c} \tag{8.47}$$

According to Kirchhoff's law Eq. (8.31) for the source function holds for $\varepsilon_L(\nu)$ and $\kappa_L'(\nu)$.

### 8.2.5. Effective Radiative Lifetime of an Excited State

In Section 8.2.1 the radiative lifetime of an excited state was defined as the reciprocal of the spontaneous transition probability $A_{ul}$. However, an excited state $u$ can be depopulated by emission of radiation to a family of lower states $l$ or by absorption of radiation to a set of higher states $u'$ (see Fig. 8.4). The sum of the transition probabilities must be equal to the reciprocal of the lifetime of the state, assuming superelastic collisions to be negligible. Thus

$$\frac{1}{\tau_u} = \sum_l A_{ul} + \sum_l B_{ul} u^0(\nu_{ul}) + \sum_{u'} B_{uu'} u^0(\nu_{uu'}) \tag{8.48}$$

because

$$\frac{A_{ul}}{B_{ul} u^0(\nu_{ul})} = \exp\left(\frac{h\nu_{ul}}{kT}\right) - 1 \tag{8.49}$$

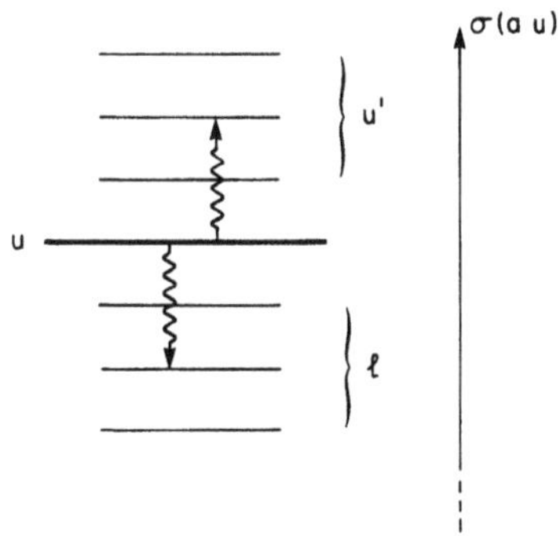


FIG. 8.4. Atomic transitions from a state $u$.

It follows that

$$\frac{1}{\tau_u} = \sum_l A_{ul}\left\{1 + \left[\exp\left(\frac{h\nu_{ul}}{kT}\right) - 1\right]^{-1}\right\} + \sum_{u'} A_{u'u}\frac{g_{u'}}{g_u}\left[\exp\left(\frac{h\nu_{uu'}}{kT}\right) - 1\right]^{-1} \tag{8.50}$$

The influence of the stimulated radiation appears through $\exp(h\nu/kT)$. For $T = 300$ K and for a photon of wavelength 600 nm, $\exp(h\nu/kT) \sim 5 \times 10^{36}$ and the effect of the stimulated emission is insignificant in the visible spectral region where $h\nu/kT \gg 1$. Therefore

$$\frac{1}{\tau_u} = \sum_l A_{ul} \tag{8.51}$$

However, at 9000 K, $\exp(h\nu/kT) \sim 16$, and the departures from Eq. (8.51) can become significant.

## 8.3. RADIATION EMISSION AND ABSORPTION

### 8.3.1. Classification of Emitted Radiation

The various types of spectra observed in plasmas are classified according to the particles that emit them and the degrees of freedom they set into action. Figure 8.5 [from Cabannes and Chapelle (1971)] shows a schematic of the various excited states of an atom or an ion and the corresponding bound–bound, bound–free, and free–free transitions.

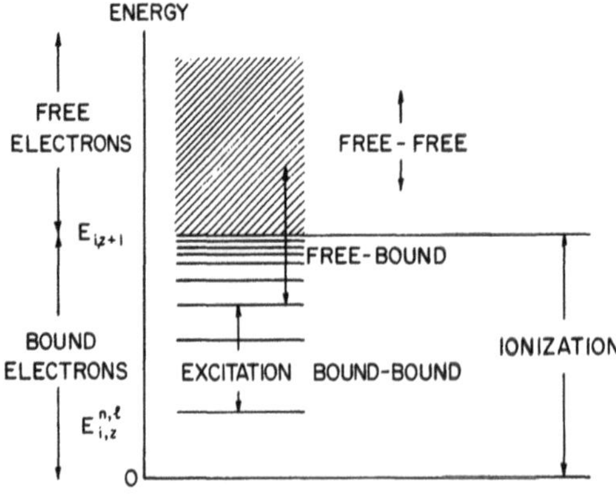


**FIG. 8.5.** Excited levels of an atom or an ion [after Cabannes and Chapelle (1971)]: $E^{n,l}_{i,z}$ = energy of an excited state; $E_{i,z+1}$ = ionization energy of the atom of species $i$ with electrical charge $ze$.

#### 8.3.1.1. Bound–Bound Transitions

Due to their electronic excitation, atoms and ions (except $H^+$) emit a spectrum of lines such that

$$E_u - E_l = h\nu_{ul} = \frac{hc}{\lambda_{ul}} \tag{8.52}$$

where $u$ and $l$ are, respectively, the upper and lower excited levels between which the transition takes place.

The radiation frequency $\nu$ (or wavelength $\lambda$) is characteristic of both the atom or ion and the emitting levels. Electrons changing their orbits remain bound to the nucleus, and this type of transition is called a "bound–bound" transition. The corresponding wavelengths extend from the infrared to the far ultraviolet.

The spectra of diatomic molecules are by far more complex than those of single atoms. Depending on the energy levels of the molecules (see Section 2.4), the following excited states have to be considered:

- electronic excitation at energies of a few eV,
- vibrational excitation at energies in the order of 0.1 eV,
- rotational excitation at energies from $10^{-4}$ eV to $10^{-2}$ eV.

Figure 8.6 [from Cabannes and Chapelle (1971)] gives the energy diagram for a CN diatomic molecule with various electronic excited states like the fundamental $X^2\Sigma^+$ state, the first excited $A^2\Pi_1$ state, and the second excited $B^2\Sigma^+$ state (see Section 2.4 for the denomination of the states).

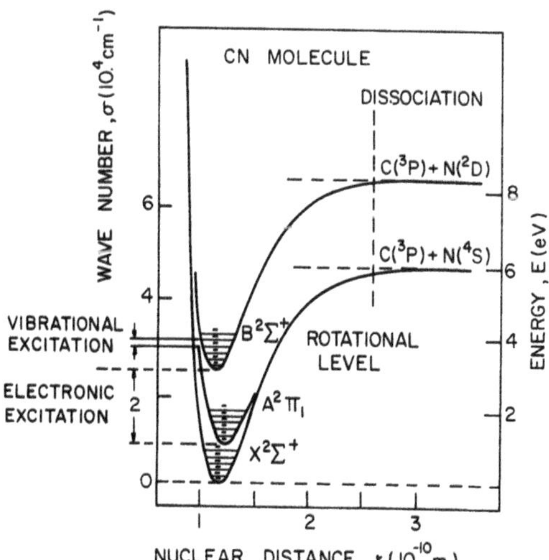


**FIG. 8.6.** Energy curves for diatomic molecule (CN) [after Cabannes and Chapelle (1971)].

Electronic transitions are modified by the presence of vibrational and rotational effects that give rise to groups of bands in band systems. For atoms, each band system corresponds to a single electronic transition.

If the energy is written (see Section 2.4.1) using spectroscopic notation,

$$E = T_e + G + F \tag{8.53}$$

where $T_e$, $G$, and $F$ are respectively, the electronic term, the vibrational term, and the rotational term expressed in $\text{cm}^{-1}$. The wave numbers of the spectral lines are given by

$$\sigma = E' - E'' = (T_e' - T_e'') + (G' - G'') + (F' - F'') \tag{8.54}$$

where the single-primed letters refer to the upper state and the double-primed letters to the lower state.

For a given electronic transition, the expression $\sigma_E = T_e' - T_e''$ is a constant.

Neglecting the rotational terms, which are very small, the variable part of the frequency is $(G' - G'')$, which gives the electronic transition of a coarse structure called the vibrational structure. Since there is no selection rule for the vibrational quantum number ($v$), a large number of bands are expected for electronic transitions. Each band is designated by the two vibrational quantum numbers $v'$ and $v''$. Some characteristic groups are observed in the band system. The groups are called sequences, and for each one $\Delta v = v' - v''$ is a constant. The separation of bands is small for a sequence because the vibrational constants $\omega_e$ and $x_e$ (see Section 2.4.3) do not change much from one electronic state to another, and thus $G' - G''$ varies slowly with $v'$ when $\Delta v$ is a constant. Figure 8.7 shows the sequences for the violet and the red systems of the diatomic molecule CN. Note that due to the significant overlapping in sequences, only a few bands are observed in each sequence.

Each band has a rotational structure due to the rotational motion of

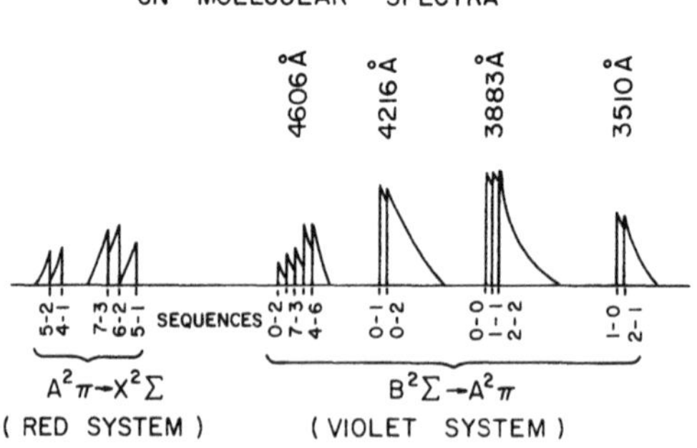


**FIG. 8.7.** Schematic spectrum of violet and red systems of the CN molecular spectra [after Cabannes and Chapelle (1971)].

the nuclei. The corresponding frequency for a given electronic band $[(T_e' - T_e'')$ and $(G' - G'')$ are then constants] is obtained from the difference $F' - F''$. These rotational spectra are themselves very complex and are classified into branches according to the value of $\Delta J = J' - J'' = 0 \pm 1$ (see Section 2.4.3). The electronic excitation of diatomic molecules gives band systems in the UV, visible, and IR regions.

If the electronic state does not change during a transition, infrared bands called the vibration–rotation bands may be observed (mainly for the fundamental X state). The wave number of the spectral lines is then

$$\sigma = (G' - G'') + (F' - F'') \tag{8.55}$$

Generally, one or only a few bands corresponding to the lowest values of the vibrational quantum number are observed.

#### 8.3.1.2. Free–Bound and Free–Free Transitions

*8.3.1.2.a. Free–Bound Transitions.* Because a free electron may assume nonquantized kinetic energies, its recombination with an ion will result in continuum radiation. For example, the recombination of free electrons and $N^+$ ions gives rise to "free–bound" transitions resulting in a continuous spectrum observable in the visible and the ultraviolet range. This same phenomenon also occurs for the recombination of free electrons with $N^{++}$, $N^{+++}$, $N_2^+$, .... The excess energy of an electron with velocity $v_e$ is converted into radiation according to the relation

$$m_e v_e^2/2 + E_{N^+}^{I} - \delta E_{N^+} - E_{N,j} = h\nu \tag{8.56}$$

where $E_{N^+}^{I}$ is the ionization energy corresponding to the reaction $N \rightarrow N^+ + e$ and $\delta E_{N^+}$ is its lowering; $E_{N,j}$ is the energy of the excited state $j$ to which the electron is trapped. A threshold value exists for the wavelength resulting from the trapping of electrons with zero velocity in the $E_{N,j}$ excited state:

$$\Delta' E_N = h\nu_{\min} \quad \text{or} \quad \lambda_{\max} = \frac{hc}{\Delta' E_N} \tag{8.57}$$

and

$$\lambda_{\max}(\mathrm{m}) = \frac{1.98647 \times 10^{-25}}{\Delta' E_N\,(\mathrm{J})} = \frac{1.23944 \times 10^{-6}}{\Delta' E_N\,(\mathrm{eV})} = \frac{1.00048 \times 10^{2}}{\Delta' E_N\,(\mathrm{cm}^{-1})} \tag{8.58}$$

where

$$\Delta E_N = E_{N^+}^{I} - \delta E_{N^+} - E_{N,j} \tag{8.59}$$

Because the electrons are distributed over a large spectrum of energies, their recombination into energy level $j$ will give rise to a spectrum between $\nu_{min}$ and large values of $\nu$.

Electrons can be trapped into every available energy state of an atom, so the number of continuous spectra coincides with the number of available energy levels.

*8.3.1.2.b. Free–Free Transitions.* Continuous spectra occur when free electrons loose part of their kinetic energy in the Coulomb field of positive ions. This loss in kinetic energy is converted into radiation reflecting the Maxwellian distribution of the electrons. The emitted energy is a continuum called "bremsstrahlung," which is typically in the infrared. Because radiation is emitted by free electrons whose kinetic energy is $\frac{1}{2}m_e v_e^2 \geqslant h\nu$, the spectral wavelength depends on the electron temperature. Free–free transition may also occur during electron–atom or electron–molecule encounters.

## 8.3.2. Line Radiation

### 8.3.2.1. Line Broadening

The absorption or emission of discrete radiation quanta implies that monochromatic radiation of frequency $\nu_{ul}$ should be observed. The examination of such radiation by a spectrometer shows, however, that the spectral line has a definite width $\delta_\nu$ or $\delta_\lambda$ around $\nu_{ul}$ or $\lambda_{ul}$. When the resolving power of the spectrometer used for line analysis is high enough (see Vol. 2), it is observed that the line width depends on the properties of the emitting source.

The spectral lines emitted by a gas discharge are found to be rather narrow at low gas pressure [Griem H. (1974)]. At higher pressures and correspondingly higher densities of atoms, ions, and electrons in the source, there will be more frequent collisions among the radiating atoms, and the resulting effects are described as *pressure broadening.* When charged particles are present in significant quantities, they generate electric microfields that perturb the normal energy levels via *Stark broadening* of the spectral lines. Note that magnetic fields may also result in the broadening of discrete energy levels through the Zeeman effect. In addition, these effects may result in a displacement of the maximum of the lines called *line shift*.

When the pressure is progressively reduced, collision and Stark broadening are also reduced and the line width asymptotically approaches a temperature-dependent finite value. The remaining width of the line is due to random motion of the radiating atoms and is described as the *Doppler effect.* When the Doppler width is reduced to negligible values (by reducing the thermal velocities of the radiating atoms), the remaining width is known as the *natural width* and is associated with the interactions of the atoms. This natural width is generally much narrower than the Doppler broadening.

In general, the frequency or wavelength distributions of emission or absorption can be described by a shape factor $P(\nu - \nu_0)$ such that

$$\int_{-\infty}^{+\infty} P(\nu - \nu_0)\, d\nu = 1 \tag{8.60}$$

with $\nu_0 = \nu_{ul}$, or

$$\int_{-\infty}^{+\infty} P(\lambda - \lambda_0)\, d\lambda = 1$$

with $\lambda_0 = \lambda_{ul}$.

These relationships mean that Eq. (8.43) can be written as

$$\varepsilon_L(\lambda) = \frac{hc}{4\pi\lambda_{ul}} A^i_{ul} n_{i,u} P(\lambda_{ul}) \tag{8.61}$$

As the integration is performed over the line contour or over the line wavelength range $\delta_\lambda$, $\varepsilon_L(\lambda)$ results from an integration (see Eq. 8.23) and is expressed in W/ster m$^3$ (not in W/ster m$^4$, as already emphasized in Section 8.1.3). Equation (8.61) is then written as

$$\varepsilon_L(\lambda_{ul}) = \int_{\text{line}} \varepsilon_\lambda\, d\lambda = \frac{1}{4\pi} n_{i,u} A^i_{ul} \frac{hc}{\lambda_{ul}} \tag{8.62}$$

which is the line emission coefficient.

Figure 8.8 shows the characteristic parameters of a line profile: $2\delta$ is the width at half of the maximum intensity, $\lambda_0$ is the central wavelength of the line, and $\Delta$ is the shift. In the following discussion we will restrict

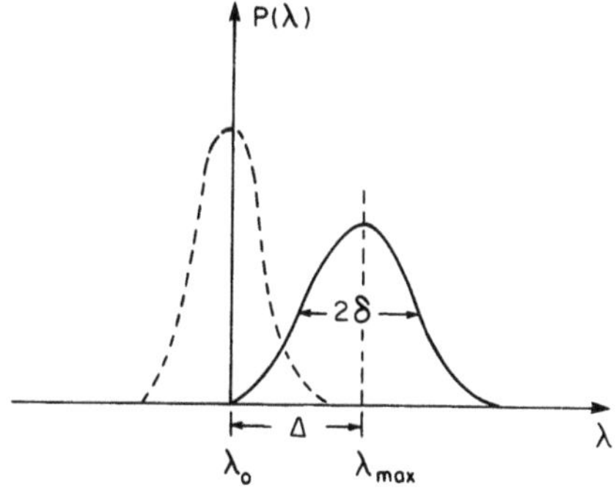


**FIG. 8.8.** Line profile and line shift.

our description to the natural, Doppler, and Stark broadenings and to the resulting profiles. It must be emphasized that when LTE prevails, the emitted line profile is identical to its absorption profile according to Kirchhoff's law. Since the self-absorption coefficient varies across the line width, the emission line profile reflecting this absorption will be more distorted at the line center compared to its wings.

*8.3.2.1.a. Natural Line Width.* The fundamental broadening of a spectral line stems from the hypothetical situation in which an atom is motionless and is shielded from electric and magnetic fields. Under these conditions the line profile is Lorentzian as defined by

$$P_N(\lambda) = \frac{\delta}{\pi[(\lambda - \lambda_0)^2 + \delta^2]} \tag{8.63}$$

The half-width is given by

$$\delta = \delta_m + \delta_n \quad \text{with} \quad \delta_m = \sum_{n=1}^{m-1} A_{mn}/4\pi \quad \text{and} \quad \delta_n = \sum_{m=1}^{n-1} A_{nm}/4\pi \tag{8.64}$$

In most practical cases for thermal plasmas this broadening is negligible compared to Stark and Doppler broadening and may be neglected in the calculations.

*8.3.2.1.b. Doppler Line Width.* If a monochromatic light source is moving with a velocity $v$ in the direction of the line of sight, the emitted light frequency appears to be shifted by an amount $v\nu_0/c$ compared to the frequency $\nu_0$ of the source at rest. The random motion of the atoms in a source causes the spectral lines emitted or absorbed by these atoms to be broadened. If the velocity distribution is Maxwellian, the resulting line profile is

$$P_D(\lambda) = \frac{1}{\Delta\lambda_p \pi^{1/2}} \exp\left[-\left(\frac{\lambda - \lambda_0}{\Delta\lambda_p}\right)^2\right] \tag{8.65}$$

where

$$\Delta\lambda_p = \lambda_0 \left(\frac{2kT}{mc^2}\right)^{1/2} \tag{8.66}$$

and $m$ is the mass of the emitting atom. The width at half the maximum intensity is given by

$$\begin{aligned} 2\delta_D &= 2\,\Delta\lambda_p(\ln 2)^{1/2} = 2\lambda_0\left(\frac{2kT\ln 2}{mc^2}\right)^{1/2} \\ &= 7.1610^{-7}\lambda_0\sqrt{T/M} \end{aligned} \tag{8.67}$$

where $M$ is the atomic mass (in g) and $\delta_D$ is expressed in the units of $\lambda_0$.

In thermal plasmas the Doppler effect usually results in a small contribution to the line broadening. For example, at $T = 10{,}000$ K and $p = 100$ kPa $2\delta_D = 3.5 \times 10^{-2}$ nm for the $H_\beta$ line at 486.1 nm, while $2\delta_D = 9 \times 10^{-3}$ nm for the NI nitrogen atomic line at 493.5 nm, and $2\delta_D = 2.5 \times 10^{-3}$ nm for the ArI atomic line at 696.5 nm. Equation (8.65) can be rewritten in terms of $\delta_D$ as follows:

$$P_D(\lambda) = \frac{(\ln 2)^{1/2}}{\delta_D \pi^{1/2}} \exp\left[-\ln 2\left(\frac{\lambda - \lambda_0}{\delta_D}\right)^2\right] \tag{8.68}$$

*8.3.2.1.c. Stark Broadening.* The Stark effect becomes predominant as soon as the rate of ionization exceeds 1%. The perturbation time resulting from a charged particle moving with a relative mean velocity $\bar{v}$ with respect to the emitting particle is given by

$$\tau_s = b/\bar{v} \tag{8.69}$$

where $b$ is the impact parameter (see Section 7.3.4). Because of the large difference between the velocities of electrons and/or atoms, the collision time $\tau_s$ for electrons can be regarded as short compared with the time between successive collisions, and the broadening is determined on the basis of the *impact approximation.* In this model the emitting system is virtually unperturbed most of the time, and broadening is described in terms of impacts that are well separated in time. For the slowly moving ions, the perturbation is practically constant over the time of interest ($1/\Delta\nu$), and thus their motion can be neglected completely. Their effect is calculated using the *quasi-static approximation* in which a static Stark effect is assumed, taking into account the statistical distribution of the electric fields. Griem (1974) and Traving (1968) have given expressions for the half-line Stark width $\delta_{e,i}$ and the Stark shift $\Delta_{e,i}$, for which the electron density $n_e$ is the most important parameter.

For example, Griem (1974) has expressed $\delta_{e,i}$ through the electron density and the parameters $w$, $d$, and $\alpha$, which are tabulated for different lines, temperatures, and electron densities in forms such as

$$2\delta_{e,i} = 2[1 + 1.75 \times 10^{-4} n_e^{1/4} \alpha(1 - 0.068 n_e^{1/6} T_e^{1/2})] 10^{17} n_e w \tag{8.70}$$

where $\delta_{e,i}$ is expressed in nm and $n_e$ in cm$^{-3}$; the relationship holds for $0.05 \le 10^{-4} n_e^{1/4}$ and $\alpha \leqslant 0.5$.

The resulting line profile is Lorentzian, for example,

$$P(\lambda) = \frac{1}{\pi} \frac{\delta_e}{[(\lambda - \lambda_0 - \Delta_{e,i})^2 + \delta_{e,i}^2]^{1/2}} \tag{8.71}$$

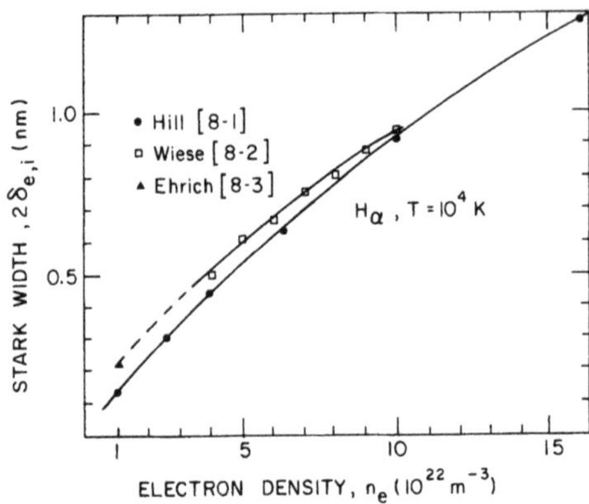


**FIG. 8.9.** The Stark width versus electron density at half maximum of the $H_\alpha$ line at 10,000 K, calculated in Ref. [6] from different sources.

For the nitrogen lines at $T = 13{,}200$ K, $n_e = 1.05 \times 10^{17}$ cm$^{-3}$, and $2\delta_{e,i} = 0.269$ nm for the NI line at 493.5 nm and 0.124 nm for the NI line at 746.8 nm. At such temperatures, the Doppler broadening, which is $2\delta_D = 0.011$ nm for NI at 493.5 nm and 0.0164 nm for NI at 746.8 nm, is less than 13% of the Stark broadening. Hydrogen atomic lines are often used to determine the electron density ($\delta_{ei}$ is by far more sensitive to $n_e$ than to $T_e$). Figure 8.9 represents, for example, the $H_\alpha$ line broadening at $T = 10{,}000$ K versus $n_e$; $2\delta_{e,i}$ is calculated with the parameters determined by Hill.[1] The experimental values of Wiese[2] and Ehrich[3] are in rather good agreement with the calculated values.

The coefficients used to calculate $\delta_{e,i}$ should be checked carefully; the relative error between the calculated and measured values, especially for species such as Fe, Cu, F, and S, can reach values up to ±30%.[4,5]

*8.3.2.1.d. Resulting Profiles.* In thermal plasmas, the line profile represents the convolution of the Doppler (Gaussian) and Stark (Lorentzian) profiles. Assuming that the two processes are independent, the result is a Voigt profile defined as

$$P(a, b) = \frac{1}{2\pi^{1/2}\delta_D} \int_{-\infty}^{+\infty} \frac{a}{\pi} \frac{\exp(-y^2)}{(b - y)^2 + a^2} dy \tag{8.72}$$

with

$$a = \frac{\delta_L}{\delta_D} (\ln 2)^{1/2} \tag{8.73}$$

and

$$b = \frac{(\lambda - \lambda_0 - \Delta)}{\delta_D} (\ln 2) \tag{8.74}$$

$\Delta$ being the line shift.

### 8.3.2.2. Volumetric Spectral Emission Coefficient Neglecting Absorption

When absorption is neglected, the line intensity is calculated according to Eq. (8.62). Assuming a Boltzmann distribution for $n_{i,u}$, it follows that

$$\varepsilon_L(\lambda_{ul}) = \int \varepsilon_\lambda \, d\lambda = \frac{1}{4\pi} n_i g_{i,u} \exp\left(-\frac{E_{i,u}}{kT}\right) \frac{1}{Q^i_{el}(T)} A^i_{ul} \frac{hc}{\lambda_{ul}} \tag{8.75}$$

where index $i$ stands for the chemical species $i$ and $Q^i_{el}(T)$ is the corresponding electronic partition function.

With the densities $n_i$ in $m^{-3}$, the transition probabilities in $s^{-1}$, and the wavelengths in m, the coefficient $hc/4\pi = 1.58078 \times 10^{-26}$. In Eq. (8.75), the temperature appears explicitly only in the exponential term, but the density of the species $n_i$ is also strongly dependent on temperature, while $Q^i_{el}$, the electronic partition function, is rather weakly temperature-dependent. Thus, for a given pressure, $\varepsilon_L$ depends strongly on $T$, mainly through the exponential term and to a lesser degree (except when dissociation is not completed or ionization important) through $n_i$.

For example, in Fig. 8.10 (from Ref. 6), which shows the volumetric

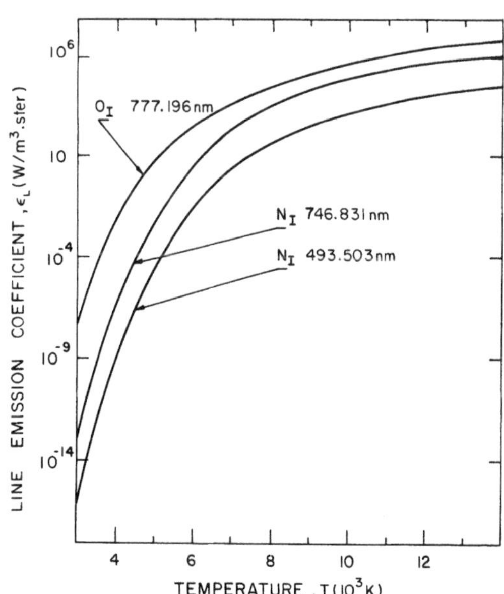


**FIG. 8.10.** Volumetric emission coefficient versus temperature at atmospheric pressure (W/m$^3$ ster) of the NI lines at 493.503 nm and 747.831 nm and the OI line at 777.196 nm.[6]

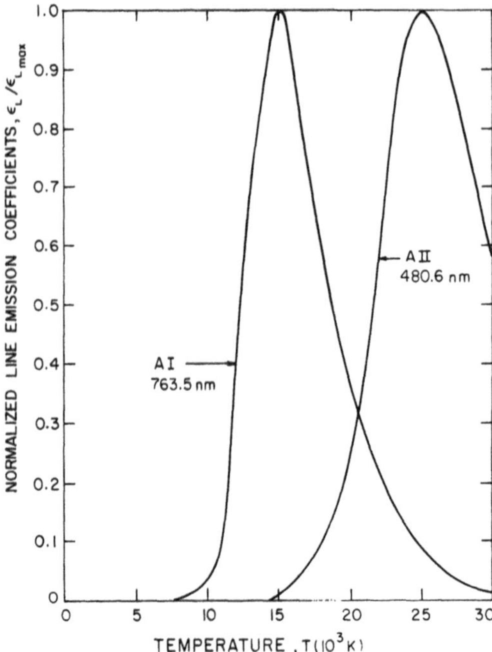


**FIG. 8.11.** Relative emission coefficients of two argon lines, ArI 763.5 nm and ArII 480.6 nm, as a function of temperature at atmospheric pressure.[7]

emission coefficient $\varepsilon_L(\lambda)$ as a function of temperature at $p = 10^5$ Pa for NI and OI lines, it is clear that the emission coefficient of NI at 493.5 nm is negligible as long as dissociation is not completed. The same is true for OI at 777.2 nm. This figure illustrates a very important experimental fact: most detectors (photomultipliers, photodides, etc.) can detect variations in a range of 3 to 4 orders of magnitude but not more. Thus, according to Fig. 8.10, temperature measurements through the absolute values of $\varepsilon_L(\lambda)$ will be restricted to temperature ranges between about 8000 and 16,000 K. In Fig. 8.11 (from Ref. 7), which illustrates the emission coefficients (normalized to the maximum intensity) of two argon lines, ArI (neutral atom) and ArII (first ion), it is clear that the neutral line will show up in the spectrum only in the temperature interval from 10,000 K to 25,000 K, since the density of argon atoms is very low beyond 25,000 K (see Fig. 5.4). The normalized emission coefficient is the emission coefficient divided by its maximum value (which occurs at approximately 15,500 K for ArI and 25,500 K for ArII). According to Eq. (8.75), the emission coefficient of an atomic or ionic line of the species $i$ will increase with pressure for a given temperature (about two orders of magnitude if the pressure is raised from 100 kPa to 10 MPa) due to the change in density, $n_i$. The maximum line emission coefficient is also shifted to higher temperatures because dissociation occurs at a higher temperature when the pressure is raised, thus shifting the position of the maximum density of atoms.

Similar results are obtained for diatomic molecules, whose emission coefficients can be given[8] by

$$\varepsilon_L(\lambda_0) = \int_{\text{line}} \varepsilon_\lambda \, d\lambda = \frac{hc}{4\pi\lambda} A^{n',v',K',J'}_{n'',v'',K'',J''} n_i(n', v', K', J') \tag{8.76}$$

where $n_i(n', v', K', J')$ is the density of the particles of species $i$ in the emitting state; their quantum numbers are $n'$ for the electronic term, $v'$ for the vibrational term, and $K'$ and $J'$ for the rotational term. Quantity $n_i(n', v', K', J')$ is given by the Boltzmann equation. Figure 8.12 represents the emission coefficient of the 0–0, 0–1, and 0–2 band heads of the $B^2\Sigma_u^+ \rightarrow X^2\Sigma_g^+$ transitions for $N_2^+$. Here also the maximum corresponds to the maximum density of $N_2^+$ (see Fig. 6.4).

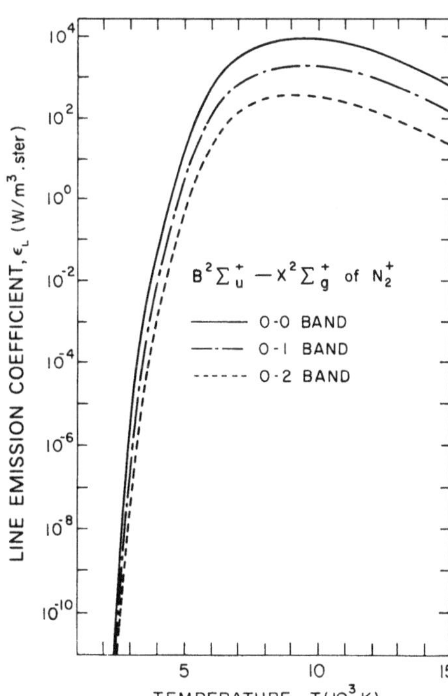


**FIG. 8.12.** Absolute volumetric emission coefficients (W/m$^3$ ster) of the band heads 0.0, 0.1, and 0.2 of the system $B^2\Sigma_u \rightarrow X^2\Sigma_g$ of $N_{2^+}$, as a function of temperature at atmospheric pressure.[8]

### 8.3.3. Continuum Radiation

#### 8.3.3.1. General Relationships

According to Griem (1964), the absorption coefficient for a particle of species $i$ and electrical charge $ze$ (where $z = 0$ for an atom, $z = 1$ for its first ion, etc.) is given by

$$\kappa_{i,z+1}(\nu, T) = \sum_{n,l} n_{i,z}^{n,l} \sigma_{i,z}^{n,l} \tag{8.77}$$

where $n_{i,z}^{n,l}$ is the density of the chemical species $i$, $ze$ is the electrical charge of the excited state defined by the principal quantum number $n$ and the azimuthal quantum number $l$, and $\sigma_{i,z}^{n,l}$ is the cross section for photoionization by a photon $h\nu$ as calculated by Bates.[9] The emission or absorption coefficient of the atom of electrical charge $ze$ is denoted by the index $z + 1$ because of its correspondence to the recombination of electrons with ions. For an optically thin plasma, we can substitute Eq. (8.77) into Eq. (8.33), and then, using Eq. (8.10), it can be shown that at equilibrium

$$\varepsilon_{i,z+1}^{n,l} = \frac{2h\nu^3}{c^2}\left[\exp\left(\frac{h\nu}{kT}\right) - 1\right]^{-1} n_{i,z}^{n,l}\sigma_{i,z}^{n,l} \tag{8.78}$$

Using the Saha equation (6.55), $n_{i,z}$ can be expressed in terms of the electron and parent ion densities as follows:

$$\frac{n_e n_{i,z+1}}{n_{i,z}} = 2\left(\frac{2\pi m_e kT}{h^2}\right)^{3/2} \frac{Q_{i,z+1}^{\text{el}}}{Q_{i,z}^{\text{el}}} \exp\left(-\frac{E_{i,z+1}^{I} - \delta E_{i,z+1}}{kT}\right) \tag{8.79}$$

where $E_{i,z+1}^{\text{I}}$ is the first ionization energy for the reaction $X_{i,z} \rightleftarrows X_{i,z+1} + e$, $\delta E_{i,z+1}$ is its lowering (see Section 4.4.2), and $Q_{i,z}^{\text{el}}$ is the electronic partition function of species $i$, whose electrical charge is $ze$. As in most cases the ion is in its ground state (denoted by the superscript 1); using the Boltzmann equation, Eq. (6.2), to express $n_{i,z+1}$ versus $n_{i,z+1}^{1}$, it follows that

$$\varepsilon_{i,z+1}^{n,l}(\nu) = \frac{h^4\nu^3}{c^2(2\pi m_e kT)^{3/2}} \frac{\sigma_{i,z}^{n,l} g_{i,z}^{n,l} n_e n_{i,z+1}^{1}}{g_{i,z}^{1}} \times \exp\left(\frac{E_{i,z+1}^{\text{I}} - \delta E_{i,z+1} - E_{i,z}^{n,l}}{kT}\right)\left[\exp\left(\frac{h\nu}{kT}\right) - 1\right]^{-1} \tag{8.80}$$

As is the case for line emission, one must be careful with the units of Eq.

(8.80), which gives $\varepsilon_{i,z+1}^{n,l}(\nu)$ in J/m$^3$ ster. If $\varepsilon$ is expressed in terms of the wavelength, $\varepsilon_{i,z+1}^{n,l}(\lambda)$ is in W/m$^4$ ster, since

$$\varepsilon(\lambda) = \frac{c}{\lambda^2}\varepsilon(\nu)$$

### 8.3.3.2. Free–Bound Transitions

In this case the emission for the different atoms of species $i$ and their parent ions of charge $z + 1$ is given by

$$\varepsilon_{\text{fb}} = \sum_i \sum_{z=0}^{z_{\max}} \sum_{n,l}^{n_{\max}} \varepsilon_{i,z+1}^{n,l} \tag{8.81}$$

In the following paragraphs the notation $\sum_{z=0}^{z_{\max}}$ will be abbreviated to $\sum_z$.

The summation is performed for all atoms and ions involved and for each of the levels $n, l$ compatible with frequencies in excess of $\nu_0$ [corresponding to electrons trapped with close to zero velocity (see Section 8.3.1)]. Therefore,

$$h\nu_0 = E_{i,z+1}^{\text{I}} - \delta E_{i,z+1} - E_{i,z}^{n,l} \tag{8.82}$$

If the resulting ion is in an excited state $n', l'$ and not in its fundamental ground state, the energy of this excited state $E_{i,z}^{n',l'}$ has to be added to the ionization energy $E_{i,z+1}$. Equation (8.82) shows clearly that the quantum states to be taken into account depend on the frequency or the wavelength at which the calculation is performed. Moreover, a limit $n_{\max}$ has to be imposed on the principal quantum number. According to Griem (1964), the levels with high values of $n$ can be considered to be hydrogenic, i.e.,

$$n_{i,z}^{\max} \simeq \left(\frac{(z + 1)^2 E_{\text{H}^+}}{\delta E_{i,z+1}}\right)^{1/2} \tag{8.83}$$

where $E_{\text{H}^+}$ is the ionization energy of hydrogen.

Equation (8.83) implies that the lowering of the ionization potential is identical to the limit of the discrete energy levels (i.e., the point at which they overlap into a continuum close to the ionization limit). Two different approaches will be used for the calculation of the photoionization cross sections:

*8.3.3.2.a. Hydrogenic Levels.* For hydrogenic levels (levels with high $n$ values), the classical photoionization cross section defined by Kramers[10] and corrected by Gaunt[11] is used:

$$(\sigma_{i,z}^{n,l})_{\text{class}} = \frac{\sigma h\alpha}{3^{3/2}}\pi r_1^2\left(\frac{E_{\text{H}^+}}{h\nu}\right)^3 \frac{(z + 1)^4}{n^5} G_{i,z}^n(\nu) \tag{8.84}$$

where $G_{i,z}^{n}$ is the Gaunt factor, which depends weakly on $\nu$ and which is generally close to unity; $\alpha = 2\pi e^2/hc$ is the constant of fine structure, and $r_1$ is the Bohr radius [see Eq. (2.5) for $n = 1$]. The relative cross section defined by Eq. (8.85) is frequently used instead of $\sigma_{i,z}^{n,l}$ [see Griem (1964)]:

$$\bar{G}_{i,z}^{n,l} = \frac{\sigma_{i,z}^{n,l} g_{i,z}^{n,l}}{(\sigma_{i,z}^{n,l})_{\text{class}}\, g_{i,z+1}^{1} 2n^2} \tag{8.85}$$

From Eq. (8.78), it can be shown that

$$\varepsilon_{i,z+1}^{n,l} = \left[\frac{16\alpha'^3 E_{\mathrm{H}^+}}{3^{3/2}\pi}\right] \frac{(z+1)^4 g_{i,z+1}^{1} n_{i,z}}{Q_{i,z}} \frac{\bar{G}_{i,z}^{n,l}}{n^3} \times \left[\exp\left(\frac{h\nu}{kT}\right) - 1\right]^{-1} \exp\left(-\frac{E_{i,z}^{n,l}}{kT}\right) \tag{8.86}$$

The constant in brackets is the right-hand side of Eq. (8.86) and is equal to $3.0077 \times 10^{-26}$ J/ster; $\varepsilon_{\text{fb}}(\nu)$ is calculated in J/m$^3$ ster.

For high quantum numbers $n'$, the levels are very close to each other and the summation in Eq. (8.81) can be replaced by an integral. At frequencies $\nu < (z+1)^2 E_{\mathrm{H}^+}/(hn'^2)$, $\bar{G}_{i,z+1}^{n,l} = 1$. According to Cabannes and Chapelle (1971), $\varepsilon_{\text{fb}}$ can be rewritten in the form

$$\varepsilon_{\text{fb}}(\nu) = \left[\frac{16\pi e^6}{3c^3(6\pi m^3 k)^{1/2}(4\pi\varepsilon_0)^3}\right] \frac{n_e}{T^{1/2}} \left[1 - \exp\left(-\frac{h\nu}{kT}\right)\right] \sum_i \sum_{z=1}^{z_{\max}} z^2 n_{i,z} \tag{8.87}$$

The constant in brackets is equal to $5.44692 \times 10^{-52}$ J m$^3$ K$^{1/2}$/ster. For a single ion $\varepsilon_{\text{fb}}$ is proportional to $n_e^2$. Other calculations are also possible using the Gaunt factors given by Menzel and Pekeris[12] and Peach.[13]

*8.3.3.2.b. Nonhydrogenic Atoms and Ions.* For nonhydrogenic atoms and ions the photoionization cross section in Eq. (8.80) can be evaluated for each level $n$ using the quantum defect method of Burgess and Seaton[14]:

$$\sigma_{i,z}^{n,l} = \frac{4\alpha'}{3} \pi r_1^2 \frac{h\nu}{E_{\mathrm{H}^+}} \frac{n_l^*}{(z+1)^4} \sum_{l'=l\pm 1} C_{l'} \, |g(n_l^*, l')|^2 \tag{8.88}$$

where $g$ is a complex expression given in Ref. 14 and where $n_l^*$ is the effective quantum number, defined as

$$n_l^* = \left(\frac{(z+1)^2 E_{\mathrm{H}^+}}{E_{i,z+1} - E_{i,z}^{n,l}}\right) \tag{8.89}$$

with

$$C_{l'} = \frac{l+1}{2l+1} \qquad \text{if } l' = l+1$$
$$C_{l'} = \frac{l}{2l+1} \qquad \text{if } l' = l-1 \tag{8.90}$$

An expression similar to Eq. (8.87) has also been proposed by Cabannes and Chapelle (1971):

$$\varepsilon_{\text{fb}} = C_1 \frac{n_e}{T^{1/2}}\left[1 - \exp\left(-\frac{h\nu}{kT}\right)\right] \sum_i \sum_z (z+1)^2 n_{i,z+1} \times \frac{g^1_{i,z+1}}{Q^{\text{el}}_{i,z-1}} \xi_{i,z}(\nu, T) \tag{8.91}$$

where $\xi$ is the Biberman factor, which takes into account the electronic structure of the atoms, and depends strongly on the frequency $\nu$ and weakly on the temperature. For example, Fig. 8.13 shows the Biberman factor for an argon plasma at 8000 K at a pressure of 105 kPa. The first peak is observed at $\lambda = 87.6$ nm and corresponds to the photoionization of the fundamental state of argon.

### 8.3.3.3. Free–Free Transitions

Free–free transitions can be calculated for thermal plasmas using a hydrogenic approximation.[10] Equation (8.80) is used with

$$\sigma^{n,l}_{i,z} = (\sigma^{n,l}_{i,z})_{\text{class}} G^{i,z}_{\text{ff}}(\nu, T) \tag{8.92}$$

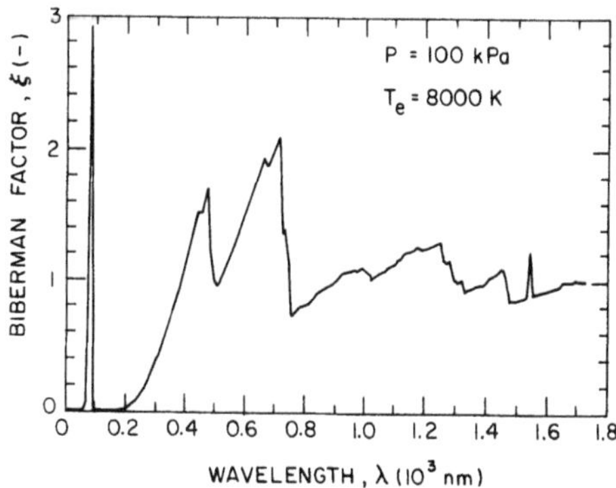


**FIG. 8.13.** Biberman factor $\xi(T, \lambda)$ for an argon plasma at a temperature of 8000 K and a pressure of 105 kPa.[24]

where $G_{ff}^{i,z}$ is the free–free Gaunt factor and, assuming hydrogenic behavior,

$$\sigma_{i,z}^{n,l} = 2n^2\sigma_{i,z+1}^{1} \tag{8.93}$$

and

$$E_{i,z+1}^{\mathrm{I}} - E_{i,z}^{n,l} = \frac{(z+1)E_{\mathrm{H}^+}}{n^2} \tag{8.94}$$

It follows that

$$\varepsilon_{ff}^{e,i} = \sum_{i,z}\sum_{n} \frac{2h}{c^2(2\pi m_e kT)^{3/2}} \frac{64\alpha'}{3^{3/2}} \pi r_1^2 E_{\mathrm{H}^+}^3 \frac{(z+1)^2}{n^3} G_{ff}^{i,z} \times n_e n_{i,z+1} \exp\left(\frac{\dfrac{(z+1)^2 E_{\mathrm{H}}}{n^2} - \delta E_{i,z+1}}{kT}\right)\left[\exp\left(\frac{h\nu}{kT}\right) - 1\right]^{-1} \tag{8.95}$$

where the superscript $e, i$ indicates that the Bremsstrahlung is due to the field of the ions.

The summation over $n$ is replaced by integration over the quasi-continuum, i.e., from $n_1$ as defined by Eq. (8.83) to $n_2 = 0$. For high frequencies, $G_{ff}^{i,z} \sim 1$ and

$$\left[\exp\left(\frac{h\nu}{kT}\right) - 1\right]^{-1} \cong \exp\left(-\frac{h\nu}{kT}\right)$$

Equation (8.95) then simplifies to

$$\varepsilon_{ff}^{e,i} = C_1 \frac{n_e}{T^{1/2}} \exp\left(-\frac{h\nu}{kT}\right) \sum_{i,z} (z+1)^2 n_{i,z+1} G_{ff}^{i,z} \tag{8.96}$$

The Gaunt factors $G_{ff}^{i,z}$ have been tabulated by Karsas and Letter.[15] Equation (8.96) shows clearly that $\varepsilon_{ff}$ is proportional to $n_e^2$ (assuming a singly ionized plasma) and the $\varepsilon_{ff}$ will have significant values at atmospheric pressure only for $n_e > 10^{21}\,\mathrm{m}^{-3}$ (about $T > 9000$ K for Ar, $H_2$, $N_2$, and $O_2$).

For low frequencies Eq. (8.95) can be written as [see Cabannes and Chapelle (1971)]

$$\varepsilon_{ff}^{e,i} = C_1 \frac{n_e}{T^{1/2}} \sum_{1,z} n_{i,z+1}(z+1)^2 G_{i,z}(\nu, T) \tag{8.97}$$

with

$$G_{i,z}(\nu, T) = \frac{\sqrt{3}}{\pi}\left[\ln\left(2.1 \times 10^8 \frac{T_e^{3/2}}{2\nu}\right) - \frac{5}{2}\gamma\right] \tag{8.98}$$

where $\gamma = 0.577$ (Euler's constant).

When the electron density is low, the free–free radiation due to collisions with neutral atoms of density $n_a$ must also be taken into account [see Cabannes and Chapelle (1971)]. Assuming only elastic collisions, it follows that

$$\varepsilon_{\text{ff}}^{ea} = \left[\frac{32e^2}{3c^3}\left(\frac{k}{2\pi m}\right)^{3/2}\frac{1}{4\pi\varepsilon_0}\right] n_e T^{3/2} \sum_a n_a G_a(\nu, T) \tag{8.99}$$

where the constant $C_2$ [in brackets on the right-hand side of Eq. (8.99)] is $3.4213 \times 10^{-43}$ J m K$^{-3/2}$ ster. The number densities $n_a$ and $n_e$ are expressed in m$^{-3}$. The Gaunt factor for the neutrals, $G_a(\nu, T)$, is given by

$$G_a(\nu, T) = \int_{x_0}^{\infty} \sigma_{ea}(x) x^2 \exp(-x)\, dx \tag{8.100}$$

with $x = mv^2/2kT_e$ and $x_0 = h\nu/kT_e$; $\sigma_{ea}(x)$ is the electron–neutral impact elastic cross section (m$^2$), which is a function of the electron velocity $v_e$ (neglecting the neutral atoms' velocity compared with that of the electrons). Equation (8.100) can be simplified by choosing a constant mean (average) value for $\sigma_{ea}$ [see Cabannes and Chapelle (1971)]:

$$G_a(\nu, T_e) = \bar{\sigma}_{ea}\left[1 + \left(1 + \frac{h\nu}{kT}\right)^2\right] \exp\left(-\frac{h\nu}{kT}\right) \tag{8.101}$$

At high frequencies ($h\nu/kT_e \gg 1$),

$$\varepsilon_{\text{ff}}^{ea}(\nu, T_e) = C_2 n_e T_e^{3/2} \left(\frac{h\nu}{kT}\right)^2 \exp\left(-\frac{h\nu}{kT}\right) \sum_a n_a \bar{\sigma}_{ea} \tag{8.102}$$

At low frequencies ($h\nu/kT_e \ll 1$),

$$\varepsilon_{\text{ff}}^{ea}(\nu, T_e) = 2C_2 n_e T^{3/2} \sum_a n_a \bar{\sigma}_{ea} \tag{8.103}$$

In general $\varepsilon_{\text{ff}}^{ea}$ is more important than $\varepsilon_{\text{ff}}^{ei}$ at low temperatures. For example, in the case of an argon plasma at atmospheric pressure, $\varepsilon_{\text{ff}}^{ea} = \varepsilon_{\text{ff}}^{ei}$ at $\lambda = 300$ nm for $n_e/n_a \sim 3 \times 10^{-3}$, which corresponds to a temperature $T_e \sim 8500$ K.

### 8.3.3.4. Total Continum Radiation

The total continuum radiation includes the free–free [Eq. (8.96) or (8.99)] and the free–bound [Eq. (8.86) or (8.91)] radiation. For high frequencies ($h\nu/kT_e \gg 1$) the continuum is reduced to the free–bound radiation, while for low frequencies ($h\nu/kT \ll 1$) the continuum is due mainly to free–free radiation.

In two-temperature plasmas, the highly excited levels are due mainly to the electrons. Expressions for the calculation of the continuum radiation have to be written in terms of $T_e$ rather than $T$.

### 8.3.3.5. Other Contributions

*8.3.3.5.a. Negative Ions.* Atoms or neutral molecules such as H, N, O, C, Cl, S, $O_2$, and $C_2$[16] can capture a free electron to form a negative ion with the emission of a continuum similar to that of recombination processes ($\varepsilon_{\text{fb}}$):

$$X + e \rightleftarrows X^- + h\nu \tag{8.104}$$

The corresponding expression for $\varepsilon_{\text{fb}}^{X^-}$ is (see Ref 17):

$$\varepsilon_{\text{fb}}^{X^-} = \frac{2h\nu^3}{c^2}\exp\left(-\frac{h\nu}{kT}\right)n_{X^-}\sigma_{X^-} \tag{8.105}$$

where $\sigma_{X^-}$ is the photoattachment cross section. It is zero for frequencies below $\nu_0$, where $h\nu_0 = E_a$ ($E_a$ is the attachment energy, see Table 2.4); $n_{X^-}$ is calculated with the help of the Saha equation for the negative ions. Equation (8.105) becomes

$$\varepsilon_{\text{fb}}^{X^-} = \frac{h^4}{c^2(2\pi mk)^{3/2}}\frac{n_e n_X}{T^{3/2}}\frac{g_{X^-}}{Q_X^{el}}\exp\left(\frac{E_a}{kT}\right)\nu^3\exp\left(-\frac{h\nu}{kT}\right) \tag{8.106}$$

where $g_{X^-}$ is the statistical weight of the ground state of the negative ion (usually negative ions have only one stable state). For example, in the case of $H^-$, $g_{X^-} = 1$. This contribution of the negative ion is proportional to $n_{X^-}$, the density of the neutral atom, and to $n_e$. It can become more important for free–free radiation or even free–bound radiation, which are proportional to the square of the electron density in weakly ionized plasmas.

*8.3.3.5.b. Pseudocontinuum.* In plasmas, lines that are broadened by the Stark effect can overlap each other and produce a continuum.

This effect is of particular importance in the case of hydrogen, where the lines are strongly broadened due to the linear Stark effect [see Cabannes and Chapelle (1971)].

### 8.3.3.6. Examples of Continuum Radiation

For an example of continuous radiation, we will consider the radiation of nitrogen plasmas, basing our discussion on the results of Bayard.[17] Figure 8.14 shows the total continuum $\varepsilon_{\text{cont}}(0)$ (full lines) expressed in (J/m$^3$ ster) and the continuum without the contribution of the electron recombination (dotted lines) for various temperatures and at atmospheric pressure. These curves underline the important role of the negative ions, especially in the temperature range between 7000 and 12,000 K, as already observed by Krey and Morris[18] and Morris and

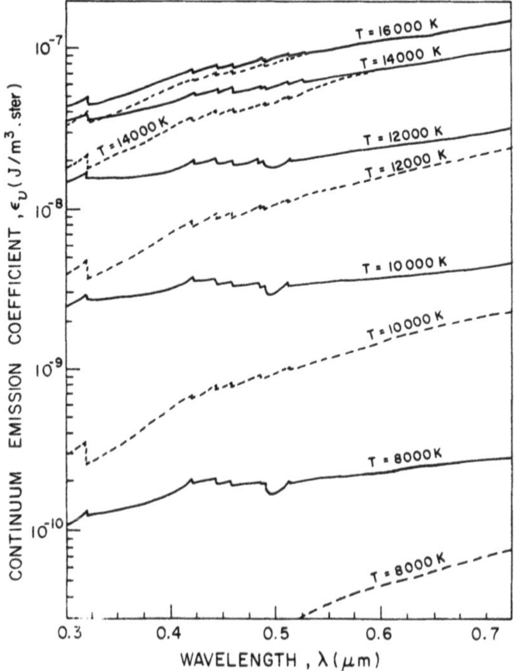


**FIG. 8.14.** Evolution of the total continuum volumetric emission coefficient (J/m$^3$ ster) in the wavelength range $0.3 \leq \lambda \leq 0.7$ nm at various temperatures and at atmospheric pressure of nitrogen (solid line) and the same coefficient minus the free bound coefficient $\varepsilon_{\text{fb}}$ (dashed line).[17]

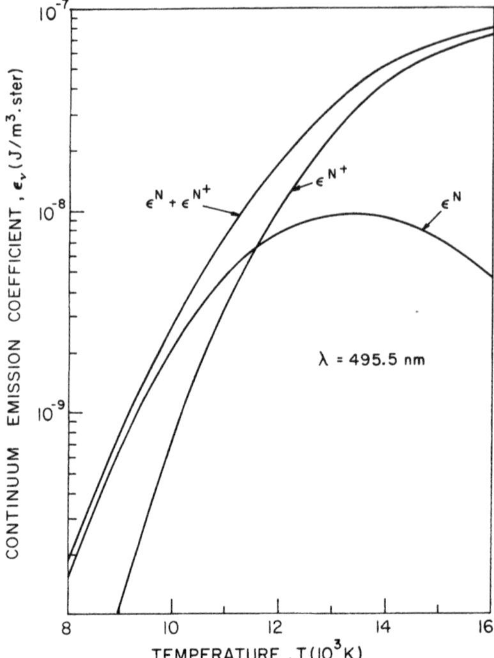


**FIG. 8.15.** Relative contributions to the continuum volumetric emission coefficient ($J/m^3$ ster) for the wavelength 495.5 nm at atmospheric pressure from 8000 K to 15,000 K.[17]

Yos.[19] The sharp variations of the continuum with the wavelength correspond to the recombination to the excited states of $N(\varepsilon_{\mathrm{fb}}^{\mathrm{N}})$ and $N^{+}(\varepsilon_{\mathrm{fb}}^{\mathrm{N}^{+}})$.

Figure 8.15 shows the relative contributions of the atoms $\varepsilon_{\nu}^{\mathrm{N}^{+}} = \varepsilon_{\mathrm{fb}}^{\mathrm{N}} + \varepsilon_{\mathrm{ff}}^{\mathrm{N}}$ and the ions $\varepsilon_{v}^{\mathrm{N}^{+}} = \varepsilon_{\mathrm{fb}}^{\mathrm{N}} + \varepsilon_{\mathrm{ff}}^{\mathrm{N}^{+}}$ to the continuum intensity as a function of temperature, at a wavelength of 495.5 nm (one of the wavelengths of the nitrogen spectra where the continuum is not overlapped by atomic or molecular spectra). In this wavelength range, the contribution of the free–free radiation is generally smaller (between 3 and 5 times) than the contribution of the free–bound radiation. The curves follow the evolution of the densities of N and $N^{+}$. At lower temperatures (see Fig. 8.16) the contribution of the free–free radiation due to molecules becomes important.

Note the sharp variation of the continuum intensity with temperature: from $10^{-20}$ $J/m^3$ ster at 4000 K to $10^{-7}$ $J/m^3$ ster at 16,000 K (a change of more than $10^{13}$ times for a temperature rise of 12,000 K at $\lambda = 495.5$ nm). Because of the inherent limits of the dynamic range of the measuring instruments, measurements will be possible only in a small temperature range, for example, between 8000 K and 16,000 K; continuum measurements are particularly difficult under 8000 K due to the

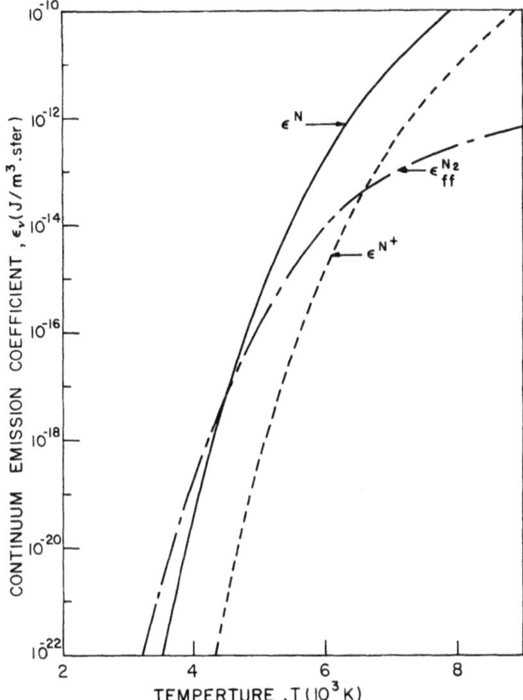


**FIG. 8.16.** Relative contributions to the continuum volumetric emission coefficient ($J/m^3$ ster) for the wavelength 495.5 nm at atmospheric pressure from 2000 K to 8000 K.[17]

very low intensity of the continuum, in comparison with the radiation emitted from the hotter regions of the plasma. The important influence of the electron density on the continuum radiation can be seen in Fig. 8.17, which gives the nitrogen continuum $\varepsilon_{\text{cont}}(\lambda)$ as a function of $n_e^2/\sqrt{T_e}$ at

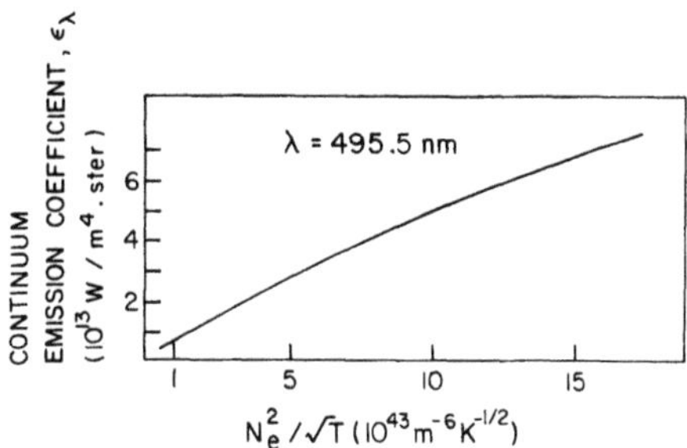


**FIG. 8.17.** Continuum volumetric emission coefficient of nitrogen ($W/m^4$ ster) as a function of $n_e^2/\sqrt{T}$ at atmospheric pressure for a wavelength of 495.5 nm.[6]

495.5 nm. In order to overcome the problems with units, this curve has been represented in units of W/m$^4$ ster and not in J/m$^3$ ster. The conversion factor $(c/\lambda^2)$ is $1.22105 \times 10^{21}$ for a wavelength of 493.5 nm:

$$\varepsilon_\lambda \quad (\text{W/m}^4\,\text{ster}) = 1.22105 \times 10^{21}\varepsilon_\nu \quad (\text{W/m}^3\,\text{ster})$$

For comparison, the continuum radiation at $\lambda = 493.5$ nm, with $\delta_\lambda = 0.3$ nm at 13,500 K, is only $1.5 \times 10^4$ W/m$^3$ ster, which is smaller than the corresponding line radiation at the same wavelength and temperature with a line width of about 0.3 nm ($3.2 \times 10^4$ W/m$^3$ ster). In most cases a ratio between one and three orders of magnitude exists between the line radiation and the continuum radiation.

### 8.3.4. Total Effective Radiation of Plasmas

If light diffusion is neglected and LTE is assumed, the radiative transfer equation (8.28) can be written[20]

$$\vec{n} \cdot \vec{\nabla} I_\nu(\vec{r}, \vec{n}) = \kappa'_\nu(B_\nu - I_\nu) \tag{8.107}$$

where $I_\nu(\nu, \theta, \varphi)$ is the specific intensity of the radiation field [see Eq. (8.1)], $\kappa'_\nu$ is the absorption coefficient [see Eq. (8.24)], $B_\nu$ is the blackbody radiation [see Eq. (8.10)], and $\vec{n}$ is the unit vector defining the radiation direction. If the radiation is assumed isotropic, then when both sides of Eq. (8.107) are multiplied by the solid angle element $d\Omega$ and integrated over $\Omega$, it follows[21] that

$$\vec{\nabla} \cdot \vec{F}_{R\nu} = 4\pi(\varepsilon_\nu - \kappa'_\nu J_\nu) \tag{8.108}$$

where $J_\nu$ and $\vec{F}_{R\nu}$ are, respectively, the mean radiation intensity and the radiation flux defined by

$$J_\nu = \frac{1}{4\pi}\int_{4\pi} I_\nu \, d\Omega \tag{8.109}$$

and

$$\vec{F}_{R\nu} = \int_{4\pi} I_\nu \cdot \vec{n} \, d\Omega \tag{8.110}$$

Equation (8.108) states that the divergence of the radiation flux per unit frequency is equal to the net radiation emitted, which is the difference between the total volumetric emission $4\pi\varepsilon_\nu$ and the total volumetric absorption $4\pi\kappa'_\nu J_\nu$. This equation, if rigorous, cannot be solved as it is.

Approximate solutions can be found, however, by making the assumptions discussed in the following paragraphs.

### 8.3.4.1. Optically Thin Plasma

Equation (8.108) reduces to

$$\vec{\nabla} \cdot \vec{F} = 4\pi \int_0^\infty \kappa_\nu' B_\nu(T)\, d\nu \tag{8.111}$$

if the energy radiated or reflected by the walls is negligible. Unfortunately, even for the classical plasma gases this approximation does not hold for all lines, especially resonance lines.

### 8.3.4.2. Gray Body Approximation

A gray body is a medium in which the absorption coefficient is independent of the wavelength. In this case simplified equations[20] can be derived for the radiation flux. However, for plasmas it is not possible to consider the absorption coefficient constant over all frequencies, and thus the spectrum must be divided into two or more frequency bands of width $\Delta\nu$, where $\kappa_{\Delta\nu}'$ is kept constant.[22] However, the variations of $\kappa_{\Delta\nu}'$ with temperature and pressure must be accounted for. Siegel and Howell[23] have published a rather extensive study of the literature related to this method.

### 8.3.4.3. Diffusion Approximation

A detailed calculation of this approximation is given by Siegel and Howell.[23] The radiative flux is written

$$\vec{F}_{R\nu} = -\frac{4\pi}{3\kappa_\nu'} \vec{\nabla} J_\nu \tag{8.112}$$

and integration over all the frequencies of Eqs. (8.108) and (8.112) results in

$$\vec{\nabla} \cdot \vec{F}_{R\nu} = 4\pi(\varepsilon_\nu - \kappa_i' J_R) \tag{8.113}$$

with

$$\vec{F}_{R\nu} = -\frac{4\pi}{3\kappa_2'} \vec{\nabla} J \tag{8.114}$$

The two mean absorption coefficients $\kappa_1$ and $\kappa_2$ are defined by

$$\kappa_1' = \frac{1}{J}\int_0^\infty J_\nu \kappa_\nu' \, d\nu \tag{8.115}$$

and

$$\kappa_2' = \frac{1}{F_R}\int_0^\infty J_\nu \kappa_\nu' \, d\nu \tag{8.116}$$

with

$$J_R = \int_0^\infty J_\nu \, d\nu \tag{8.117}$$

These coefficients allow the determination of the radiative transfer[21] with certain boundary conditions.

The diffusion approximation is the most rigorous one. However, the lines make the solution highly complex.[24] This complexity of the solution of Eqs. (8.112) to (8.117) for the lines is related to the following considerations:[5]

- Lines have very often an extremely narrow width, implying that we need to consider very small frequency intervals compared to the whole spectrum.
- Lines may be broadened according to local temperatures and the resulting species number densities and their absorption coefficients then vary along the line width[26] (see Fig. 8.18, for example).

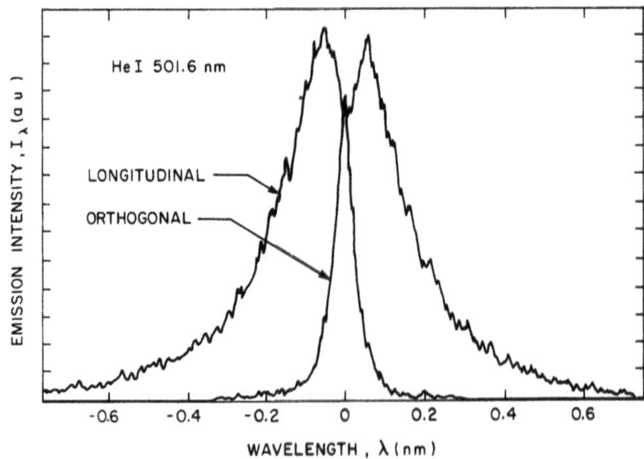


**FIG. 8.18.** HeI 501.6 nm line profile measured orthogonally (⊥) and longitudinally (∥) to a He discharge ($p = 92$ kPa) with thickness of 1 mm and length of 230 mm.[26]

- Lines may be shifted according to the local temperature, and this shift may also induce nonnegligible variation of the absorption coefficient for a given frequency $\nu$.

### 8.3.4.4. Effective Emission Coefficient of Lowke

If the diffusion approximation is used for lines, it is mandatory to make assumptions about the dependence of the plasma temperature on space variables such as the plasma jet axis $z$ and radius $r$. In most cases in DC plasma jets or RF plasma jets (downstream of the induction coil), the temperature distribution along the jet radius at a distance $z$ from the torch exit can be represented by a function of the type

$$\frac{T}{T_{\max}} = \left(1 - \frac{r}{R}\right)^n \tag{8.118}$$

where $T_{\max}$ is the temperature at the axis of the jet ($r = 0$), $R$ is the jet radius at position $z$, and $n$ is an exponent. Using such a distribution to calculate the effective emission according to Eqs. (8.112) to (8.117) is very complex. That is why in most cases the distribution given by Eq. (8.118) is replaced by a rectangular profile: $T = T_{\max}$ for $r < R$ and $T = 0$ for $r \geq R$ (this approximation is often sufficient[5]). The problem is now that of radiative heat transfer in an isothermal cylinder of temperature $T_{\max}$ and radius $R$. Lowke[25] has shown that under such conditions it is possible to use an effective emission coefficient $\varepsilon_E$ corresponding to the effective radiation emitted from the axis of the isothermal cylinder (i.e., the fraction of the total power that is radiated per unit volume and unit solid angle into a volume element surrounding the axis of the cylinder and escaping from the cylinder after crossing a thickness $R$ of the isothermal plasma). Here $\varepsilon_E$ is given by

$$\varepsilon_E = \int_0^\infty B_\nu \kappa'_\nu G_1(\kappa'_\nu, R)\, d\nu \tag{8.119}$$

where $G_1$ is a function accounting for the cylindrical geometry of the plasma. More recently, Liebermann and Lowke[27] have shown that the isothermal cylinder could be approximated by an isothermal sphere with a precision better than 90%. In this case Eq. (8.119) can be simplified to

$$\varepsilon_E = \int_0^\infty B_\nu \kappa'_\nu \exp(-\kappa'_\nu B_\nu)\, d\nu \tag{8.120}$$

*8.3.4.4.a. Effective Line Radiation.* The main difficulty with Eq. (8.120) is determining the variation of the absorption coefficient $\kappa'_\nu$ with frequency $\nu$, taking into account the mechanisms related to the continuum emission on one hand and to the line profiles on the other. To further simplify the calculation, another assumption must be made: the lines are isolated, i.e., overlapping of lines has a negligible effect on the radiative transfer of the lines. In this way, each line can be dealt with separately. This assumption introduces the notion of the escape factor $\Lambda$ (see Refs. 5 and 28). The escape factor $\Lambda$ is the ratio of the radiation escaping from the plasma to that escaping from a transparent plasma. The relevance of the escape factor is that it depends only on a limited number of parameters.

In the general case, Eq. (8.119) can be written as

$$\varepsilon_E(T) = \int_0^\infty B_\nu(T)[\kappa_c' + \kappa_0' P(\nu)]\left[1 - \exp\left(-\frac{h\nu}{kT}\right)\right]$$
$$\times \exp\left\{-[\kappa_c' + \kappa_0' P(\nu)]\left[1 - \exp\left(-\frac{h\nu}{kT}\right)R\right]\right\} d\nu \quad (8.121)$$

where $\kappa_c'$ is the absorption coefficient of the continuum close to the line, $\kappa_0'$ is the absorption coefficient at the line center (frequency $\nu_0 = \nu_{ul}$), and $P(\nu)$ characterizes the line profile [in most cases given by Eq. (8.72)]. The factor $[1 - \exp(-h\nu/kT)]$ has been introduced to account for induced emission.

The spectral band $\Delta\nu$ associated with one line of central frequency $\nu_0$ is very narrow compared to the whole spectrum, so the Planck function $B_\nu$ and $\exp(-h\nu/kT)$ can be considered constant within this band. Moreover, because the $\varepsilon_E$ related to the continuum and due only to the continuum is calculated independently, it does not need to be accounted for in Eq. (8.121) as an emission term. However, it plays a role in the absorption of the line (in the exponential term). Thus, we can finally write Eq. (8.121) follows:

$$\varepsilon_{E\nu_0} = B_{\nu_0}(T)\left[1 - \exp\left(-\frac{h\nu_0}{kT}\right)\right]\kappa_0' \int_0^\infty P(\nu)$$
$$\times \exp\left\{-[\kappa_c' + \kappa_0' P(\nu)]\left[1 - \exp\left(-\frac{h\nu}{kT}\right)\right]R\right\} d\nu \quad (8.122)$$

The subscript $\nu_0$ indicates that $\varepsilon_{E\nu_0}$ is related to the line centered at the frequency $\nu_0$.

To introduce the escape factor, a new assumption must be made:[5] in the induced emission expression $[1 - \exp(-h\nu/kT)]$ the exponential term approaches zero. There are two justifications for this assumption:

- The more intense lines, which are also the more absorbed ones, are generally situated at wavelengths below 200 nm. For the temperature conditions generally encountered in thermal plasmas, $\exp(-h\nu/kT)$ is very small and the assumption is justified.
- For $\lambda > 200$ nm, the induced emission correction is more important, but the lines have a lower intensity (they are no longer resonance lines) and are more broadened (due to the Stark effect). In this case the total absorption term $\exp(-\kappa' R)$ is close to 1, because $\kappa' R$ is small and the influence of the induced emission on the exponent is negligible.

Finally, having eliminated the induced emission coefficient, if we assume that the continuum absorption coefficient $\kappa'_c$ is constant within the line spectral band width, the effective emission coefficient of the line becomes

$$\varepsilon_{E\nu_0} = B_{\nu_0}(T)\left[1 - \exp\left(-\frac{h\nu_0}{kT}\right)\right]\kappa'_0 \exp(-\kappa'_c R)\int_0^\infty P(\nu)\exp[-\kappa'_0 P(\nu)R]\,d\nu \tag{8.123}$$

In Eq. (8.123) the integral term

$$\Lambda_{ul} = \int_0^\infty P(\nu)\exp[-\kappa'_0 P(\nu)R]\,d\nu \tag{8.124}$$

is the escape factor for the transition $\nu_{ul}$. The escape factor can be calculated[29,30] provided the line profiles and the temperature distributions $T(r, z)$ are known. For example, in the case considered here (a cylindrical plasma with uniform temperature) the escape profiles according to Essoltani[24] are as follows:

- For a Gaussian profile

$$\Lambda_{ul}^G = 2\ln(\kappa'_0 D)\frac{1 + \dfrac{\kappa'_0 D}{2 + (\kappa'_0 D)^2}}{1 + \kappa'_0 D[\pi \ln(1 + \kappa'_0 D)]^{1/2}} \tag{8.125}$$

where $\kappa'_0$ is the absorption at the line center and $D$ the plasma cylinder radius.

• For a Lorentzian profile

$$\Lambda_{ul}^{L} = 1.95 \times \frac{1 + \dfrac{\kappa_0' D}{2 + (\kappa_0' D)^2}}{1 + (\pi \kappa_0' D)^{1/2}} \tag{8.126}$$

• For a Voigt profile the calculation of $\Lambda_{ul}^{V}$ is not possible. However, Drawin[29] has proposed a semiempirical method based on the equation

$$\Lambda_{ul}^{V} = \Lambda_{ul}^{G} + f(\alpha, \tau_0)\Lambda_{ul}^{L} \tag{8.127}$$

where

$$\alpha = \frac{\delta_L}{\delta_G} = \frac{\text{Lorentzian profile width}}{\text{Gaussian profile width}}$$

$\tau_0$ is the optical depth [see Eq. (8.27)] and $f(\alpha, \tau_0)$ is a function given by Drawin.[29]

Equations (8.125) to (8.127) can be used to tabulate the different escape factors versus $\kappa_0'$, and $R$ and Eq. (8.123) can be calculated for the different lines.

Note that $\Lambda$ can also be calculated as a function of the optical depth $\tau_0$; for example, for a Lorentzian profile Eq. (8.124) can be written as

$$\Lambda_{ul} = \int_{\nu} P(\nu) \exp\left(-\tau_0 \frac{\delta_\nu^2}{\delta_\nu^2 + (\Delta\nu)^2}\right) d\nu \tag{8.128}$$

where $\delta_\nu$ is the half width of the line at half the maximum intensity. In the case of a plasma cylinder at uniform temperature, $\tau_0$ is given by

$$\tau_0 = \frac{1}{\delta_\nu} \frac{h\nu_0}{\pi} B_{lh} n_l R \tag{8.129}$$

where $B_{lh}$ is the Einstein absorption coefficient [see Eq. (8.37)].

Figure 8.18 illustrates the importance of the plasma thickness on the line shape. It shows the HeI 501.6 nm line obtained in a linear discharge at 700 Torr; its thickness is 1 mm and length 230 mm. The measurement performed orthogonally to the discharge results in a rather narrow line (Fig. 8.18), while a measurement performed longitudinally (Fig. 8.18) is broadened with a dip in its central part. The theoretical profile (in Fig. 8.18) calculated using the equation

$$I(\lambda) = \frac{\varepsilon_\lambda}{\kappa_\lambda'} [1 - \exp(-\kappa_\lambda' l)] \tag{8.130}$$

is in fairly good agreement with the experimental profile.

Even with all the simplifying assumptions, the calculation of the lines is long and tedious (for example, in an Ar–Fe plasma more than 4300 lines have to be considered[24]). Thus in most cases it is assumed that:

- All the lines with low spontaneous emission coefficients $A_{ul}$ ($A_{ul} < 10^4\,\text{s}^{-1}$) are optically thin.
- All the lines with absorption coefficients $\kappa_0'$ below $0.01\,\text{cm}^{-1}$ are optically thin. However, because $\kappa_0'$ varies with temperature (or, more precisely, with the density of the lower level population), $\kappa_0'$ has to be calculated for a rather wide temperature range (6000–20,000 K, for example).

*8.3.4.4.b. Effective Continuum Radiation.* To calculate the effective emission of the continuum, Eq. (8.120) is used and $\kappa_\nu'$ is calculated according to Eq. (8.47). Figure 8.19 (from Ref. 5) shows the evolution of $\kappa_\nu'$ as a function of frequency for a nitrogen plasma at atmospheric pressure and 10,000 K. Roughly three zones can be distinguished:

- For high frequencies (UV or far-UV, zone 1: $\nu > \nu_0$ with $\nu_0 = 3.52 \times 10^{15}$ Hz) $\kappa_\nu'$ has a rather high value, varying from about 1 to $10\,\text{cm}^{-1}$. In this high-frequency zone the continuum is mainly due to free–bound transitions; in fact the frequency $\nu_0$ corresponds to the ionization energy of the fundamental state of nitrogen atom: $E^{\text{I}}_{\text{N}^+} = h\nu_{\text{ON}}$ with $E^{\text{I}}_{\text{N}^+} = 14.55\,\text{eV}$. Any photon whose frequency is higher than $\nu_0$ can induce the photoionization of the ground state according to the reaction

$$X_g + h\nu \rightarrow \text{X}^+ + e^-$$

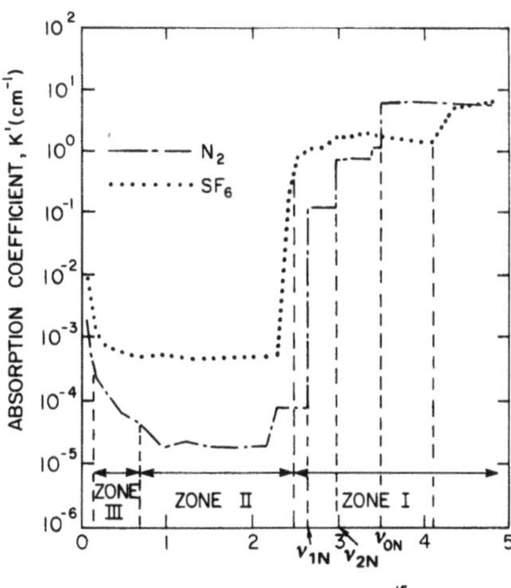


**FIG. 8.19.** The nitrogen continuum absorption coefficient $\kappa_\nu'$ as a function of frequency at $T = 10{,}000$ K and $p = 100$ kPa.[5]

Because the ground state is highly populated and the photoionization cross section of the reaction is high, $\kappa'_\nu$ is also high. In fact, for the nitrogen atom the ground state can be considered to correspond to three levels with, respectively, values of 0 eV, 0.238 eV, and 3.76 eV. The ionization energies of the last two levels are, respectively, 12.17 eV and 10.79 eV, corresponding to the threshold frequencies $\nu_{1N} = 2.94 \times 10^{15}$ Hz and $\nu_{2N} = 2.65 \times 10^{15}$ Hz.

- In the second zone ($10^{14} < \nu < 2.65 \times 10^{15}$ Hz, corresponding to the near-IR, visible, and near-UV spectra) $\kappa'_\nu$ varies very little. Its value is between $10^{-4}$ and $10^{-5}$ cm$^{-1}$.
- In the far-IR (zone III, $\nu < 10^{14}$ Hz) the continuum is mainly due to free–free radiation, and $\kappa'_\nu$ increases when the frequency decreases down to the plasma frequency.

When temperature increases, the behavior of $\kappa'_\nu$ varies in each zone depending on the phenomena controlling $\kappa'_\nu$. When temperature increases, $\kappa'_\nu$ decreases in zone 1 (fewer atoms in the ground states), it is almost constant in zone 2, and increases in zone 3 (with electron density). Variations with pressure are closely linked to the plasma composition, especially the concentration of electrons and ionized species.

## 8.4. EXAMPLES OF RESULTS

### 8.4.1. Classical Plasma Gases

#### 8.4.1.1. Argon Plasma

The argon plasma is probably the most intensively studied plasma. Figure 8.20 shows the evolution with temperature of the total volumetric emission coefficient ($\varepsilon_T = \int_0^\infty \varepsilon_\lambda \, d_\lambda = \int_0^\infty \varepsilon_\nu \, d_\nu$ expressed in W/m$^3$). The data, compiled by Boulos,[31] are for a temperature range of 8000–24,000 K and an optically thin plasma; the resonance lines are neglected. The data from different sources[18,32–35] are relatively dispersed, especially at high temperatures, where there is up to an order of magnitude difference between the extrema. A more recent compilation (of data from Refs. 18, 32, 33, and 35–38) by Wilbers *et al.*[39] shows less widely dispersed results (see Fig. 8.21). It is worth emphasizing that the results of Fig. 8.21 are expressed in W/m$^3$ ster, while those of Fig. 8.20 are in

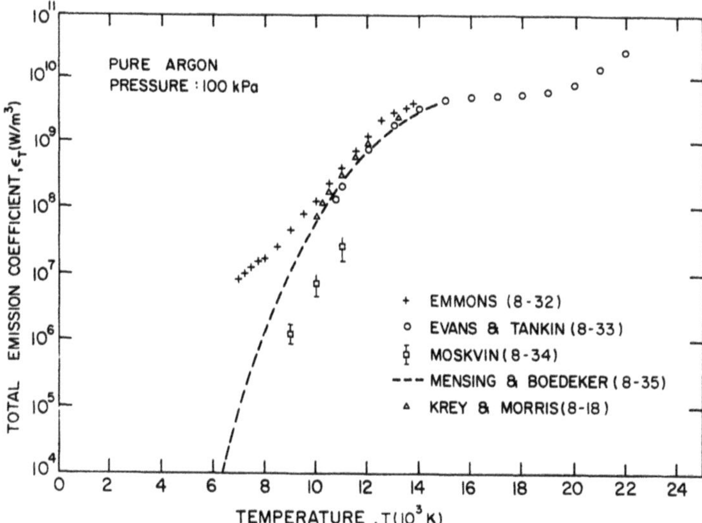


**FIG. 8.20.** Total volumetric emission coefficient of an argon plasma versus temperature if the plasma is assumed to be optically thin and the resonance lines are neglected.[31]

W/m$^3$. Correspondence between the two can be obtained by multiplying the results in W/m$^3$ ster by $4\pi$. Only Yabukov,[38] Owano *et al.*,[37] Emmons,[32] and Wilbers *et al.*[39] have extended their calculations to lower temperatures (down to 5000 K). For example, the more recent results of

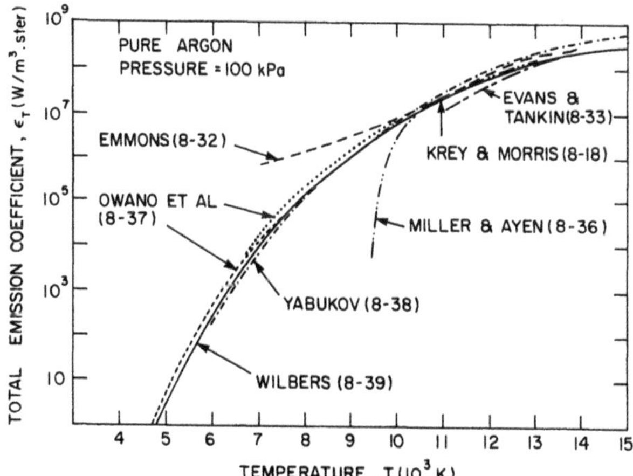


**FIG. 8.21.** Total volumetric emission coefficient of an argon plasma versus temperature of the plasma which is assumed to be optically thin, and the resonance lines are neglected.[39]

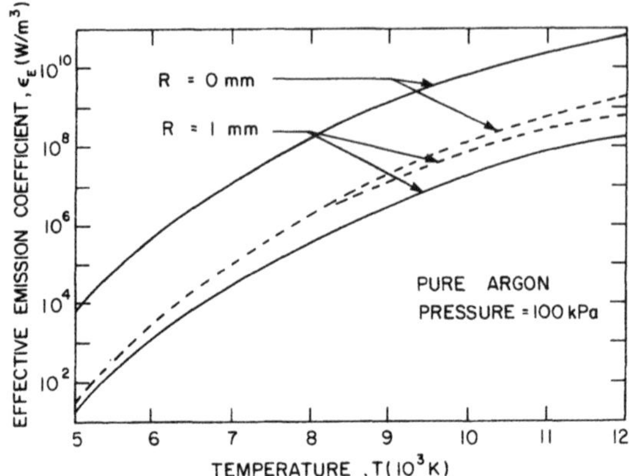


**FIG. 8.22.** Evolution of the effective volumetric emission coefficient, $\varepsilon_E$, versus temperature at 100 kPa for an argon plasma assumed to be either optically thin ($R = 0$) or partially absorbing ($R = 1$ mm). The radiation from the resonance lines (full line) separated from that of excited states lines (dotted line).[24]

Wilbers *et al.* show higher values than those of Yabukov. According to Wilbers *et al.*, the difference occurs because their spectral range (100 nm to $10^5$ nm) is larger than that of Yabukov. Similar considerations hold when comparing the results of Wilbers to those of Owano *et al.*, who conducted measurements from 250 to 2500 nm. The latter did not account for the resonance lines below 200 nm. Also, the continuum emissivity between 100 nm and 250 nm and above 2500 nm was not taken into account. For their measurements, the deviations between the measured values and the upper limits increase with increasing temperature. This result is probably due to the increasing influence of the continuum emissivity in the omitted wavelength regions.

Of course, very different results are obtained when considering resonance lines and absorption, as illustrated in Fig. 8.22 from Essoltani[24] (note that the effective emission coefficient is expressed in the same units as the total volumetric emission coefficient). The evolution of the resonance and excited-state lines at atmospheric pressure as a function of temperature is shown both without ($R = 0$) and with ($R = 1$ mm) absorption. It is clear from this figure that in optically thin plasmas the resonance lines make the main contribution (almost two orders of magnitude higher than that of the other lines). However, as soon as absorption is taken into account, the emission of the resonance lines along a radius of 1 mm is drastically reduced (almost three orders of magnitude), while that of the other excited lines is rather unaffected. The same tendencies with $\varepsilon_E$ expressed in W/m³ ster instead of W/m³ are

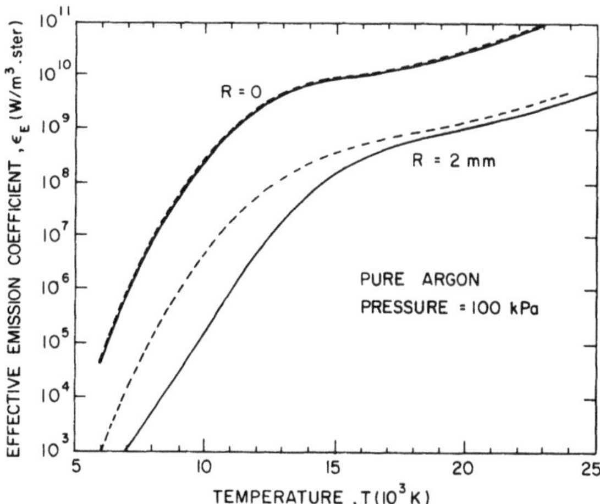


**FIG. 8.23.** Evolution of the effective volumetric emission coefficient versus temperature at 100 kPa for an argon plasma assumed to be either optically thin ($R = 0$) or partially absorbing ($R = 2$ mm). The radiation from the resonance (full line) is separated from that of excited states (dotted line).[28]

shown in Fig. 8.23 from Gleizes *et al.*[28] Here again, very strong absorption of the resonance line is observed: for a plasma radius of 2 mm, the escape factor is lower than $10^{-3}$ when $T \leqslant 10{,}000$ K and is equal to about 0.016 when $T = 15{,}000$ K. The escape factor increases with temperature because of the decrease in the density of the ground-state atoms and of broadening of the lines due to the Stark effect. As shown in Fig. 8.22, the absorption of the other lines is rather small.

### 8.4.1.2. Nitrogen Plasma

Figure 8.24 from Boulos *et al.*[31] shows the results for a nitrogen plasma in $W/m^3$ obtained by various authors,[18,40–43] assuming optically thin plasmas. Here again, a certain dispersion of the data can be observed. Figure 8.25 compares the results in $W/m^3$ ster of the calculations of Rahmani[5] with those of Hermann and Schade[44] and Allen[45] and with the experimental results of Ernst *et al.*[46] The experimental values are in rather good agreement with predictions up to 20,000 K.

When considering absorption and the effective emission coefficient, as we did for argon, the results are rather different, as illustrated in Fig. 8.26 for a plasma radius of 20 mm. It is interesting to note that for $T \leq 17{,}000$ K, the continuum radiation is almost equivalent to that of

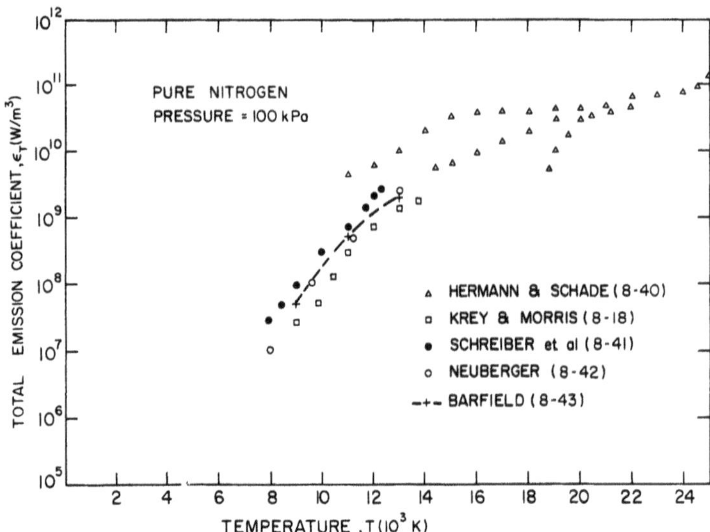


**FIG. 8.24.** Total volumetric emission coefficient of a nitrogen plasma assumed to be optically thin as a function of temperature at 100 kPa.[31]

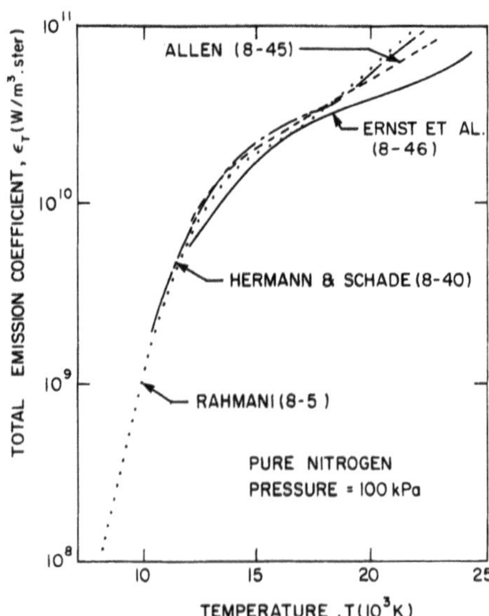


**FIG. 8.25.** Total volumetric emission coefficient of a nitrogen plasma assumed to be optically thin as a function of temperature at 100 kPa.[5]

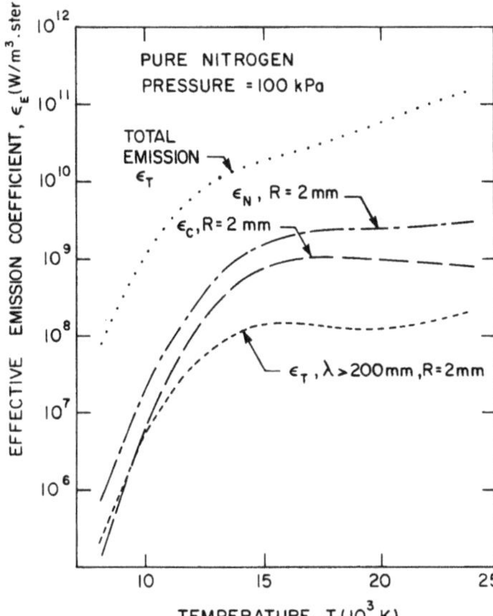


**FIG. 8.26.** Evolution of the effective volumetric emission coefficient of a nitrogen plasma with a radius of 20 mm, as a function of temperature at 100 kPa,[5] total coefficient with no absorption, effective total emission coefficient, effective emission coefficient due to continuum, and effective emission coefficient for $\lambda > 200$ nm (no resonance lines).

the lines in $\varepsilon_E$. Figure 8.27 shows the evolution of the effective emission coefficient for a nitrogen plasma for different plasma radii at atmospheric pressure. As we saw for argon, the radiation is absorbed (almost 95%) in the first millimeter of plasma. For $T > 18{,}000$ K the absorption becomes more important. It is also important to note that $\varepsilon_E$ becomes almost

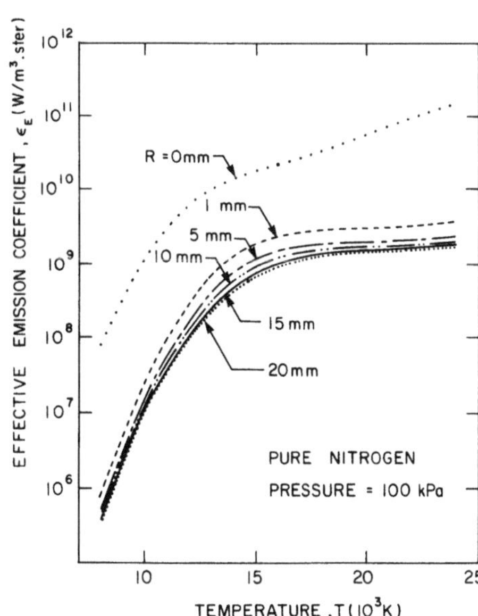


**FIG. 8.27.** Effective volumetric emission coefficient of a nitrogen plasma for different plasma radii $R$ ($R = 0$ corresponds to optically thin plasma) as a function of temperature at 100 kPa.[5]

constant for $T > 16{,}000$ K because the electron density becomes almost constant for higher temperatures (see Fig. 6.2). Thus the Stark line width $\delta_\nu$ also reaches a limiting value, and because the optical depth is inversely proportional to $\delta_\nu$ [see Eq. (8.129)] absorption no longer varies. Moreover, the Stark width of the ionic lines is also less than that of the atomic lines, thus the atomic lines (for the same absorption coefficient) will be more absorbed. These two facts explain the relative increase of absorption for $T > 18{,}000$ K.

### 8.4.2. Plasma Seeded with Metallic Vapors

The presence of small amounts of metallic vapors substantially increases the radiative losses of a thermal plasma and modifies, for example, the heat transfer to particles when they evaporate.[47]

#### 8.4.2.1. Argon and Iron

The calculations presented here have been developed by Essoltani.[24] Figure 8.28 shows the total volumetric emission coefficients for iron and argon in an optically thin plasma at 100 kPa. In spite of the low percent of iron (1 mol%) iron radiates much more than argon does, especially at low temperatures ($T \leqslant 10{,}000$ K), due to two factors:

1. the great number of iron lines (3226 lines in the iron calculation, compared with less than 500 for argon),
2. the low ionization potential of iron ($E^{\mathrm{I}}_{\mathrm{Fe}^+} = 7.9$ eV) and the low

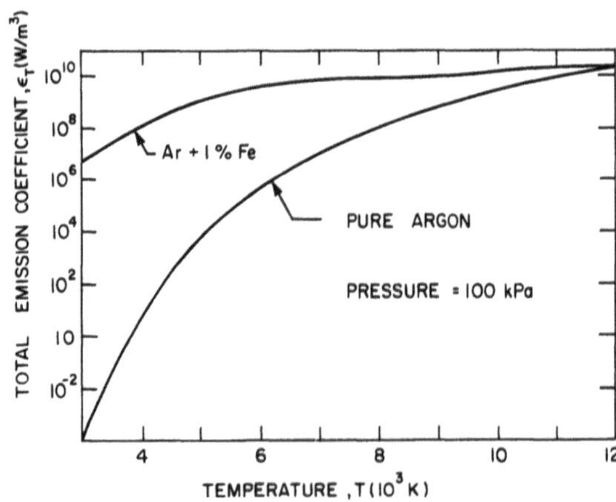


FIG. 8.28. Total volumetric emission coefficient of an Ar–Fe plasms (1 mol%) assumed to be optically thin, as a function of temperature at 100 kPa.[24]

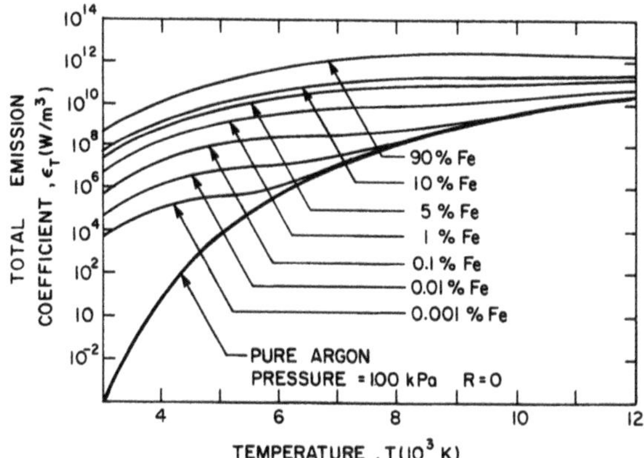


**FIG. 8.29.** Total volumetric emission coefficient of an Ar–Fe plasma assumed to be optically thin, for different mole percents of iron as a function of temperature at 100 kPa.[24]

energy of the first excited level of iron ($E_1 = 0.85$ eV). The corresponding values for Ar are 15.75 and 11.54 eV, respectively. For $T > 12{,}000$ K the argon emissivity exceeds that of iron.

The importance of iron for the total volumetric emissivity $\varepsilon_T$ is illustrated in Fig. 8.29, where the evolution of $\varepsilon_T$ versus temperature for the Ar–Fe mixture has been plotted for various iron mole percents. Below 7000 K even 0.001% of iron results in a dramatic increase in $\varepsilon_T$ (eight orders of magnitude at 3000 K). When considering the effective volumetric emission coefficient $\varepsilon_{EFe}$ of 1 mol% of Fe in Ar at different plasma radii (see Fig. 8.30), we can see that:

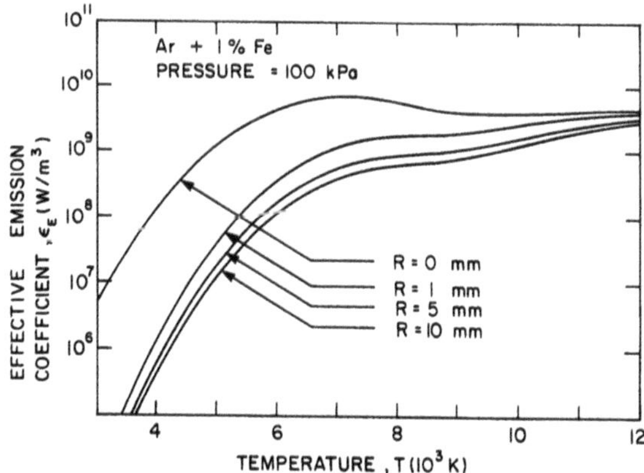


**FIG. 8.30.** Effective volumetric emission coefficient of an Ar–Fe plasma (1 mol%) for different plasma radii as a function of temperature at 100 kPa.[24]

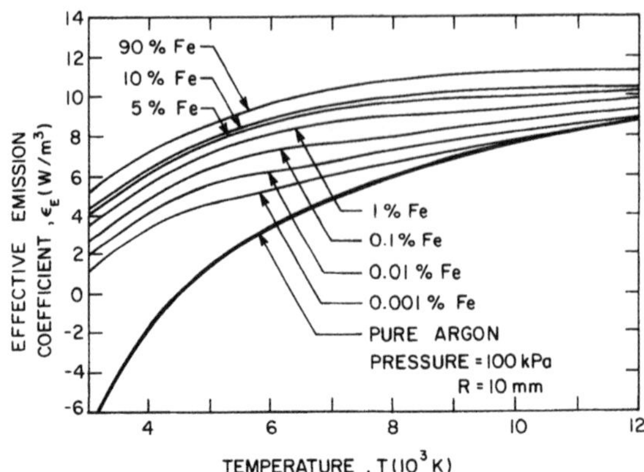


**FIG. 8.31.** Effective volumetric emission coefficient of an Ar–Fe plasma as a function of temperature at 100 kPa for different mole percents of iron and a plasma radius of 10 mm.[24]

- With no absorption ($R = 0$) $\varepsilon_{\mathrm{EFe}}$ increases dramatically with temperature from 3000 to 7000 K in connection with the high densities of the excited levels in this temperature range. Due to the low value of the iron ionization potential ($E^{\mathrm{I}}_{\mathrm{Fe^+}} = 7.9$ eV) for $T > 7000$ K, ions are produced and the neutral atoms' densities decrease, slightly reducing $\varepsilon_{\mathrm{EFe}}$.
- Due to the important contribution of the resonance lines to the total emissivity, $\varepsilon_{\mathrm{EFe}}$ decreases very rapidly within the first millimeter of plasma and much more slowly at radii greater than 1 mm. When comparing Fig. 8.31, which gives $\varepsilon_E$ for various iron percents and a plasma radius of 10 mm, to Fig. 8.29, which gives $\varepsilon_T$ for an optically thin plasma, the importance of the absorption of the argon resonance lines in the whole temperature range is obvious, while the resonance lines of iron mainly absorb at temperatures below 9000 K. In both cases, however, it is clear that the presence of iron vapor, even at low percentages, dramatically increases the plasma emission below 10,000 K.

### 8.4.2.2. Argon and Copper

The calculations for an argon and copper plasma have been performed by Gleizes *et al.*[28] Figure 8.32 compares the effective volumetric emission coefficients for an argon plasma containing 1 mol% iron and one containing 1 mol% copper; the arc radius is 10 mm. Because of the lower number of Cu lines compared to those of Fe, $\varepsilon_{\mathrm{EFe}}$ is almost one order of magnitude higher than $\varepsilon_{\mathrm{ECu}}$. Note that (as already emphasized

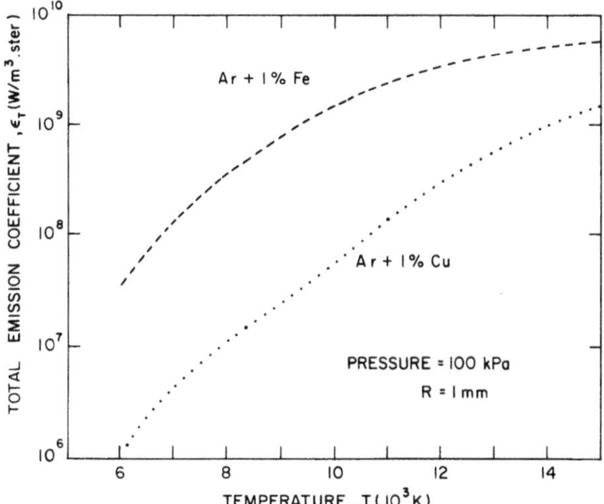


**FIG. 8.32.** Effective volumetric emission coefficients of Ar–Fe and Ar–Cu plasmas (both with 1 mol% of metallic vapor) with a plasma radius of 10 mm as a function of temperature at 100 kPa.[28]

for Ar–Fe plasma) the absorption of argon in the Ar–Cu plasma is higher than that of Cu because of the strong absorption of Ar resonance lines, especially at high temperatures (see Fig. 8.33 from Ref. 28).

For a given percentage of Cu (1 mol%), the influence of thickness of

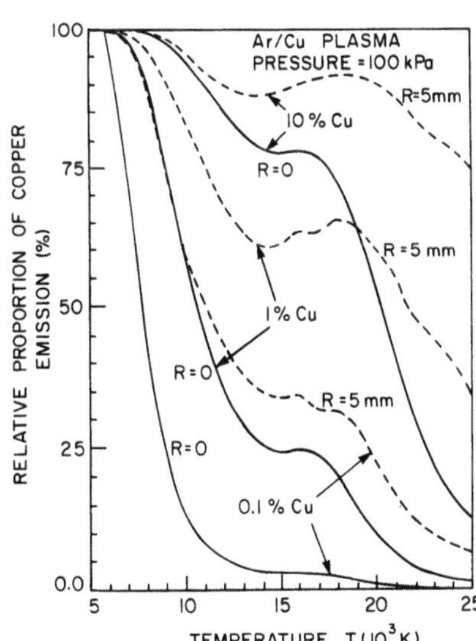


**FIG. 8.33.** Relative proportion of copper emission in Ar–Cu plasma, at 100 kPa; solid line, $R = 0$ (no absorption); dotted line, $R = 5$ mm.[28]

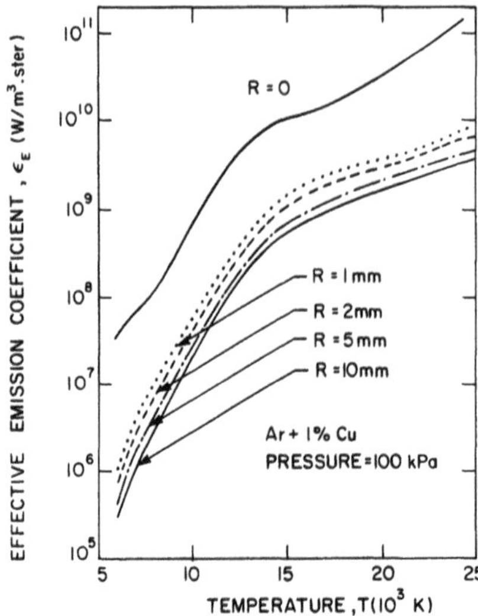


**FIG. 8.34.** Influence of plasma thickness on the effective emission coefficient of an Ar–Cu plasma (1 mol% Cu) at 100 kPa.[28]

the plasma is illustrated in Fig. 8.34 (from Ref. 28). As we saw for iron, overall there is a strong absorption at low temperatures corresponding to high densities of absorbing atoms. Also note that the values of $\varepsilon_E$ are hardly dependent on the plasma radius as soon as it is greater than 1–2 mm. Finally, the influence of the copper mole percent is illustrated in Fig. 8.35, which shows the evolution of $\varepsilon_E$ with temperature (at 100 kPa) for a plasma radius of 2 mm.

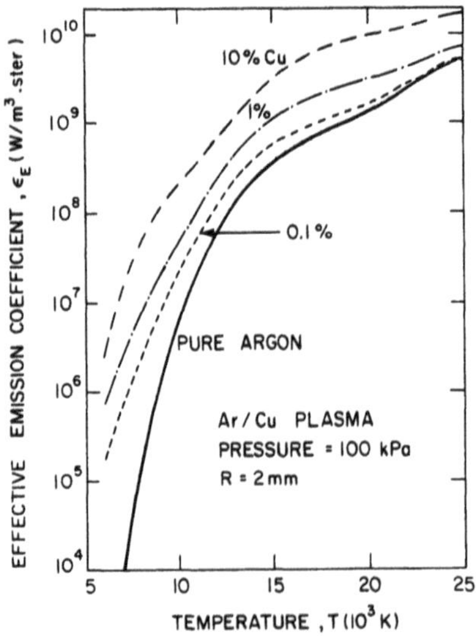


**FIG. 8.35.** Influence of the copper concentration on the effective emission coefficient of an Ar–Cu plasma at $p = 100$ kPa and for a plasma radius of 2 mm.[28]

## 8.5. BLACKBODY RADIATION OF HIGH-TEMPERATURE GASES

The emission coefficients described in the previous paragraphs are all based on the assumption that the plasma is optically thin, with all the radiation escaping from the plasma. This assumption may fail for the following two situations.

(a) In the first case very strong absorption occurs for resonance lines (transition between the first excited state and the fundamental level) for which the absorption coefficient $\kappa_L'(\nu)$ [see Eq. (8.47)] is so high that a layer of thickness $L$ of a fraction of a millimeter is already sufficient for complete absorption.[48] (see also the results presented in Section 8.4). In the immediate neighborhood of such a resonance line, the absorption coefficient is usually many orders of magnitude smaller. For lines corresponding to lower-level transitions, the absorption coefficient may be such that in a layer of a few millimeters self-absorption will be responsible for a reduction of 20 to 50% of the line intensity. This reduction may lead to significant errors when such lines are used for spectroscopic temperature measurements with no corrections for this absorption.

(b) The optically thin assumption may also fail for high-pressure plasmas ($p > 100\,\mathrm{kPa}$). Finkelnburg and Peters[48] have calculated the conditions for which the continuous radiation of a laboratory plasma approaches that of blackbody radiation ($\varepsilon_\nu \geq 0.9\varepsilon_\nu^0$) in the visible range of the spectrum (500.0 nm). Figure 8.36 shows the results of their calculation for different gases. By considering only the singly ionized

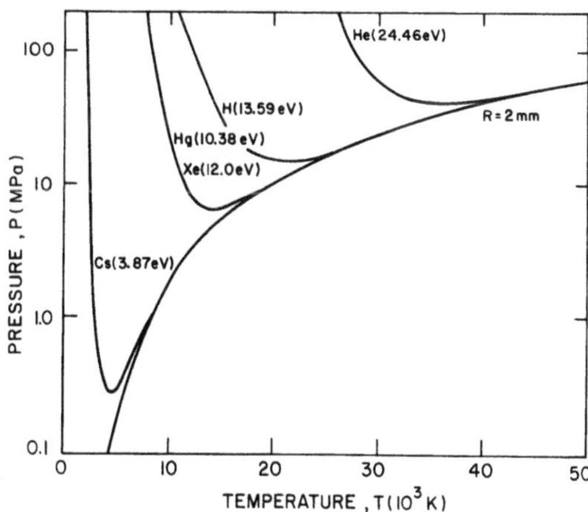


**FIG. 8.36.** Regime in which a plasma control volume with a radius of 2 mm can be considered to be a blackbody.[7]

species of these gases, all curves merge into a common curve that corresponds to an ionization degree of 100%. Above this common curve, a laboratory plasma with a layer thickness of 2 mm or larger would become a blackbody radiator at the plasma temperature. Argon, nitrogen, oxygen, which have ionization energies between 14 and 16 eV, would fall between the curves for helium and hydrogen, i.e., become blackbody radiators for temperatures $T > 2 \times 10^4$ K and pressures $p > 20$ MPa. Cesium, with the lowest ionization potential, would require a minimum pressure of 0.5 MPa to become a blackbody radiator at 5000 K.

## LIST OF SYMBOLS

| | |
|---|---|
| $A^i_{ul}$ | transition probability ($s^{-1}$) for spontaneous emission |
| $B_{lu}$ | transition probability for absorption ($m^3/J\,s^2$) |
| $B_\lambda$ | blackbody monochromatic radiation intensity ($W/m^3$ ster) |
| $B_\nu$ | blackbody monochromatic radiation intensity ($J/m^2$ ster) |
| $B_{\mathrm{ul}}$ | transition probability for induced emission ($m^3/J\,s^2$) |
| $b$ | impact parameter |
| $c$ | velocity of light ($2.998 \times 10^8$ m/s) |
| $E$ | energy |
| $E^{\mathrm{I}}_{\mathrm{H}^+}$ | ionization energy of the hydrogen atom (13.6 eV) |
| $E_{i,u}$ | energy of the excited state $u$ of the chemical species $i$ |
| $E^{\mathrm{I}}_{\mathrm{X}^+}$ | ionization energy of the atom X |
| $E_{\mathrm{X}_j}$ | excited state $j$ of the atom X |
| $E_\nu$ | monochromatic radiation energy |
| $e$ | charge of the electron ($-1.6 \times 10^{-19}$ As) |
| $F$ | energy of the rotational excited state in($cm^{-1}$) |
| $F_{R\nu}$ | monochromatic radiation flux |
| $G$ | energy of the vibrational excited state (expressed in $cm^{-1}$) |
| $G_1$ | function accounting for the cylindrical geometry of the plasma |
| $G^n_{i,z}$ | Gaunt factor |
| $g_u$ | statistical weight or degeneracy |
| $H^+$ | total radiation flux in positive direction ($W/m^2$) |
| $H^-$ | total radiation flux in negative direction ($W/m^2$) |
| $H^0$ | total flux (intensity emitted per unit surface per unit time into the half sphere) for a blackbody |
| $h$ | Planck constant ($6.6 \times 10^{-34}$ W $s^2$) |
| $I_\nu(\theta, \varphi)$ | monochromatic radiation intensity ($J/m^2$ ster) |

| | |
|---|---|
| $I_\lambda(\theta, \varphi)$ | monochromatic radiation intensity ($W/m^3$ ster) |
| $I(\theta, \varphi)$ | directional total radiation intensity ($W/m^2$ ster) |
| $I$ | total radiation intensity ($W/m^2$) |
| $I_\nu$ | monochromatic radiation intensity ($J/m^2$ ster) |
| $J$ | rotational quantum number |
| $J_R$ | total radiation flux |
| $k$ | Boltzmann constant ($1.38 \times 10^{-23}$ J/K part) |
| $l$ | azimuthal quantum number |
| $M$ | atomic mass (g) |
| $m_e$ | mass of the electron ($9.11 \times 10^{-31}$ kg) |
| $N_u(t)$ | population of excited state $u$ |
| $n$ | principal quantum number |
| $\vec{n}$ | surface normal |
| $n_e$ | electron density |
| $n_{i,u}$ | density of the excited state $u$ of the chemical species $i$ ($m^{-3}$) |
| $n_r$ | refractive index |
| $n_{i,z}^{n,l}$ | density of the chemical species $i$; $ze$ is the electrical charge of the excited state defined by the principal quantum number $n$ and the azimuthal quantum number $l$ |
| $n_l^*$ | effective quantum number |
| $p$ | pressure (Pa) |
| $P(\nu - \nu_0)$ | shape factor of a spectral line (s) |
| $P(\lambda - \lambda_0)$ | shape factor of a spectral line ($m^{-1}$) |
| $Q_{i,z}^{el}$ | electronic partition function of the chemical species $i$ with electrical charge $ze$ |
| $r_1$ | Bohr radius ($5.3 \times 10^{-11}$ m) of the ground state |
| $R$ | radius of the elemental plasma control volume (m) |
| $S_\nu$ | source function: $S_\nu = \varepsilon_\nu / \kappa'_\nu n_r$ ($J/m^2$ ster) |
| $S$ | cross section ($m^2$) |
| $T_e$ | energy of the electronic excited state (expressed in $cm^{-1}$) |
| $t$ | time (s) |
| $u$ | total radiation density ($J/m^3$) |
| $u_\nu(\theta, \varphi)$ | monochromatic radiation density (J s/$m^3$ ster) |
| $u_\nu^0(T)$ | blackbody monochromatic radiation density (Planck law) (J s/$m^3$ ster) |
| $v$ | vibrational quantum number |
| $\bar{v}$ | mean velocity of an atom or an ion |
| $v_e$ | velocity of the electron |

## Greek Symbols

| | |
|---|---|
| $\alpha$ | constant of fine structure ($\alpha = 2\pi e^2/hc$) |
| $\delta_{e,i}$ | Stark width at half maximum intensity (nm) |

| | |
|---|---|
| $\delta_{E_{X^+}}$ | lowering of ionization energy of the atom X (eV) |
| $\delta_\lambda$ | width of the spectral line (nm) |
| $\delta_0$ | doppler width at half the maximum intensity (nm) |
| $\Delta J$ | $\Delta J = J' - J''$ difference in rotational quantum numbers related respectively to the upper ′ and lower ″ states |
| $\Delta v$ | $\Delta v = v' - v''$ difference in vibrational quantum numbers related respectively to the upper ′ and lower ″ states |
| $\varepsilon_E$ | effective emission coefficient ($W/m^3$ ster) |
| $\varepsilon_{fb}$ | emission coefficient for free–bound transition |
| $\varepsilon_{ff}$ | emission coefficient for free–free transition |
| $\varepsilon_{ff}^{e,i}$ | emission coefficient for free–free transitions due to the field of ions |
| $\varepsilon_L$ | line emission coefficient ($W/m^3$ ster) |
| $\varepsilon_{ff}^{e,a}$ | emission coefficient for the free–free transitions due to elastic collisions |
| $\varepsilon_{i,z+1}^{nl}$ | emission coefficient of particles of chemical species $i$ with electrical charge $ze$; the excited state is defined by the quantum number $n$ and $l$ |
| $\varepsilon_T$ | total emission coefficient ($W/m^3$ ster) |
| $\varepsilon_\lambda$ | monochromatic emission coefficient ($W/m^4$ ster) |
| $\varepsilon_\nu$ | monochromatic emission coefficient ($J/m^3$ ster) |
| $\zeta_{i,z}(\nu, T)$ | Biberman factor |
| $\theta$ | angle with respect to the surface normal $\vec{n}$ |
| $\kappa_\nu$ | monochromatic absorption coefficient including induced emission ($cm^{-1}$) |
| $\kappa'_\nu$ | monochromatic absorption coefficient per unit length without induced emission ($cm^{-1}$) |
| $\kappa_{i,z+1}$ | absorption coefficient for a particle of species $i$ and electrical charge $z$ ($z = 0$ for an atom, $z = 1$ for its first ion) ($cm^{-1}$) |
| $\kappa'_L(\nu)$ | integrated absorption coefficient over the line profile ($cm^{-1}\,s^{-1}$) |
| $\lambda$ | wavelength (nm) |
| $\lambda_{max}$ | wavelength giving the maximum value of $B_v$ at a given temperature |
| $\Lambda$ | escape factor |
| $\nu$ | radiation frequency |
| $\sigma_{ul}$ | wave number for the transition between the state $u$ and the state $l$ ($m^{-1}$) |
| $\sigma_{i,z}^{n,l}$ | cross section for photoionization |
| $\sigma_{i,z}^{n,l}$class | classical photoionization cross section |
| $\tau_s$ | perturbation time resulting from a charged particle moving with a relative velocity $\bar{v}$ |

$\tau_{ul}$ lifetime of the excited state $u$ (s)
$\tau_\nu$ optical depth (dimensionless)
$\varphi$ azimuthal angle with respect to the normal $\vec{n}$
$\Omega$ solid angle

## Superscripts

$'$ upper energy level
$''$ lower energy level
$n$ principal quantum number
$l$ azimuthal quantum number

## Subscripts

$i$ chemical species $i$
$l$ lower excited state
$u$ upper excited state
$z$ electrical charge of the particle

## GENERAL BIBLIOGRAPHY

Cabannes, F. and J. Chapelle, "Spectroscopic Plasma Diagnostic," Chapter 7 in *Reactions Under Plasma Conditions,* Vol. 1, New York: Wiley Interscience, 1971.

Griem, H. R., *Plasma Spectroscopy,* New York: McGraw-Hill, 1964.

Griem, H. R., *Spectral Broadening by Plasma,* New York and London: Academic Press, 1974.

Herzberg, G., *Atomic Spectra and Atomic Structure,* New York: Dover, 1944.

Herzberg, G., *Spectra of Diatomic Molecules,* New York: D. van Nostrand, 1969.

Pecker-Wimel, C., *Introduction à la spectroscopie des plasmas,* London: Gordon and Breach, 1967.

Traving, G., *Plasma Diagnostics,* Chapter II; (Lochte-Holtgreven, ed.), Wiley, New York, 1968.

## REFERENCES

1. R. A. Hill, *J. Quant. Spectrosc. Radiat. Transfer* **7** (1963): 82.
2. W. E. Wiese, P. E. Kelleher, and V. Helbig, *Phys. Rev. A* **11** (1975): 1854.
3. H. Ehrich and M. J. Kusch, *Z. Naturforsch., A* **28** (1973): 1794.
4. R. Konjevic and N. Konjevic, Fysika **18** (1986): 327.
5. B. Rahmani, "Calcul de l'émission nette du rayonnement des arcs dans $SF_6$ et dans les mélanges $SF_6$–$N_2$," Thèse de Doc. Ing. (Univ. of Toulouse, France, Feb., 1989).
6. J. M. Baronnet, "Contribution à l'étude spectroscopique des plasmas d'azote produits par un générateur à arc soufflé; application à la chimie des plasmas: synthèse des oxydes d'azote," Thèse doctorat d'État (Université de Limoges, France, Nov. 1978).
7. E. Pfender, "Diagnostic Techniques," in *Continuing Education: Plasma Technology and Applications,* (2nd World Congress of Chemical Engineering and World Chemical Montreal, 4–9 Oct., 1981).
8. P. Fauchais, K. Lapworth, and J. M. Baronnet, "First report on measurement of temperature and concentration of excited species in optically thin plasmas," IUPAC

Subcommittee on Plasma Chemistry, P. Fauchais, ed. (Univ. of Limoges, France, April 1974).
9. D. R. Bates, *Atomic and Molecular Processes* (New York: Academic Press, 1962).
10. H. A. Kramers, *Philos. Mag.* **46** (1923): 836.
11. J. Gaunt, *Philos. Trans. R. Soc. London, Ser. A* **229** (1930): 163.
12. D. H. Menzel and C. L. Pekeris, *Mon. Nat. R. Astron. Soc.* **96** (1935): 77.
13. G. Peach, *Mon. Nat. R. Astron. Soc.* **124** (1962): 371.
14. A. Burgess and M. J. Seaton, *Mon. Nat. R. Astron. Soc.* **120** (1960): 121.
15. W. J. Karsas and R. Letter, *Astron. J. Suppl. Sci.* **6** (1961): 167.
16. W. H. Soon and J. A. Kunc, *Phys. Rev. A* **43** (1991): 723.
17. S. Bayard, "Contribution au calcul des fonctions de partition des plasmas azote-silicium-aluminium et détermination des températures à partir du fond continu de l'azote," Thèse de doctorat de 3e cycle (University of Limoges, France, 30 April 1974).
18. R. U. Krey and J. C. Morris, *Phys. Fluids* **13** (1970): 1483.
19. J. L. Morris and J. M. Yos, *Radiation Studies of Arc Heated Plasmas* (ARL 71–0317 AFSC-0390-41 CR).
20. D. H. Sampson, *Radiative Contributions to Energy and Momentum Transport in Gas* (New York: Interscience, 1965).
21. J. J. Lowke and E. R. Capriotti, *J. Quant. Spectrosc. Radiat. Transfer* **9** (1969): 107.
22. N. E. Nicolet, C. E. Shepard, K. J. Clark, A. Balakushnan, J. P. Kesseling, K. E. Suchsland, and J. J. Reese Jr., *Analysis and Design Study for a High Pressure, High Enthalpy Constricted Arc Heater* (Rep. AEDC-TR-75-47, 1975).
23. R. Siegel and J. R. Howell, *Thermal Radiation Heat Transfer* (New York: McGraw-Hill 1981).
24. A. Essoltani, "Étude du rayonnement émis par un plasma d'argon en présence de vapeur métallique,". Thèse de doctorat es Sciences Appliquées, Spécialité Génie Chimique (Université de Sherbrooke, Sherbrooke, Québec, CN, May 1991).
25. J. J. Lowke, *J. Quant. Spectrosc. Radiat. Transfer* **14** (1974): 111.
26. M. Gand, "Relaxation d'un plasma d'hélium créé par claquage rapide," (Thèse 3$^{e}$ cycle (Univ. of Orleans, France, July, 1978).
27. R. W. Liebermann and J. J. Lowke, *J. Quant. Spectrosc. Radiat. Transfer* **16** (1976): 253.
28. A. Gleizes, J. J. Gonzalez, B. Liani, and B. Rahmani, *Journal de Physique* **51** C5 (1990): 213.
29. A. W. Drawin and F. Emard, *Beiträge aus der Plasma Physik* **13** (1973): 143.
30. F. E. Irons, *J. Quant. Spectrosc. Radiat. Transfer* **22** (1979): 1.
31. M. Boulos, "Thermodynamic and transport properties of argon, nitrogen and oxygen at atmospheric pressure over the temperature range 3000–20,000 K," Internal report (Univ. of Sherbrooke, CN, May 1984).
32. M. W. Emmons, *Phys. Fluids* **10** (1967): 1125.
33. D. L. Evans and R. S. Tankin, *Phys. Fluids* **10** (1967): 1137.
34. Yu. V. Moskvin, *Teplofizika Vysokikh Temperatur* **6** (1968): 1.
35. A. E. Mensing and L. R. Boedeker, *Theoretical Investigations of RF Induction Heated Plasmas* (NASA-CR-1312, 1969).
36. R. C. Miller and R. J. Ayen, *J. Appl. Phys.* **40** (1990): 5260.
37. T. G. Owano, M. H. Gordon, and C. H. Kruger, *Phys. Fluids* **B2** (1990): 3184.
38. I. T. Yabukov, *Opt. Spectrosc.* **19** (1965): 277.
39. A. T. M. Wilbers, J. J. Beulens, and D. C. Schram, *International Symposium on Plasma Chemistry 10* 1 1.1-4, (U. Ehlemann *et al.*, eds.) (Univ. of Bocham, Germany, 1991).
40. W. Hermann and E. Schade, Z. *Phys.* **233** (1970): 333.

41. P. W. Schreiber, A. M. Hunter, and K. R. Benedetto, *AIAA J.* **10** (1972): 670.
42. A. W. Neuberger (AIAA Paper 73–744, delivered at AIAA 8th Thermophysics Conference, Palm Springs, CA, 1973).
43. W. P. Barfield, *J. Quant. Spectrosc. Radiat.* **17** (1977): 471.
44. W. Hermann and E. Schade, *J. Quant. Spectrosc. Radiat. Transfer* **12** (1972): 1257.
45. R. A. Allen, N.A.S.A. Contractor Report, CR 557.
46. K. A. Ernst, J. G. Kopainsky, and H. H. Maecker, *IEEE Trans. Plasma Sci.* **1**, 3 (1973).
47. A. Essoltani, P. Proulx, M. Boulos, and A. Gleizes, *International Symposium on Plasma Chemistry 10* 1 1.1–7, (U. Ehlemann *et al.*, eds.) (Univ. of Bochum, Germany, 1991).
48, A. Finkelnburg and Th. Peters, "Kontinuierliche Spektren," In *Encyclopedia of Physics,* Vol. 28, *Spectroscopy II* (Berlin: Springer-Verlag, 1957).

Appendix

# Thermodynamic and Transport Properties of Pure Gases and Their Mixtures at Atmospheric Pressure Over the Temperature Range 500–24,000 K

## A.1. INTRODUCTION

Thermodynamic and transport properties of plasmas are a prerequisite for any plasma modeling work. Compared to calculations, for ordinary gases, plasmas impose additional difficulties due to the large number of chemical species at elevated temperatures (including charged particles) and the chemical reactions taking place in plasmas. As shown in Chapters 6, 7, and 8, the calculation of plasma properties can be a formidable task, especially as far as transport properties are concerned. Collision cross-sections required for those calculations suffer from relatively large uncertainties associated with the assumptions which have to be introduced for the interaction potentials. Experimental data, on the other hand, are only available for a limited number of collision processes. The data presented in the following tables have been computed from the most reliable sources presently available (see Acknowledgments and References). They have been computed at the Université de Limoges in France, and thoroughly validated against the data banks available at the University of Minnesota, U.S.A., and the Université de Sherbrooke, Canada. The former were derived from an extensive computer program for thermodynamic and transport properties of plasma developed at the University of Minnesota. The Université de Sherbrooke data bank, on the other hand, is based on a compilation of a large number of literature data which include both experimental and theoretical studies published over the past forty years.

## A.2. CALCULATION METHOD

### A.2.1. Thermodynamic Properties

Depending on the temperature, pressure, and initial mole fractions, a set of $K$-equations ($K$ being the number of chemical species including ions and electrons present in the mixture) are solved simultaneously in order to determine the densities of the different species. The method consists of minimizing the Gibbs free energy, $G$, with mass-conservation and electrical-neutrality equations. The minimization scheme is based on the use of Lagrange multipliers, and the solution of the corresponding set of equations is by the steeper decent method according to White and Dantzig.[1] The developed computer code[2] takes into account, when necessary, the virial expression of the equation of state at high pressure as well as the Debye corrections related to the high densities of charged particles. Knowledge of the partition functions of the different species is a prerequisite to the calculation. These were determined by using the electronic energy levels and spectroscopic data from the literature.[3] The limitation of the atomic and ionic partition functions was calculated assuming that the lowering of the ionization potential[4] is given by

$$\delta E_i = (z_i + 1)e^2/4\pi\varepsilon_0 d \tag{A.1}$$

and

$$d^2 = \varepsilon_0 kTV \Big/ \left(e^2 \sum z_i^2 N_i\right) \tag{A.2}$$

$V$ being the volume, $e$ the charge of the electron, $\varepsilon_0$ the vacuum dielectric constant, $k$ Boltzmann's constant, $z_i$ the charge number of the chemical species $i$, and $N_i$ the number of particles of the chemical species $i$. For molecules, the partition function was calculated by looking for the $v$, $J$ values limiting the rotational–vibrational partition function.[5] The data used for diatomic species were those of Herzberg.[6]

The calculated densities, partition functions, and their derivatives were used to calculate the other thermodynamic properties such as total specific enthalpy and the specific heat of the plasma for temperatures up to 24,000 K.

### A.2.2. Transport Properties

The transport properties were computed using Sonine's polynomial expansion of the first order Chapman–Enskog approximation of the Boltzmann equation.[7,8] Since the electron mass is smaller than that of the heavy species, the heavy-species Boltzmann equation was decoupled from that of the electrons. The electron and heavy-particle transport properties were calculated independently following the method of Devoto[9] developed at Limoges successively by Gorse,[10] Bonnefoi,[11,12] and Aubreton.[13] According to this method, the transport properties are given as a function of temperature, densities of the different species, and collision integrals obtained either from the literature or by estimating them using interaction potentials. The calculated collision integrals were fitted by polynomials of the following form as a function of temperature:

$$\Omega(l, s) = a(1) \log T + \sum_{i=2}^{8} a(i) T^{m(i)} \tag{A.3}$$

The number of coefficients $i$ was limited to 9, and the exponents $m(i)$ may vary between $-9$ and $+9$. The different coefficients and exponents were stored in files. The data were tabulated over the temperature range between 300 and 24,000 K with a 100-K step. A literature search was made to find the most reliable data for the interaction potentials.

## A.3. PLASMA TABLES

The following tables (A.1–A.7; beginning on page 388) provide data on the thermodynamic and transport properties of selected gases and mixtures at atmospheric pressure over the temperature range 500–24,000 K.

# TABLE A.1. Thermodynamic and Transport Properties

| $T$ (K) | Density (kg/m$^3$) | Enthalpy (J/kg) | Sp heat (J/kg · K) | Viscosity (kg/m · s) | Therm cond (W/m · K) | Elec cond (A/V · m) |
|---|---|---|---|---|---|---|
| | | | **ARGON** | | | |
| 500 | 9.7353E-01 | 1.0499E+05 | 5.2033E+02 | 3.4224E-05 | 2.6712E-02 | 3.0784E-23 |
| 600 | 8.1121E-01 | 1.5705E+05 | 5.2063E+02 | 3.9245E-05 | 3.0631E-02 | 2.5610E-23 |
| 700 | 6.9530E-01 | 2.0910E+05 | 5.2053E+02 | 4.4858E-05 | 3.5011E-02 | 3.7555E-23 |
| 800 | 6.0839E-01 | 2.6115E+05 | 5.2048E+02 | 4.9742E-05 | 3.8823E-02 | 3.7903E-23 |
| 900 | 5.4079E-01 | 3.1319E+05 | 5.2044E+02 | 5.3921E-05 | 4.2085E-02 | 3.9136E-23 |
| 1000 | 4.8672E-01 | 3.6524E+05 | 5.2041E+02 | 5.7633E-05 | 4.4982E-02 | 3.9216E-23 |
| 1100 | 4.4248E-01 | 4.1727E+05 | 5.2039E+02 | 6.1076E-05 | 4.7669E-02 | 3.8200E-23 |
| 1200 | 4.0561E-01 | 4.6931E+05 | 5.2038E+02 | 6.4374E-05 | 5.0243E-02 | 3.6620E-23 |
| 1300 | 3.7441E-01 | 5.2135E+05 | 5.2037E+02 | 6.7598E-05 | 5.2760E-02 | 3.5018E-23 |
| 1400 | 3.4767E-01 | 5.7339E+05 | 5.2036E+02 | 7.0787E-05 | 5.5249E-02 | 7.4108E-21 |
| 1500 | 3.2450E-01 | 6.2542E+05 | 5.2036E+02 | 7.3960E-05 | 5.7725E-02 | 1.7575E-18 |
| 1600 | 3.0422E-01 | 6.7746E+05 | 5.2035E+02 | 7.7125E-05 | 6.0195E-02 | 8.8694E-17 |
| 1700 | 2.8633E-01 | 7.2949E+05 | 5.2035E+02 | 8.0284E-05 | 6.2661E-02 | 2.9060E-15 |
| 1800 | 2.7042E-01 | 7.8152E+05 | 5.2034E+02 | 8.3437E-05 | 6.5122E-02 | 6.7821E-14 |
| 1900 | 2.5619E-01 | 8.3356E+05 | 5.2034E+02 | 8.6582E-05 | 6.7576E-02 | 1.1967E-12 |
| 2000 | 2.4338E-01 | 8.8559E+05 | 5.2034E+02 | 8.9715E-05 | 7.0022E-02 | 1.6592E-11 |
| 2100 | 2.3179E-01 | 9.3763E+05 | 5.2034E+02 | 9.2834E-05 | 7.2456E-02 | 1.8351E-10 |
| 2200 | 2.2126E-01 | 9.8966E+05 | 5.2034E+02 | 9.5935E-05 | 7.4877E-02 | 1.6042E-09 |
| 2300 | 2.1164E-01 | 1.0417E+06 | 5.2033E+02 | 9.9017E-05 | 7.7282E-02 | 1.0861E-08 |
| 2400 | 2.0282E-01 | 1.0937E+06 | 5.2033E+02 | 1.0208E-04 | 7.9669E-02 | 5.6919E-08 |
| 2500 | 1.9471E-01 | 1.1458E+06 | 5.2033E+02 | 1.0511E-04 | 8.2038E-02 | 2.3956E-07 |
| 2600 | 1.8722E-01 | 1.1978E+06 | 5.2033E+02 | 1.0812E-04 | 8.4387E-02 | 8.5541E-07 |
| 2700 | 1.8029E-01 | 1.2498E+06 | 5.2033E+02 | 1.1110E-04 | 8.6714E-02 | 2.7145E-06 |
| 2800 | 1.7385E-01 | 1.3019E+06 | 5.2033E+02 | 1.1405E-04 | 8.9019E-02 | 7.8901E-06 |
| 2900 | 1.6786E-01 | 1.3539E+06 | 5.2033E+02 | 1.1698E-04 | 9.1301E-02 | 2.1380E-05 |
| 3000 | 1.6226E-01 | 1.4059E+06 | 5.2033E+02 | 1.1987E-04 | 9.3561E-02 | 5.4577E-05 |
| 3100 | 1.5703E-01 | 1.4580E+06 | 5.2033E+02 | 1.2274E-04 | 9.5798E-02 | 1.3214E-04 |
| 3200 | 1.5212E-01 | 1.5100E+06 | 5.2033E+02 | 1.2557E-04 | 9.8014E-02 | 3.0489E-04 |
| 3300 | 1.4751E-01 | 1.5620E+06 | 5.2033E+02 | 1.2837E-04 | 1.0021E-01 | 6.7299E-04 |
| 3400 | 1.4317E-01 | 1.6141E+06 | 5.2033E+02 | 1.3114E-04 | 1.0240E-01 | 1.4257E-03 |
| 3500 | 1.3908E-01 | 1.6661E+06 | 5.2033E+02 | 1.3389E-04 | 1.0458E-01 | 2.9073E-03 |
| 3600 | 1.3522E-01 | 1.7181E+06 | 5.2033E+02 | 1.3660E-04 | 1.0679E-01 | 5.7210E-03 |
| 3700 | 1.3157E-01 | 1.7702E+06 | 5.2033E+02 | 1.3928E-04 | 1.0904E-01 | 1.0891E-02 |
| 3800 | 1.2810E-01 | 1.8222E+06 | 5.2033E+02 | 1.4193E-04 | 1.1140E-01 | 2.0098E-02 |
| 3900 | 1.2482E-01 | 1.8742E+06 | 5.2033E+02 | 1.4455E-04 | 1.1395E-01 | 3.6037E-02 |
| 4000 | 1.2170E-01 | 1.9263E+06 | 5.2033E+02 | 1.4715E-04 | 1.1682E-01 | 6.2885E-02 |
| 4100 | 1.1873E-01 | 1.9783E+06 | 5.2033E+02 | 1.4972E-04 | 1.2019E-01 | 1.0700E-01 |
| 4200 | 1.1590E-01 | 2.0303E+06 | 5.2033E+02 | 1.5226E-04 | 1.2435E-01 | 1.7783E-01 |
| 4300 | 1.1321E-01 | 2.0824E+06 | 5.2033E+02 | 1.5477E-04 | 1.2970E-01 | 2.8904E-01 |
| 4400 | 1.1064E-01 | 2.1344E+06 | 5.2034E+02 | 1.5726E-04 | 1.3680E-01 | 4.6013E-01 |
| 4500 | 1.0818E-01 | 2.1864E+06 | 5.2034E+02 | 1.5972E-04 | 1.4639E-01 | 7.1816E-01 |
| 4600 | 1.0583E-01 | 2.2385E+06 | 5.2035E+02 | 1.6216E-04 | 1.5953E-01 | 1.1010E+00 |
| 4700 | 1.0358E-01 | 2.2905E+06 | 5.2036E+02 | 1.6458E-04 | 1.7754E-01 | 1.6587E+00 |
| 4800 | 1.0142E-01 | 2.3425E+06 | 5.2038E+02 | 1.6697E-04 | 2.0219E-01 | 2.4584E+00 |
| 4900 | 9.9348E-02 | 2.3946E+06 | 5.2041E+02 | 1.6934E-04 | 2.3575E-01 | 3.5882E+00 |
| 5000 | 9.7361E-02 | 2.4466E+06 | 5.2045E+02 | 1.7169E-04 | 2.8105E-01 | 5.1615E+00 |

TABLE A.1. (*Continued*)

| $T$ (K) | Density (kg/m$^3$) | Enthalpy (J/kg) | Sp heat (J/kg · K) | Viscosity (kg/m · s) | Therm cond (W/m · K) | Elec cond (A/V · m) |
|---|---|---|---|---|---|---|
| | | | **ARGON** | | | |
| 5100 | 9.5452E-02 | 2.4987E+06 | 5.2050E+02 | 1.7402E-04 | 1.3604E-01 | 7.0610E+00 |
| 5200 | 9.3617E-02 | 2.5507E+06 | 5.2057E+02 | 1.7632E-04 | 1.3793E-01 | 9.7731E+00 |
| 5300 | 9.1850E-02 | 2.6028E+06 | 5.2066E+02 | 1.7861E-04 | 1.3985E-01 | 1.3321E+01 |
| 5400 | 9.0149E-02 | 2.6549E+06 | 5.2079E+02 | 1.8088E-04 | 1.4179E-01 | 1.7883E+01 |
| 5500 | 8.8510E-02 | 2.7070E+06 | 5.2095E+02 | 1.8314E-04 | 1.4378E-01 | 2.3647E+01 |
| 5600 | 8.6930E-02 | 2.7591E+06 | 5.2116E+02 | 1.8537E-04 | 1.4582E-01 | 3.0805E+01 |
| 5700 | 8.5404E-02 | 2.8112E+06 | 5.2143E+02 | 1.8759E-04 | 1.4794E-01 | 3.9542E+01 |
| 5800 | 8.3932E-02 | 2.8634E+06 | 5.2178E+02 | 1.8979E-04 | 1.5015E-01 | 5.0028E+01 |
| 5900 | 8.2509E-02 | 2.9156E+06 | 5.2221E+02 | 1.9198E-04 | 1.5248E-01 | 6.2412E+01 |
| 6000 | 8.1133E-02 | 2.9679E+06 | 5.2276E+02 | 1.9415E-04 | 1.5494E-01 | 7.6821E+01 |
| 6100 | 7.9803E-02 | 3.0202E+06 | 5.2344E+02 | 1.9631E-04 | 1.5756E-01 | 9.3364E+01 |
| 6200 | 7.8515E-02 | 3.0727E+06 | 5.2427E+02 | 1.9845E-04 | 1.6037E-01 | 1.1214E+02 |
| 6300 | 7.7268E-02 | 3.1252E+06 | 5.2529E+02 | 2.0058E-04 | 1.6338E-01 | 1.3324E+02 |
| 6400 | 7.6059E-02 | 3.1779E+06 | 5.2652E+02 | 2.0270E-04 | 1.6663E-01 | 1.5679E+02 |
| 6500 | 7.4888E-02 | 3.2307E+06 | 5.2801E+02 | 2.0481E-04 | 1.7012E-01 | 1.8290E+02 |
| 6600 | 7.3752E-02 | 3.2836E+06 | 5.2980E+02 | 2.0690E-04 | 1.7388E-01 | 2.1174E+02 |
| 6700 | 7.2649E-02 | 3.3368E+06 | 5.3193E+02 | 2.0899E-04 | 1.7793E-01 | 2.4348E+02 |
| 6800 | 7.1578E-02 | 3.3903E+06 | 5.3445E+02 | 2.1106E-04 | 1.8229E-01 | 2.7831E+02 |
| 6900 | 7.0538E-02 | 3.4440E+06 | 5.3742E+02 | 2.1312E-04 | 1.8696E-01 | 3.1644E+02 |
| 7000 | 6.9527E-02 | 3.4981E+06 | 5.4091E+02 | 2.1517E-04 | 1.9198E-01 | 3.5808E+02 |
| 7100 | 6.8544E-02 | 3.5526E+06 | 5.4498E+02 | 2.1721E-04 | 1.9734E-01 | 4.0342E+02 |
| 7200 | 6.7587E-02 | 3.6076E+06 | 5.4971E+02 | 2.1925E-04 | 2.0308E-01 | 4.5261E+02 |
| 7300 | 6.6656E-02 | 3.6631E+06 | 5.5518E+02 | 2.2127E-04 | 2.0922E-01 | 5.0579E+02 |
| 7400 | 6.5749E-02 | 3.7192E+06 | 5.6148E+02 | 2.2328E-04 | 2.1578E-01 | 5.6304E+02 |
| 7500 | 6.4865E-02 | 3.7761E+06 | 5.6859E+02 | 2.2529E-04 | 2.2278E-01 | 6.2440E+02 |
| 7600 | 6.4003E-02 | 3.8338E+06 | 5.7678E+02 | 2.2729E-04 | 2.3026E-01 | 6.8984E+02 |
| 7700 | 6.3162E-02 | 3.8924E+06 | 5.8587E+02 | 2.2927E-04 | 2.3823E-01 | 7.5930E+02 |
| 7800 | 6.2342E-02 | 3.9520E+06 | 5.9654E+02 | 2.3125E-04 | 2.4675E-01 | 8.3264E+02 |
| 7900 | 6.1540E-02 | 4.0129E+06 | 6.0840E+02 | 2.3322E-04 | 2.5586E-01 | 9.0972E+02 |
| 8000 | 6.0757E-02 | 4.0750E+06 | 6.2133E+02 | 2.3518E-04 | 2.6555E-01 | 9.9034E+02 |
| 8100 | 5.9991E-02 | 4.1386E+06 | 6.3655E+02 | 2.3713E-04 | 2.7592E-01 | 1.0742E+03 |
| 8200 | 5.9241E-02 | 4.2040E+06 | 6.5319E+02 | 2.3906E-04 | 2.8698E-01 | 1.1612E+03 |
| 8300 | 5.8507E-02 | 4.2711E+06 | 6.7170E+02 | 2.4099E-04 | 2.9878E-01 | 1.2510E+03 |
| 8400 | 5.7788E-02 | 4.3405E+06 | 6.9315E+02 | 2.4290E-04 | 3.1142E-01 | 1.3433E+03 |
| 8500 | 5.7083E-02 | 4.4120E+06 | 7.1515E+02 | 2.4480E-04 | 3.2480E-01 | 1.4378E+03 |
| 8600 | 5.6391E-02 | 4.4860E+06 | 7.4024E+02 | 2.4668E-04 | 3.3902E-01 | 1.5343E+03 |
| 8700 | 5.5711E-02 | 4.5628E+06 | 7.6784E+02 | 2.4855E-04 | 3.5411E-01 | 1.6326E+03 |
| 8800 | 5.5043E-02 | 4.6426E+06 | 7.9813E+02 | 2.5040E-04 | 3.7011E-01 | 1.7324E+03 |
| 8900 | 5.4387E-02 | 4.7259E+06 | 8.3359E+02 | 2.5222E-04 | 3.8722E-01 | 1.8336E+03 |
| 9000 | 5.3740E-02 | 4.8127E+06 | 8.6785E+02 | 2.5402E-04 | 4.0512E-01 | 1.9358E+03 |
| 9100 | 5.3103E-02 | 4.9035E+06 | 9.0733E+02 | 2.5579E-04 | 4.2400E-01 | 2.0390E+03 |
| 9200 | 5.2476E-02 | 4.9988E+06 | 9.5381E+02 | 2.5754E-04 | 4.4416E-01 | 2.1430E+03 |
| 9300 | 5.1856E-02 | 5.0986E+06 | 9.9724E+02 | 2.5924E-04 | 4.6510E-01 | 2.2476E+03 |
| 9400 | 5.1244E-02 | 5.2039E+06 | 1.0530E+03 | 2.6090E-04 | 4.8746E-01 | 2.3528E+03 |
| 9500 | 5.0639E-02 | 5.3141E+06 | 1.1027E+03 | 2.6252E-04 | 5.1052E-01 | 2.4584E+03 |
| 9600 | 5.0040E-02 | 5.4310E+06 | 1.1687E+03 | 2.6408E-04 | 5.3517E-01 | 2.5643E+03 |
| 9700 | 4.9447E-02 | 5.5535E+06 | 1.2254E+03 | 2.6559E-04 | 5.6042E-01 | 2.6705E+03 |

(*continued*)

TABLE A.1. *(Continued)*

| $T$ (K) | Density (kg/m$^3$) | Enthalpy (J/kg) | Sp heat (J/kg · K) | Viscosity (kg/m · s) | Therm cond (W/m · K) | Elec cond (A/V · m) |
|---|---|---|---|---|---|---|
| | | | **ARGON** | | | |
| 9800 | 4.8859E-02 | 5.6838E+06 | 1.3030E+03 | 2.6702E-04 | 5.8741E-01 | 2.7769E+03 |
| 9900 | 4.8276E-02 | 5.8206E+06 | 1.3671E+03 | 2.6838E-04 | 6.1489E-01 | 2.8834E+03 |
| 10000 | 4.7696E-02 | 5.9663E+06 | 1.4577E+03 | 2.6965E-04 | 6.4428E-01 | 2.9900E+03 |
| 10100 | 4.7120E-02 | 6.1193E+06 | 1.5294E+03 | 2.7083E-04 | 6.7400E-01 | 3.0966E+03 |
| 10200 | 4.6546E-02 | 6.2827E+06 | 1.6345E+03 | 2.7189E-04 | 7.0582E-01 | 3.2033E+03 |
| 10300 | 4.5975E-02 | 6.4560E+06 | 1.7333E+03 | 2.7284E-04 | 7.3894E-01 | 3.3099E+03 |
| 10400 | 4.5406E-02 | 6.6376E+06 | 1.8161E+03 | 2.7365E-04 | 7.7204E-01 | 3.4165E+03 |
| 10500 | 4.4838E-02 | 6.8323E+06 | 1.9469E+03 | 2.7431E-04 | 8.0759E-01 | 3.5231E+03 |
| 10600 | 4.4271E-02 | 7.0389E+06 | 2.0656E+03 | 2.7481E-04 | 8.4446E-01 | 3.6295E+03 |
| 10700 | 4.3704E-02 | 7.2548E+06 | 2.1591E+03 | 2.7514E-04 | 8.8084E-01 | 3.7359E+03 |
| 10800 | 4.3137E-02 | 7.4868E+06 | 2.3198E+03 | 2.7527E-04 | 9.2007E-01 | 3.8422E+03 |
| 10900 | 4.2570E-02 | 7.7328E+06 | 2.4603E+03 | 2.7519E-04 | 9.6057E-01 | 3.9484E+03 |
| 11000 | 4.2003E-02 | 7.9937E+06 | 2.6087E+03 | 2.7489E-04 | 1.0023E+00 | 4.0544E+03 |
| 11100 | 4.1434E-02 | 8.2650E+06 | 2.7132E+03 | 2.7435E-04 | 1.0428E+00 | 4.1603E+03 |
| 11200 | 4.0864E-02 | 8.5572E+06 | 2.9221E+03 | 2.7355E-04 | 1.0867E+00 | 4.2660E+03 |
| 11300 | 4.0293E-02 | 8.8666E+06 | 3.0941E+03 | 2.7248E-04 | 1.1317E+00 | 4.3716E+03 |
| 11400 | 3.9720E-02 | 9.1940E+06 | 3.2745E+03 | 2.7113E-04 | 1.1779E+00 | 4.4770E+03 |
| 11500 | 3.9145E-02 | 9.5326E+06 | 3.3854E+03 | 2.6947E-04 | 1.2217E+00 | 4.5822E+03 |
| 11600 | 3.8568E-02 | 9.8976E+06 | 3.6503E+03 | 2.6751E-04 | 1.2696E+00 | 4.6872E+03 |
| 11700 | 3.7990E-02 | 1.0283E+07 | 3.8548E+03 | 2.6524E-04 | 1.3184E+00 | 4.7920E+03 |
| 11800 | 3.7410E-02 | 1.0690E+07 | 4.0676E+03 | 2.6265E-04 | 1.3680E+00 | 4.8965E+03 |
| 11900 | 3.6828E-02 | 1.1119E+07 | 4.2884E+03 | 2.5973E-04 | 1.4183E+00 | 5.0008E+03 |
| 12000 | 3.6243E-02 | 1.1558E+07 | 4.3942E+03 | 2.5646E-04 | 1.4650E+00 | 5.1049E+03 |
| 12100 | 3.5658E-02 | 1.2032E+07 | 4.7392E+03 | 2.5289E-04 | 1.5162E+00 | 5.2086E+03 |
| 12200 | 3.5072E-02 | 1.2530E+07 | 4.9812E+03 | 2.4900E-04 | 1.5679E+00 | 5.3120E+03 |
| 12300 | 3.4485E-02 | 1.3053E+07 | 5.2298E+03 | 2.4480E-04 | 1.6198E+00 | 5.4150E+03 |
| 12400 | 3.3897E-02 | 1.3602E+07 | 5.4841E+03 | 2.4030E-04 | 1.6719E+00 | 5.5176E+03 |
| 12500 | 3.3310E-02 | 1.4176E+07 | 5.7436E+03 | 2.3552E-04 | 1.7239E+00 | 5.6199E+03 |
| 12600 | 3.2723E-02 | 1.4777E+07 | 6.0073E+03 | 2.3048E-04 | 1.7758E+00 | 5.7216E+03 |
| 12700 | 3.2134E-02 | 1.5383E+07 | 6.0645E+03 | 2.2516E-04 | 1.8220E+00 | 5.8233E+03 |
| 12800 | 3.1549E-02 | 1.6036E+07 | 6.5252E+03 | 2.1966E-04 | 1.8729E+00 | 5.9242E+03 |
| 12900 | 3.0967E-02 | 1.6715E+07 | 6.7942E+03 | 2.1396E-04 | 1.9230E+00 | 6.0245E+03 |
| 13000 | 3.0387E-02 | 1.7421E+07 | 7.0625E+03 | 2.0811E-04 | 1.9722E+00 | 6.1243E+03 |
| 13100 | 2.9810E-02 | 1.8154E+07 | 7.3286E+03 | 2.0212E-04 | 2.0203E+00 | 6.2235E+03 |
| 13200 | 2.9238E-02 | 1.8913E+07 | 7.5907E+03 | 1.9603E-04 | 2.0670E+00 | 6.3220E+03 |
| 13300 | 2.8672E-02 | 1.9698E+07 | 7.8469E+03 | 1.8988E-04 | 2.1122E+00 | 6.4199E+03 |
| 13400 | 2.8111E-02 | 2.0507E+07 | 8.0952E+03 | 1.8368E-04 | 2.1556E+00 | 6.5170E+03 |
| 13500 | 2.7556E-02 | 2.1341E+07 | 8.3336E+03 | 1.7748E-04 | 2.1970E+00 | 6.6134E+03 |
| 13600 | 2.7010E-02 | 2.2197E+07 | 8.5599E+03 | 1.7130E-04 | 2.2362E+00 | 6.7091E+03 |
| 13700 | 2.6466E-02 | 2.3040E+07 | 8.4306E+03 | 1.6504E-04 | 2.2684E+00 | 6.8047E+03 |
| 13800 | 2.5936E-02 | 2.3935E+07 | 8.9539E+03 | 1.5897E-04 | 2.3028E+00 | 6.8988E+03 |
| 13900 | 2.5416E-02 | 2.4848E+07 | 9.1321E+03 | 1.5300E-04 | 2.3345E+00 | 6.9920E+03 |
| 14000 | 2.4906E-02 | 2.5777E+07 | 9.2898E+03 | 1.4715E-04 | 2.3633E+00 | 7.0844E+03 |
| 14100 | 2.4408E-02 | 2.6720E+07 | 9.4252E+03 | 1.4145E-04 | 2.3892E+00 | 7.1758E+03 |
| 14200 | 2.3922E-02 | 2.7674E+07 | 9.5364E+03 | 1.3591E-04 | 2.4121E+00 | 7.2664E+03 |
| 14300 | 2.3448E-02 | 2.8636E+07 | 9.6220E+03 | 1.3054E-04 | 2.4320E+00 | 7.3559E+03 |

TABLE A.1. (*Continued*)

| $T$ (K) | Density (kg/m$^3$) | Enthalpy (J/kg) | Sp heat (J/kg · K) | Viscosity (kg/m · s) | Therm cond (W/m · K) | Elec cond (A/V · m) |
|---|---|---|---|---|---|---|
| | | | **ARGON** | | | |
| 14400 | 2.2987E-02 | 2.9604E+07 | 9.6808E+03 | 1.2536E-04 | 2.4488E+00 | 7.4446E+03 |
| 14500 | 2.2539E-02 | 3.0575E+07 | 9.7118E+03 | 1.2038E-04 | 2.4625E+00 | 7.5322E+03 |
| 14600 | 2.2105E-02 | 3.1546E+07 | 9.7145E+03 | 1.1561E-04 | 2.4733E+00 | 7.6189E+03 |
| 14700 | 2.1685E-02 | 3.2515E+07 | 9.6888E+03 | 1.1105E-04 | 2.4813E+00 | 7.7047E+03 |
| 14800 | 2.1279E-02 | 3.3479E+07 | 9.6348E+03 | 1.0670E-04 | 2.4866E+00 | 7.7894E+03 |
| 14900 | 2.0887E-02 | 3.4434E+07 | 9.5531E+03 | 1.0257E-04 | 2.4894E+00 | 7.8732E+03 |
| 15000 | 2.0510E-02 | 3.5379E+07 | 9.4447E+03 | 9.8656E-05 | 2.4899E+00 | 7.9561E+03 |
| 15100 | 2.0147E-02 | 3.6310E+07 | 9.3111E+03 | 9.4960E-05 | 2.4884E+00 | 8.0380E+03 |
| 15200 | 1.9797E-02 | 3.7225E+07 | 9.1538E+03 | 9.1476E-05 | 2.4851E+00 | 8.1189E+03 |
| 15300 | 1.9462E-02 | 3.8123E+07 | 8.9748E+03 | 8.8201E-05 | 2.4803E+00 | 8.1990E+03 |
| 15400 | 1.9140E-02 | 3.9000E+07 | 8.7762E+03 | 8.5131E-05 | 2.4742E+00 | 8.2782E+03 |
| 15500 | 1.8832E-02 | 3.9856E+07 | 8.5605E+03 | 8.2258E-05 | 2.4671E+00 | 8.3565E+03 |
| 15600 | 1.8537E-02 | 4.0689E+07 | 8.3302E+03 | 7.9577E-05 | 2.4593E+00 | 8.4340E+03 |
| 15700 | 1.8254E-02 | 4.1498E+07 | 8.0876E+03 | 7.7081E-05 | 2.4511E+00 | 8.5107E+03 |
| 15800 | 1.7984E-02 | 4.2282E+07 | 7.8353E+03 | 7.4762E-05 | 2.4427E+00 | 8.5866E+03 |
| 15900 | 1.7725E-02 | 4.3039E+07 | 7.5760E+03 | 7.2613E-05 | 2.4343E+00 | 8.6617E+03 |
| 16000 | 1.7477E-02 | 4.3770E+07 | 7.3119E+03 | 7.0625E-05 | 2.4262E+00 | 8.7361E+03 |
| 16100 | 1.7241E-02 | 4.4475E+07 | 7.0453E+03 | 6.8791E-05 | 2.4186E+00 | 8.8099E+03 |
| 16200 | 1.7014E-02 | 4.5153E+07 | 6.7783E+03 | 6.7103E-05 | 2.4116E+00 | 8.8829E+03 |
| 16300 | 1.6798E-02 | 4.5804E+07 | 6.5129E+03 | 6.5553E-05 | 2.4054E+00 | 8.9554E+03 |
| 16400 | 1.6590E-02 | 4.6429E+07 | 6.2507E+03 | 6.4134E-05 | 2.4000E+00 | 9.0272E+03 |
| 16500 | 1.6392E-02 | 4.7028E+07 | 5.9933E+03 | 6.2838E-05 | 2.3958E+00 | 9.0985E+03 |
| 16600 | 1.6202E-02 | 4.7603E+07 | 5.7421E+03 | 6.1657E-05 | 2.3926E+00 | 9.1692E+03 |
| 16700 | 1.6020E-02 | 4.8152E+07 | 5.4981E+03 | 6.0585E-05 | 2.3906E+00 | 9.2394E+03 |
| 16800 | 1.5845E-02 | 4.8679E+07 | 5.2621E+03 | 5.9616E-05 | 2.3898E+00 | 9.3090E+03 |
| 16900 | 1.5677E-02 | 4.9182E+07 | 5.0353E+03 | 5.8741E-05 | 2.3902E+00 | 9.3782E+03 |
| 17000 | 1.5516E-02 | 4.9664E+07 | 4.8178E+03 | 5.7957E-05 | 2.3920E+00 | 9.4470E+03 |
| 17100 | 1.5361E-02 | 5.0125E+07 | 4.6104E+03 | 5.7255E-05 | 2.3950E+00 | 9.5153E+03 |
| 17200 | 1.5211E-02 | 5.0566E+07 | 4.4132E+03 | 5.6632E-05 | 2.3993E+00 | 9.5831E+03 |
| 17300 | 1.5068E-02 | 5.0989E+07 | 4.2265E+03 | 5.6081E-05 | 2.4048E+00 | 9.6506E+03 |
| 17400 | 1.4929E-02 | 5.1394E+07 | 4.0503E+03 | 5.5597E-05 | 2.4116E+00 | 9.7176E+03 |
| 17500 | 1.4795E-02 | 5.1782E+07 | 3.8846E+03 | 5.5177E-05 | 2.4197E+00 | 9.7843E+03 |
| 17600 | 1.4666E-02 | 5.2155E+07 | 3.7295E+03 | 5.4815E-05 | 2.4288E+00 | 9.8505E+03 |
| 17700 | 1.4540E-02 | 5.2514E+07 | 3.5848E+03 | 5.4506E-05 | 2.4392E+00 | 9.9164E+03 |
| 17800 | 1.4419E-02 | 5.2859E+07 | 3.4503E+03 | 5.4248E-05 | 2.4506E+00 | 9.9818E+03 |
| 17900 | 1.4301E-02 | 5.3191E+07 | 3.3258E+03 | 5.4035E-05 | 2.4631E+00 | 1.0047E+04 |
| 18000 | 1.4187E-02 | 5.3513E+07 | 3.2111E+03 | 5.3865E-05 | 2.4767E+00 | 1.0111E+04 |
| 18100 | 1.4076E-02 | 5.3823E+07 | 3.1061E+03 | 5.3734E-05 | 2.4912E+00 | 1.0176E+04 |
| 18200 | 1.3968E-02 | 5.4124E+07 | 3.0105E+03 | 5.3639E-05 | 2.5066E+00 | 1.0240E+04 |
| 18300 | 1.3863E-02 | 5.4417E+07 | 2.9240E+03 | 5.3576E-05 | 2.5230E+00 | 1.0303E+04 |
| 18400 | 1.3760E-02 | 5.4701E+07 | 2.8465E+03 | 5.3543E-05 | 2.5402E+00 | 1.0366E+04 |
| 18500 | 1.3660E-02 | 5.4979E+07 | 2.7777E+03 | 5.3537E-05 | 2.5582E+00 | 1.0428E+04 |
| 18600 | 1.3562E-02 | 5.5251E+07 | 2.7175E+03 | 5.3554E-05 | 2.5769E+00 | 1.0490E+04 |
| 18700 | 1.3466E-02 | 5.5517E+07 | 2.6657E+03 | 5.3592E-05 | 2.5964E+00 | 1.0551E+04 |
| 18800 | 1.3372E-02 | 5.5780E+07 | 2.6221E+03 | 5.3649E-05 | 2.6166E+00 | 1.0612E+04 |
| 18900 | 1.3280E-02 | 5.6038E+07 | 2.5865E+03 | 5.3722E-05 | 2.6375E+00 | 1.0672E+04 |
| 19000 | 1.3207E-02 | 5.6273E+07 | 2.3449E+03 | 5.4238E-05 | 2.6584E+00 | 1.0728E+04 |

(*continued*)

**TABLE A.1.** *(Continued)*

| $T$ (K) | Density (kg/m$^3$) | Enthalpy (J/kg) | Sp heat (J/kg · K) | Viscosity (kg/m · s) | Therm cond (W/m · K) | Elec cond (A/V · m) |
|---|---|---|---|---|---|---|
| | | | **ARGON** | | | |
| 19100 | 1.3118E-02 | 5.6526E+07 | 2.5353E+03 | 5.4333E-05 | 2.6805E+00 | 1.0787E+04 |
| 19200 | 1.3030E-02 | 5.6779E+07 | 2.5239E+03 | 5.4437E-05 | 2.7031E+00 | 1.0845E+04 |
| 19300 | 1.2944E-02 | 5.7031E+07 | 2.5201E+03 | 5.4547E-05 | 2.7263E+00 | 1.0902E+04 |
| 19400 | 1.2859E-02 | 5.7283E+07 | 2.5240E+03 | 5.4662E-05 | 2.7499E+00 | 1.0958E+04 |
| 19500 | 1.2775E-02 | 5.7537E+07 | 2.5358E+03 | 5.4778E-05 | 2.7740E+00 | 1.1014E+04 |
| 19600 | 1.2692E-02 | 5.7792E+07 | 2.5552E+03 | 5.4893E-05 | 2.7986E+00 | 1.1068E+04 |
| 19700 | 1.2610E-02 | 5.8050E+07 | 2.5825E+03 | 5.5005E-05 | 2.8235E+00 | 1.1122E+04 |
| 19800 | 1.2529E-02 | 5.8312E+07 | 2.6176E+03 | 5.5113E-05 | 2.8489E+00 | 1.1174E+04 |
| 19900 | 1.2448E-02 | 5.8578E+07 | 2.6606E+03 | 5.5213E-05 | 2.8746E+00 | 1.1225E+04 |
| 20000 | 1.2369E-02 | 5.8849E+07 | 2.7116E+03 | 5.5304E-05 | 2.9006E+00 | 1.1275E+04 |
| 20100 | 1.2289E-02 | 5.9126E+07 | 2.7707E+03 | 5.5382E-05 | 2.9270E+00 | 1.1323E+04 |
| 20200 | 1.2211E-02 | 5.9410E+07 | 2.8380E+03 | 5.5447E-05 | 2.9536E+00 | 1.1371E+04 |
| 20300 | 1.2132E-02 | 5.9702E+07 | 2.9136E+03 | 5.5496E-05 | 2.9805E+00 | 1.1416E+04 |
| 20400 | 1.2054E-02 | 6.0001E+07 | 2.9976E+03 | 5.5527E-05 | 3.0076E+00 | 1.1460E+04 |
| 20500 | 1.1977E-02 | 6.0310E+07 | 3.0901E+03 | 5.5537E-05 | 3.0350E+00 | 1.1502E+04 |
| 20600 | 1.1899E-02 | 6.0630E+07 | 3.1912E+03 | 5.5525E-05 | 3.0626E+00 | 1.1543E+04 |
| 20700 | 1.1822E-02 | 6.0960E+07 | 3.3011E+03 | 5.5489E-05 | 3.0904E+00 | 1.1582E+04 |
| 20800 | 1.1745E-02 | 6.1302E+07 | 3.4197E+03 | 5.5428E-05 | 3.1183E+00 | 1.1619E+04 |
| 20900 | 1.1667E-02 | 6.1656E+07 | 3.5472E+03 | 5.5338E-05 | 3.1464E+00 | 1.1654E+04 |
| 21000 | 1.1590E-02 | 6.2025E+07 | 3.6837E+03 | 5.5219E-05 | 3.1747E+00 | 1.1687E+04 |
| 21100 | 1.1513E-02 | 6.2408E+07 | 3.8290E+03 | 5.5069E-05 | 3.2030E+00 | 1.1718E+04 |
| 21200 | 1.1435E-02 | 6.2806E+07 | 3.9833E+03 | 5.4888E-05 | 3.2315E+00 | 1.1746E+04 |
| 21300 | 1.1357E-02 | 6.3221E+07 | 4.1465E+03 | 5.4673E-05 | 3.2600E+00 | 1.1773E+04 |
| 21400 | 1.1279E-02 | 6.3652E+07 | 4.3186E+03 | 5.4424E-05 | 3.2886E+00 | 1.1797E+04 |
| 21500 | 1.1201E-02 | 6.4102E+07 | 4.4994E+03 | 5.4140E-05 | 3.3173E+00 | 1.1819E+04 |
| 21600 | 1.1122E-02 | 6.4571E+07 | 4.6887E+03 | 5.3820E-05 | 3.3460E+00 | 1.1838E+04 |
| 21700 | 1.1043E-02 | 6.5060E+07 | 4.8865E+03 | 5.3465E-05 | 3.3747E+00 | 1.1856E+04 |
| 21800 | 1.0963E-02 | 6.5569E+07 | 5.0924E+03 | 5.3074E-05 | 3.4035E+00 | 1.1871E+04 |
| 21900 | 1.0883E-02 | 6.6100E+07 | 5.3062E+03 | 5.2648E-05 | 3.4322E+00 | 1.1883E+04 |
| 22000 | 1.0802E-02 | 6.6653E+07 | 5.5275E+03 | 5.2186E-05 | 3.4610E+00 | 1.1893E+04 |
| 22100 | 1.0721E-02 | 6.7228E+07 | 5.7559E+03 | 5.1690E-05 | 3.4897E+00 | 1.1901E+04 |
| 22200 | 1.0640E-02 | 6.7827E+07 | 5.9910E+03 | 5.1160E-05 | 3.5185E+00 | 1.1907E+04 |
| 22300 | 1.0558E-02 | 6.8450E+07 | 6.2322E+03 | 5.0598E-05 | 3.5472E+00 | 1.1911E+04 |
| 22400 | 1.0475E-02 | 6.9098E+07 | 6.4791E+03 | 5.0005E-05 | 3.5758E+00 | 1.1912E+04 |
| 22500 | 1.0392E-02 | 6.9771E+07 | 6.7309E+03 | 4.9382E-05 | 3.6044E+00 | 1.1912E+04 |
| 22600 | 1.0308E-02 | 7.0470E+07 | 6.9871E+03 | 4.8732E-05 | 3.6330E+00 | 1.1909E+04 |
| 22700 | 1.0224E-02 | 7.1195E+07 | 7.2468E+03 | 4.8056E-05 | 3.6616E+00 | 1.1905E+04 |
| 22800 | 1.0140E-02 | 7.1946E+07 | 7.5094E+03 | 4.7357E-05 | 3.6900E+00 | 1.1899E+04 |
| 22900 | 1.0055E-02 | 7.2723E+07 | 7.7741E+03 | 4.6636E-05 | 3.7185E+00 | 1.1891E+04 |
| 23000 | 9.9693E-03 | 7.3527E+07 | 8.0399E+03 | 4.5896E-05 | 3.7468E+00 | 1.1882E+04 |
| 23100 | 9.8835E-03 | 7.4358E+07 | 8.3059E+03 | 4.5139E-05 | 3.7751E+00 | 1.1872E+04 |
| 23200 | 9.7973E-03 | 7.5215E+07 | 8.5714E+03 | 4.4368E-05 | 3.8034E+00 | 1.1860E+04 |
| 23300 | 9.7200E-03 | 7.6059E+07 | 8.4441E+03 | 4.3898E-05 | 3.8331E+00 | 1.1841E+04 |
| 23400 | 9.6331E-03 | 7.6968E+07 | 9.0915E+03 | 4.3100E-05 | 3.8613E+00 | 1.1828E+04 |
| 23500 | 9.5460E-03 | 7.7903E+07 | 9.3495E+03 | 4.2296E-05 | 3.8895E+00 | 1.1813E+04 |
| 23600 | 9.4588E-03 | 7.8864E+07 | 9.6029E+03 | 4.1487E-05 | 3.9176E+00 | 1.1798E+04 |
| 23700 | 9.3716E-03 | 7.9849E+07 | 9.8509E+03 | 4.0676E-05 | 3.9456E+00 | 1.1783E+04 |
| 23800 | 9.2844E-03 | 8.0858E+07 | 1.0092E+04 | 3.9864E-05 | 3.9736E+00 | 1.1767E+04 |
| 23900 | 9.1973E-03 | 8.1891E+07 | 1.0327E+04 | 3.9055E-05 | 4.0016E+00 | 1.1751E+04 |
| 24000 | 9.1103E-03 | 8.2946E+07 | 1.0553E+04 | 3.8250E-05 | 4.0294E+00 | 1.1735E+04 |

**TABLE A.1.** (*Continued*)

| $T$ (K) | Density (kg/m$^3$) | Enthalpy (J/kg) | Sp heat (J/kg · K) | Viscosity (kg/m · s) | Therm cond (W/m · K) | Elec cond (A/V · m) |
|---|---|---|---|---|---|---|
| | | | **HELIUM** | | | |
| 500 | 9.7557E-02 | 4.1449E+06 | 5.1931E+03 | 2.8964E-05 | 2.2562E-01 | 2.9209E-24 |
| 600 | 8.1297E-02 | 4.6642E+06 | 5.1931E+03 | 3.3144E-05 | 2.5818E-01 | 2.6384E-24 |
| 700 | 6.9683E-02 | 5.1835E+06 | 5.1931E+03 | 3.7170E-05 | 2.8954E-01 | 2.4200E-24 |
| 800 | 6.0973E-02 | 5.7028E+06 | 5.1931E+03 | 4.1075E-05 | 3.1996E-01 | 2.2446E-24 |
| 900 | 5.4198E-02 | 6.2221E+06 | 5.1931E+03 | 4.4883E-05 | 3.4962E-01 | 2.0997E-24 |
| 1000 | 4.8778E-02 | 6.7415E+06 | 5.1931E+03 | 4.8608E-05 | 3.7864E-01 | 1.9776E-24 |
| 1100 | 4.4344E-02 | 7.2608E+06 | 5.1931E+03 | 5.2263E-05 | 4.0711E-01 | 1.8728E-24 |
| 1200 | 4.0649E-02 | 7.7801E+06 | 5.1931E+03 | 5.5856E-05 | 4.3510E-01 | 1.7818E-24 |
| 1300 | 3.7522E-02 | 8.2994E+06 | 5.1931E+03 | 5.9394E-05 | 4.6266E-01 | 1.7018E-24 |
| 1400 | 3.4842E-02 | 8.8187E+06 | 5.1931E+03 | 6.2883E-05 | 4.8984E-01 | 1.6308E-24 |
| 1500 | 3.2519E-02 | 9.3380E+06 | 5.1931E+03 | 6.6327E-05 | 5.1667E-01 | 1.5674E-24 |
| 1600 | 3.0486E-02 | 9.8573E+06 | 5.1931E+03 | 6.9731E-05 | 5.4318E-01 | 1.5102E-24 |
| 1700 | 2.8693E-02 | 1.0377E+07 | 5.1931E+03 | 7.3099E-05 | 5.6942E-01 | 1.4584E-24 |
| 1800 | 2.7099E-02 | 1.0896E+07 | 5.1931E+03 | 7.6432E-05 | 5.9538E-01 | 1.4112E-24 |
| 1900 | 2.5673E-02 | 1.1415E+07 | 5.1931E+03 | 7.9735E-05 | 6.2111E-01 | 1.3679E-24 |
| 2000 | 2.4389E-02 | 1.1935E+07 | 5.1931E+03 | 8.3008E-05 | 6.4661E-01 | 3.2311E-24 |
| 2100 | 2.3228E-02 | 1.2454E+07 | 5.1931E+03 | 8.6255E-05 | 6.7190E-01 | 9.7793E-23 |
| 2200 | 2.2172E-02 | 1.2973E+07 | 5.1931E+03 | 8.9477E-05 | 6.9700E-01 | 2.1741E-21 |
| 2300 | 2.1208E-02 | 1.3493E+07 | 5.1931E+03 | 9.2676E-05 | 7.2192E-01 | 3.6959E-20 |
| 2400 | 2.0324E-02 | 1.4012E+07 | 5.1931E+03 | 9.5854E-05 | 7.4667E-01 | 4.9677E-19 |
| 2500 | 1.9511E-02 | 1.4531E+07 | 5.1931E+03 | 9.9011E-05 | 7.7126E-01 | 5.4303E-18 |
| 2600 | 1.8761E-02 | 1.5050E+07 | 5.1931E+03 | 1.0215E-04 | 7.9571E-01 | 4.9438E-17 |
| 2700 | 1.8066E-02 | 1.5570E+07 | 5.1931E+03 | 1.0527E-04 | 8.2001E-01 | 3.8253E-16 |
| 2800 | 1.7421E-02 | 1.6089E+07 | 5.1931E+03 | 1.0837E-04 | 8.4418E-01 | 2.5598E-15 |
| 2900 | 1.6820E-02 | 1.6608E+07 | 5.1931E+03 | 1.1146E-04 | 8.6823E-01 | 1.5038E-14 |
| 3000 | 1.6259E-02 | 1.7128E+07 | 5.1931E+03 | 1.1453E-04 | 8.9216E-01 | 7.8572E-14 |
| 3100 | 1.5735E-02 | 1.7647E+07 | 5.1931E+03 | 1.1759E-04 | 9.1598E-01 | 3.6927E-13 |
| 3200 | 1.5243E-02 | 1.8166E+07 | 5.1931E+03 | 1.2063E-04 | 9.3970E-01 | 1.5766E-12 |
| 3300 | 1.4781E-02 | 1.8686E+07 | 5.1931E+03 | 1.2366E-04 | 9.6331E-01 | 6.1688E-12 |
| 3400 | 1.4347E-02 | 1.9205E+07 | 5.1931E+03 | 1.2668E-04 | 9.8683E-01 | 2.2289E-11 |
| 3500 | 1.3937E-02 | 1.9724E+07 | 5.1931E+03 | 1.2969E-04 | 1.0103E+00 | 7.4881E-11 |
| 3600 | 1.3550E-02 | 2.0244E+07 | 5.1931E+03 | 1.3269E-04 | 1.0336E+00 | 2.3532E-10 |
| 3700 | 1.3183E-02 | 2.0763E+07 | 5.1931E+03 | 1.3567E-04 | 1.0569E+00 | 6.9552E-10 |
| 3800 | 1.2836E-02 | 2.1282E+07 | 5.1931E+03 | 1.3865E-04 | 1.0800E+00 | 1.9427E-09 |
| 3900 | 1.2507E-02 | 2.1802E+07 | 5.1931E+03 | 1.4162E-04 | 1.1031E+00 | 5.1504E-09 |
| 4000 | 1.2195E-02 | 2.2321E+07 | 5.1931E+03 | 1.4457E-04 | 1.1262E+00 | 1.3011E-08 |
| 4100 | 1.1897E-02 | 2.2840E+07 | 5.1931E+03 | 1.4752E-04 | 1.1492E+00 | 3.1429E-08 |
| 4200 | 1.1614E-02 | 2.3359E+07 | 5.1931E+03 | 1.5046E-04 | 1.1721E+00 | 7.2830E-08 |
| 4300 | 1.1344E-02 | 2.3879E+07 | 5.1931E+03 | 1.5339E-04 | 1.1949E+00 | 1.6236E-07 |
| 4400 | 1.1086E-02 | 2.4398E+07 | 5.1931E+03 | 1.5632E-04 | 1.2177E+00 | 3.4914E-07 |
| 4500 | 1.0840E-02 | 2.4917E+07 | 5.1931E+03 | 1.5924E-04 | 1.2404E+00 | 7.2595E-07 |
| 4600 | 1.0604E-02 | 2.5437E+07 | 5.1931E+03 | 1.6215E-04 | 1.2631E+00 | 1.4627E-06 |
| 4700 | 1.0378E-02 | 2.5956E+07 | 5.1931E+03 | 1.6505E-04 | 1.2857E+00 | 2.8615E-06 |
| 4800 | 1.0162E-02 | 2.6475E+07 | 5.1931E+03 | 1.6795E-04 | 1.3082E+00 | 5.4454E-06 |
| 4900 | 9.9548E-03 | 2.6995E+07 | 5.1931E+03 | 1.7084E-04 | 1.3308E+00 | 1.0097E-05 |
| 5000 | 9.7557E-03 | 2.7514E+07 | 5.1931E+03 | 1.7372E-04 | 1.3532E+00 | 1.8272E-05 |

(*continued*)

**TABLE A.1.** *(Continued)*

| $T$ (K) | Density (kg/m$^3$) | Enthalpy (J/kg) | Sp heat (J/kg · K) | Viscosity (kg/m · s) | Therm cond (W/m · K) | Elec cond (A/V · m) |
|---|---|---|---|---|---|---|
| | | | **HELIUM** | | | |
| 5100 | 9.5644E-03 | 2.8033E+07 | 5.1931E+03 | 1.7660E-04 | 1.3757E+00 | 3.2315E-05 |
| 5200 | 9.3805E-03 | 2.8553E+07 | 5.1931E+03 | 1.7947E-04 | 1.3980E+00 | 5.5926E-05 |
| 5300 | 9.2035E-03 | 2.9072E+07 | 5.1931E+03 | 1.8234E-04 | 1.4204E+00 | 9.4831E-05 |
| 5400 | 9.0330E-03 | 2.9591E+07 | 5.1931E+03 | 1.8520E-04 | 1.4427E+00 | 1.5773E-04 |
| 5500 | 8.8688E-03 | 3.0111E+07 | 5.1931E+03 | 1.8806E-04 | 1.4649E+00 | 2.5761E-04 |
| 5600 | 8.7104E-03 | 3.0630E+07 | 5.1931E+03 | 1.9091E-04 | 1.4872E+00 | 4.1352E-04 |
| 5700 | 8.5576E-03 | 3.1149E+07 | 5.1931E+03 | 1.9376E-04 | 1.5094E+00 | 6.5303E-04 |
| 5800 | 8.4101E-03 | 3.1668E+07 | 5.1931E+03 | 1.9660E-04 | 1.5316E+00 | 1.0154E-03 |
| 5900 | 8.2675E-03 | 3.2188E+07 | 5.1931E+03 | 1.9944E-04 | 1.5537E+00 | 1.5556E-03 |
| 6000 | 8.1297E-03 | 3.2707E+07 | 5.1931E+03 | 2.0228E-04 | 1.5759E+00 | 2.3503E-03 |
| 6100 | 7.9965E-03 | 3.3226E+07 | 5.1931E+03 | 2.0511E-04 | 1.5981E+00 | 3.5040E-03 |
| 6200 | 7.8675E-03 | 3.3746E+07 | 5.1931E+03 | 2.0793E-04 | 1.6202E+00 | 5.1579E-03 |
| 6300 | 7.7426E-03 | 3.4265E+07 | 5.1931E+03 | 2.1076E-04 | 1.6424E+00 | 7.5016E-03 |
| 6400 | 7.6216E-03 | 3.4784E+07 | 5.1932E+03 | 2.1358E-04 | 1.6647E+00 | 1.0786E-02 |
| 6500 | 7.5044E-03 | 3.5304E+07 | 5.1932E+03 | 2.1639E-04 | 1.6871E+00 | 1.5338E-02 |
| 6600 | 7.3907E-03 | 3.5823E+07 | 5.1932E+03 | 2.1921E-04 | 1.7095E+00 | 2.1584E-02 |
| 6700 | 7.2804E-03 | 3.6342E+07 | 5.1932E+03 | 2.2202E-04 | 1.7321E+00 | 3.0068E-02 |
| 6800 | 7.1733E-03 | 3.6862E+07 | 5.1932E+03 | 2.2482E-04 | 1.7550E+00 | 4.1494E-02 |
| 6900 | 7.0693E-03 | 3.7381E+07 | 5.1933E+03 | 2.2762E-04 | 1.7781E+00 | 5.6736E-02 |
| 7000 | 6.9683E-03 | 3.7900E+07 | 5.1933E+03 | 2.3042E-04 | 1.8016E+00 | 7.6891E-02 |
| 7100 | 6.8702E-03 | 3.8420E+07 | 5.1934E+03 | 2.3322E-04 | 1.8256E+00 | 1.0334E-01 |
| 7200 | 6.7748E-03 | 3.8939E+07 | 5.1935E+03 | 2.3602E-04 | 1.8501E+00 | 1.3780E-01 |
| 7300 | 6.6820E-03 | 3.9458E+07 | 5.1936E+03 | 2.3881E-04 | 1.8755E+00 | 1.8231E-01 |
| 7400 | 6.5917E-03 | 3.9978E+07 | 5.1937E+03 | 2.4160E-04 | 1.9017E+00 | 2.3941E-01 |
| 7500 | 6.5038E-03 | 4.0497E+07 | 5.1939E+03 | 2.4438E-04 | 1.9292E+00 | 3.1217E-01 |
| 7600 | 6.4182E-03 | 4.1016E+07 | 5.1941E+03 | 2.4717E-04 | 1.9580E+00 | 4.0426E-01 |
| 7700 | 6.3348E-03 | 4.1536E+07 | 5.1943E+03 | 2.4995E-04 | 1.9886E+00 | 5.2009E-01 |
| 7800 | 6.2536E-03 | 4.2055E+07 | 5.1946E+03 | 2.5273E-04 | 1.9689E+00 | 6.6395E-01 |
| 7900 | 6.1745E-03 | 4.2575E+07 | 5.1950E+03 | 2.5550E-04 | 1.9905E+00 | 8.4338E-01 |
| 8000 | 6.0973E-03 | 4.3094E+07 | 5.1955E+03 | 2.5828E-04 | 2.0122E+00 | 1.0650E+00 |
| 8100 | 6.0220E-03 | 4.3614E+07 | 5.1960E+03 | 2.6105E-04 | 2.0339E+00 | 1.3372E+00 |
| 8200 | 5.9486E-03 | 4.4134E+07 | 5.1967E+03 | 2.6382E-04 | 2.0556E+00 | 1.6697E+00 |
| 8300 | 5.8769E-03 | 4.4653E+07 | 5.1975E+03 | 2.6659E-04 | 2.0772E+00 | 2.0738E+00 |
| 8400 | 5.8069E-03 | 4.5173E+07 | 5.1985E+03 | 2.6935E-04 | 2.0989E+00 | 2.5624E+00 |
| 8500 | 5.7386E-03 | 4.5693E+07 | 5.1996E+03 | 2.7212E-04 | 2.1206E+00 | 3.1504E+00 |
| 8600 | 5.6719E-03 | 4.6213E+07 | 5.2010E+03 | 2.7488E-04 | 2.1423E+00 | 3.8545E+00 |
| 8700 | 5.6067E-03 | 4.6734E+07 | 5.2025E+03 | 2.7764E-04 | 2.1641E+00 | 4.6939E+00 |
| 8800 | 5.5429E-03 | 4.7254E+07 | 5.2044E+03 | 2.8039E-04 | 2.1859E+00 | 5.6900E+00 |
| 8900 | 5.4807E-03 | 4.7775E+07 | 5.2066E+03 | 2.8315E-04 | 2.2077E+00 | 6.8669E+00 |
| 9000 | 5.4197E-03 | 4.8296E+07 | 5.2091E+03 | 2.8590E-04 | 2.2295E+00 | 8.2515E+00 |
| 9100 | 5.3602E-03 | 4.8817E+07 | 5.2120E+03 | 2.8866E-04 | 2.2514E+00 | 9.8736E+00 |
| 9200 | 5.3019E-03 | 4.9338E+07 | 5.2154E+03 | 2.9140E-04 | 2.2734E+00 | 1.1766E+01 |
| 9300 | 5.2449E-03 | 4.9860E+07 | 5.2193E+03 | 2.9415E-04 | 2.2955E+00 | 1.3966E+01 |
| 9400 | 5.1890E-03 | 5.0383E+07 | 5.2238E+03 | 2.9690E-04 | 2.3176E+00 | 1.6512E+01 |
| 9500 | 5.1344E-03 | 5.0906E+07 | 5.2289E+03 | 2.9964E-04 | 2.3399E+00 | 1.9448E+01 |
| 9600 | 5.0809E-03 | 5.1429E+07 | 5.2348E+03 | 3.0238E-04 | 2.3623E+00 | 2.2820E+01 |
| 9700 | 5.0285E-03 | 5.1953E+07 | 5.2414E+03 | 3.0512E-04 | 2.3848E+00 | 2.6680E+01 |

**TABLE A.1.** *(Continued)*

| $T$ (K) | Density (kg/m$^3$) | Enthalpy (J/kg) | Sp heat (J/kg·K) | Viscosity (kg/m·s) | Therm cond (W/m·K) | Elec cond (A/V·m) |
|---|---|---|---|---|---|---|
| | | | **HELIUM** | | | |
| 9800 | 4.9771E-03 | 5.2478E+07 | 5.2489E+03 | 3.0786E-04 | 2.4074E+00 | 3.1081E+01 |
| 9900 | 4.9268E-03 | 5.3004E+07 | 5.2574E+03 | 3.1059E-04 | 2.4303E+00 | 3.6081E+01 |
| 10000 | 4.8775E-03 | 5.3531E+07 | 5.2670E+03 | 3.1332E-04 | 2.4533E+00 | 4.1742E+01 |
| 10100 | 4.8291E-03 | 5.4058E+07 | 5.2778E+03 | 3.1605E-04 | 2.4766E+00 | 4.8129E+01 |
| 10200 | 4.7817E-03 | 5.4587E+07 | 5.2899E+03 | 3.1877E-04 | 2.5001E+00 | 5.5311E+01 |
| 10300 | 4.7352E-03 | 5.5118E+07 | 5.3034E+03 | 3.2149E-04 | 2.5238E+00 | 6.3359E+01 |
| 10400 | 4.6896E-03 | 5.5649E+07 | 5.3185E+03 | 3.2421E-04 | 2.5479E+00 | 7.2348E+01 |
| 10500 | 4.6448E-03 | 5.6183E+07 | 5.3353E+03 | 3.2693E-04 | 2.5722E+00 | 8.2357E+01 |
| 10600 | 4.6009E-03 | 5.6718E+07 | 5.3540E+03 | 3.2964E-04 | 2.5970E+00 | 9.3466E+01 |
| 10700 | 4.5577E-03 | 5.7256E+07 | 5.3746E+03 | 3.3235E-04 | 2.6221E+00 | 1.0576E+02 |
| 10800 | 4.5154E-03 | 5.7796E+07 | 5.3975E+03 | 3.3505E-04 | 2.6476E+00 | 1.1932E+02 |
| 10900 | 4.4738E-03 | 5.8338E+07 | 5.4228E+03 | 3.3775E-04 | 2.6736E+00 | 1.3424E+02 |
| 11000 | 4.4330E-03 | 5.8883E+07 | 5.4506E+03 | 3.4044E-04 | 2.7000E+00 | 1.5060E+02 |
| 11100 | 4.3928E-03 | 5.9431E+07 | 5.4812E+03 | 3.4312E-04 | 2.7270E+00 | 1.6851E+02 |
| 11200 | 4.3534E-03 | 5.9983E+07 | 5.5147E+03 | 3.4581E-04 | 2.7545E+00 | 1.8804E+02 |
| 11300 | 4.3146E-03 | 6.0538E+07 | 5.5515E+03 | 3.4848E-04 | 2.7827E+00 | 2.0930E+02 |
| 11400 | 4.2765E-03 | 6.1097E+07 | 5.5917E+03 | 3.5115E-04 | 2.8114E+00 | 2.3237E+02 |
| 11500 | 4.2390E-03 | 6.1660E+07 | 5.6355E+03 | 3.5381E-04 | 2.8409E+00 | 2.5736E+02 |
| 11600 | 4.2022E-03 | 6.2229E+07 | 5.6833E+03 | 3.5646E-04 | 2.8710E+00 | 2.8436E+02 |
| 11700 | 4.1659E-03 | 6.2802E+07 | 5.7353E+03 | 3.5910E-04 | 2.9019E+00 | 3.1346E+02 |
| 11800 | 4.1302E-03 | 6.3381E+07 | 5.7917E+03 | 3.6173E-04 | 2.9336E+00 | 3.4476E+02 |
| 11900 | 4.0951E-03 | 6.3967E+07 | 5.8529E+03 | 3.6435E-04 | 2.9662E+00 | 3.7835E+02 |
| 12000 | 4.0606E-03 | 6.4559E+07 | 5.9192E+03 | 3.6696E-04 | 2.9996E+00 | 4.1432E+02 |
| 12100 | 4.0265E-03 | 6.5158E+07 | 5.9908E+03 | 3.6955E-04 | 3.0340E+00 | 4.5276E+02 |
| 12200 | 3.9930E-03 | 6.5765E+07 | 6.0682E+03 | 3.7214E-04 | 3.0693E+00 | 4.9376E+02 |
| 12300 | 3.9599E-03 | 6.6380E+07 | 6.1515E+03 | 3.7470E-04 | 3.1057E+00 | 5.3739E+02 |
| 12400 | 3.9274E-03 | 6.7004E+07 | 6.2413E+03 | 3.7725E-04 | 3.1431E+00 | 5.8375E+02 |
| 12500 | 3.8953E-03 | 6.7638E+07 | 6.3378E+03 | 3.7979E-04 | 3.1816E+00 | 6.3289E+02 |
| 12600 | 3.8636E-03 | 6.8282E+07 | 6.4414E+03 | 3.8230E-04 | 3.2213E+00 | 6.8489E+02 |
| 12700 | 3.8324E-03 | 6.8937E+07 | 6.5525E+03 | 3.8479E-04 | 3.2622E+00 | 7.3981E+02 |
| 12800 | 3.8017E-03 | 6.9604E+07 | 6.6714E+03 | 3.8726E-04 | 3.3043E+00 | 7.9770E+02 |
| 12900 | 3.7713E-03 | 7.0284E+07 | 6.7987E+03 | 3.8971E-04 | 3.3478E+00 | 8.5862E+02 |
| 13000 | 3.7413E-03 | 7.0977E+07 | 6.9347E+03 | 3.9213E-04 | 3.3925E+00 | 9.2258E+02 |
| 13100 | 3.7117E-03 | 7.1685E+07 | 7.0798E+03 | 3.9452E-04 | 3.4387E+00 | 9.8964E+02 |
| 13200 | 3.6824E-03 | 7.2409E+07 | 7.2345E+03 | 3.9688E-04 | 3.4862E+00 | 1.0598E+03 |
| 13300 | 3.6535E-03 | 7.3149E+07 | 7.3992E+03 | 3.9921E-04 | 3.5353E+00 | 1.1331E+03 |
| 13400 | 3.6250E-03 | 7.3906E+07 | 7.5744E+03 | 4.0150E-04 | 3.5858E+00 | 1.2095E+03 |
| 13500 | 3.5968E-03 | 7.4682E+07 | 7.7605E+03 | 4.0376E-04 | 3.6379E+00 | 1.2890E+03 |
| 13600 | 3.5688E-03 | 7.5478E+07 | 7.9581E+03 | 4.0597E-04 | 3.6916E+00 | 1.3716E+03 |
| 13700 | 3.5412E-03 | 7.6295E+07 | 8.1677E+03 | 4.0813E-04 | 3.7469E+00 | 1.4572E+03 |
| 13800 | 3.5139E-03 | 7.7134E+07 | 8.3897E+03 | 4.1025E-04 | 3.8038E+00 | 1.5459E+03 |
| 13900 | 3.4868E-03 | 7.7996E+07 | 8.6246E+03 | 4.1232E-04 | 3.8625E+00 | 1.6375E+03 |
| 14000 | 3.4601E-03 | 7.8884E+07 | 8.8730E+03 | 4.1433E-04 | 3.9228E+00 | 1.7321E+03 |
| 14100 | 3.4335E-03 | 7.9797E+07 | 9.1355E+03 | 4.1627E-04 | 3.9849E+00 | 1.8295E+03 |
| 14200 | 3.4072E-03 | 8.0738E+07 | 9.4125E+03 | 4.1816E-04 | 4.0488E+00 | 1.9297E+03 |
| 14300 | 3.3811E-03 | 8.1709E+07 | 9.7046E+03 | 4.1997E-04 | 4.1144E+00 | 2.0325E+03 |
| 14400 | 3.3553E-03 | 8.2710E+07 | 1.0012E+04 | 4.2171E-04 | 4.1819E+00 | 2.1380E+03 |

*(continued)*

TABLE A.1. (*Continued*)

| $T$ (K) | Density (kg/m$^3$) | Enthalpy (J/kg) | Sp heat (J/kg · K) | Viscosity (kg/m · s) | Therm cond (W/m · K) | Elec cond (A/V · m) |
|---|---|---|---|---|---|---|
| | | | **HELIUM** | | | |
| 14500 | 3.3296E-03 | 8.3744E+07 | 1.0336E+04 | 4.2337E-04 | 4.2512E+00 | 2.2460E+03 |
| 14600 | 3.3042E-03 | 8.4811E+07 | 1.0677E+04 | 4.2494E-04 | 4.3223E+00 | 2.3565E+03 |
| 14700 | 3.2789E-03 | 8.5915E+07 | 1.1035E+04 | 4.2642E-04 | 4.3952E+00 | 2.4692E+03 |
| 14800 | 3.2538E-03 | 8.7056E+07 | 1.1412E+04 | 4.2780E-04 | 4.4700E+00 | 2.5842E+03 |
| 14900 | 3.2288E-03 | 8.8237E+07 | 1.1806E+04 | 4.2907E-04 | 4.5467E+00 | 2.7012E+03 |
| 15000 | 3.2040E-03 | 8.9459E+07 | 1.2220E+04 | 4.3024E-04 | 4.6251E+00 | 2.8202E+03 |
| 15100 | 3.1794E-03 | 9.0724E+07 | 1.2654E+04 | 4.3129E-04 | 4.7055E+00 | 2.9411E+03 |
| 15200 | 3.1549E-03 | 9.2035E+07 | 1.3108E+04 | 4.3221E-04 | 4.7876E+00 | 3.0638E+03 |
| 15300 | 3.1304E-03 | 9.3393E+07 | 1.3583E+04 | 4.3299E-04 | 4.8716E+00 | 3.1881E+03 |
| 15400 | 3.1061E-03 | 9.4801E+07 | 1.4080E+04 | 4.3364E-04 | 4.9574E+00 | 3.3139E+03 |
| 15500 | 3.0819E-03 | 9.6261E+07 | 1.4599E+04 | 4.3413E-04 | 5.0449E+00 | 3.4412E+03 |
| 15600 | 3.0578E-03 | 9.7775E+07 | 1.5141E+04 | 4.3447E-04 | 5.1342E+00 | 3.5697E+03 |
| 15700 | 3.0338E-03 | 9.9346E+07 | 1.5706E+04 | 4.3464E-04 | 5.2253E+00 | 3.6995E+03 |
| 15800 | 3.0098E-03 | 1.0098E+08 | 1.6295E+04 | 4.3464E-04 | 5.3180E+00 | 3.8303E+03 |
| 15900 | 2.9859E-03 | 1.0267E+08 | 1.6909E+04 | 4.3446E-04 | 5.4125E+00 | 3.9621E+03 |
| 16000 | 2.9620E-03 | 1.0442E+08 | 1.7549E+04 | 4.3408E-04 | 5.5086E+00 | 4.0948E+03 |
| 16100 | 2.9382E-03 | 1.0624E+08 | 1.8214E+04 | 4.3350E-04 | 5.6063E+00 | 4.2283E+03 |
| 16200 | 2.9145E-03 | 1.0813E+08 | 1.8906E+04 | 4.3272E-04 | 5.7056E+00 | 4.3625E+03 |
| 16300 | 2.8907E-03 | 1.1010E+08 | 1.9625E+04 | 4.3172E-04 | 5.8064E+00 | 4.4972E+03 |
| 16400 | 2.8670E-03 | 1.1213E+08 | 2.0372E+04 | 4.3050E-04 | 5.9088E+00 | 4.6325E+03 |
| 16500 | 2.8433E-03 | 1.1425E+08 | 2.1146E+04 | 4.2905E-04 | 6.0127E+00 | 4.7682E+03 |
| 16600 | 2.8195E-03 | 1.1644E+08 | 2.1950E+04 | 4.2737E-04 | 6.1180E+00 | 4.9043E+03 |
| 16700 | 2.7958E-03 | 1.1872E+08 | 2.2783E+04 | 4.2544E-04 | 6.2247E+00 | 5.0406E+03 |
| 16800 | 2.7721E-03 | 1.2109E+08 | 2.3646E+04 | 4.2326E-04 | 6.3329E+00 | 5.1771E+03 |
| 16900 | 2.7484E-03 | 1.2354E+08 | 2.4539E+04 | 4.2084E-04 | 6.4424E+00 | 5.3138E+03 |
| 17000 | 2.7246E-03 | 1.2608E+08 | 2.5441E+04 | 4.1815E-04 | 6.5523E+00 | 5.4505E+03 |
| 17100 | 2.7008E-03 | 1.2873E+08 | 2.6417E+04 | 4.1521E-04 | 6.6644E+00 | 5.5872E+03 |
| 17200 | 2.6770E-03 | 1.3147E+08 | 2.7403E+04 | 4.1201E-04 | 6.7779E+00 | 5.7239E+03 |
| 17300 | 2.6532E-03 | 1.3431E+08 | 2.8422E+04 | 4.0855E-04 | 6.8927E+00 | 5.8605E+03 |
| 17400 | 2.6293E-03 | 1.3726E+08 | 2.9473E+04 | 4.0483E-04 | 7.0087E+00 | 5.9970E+03 |
| 17500 | 2.6054E-03 | 1.4031E+08 | 3.0557E+04 | 4.0085E-04 | 7.1261E+00 | 6.1333E+03 |
| 17600 | 2.5814E-03 | 1.4348E+08 | 3.1674E+04 | 3.9661E-04 | 7.2447E+00 | 6.2693E+03 |
| 17700 | 2.5574E-03 | 1.4676E+08 | 3.2823E+04 | 3.9212E-04 | 7.3647E+00 | 6.4050E+03 |
| 17800 | 2.5334E-03 | 1.5016E+08 | 3.4006E+04 | 3.8739E-04 | 7.4859E+00 | 6.5405E+03 |
| 17900 | 2.5093E-03 | 1.5368E+08 | 3.5222E+04 | 3.8241E-04 | 7.6083E+00 | 6.6756E+03 |
| 18000 | 2.4851E-03 | 1.5733E+08 | 3.6471E+04 | 3.7721E-04 | 7.7321E+00 | 6.8103E+03 |
| 18100 | 2.4609E-03 | 1.6111E+08 | 3.7753E+04 | 3.7178E-04 | 7.8571E+00 | 6.9446E+03 |
| 18200 | 2.4366E-03 | 1.6501E+08 | 3.9068E+04 | 3.6614E-04 | 7.9835E+00 | 7.0785E+03 |
| 18300 | 2.4123E-03 | 1.6905E+08 | 4.0415E+04 | 3.6029E-04 | 8.1110E+00 | 7.2119E+03 |
| 18400 | 2.3880E-03 | 1.7323E+08 | 4.1794E+04 | 3.5426E-04 | 8.2398E+00 | 7.3449E+03 |
| 18500 | 2.3635E-03 | 1.7755E+08 | 4.3142E+04 | 3.4805E-04 | 8.3681E+00 | 7.4773E+03 |
| 18600 | 2.3391E-03 | 1.8201E+08 | 4.4641E+04 | 3.4169E-04 | 8.4992E+00 | 7.6092E+03 |
| 18700 | 2.3146E-03 | 1.8662E+08 | 4.6112E+04 | 3.3517E-04 | 8.6315E+00 | 7.7405E+03 |
| 18800 | 2.2901E-03 | 1.9138E+08 | 4.7612E+04 | 3.2853E-04 | 8.7648E+00 | 7.8713E+03 |
| 18900 | 2.2655E-03 | 1.9630E+08 | 4.9140E+04 | 3.2176E-04 | 8.8991E+00 | 8.0015E+03 |
| 19000 | 2.2410E-03 | 2.0137E+08 | 5.0694E+04 | 3.1490E-04 | 9.0343E+00 | 8.1311E+03 |

**TABLE A.1.** *(Continued)*

| $T$ (K) | Density (kg/m$^3$) | Enthalpy (J/kg) | Sp heat (J/kg·K) | Viscosity (kg/m·s) | Therm cond (W/m·K) | Elec cond (A/V·m) |
|---|---|---|---|---|---|---|
| | | | **HELIUM** | | | |
| 19100 | 2.2164E-03 | 2.0659E+08 | 5.2274E+04 | 3.0796E-04 | 9.1702E+00 | 8.2601E+03 |
| 19200 | 2.1918E-03 | 2.1198E+08 | 5.3877E+04 | 3.0095E-04 | 9.3068E+00 | 8.3884E+03 |
| 19300 | 2.1672E-03 | 2.1753E+08 | 5.5502E+04 | 2.9389E-04 | 9.4438E+00 | 8.5161E+03 |
| 19400 | 2.1426E-03 | 2.2325E+08 | 5.7146E+04 | 2.8679E-04 | 9.5812E+00 | 8.6431E+03 |
| 19500 | 2.1180E-03 | 2.2913E+08 | 5.8808E+04 | 2.7968E-04 | 9.7186E+00 | 8.7695E+03 |
| 19600 | 2.0934E-03 | 2.3518E+08 | 6.0485E+04 | 2.7257E-04 | 9.8558E+00 | 8.8952E+03 |
| 19700 | 2.0689E-03 | 2.4139E+08 | 6.2174E+04 | 2.6547E-04 | 9.9927E+00 | 9.0202E+03 |
| 19800 | 2.0444E-03 | 2.4778E+08 | 6.3872E+04 | 2.5839E-04 | 1.0129E+01 | 9.1444E+03 |
| 19900 | 2.0199E-03 | 2.5434E+08 | 6.5576E+04 | 2.5136E-04 | 1.0264E+01 | 9.2680E+03 |
| 20000 | 1.9955E-03 | 2.6107E+08 | 6.7283E+04 | 2.4438E-04 | 1.0398E+01 | 9.3908E+03 |
| 20100 | 1.9712E-03 | 2.6797E+08 | 6.8988E+04 | 2.3747E-04 | 1.0530E+01 | 9.5129E+03 |
| 20200 | 1.9470E-03 | 2.7503E+08 | 7.0688E+04 | 2.3063E-04 | 1.0660E+01 | 9.6342E+03 |
| 20300 | 1.9229E-03 | 2.8227E+08 | 7.2379E+04 | 2.2389E-04 | 1.0788E+01 | 9.7548E+03 |
| 20400 | 1.8989E-03 | 2.8968E+08 | 7.4056E+04 | 2.1724E-04 | 1.0913E+01 | 9.8746E+03 |
| 20500 | 1.8750E-03 | 2.9723E+08 | 7.5543E+04 | 2.1070E-04 | 1.1033E+01 | 9.9936E+03 |
| 20600 | 1.8512E-03 | 3.0497E+08 | 7.7343E+04 | 2.0428E-04 | 1.1151E+01 | 1.0112E+04 |
| 20700 | 1.8277E-03 | 3.1286E+08 | 7.8949E+04 | 1.9798E-04 | 1.1265E+01 | 1.0229E+04 |
| 20800 | 1.8042E-03 | 3.2091E+08 | 8.0523E+04 | 1.9181E-04 | 1.1374E+01 | 1.0346E+04 |
| 20900 | 1.7810E-03 | 3.2912E+08 | 8.2057E+04 | 1.8578E-04 | 1.1479E+01 | 1.0462E+04 |
| 21000 | 1.7580E-03 | 3.3747E+08 | 8.3546E+04 | 1.7988E-04 | 1.1578E+01 | 1.0577E+04 |
| 21100 | 1.7351E-03 | 3.4597E+08 | 8.4985E+04 | 1.7414E-04 | 1.1671E+01 | 1.0691E+04 |
| 21200 | 1.7125E-03 | 3.5461E+08 | 8.6367E+04 | 1.6854E-04 | 1.1758E+01 | 1.0804E+04 |
| 21300 | 1.6902E-03 | 3.6338E+08 | 8.7688E+04 | 1.6309E-04 | 1.1838E+01 | 1.0916E+04 |
| 21400 | 1.6681E-03 | 3.7227E+08 | 8.8941E+04 | 1.5779E-04 | 1.1912E+01 | 1.1028E+04 |
| 21500 | 1.6462E-03 | 3.8128E+08 | 9.0121E+04 | 1.5265E-04 | 1.1978E+01 | 1.1139E+04 |
| 21600 | 1.6246E-03 | 3.9041E+08 | 9.1222E+04 | 1.4766E-04 | 1.2036E+01 | 1.1249E+04 |
| 21700 | 1.6033E-03 | 3.9963E+08 | 9.2238E+04 | 1.4283E-04 | 1.2086E+01 | 1.1358E+04 |
| 21800 | 1.5824E-03 | 4.0895E+08 | 9.3165E+04 | 1.3815E-04 | 1.2127E+01 | 1.1466E+04 |
| 21900 | 1.5617E-03 | 4.1835E+08 | 9.3998E+04 | 1.3363E-04 | 1.2160E+01 | 1.1573E+04 |
| 22000 | 1.5413E-03 | 4.2782E+08 | 9.4731E+04 | 1.2926E-04 | 1.2185E+01 | 1.1679E+04 |
| 22100 | 1.5213E-03 | 4.3736E+08 | 9.5362E+04 | 1.2505E-04 | 1.2200E+01 | 1.1785E+04 |
| 22200 | 1.5016E-03 | 4.4694E+08 | 9.5885E+04 | 1.2098E-04 | 1.2207E+01 | 1.1890E+04 |
| 22300 | 1.4823E-03 | 4.5657E+08 | 9.6299E+04 | 1.1706E-04 | 1.2204E+01 | 1.1993E+04 |
| 22400 | 1.4633E-03 | 4.6623E+08 | 9.6599E+04 | 1.1329E-04 | 1.2193E+01 | 1.2096E+04 |
| 22500 | 1.4447E-03 | 4.7591E+08 | 9.6785E+04 | 1.0966E-04 | 1.2172E+01 | 1.2198E+04 |
| 22600 | 1.4265E-03 | 4.8560E+08 | 9.6854E+04 | 1.0618E-04 | 1.2143E+01 | 1.2300E+04 |
| 22700 | 1.4086E-03 | 4.9528E+08 | 9.6805E+04 | 1.0283E-04 | 1.2105E+01 | 1.2400E+04 |
| 22800 | 1.3911E-03 | 5.0494E+08 | 9.6639E+04 | 9.9612E-05 | 1.2058E+01 | 1.2499E+04 |
| 22900 | 1.3740E-03 | 5.1458E+08 | 9.6356E+04 | 9.6530E-05 | 1.2003E+01 | 1.2598E+04 |
| 23000 | 1.3573E-03 | 5.2417E+08 | 9.5956E+04 | 9.3576E-05 | 1.1941E+01 | 1.2696E+04 |
| 23100 | 1.3410E-03 | 5.3372E+08 | 9.5442E+04 | 9.0747E-05 | 1.1870E+01 | 1.2793E+04 |
| 23200 | 1.3250E-03 | 5.4320E+08 | 9.4816E+04 | 8.8039E-05 | 1.1793E+01 | 1.2889E+04 |
| 23300 | 1.3095E-03 | 5.5261E+08 | 9.4081E+04 | 8.5450E-05 | 1.1709E+01 | 1.2984E+04 |
| 23400 | 1.2943E-03 | 5.6193E+08 | 9.3240E+04 | 8.2975E-05 | 1.1618E+01 | 1.3079E+04 |
| 23500 | 1.2795E-03 | 5.7116E+08 | 9.2299E+04 | 8.0612E-05 | 1.1521E+01 | 1.3172E+04 |
| 23600 | 1.2651E-03 | 5.8029E+08 | 9.1262E+04 | 7.8356E-05 | 1.1420E+01 | 1.3265E+04 |
| 23700 | 1.2511E-03 | 5.8930E+08 | 9.0133E+04 | 7.6204E-05 | 1.1313E+01 | 1.3358E+04 |
| 23800 | 1.2374E-03 | 5.9819E+08 | 8.8919E+04 | 7.4152E-05 | 1.1202E+01 | 1.3449E+04 |
| 23900 | 1.2241E-03 | 6.0696E+08 | 8.7625E+04 | 7.2198E-05 | 1.1088E+01 | 1.3540E+04 |
| 24000 | 1.2112E-03 | 6.1558E+08 | 8.6258E+04 | 7.0338E-05 | 1.0970E+01 | 1.3630E+04 |

*(continued)*

TABLE A.1. (*Continued*)

| $T$ (K) | Density ($kg/m^3$) | Enthalpy (J/kg) | Sp heat (J/kg·K) | Viscosity (kg/m·s) | Therm cond (W/m·K) | Elec cond (A/V·m) |
|---|---|---|---|---|---|---|
| | | | **HYDROGEN** | | | |
| 500 | 4.8646E-02 | 3.0055E+06 | 1.4450E+04 | 1.2055E-05 | 2.5776E-01 | 1.6334E-24 |
| 600 | 4.0509E-02 | 4.5094E+06 | 1.5039E+04 | 1.3775E-05 | 2.9747E-01 | 1.4800E-24 |
| 700 | 3.4703E-02 | 6.0383E+06 | 1.5289E+04 | 1.5435E-05 | 3.3676E-01 | 1.3007E-24 |
| 800 | 3.0353E-02 | 7.5935E+06 | 1.5552E+04 | 1.7042E-05 | 3.7576E-01 | 1.1751E-24 |
| 900 | 2.6971E-02 | 9.1759E+06 | 1.5824E+04 | 1.8605E-05 | 4.1462E-01 | 1.0832E-24 |
| 1000 | 2.4267E-02 | 1.0786E+07 | 1.6105E+04 | 2.0132E-05 | 4.5350E-01 | 1.0105E-24 |
| 1100 | 2.2056E-02 | 1.2426E+07 | 1.6393E+04 | 2.1626E-05 | 4.9250E-01 | 9.4923E-25 |
| 1200 | 2.0214E-02 | 1.4094E+07 | 1.6686E+04 | 2.3094E-05 | 5.3169E-01 | 3.3235E-24 |
| 1300 | 1.8656E-02 | 1.5793E+07 | 1.6984E+04 | 2.4537E-05 | 5.7117E-01 | 2.0710E-23 |
| 1400 | 1.7321E-02 | 1.7521E+07 | 1.7286E+04 | 2.5960E-05 | 6.1114E-01 | 2.8983E-21 |
| 1500 | 1.6164E-02 | 1.9281E+07 | 1.7597E+04 | 2.7363E-05 | 6.5205E-01 | 2.4443E-19 |
| 1600 | 1.5152E-02 | 2.1074E+07 | 1.7924E+04 | 2.8748E-05 | 6.9487E-01 | 1.1868E-17 |
| 1700 | 1.4259E-02 | 2.2902E+07 | 1.8285E+04 | 3.0118E-05 | 7.4152E-01 | 3.6555E-16 |
| 1800 | 1.3464E-02 | 2.4773E+07 | 1.8711E+04 | 3.1474E-05 | 7.9535E-01 | 7.7048E-15 |
| 1900 | 1.2752E-02 | 2.6698E+07 | 1.9251E+04 | 3.2817E-05 | 8.6174E-01 | 1.1796E-13 |
| 2000 | 1.2109E-02 | 2.8696E+07 | 1.9972E+04 | 3.4148E-05 | 9.4869E-01 | 1.3757E-12 |
| 2100 | 1.1523E-02 | 3.0792E+07 | 2.0968E+04 | 3.5470E-05 | 1.0673E+00 | 1.2701E-11 |
| 2200 | 1.0986E-02 | 3.3028E+07 | 2.2354E+04 | 3.6784E-05 | 1.2323E+00 | 9.5760E-11 |
| 2300 | 1.0488E-02 | 3.5455E+07 | 2.4269E+04 | 3.8092E-05 | 1.4619E+00 | 6.0491E-10 |
| 2400 | 1.0022E-02 | 3.8142E+07 | 2.6870E+04 | 3.9399E-05 | 1.7776E+00 | 3.2696E-09 |
| 2500 | 9.5792E-03 | 4.1175E+07 | 3.0329E+04 | 4.0707E-05 | 2.2036E+00 | 1.5389E-08 |
| 2600 | 9.1543E-03 | 4.4658E+07 | 3.4832E+04 | 4.2019E-05 | 2.7655E+00 | 6.3998E-08 |
| 2700 | 8.7407E-03 | 4.8714E+07 | 4.0564E+04 | 4.3336E-05 | 3.4882E+00 | 2.3811E-07 |
| 2800 | 8.3331E-03 | 5.3485E+07 | 4.7710E+04 | 4.4656E-05 | 4.3932E+00 | 8.0105E-07 |
| 2900 | 7.9270E-03 | 5.9129E+07 | 5.6441E+04 | 4.5970E-05 | 5.4944E+00 | 2.4596E-06 |
| 3000 | 7.5189E-03 | 6.5819E+07 | 6.6895E+04 | 4.7260E-05 | 6.7934E+00 | 6.9504E-06 |
| 3100 | 7.1071E-03 | 7.3734E+07 | 7.9148E+04 | 4.8496E-05 | 8.2730E+00 | 1.8213E-05 |
| 3200 | 6.6914E-03 | 8.3050E+07 | 9.3162E+04 | 4.9635E-05 | 9.8906E+00 | 4.4568E-05 |
| 3300 | 6.2740E-03 | 9.3921E+07 | 1.0871E+05 | 5.0622E-05 | 1.1572E+01 | 1.0251E-04 |
| 3400 | 5.8588E-03 | 1.0645E+08 | 1.2528E+05 | 5.1397E-05 | 1.3208E+01 | 2.2296E-04 |
| 3500 | 5.4520E-03 | 1.2064E+08 | 1.4196E+05 | 5.1904E-05 | 1.4660E+01 | 4.6113E-04 |
| 3600 | 5.0607E-03 | 1.3639E+08 | 1.5742E+05 | 5.2110E-05 | 1.5770E+01 | 9.1166E-04 |
| 3700 | 4.6924E-03 | 1.5338E+08 | 1.6995E+05 | 5.2018E-05 | 1.6395E+01 | 1.7306E-03 |
| 3800 | 4.3540E-03 | 1.7116E+08 | 1.7777E+05 | 5.1670E-05 | 1.6440E+01 | 3.1670E-03 |
| 3900 | 4.0502E-03 | 1.8911E+08 | 1.7950E+05 | 5.1146E-05 | 1.5892E+01 | 5.6088E-03 |
| 4000 | 3.7833E-03 | 2.0657E+08 | 1.7463E+05 | 5.0545E-05 | 1.4830E+01 | 9.6400E-03 |
| 4100 | 3.5529E-03 | 2.2295E+08 | 1.6380E+05 | 4.9964E-05 | 1.3406E+01 | 1.6123E-02 |
| 4200 | 3.3562E-03 | 2.3782E+08 | 1.4864E+05 | 4.9482E-05 | 1.1802E+01 | 2.6303E-02 |
| 4300 | 3.1893E-03 | 2.5094E+08 | 1.3122E+05 | 4.9148E-05 | 1.0185E+01 | 4.1939E-02 |
| 4400 | 3.0477E-03 | 2.6229E+08 | 1.1348E+05 | 4.8984E-05 | 8.6772E+00 | 6.5466E-02 |
| 4500 | 2.9267E-03 | 2.7197E+08 | 9.6869E+04 | 4.8987E-05 | 7.3487E+00 | 1.0021E-01 |
| 4600 | 2.8224E-03 | 2.8019E+08 | 8.2223E+04 | 4.9146E-05 | 6.2253E+00 | 1.5061E-01 |
| 4700 | 2.7313E-03 | 2.8718E+08 | 6.9849E+04 | 4.9439E-05 | 5.3030E+00 | 2.2241E-01 |
| 4800 | 2.6508E-03 | 2.9315E+08 | 5.9704E+04 | 4.9845E-05 | 4.5613E+00 | 3.2337E-01 |
| 4900 | 2.5788E-03 | 2.9831E+08 | 5.1554E+04 | 5.0343E-05 | 3.9734E+00 | 4.6319E-01 |
| 5000 | 2.5135E-03 | 3.0281E+08 | 4.5094E+04 | 5.0916E-05 | 3.5120E+00 | 6.5421E-01 |

TABLE A.1. (Continued)

| $T$ (K) | Density (kg/m$^3$) | Enthalpy (J/kg) | Sp heat (J/kg · K) | Viscosity (kg/m · s) | Therm cond (W/m · K) | Elec cond (A/V · m) |
|---|---|---|---|---|---|---|
| | | | **HYDROGEN** | | | |
| 5100 | 2.4536E-03 | 3.0682E+08 | 4.0014E+04 | 5.1549E-05 | 3.1526E+00 | 9.1185E-01 |
| 5200 | 2.3983E-03 | 3.1042E+08 | 3.6037E+04 | 5.2229E-05 | 2.8740E+00 | 1.2552E+00 |
| 5300 | 2.3467E-03 | 3.1371E+08 | 3.2928E+04 | 5.2947E-05 | 2.6590E+00 | 1.7075E+00 |
| 5400 | 2.2982E-03 | 3.1676E+08 | 3.0496E+04 | 5.3695E-05 | 2.4941E+00 | 2.2971E+00 |
| 5500 | 2.2525E-03 | 3.1962E+08 | 2.8593E+04 | 5.4467E-05 | 2.3683E+00 | 3.0576E+00 |
| 5600 | 2.2092E-03 | 3.2233E+08 | 2.7099E+04 | 5.5258E-05 | 2.2732E+00 | 4.0292E+00 |
| 5700 | 2.1680E-03 | 3.2492E+08 | 2.5925E+04 | 5.6063E-05 | 2.2022E+00 | 5.2588E+00 |
| 5800 | 2.1286E-03 | 3.2742E+08 | 2.5000E+04 | 5.6882E-05 | 2.1504E+00 | 6.8011E+00 |
| 5900 | 2.0909E-03 | 3.2985E+08 | 2.4272E+04 | 5.7710E-05 | 2.1137E+00 | 8.7193E+00 |
| 6000 | 2.0547E-03 | 3.3222E+08 | 2.3701E+04 | 5.8545E-05 | 2.0890E+00 | 1.1085E+01 |
| 6100 | 2.0200E-03 | 3.3455E+08 | 2.3254E+04 | 5.9388E-05 | 2.0741E+00 | 1.3981E+01 |
| 6200 | 1.9865E-03 | 3.3684E+08 | 2.2908E+04 | 6.0235E-05 | 2.0670E+00 | 1.7498E+01 |
| 6300 | 1.9542E-03 | 3.3910E+08 | 2.2645E+04 | 6.1086E-05 | 2.0663E+00 | 2.1737E+01 |
| 6400 | 1.9230E-03 | 3.4135E+08 | 2.2452E+04 | 6.1940E-05 | 2.0709E+00 | 2.6813E+01 |
| 6500 | 1.8929E-03 | 3.4358E+08 | 2.2318E+04 | 6.2797E-05 | 2.0799E+00 | 3.2847E+01 |
| 6600 | 1.8637E-03 | 3.4580E+08 | 2.2235E+04 | 6.3656E-05 | 2.0927E+00 | 3.9973E+01 |
| 6700 | 1.8355E-03 | 3.4802E+08 | 2.2198E+04 | 6.4515E-05 | 2.1086E+00 | 4.8335E+01 |
| 6800 | 1.8081E-03 | 3.5024E+08 | 2.2204E+04 | 6.5375E-05 | 2.1274E+00 | 5.8086E+01 |
| 6900 | 1.7815E-03 | 3.5247E+08 | 2.2249E+04 | 6.6235E-05 | 2.1486E+00 | 6.9388E+01 |
| 7000 | 1.7557E-03 | 3.5470E+08 | 2.2331E+04 | 6.7094E-05 | 2.1721E+00 | 8.2411E+01 |
| 7100 | 1.7306E-03 | 3.5695E+08 | 2.2452E+04 | 6.7953E-05 | 2.1976E+00 | 9.7331E+01 |
| 7200 | 1.7063E-03 | 3.5921E+08 | 2.2597E+04 | 6.8809E-05 | 2.2248E+00 | 1.1433E+02 |
| 7300 | 1.6826E-03 | 3.6148E+08 | 2.2793E+04 | 6.9663E-05 | 2.2540E+00 | 1.3359E+02 |
| 7400 | 1.6595E-03 | 3.6379E+08 | 2.3022E+04 | 7.0514E-05 | 2.2849E+00 | 1.5530E+02 |
| 7500 | 1.6370E-03 | 3.6611E+08 | 2.3278E+04 | 7.1361E-05 | 2.3171E+00 | 1.7965E+02 |
| 7600 | 1.6151E-03 | 3.6847E+08 | 2.3595E+04 | 7.2204E-05 | 2.3514E+00 | 2.0682E+02 |
| 7700 | 1.5937E-03 | 3.7087E+08 | 2.3946E+04 | 7.3041E-05 | 2.3873E+00 | 2.3699E+02 |
| 7800 | 1.5728E-03 | 3.7330E+08 | 2.4340E+04 | 7.3872E-05 | 2.4249E+00 | 2.7033E+02 |
| 7900 | 1.5524E-03 | 3.7578E+08 | 2.4782E+04 | 7.4696E-05 | 2.4642E+00 | 3.0699E+02 |
| 8000 | 1.5325E-03 | 3.7831E+08 | 2.5305E+04 | 7.5511E-05 | 2.5061E+00 | 3.4714E+02 |
| 8100 | 1.5131E-03 | 3.8089E+08 | 2.5821E+04 | 7.6317E-05 | 2.5490E+00 | 3.9089E+02 |
| 8200 | 1.4940E-03 | 3.8353E+08 | 2.6419E+04 | 7.7112E-05 | 2.5936E+00 | 4.3834E+02 |
| 8300 | 1.4754E-03 | 3.8624E+08 | 2.7074E+04 | 7.7895E-05 | 2.6400E+00 | 4.8963E+02 |
| 8400 | 1.4571E-03 | 3.8902E+08 | 2.7789E+04 | 7.8663E-05 | 2.6882E+00 | 5.4481E+02 |
| 8500 | 1.4392E-03 | 3.9189E+08 | 2.8646E+04 | 7.9416E-05 | 2.7401E+00 | 6.0393E+02 |
| 8600 | 1.4216E-03 | 3.9483E+08 | 2.9425E+04 | 8.0151E-05 | 2.7920E+00 | 6.6702E+02 |
| 8700 | 1.4044E-03 | 3.9786E+08 | 3.0341E+04 | 8.0867E-05 | 2.8457E+00 | 7.3408E+02 |
| 8800 | 1.3874E-03 | 4.0101E+08 | 3.1455E+04 | 8.1560E-05 | 2.9039E+00 | 8.0508E+02 |
| 8900 | 1.3708E-03 | 4.0425E+08 | 3.2412E+04 | 8.2229E-05 | 2.9612E+00 | 8.7999E+02 |
| 9000 | 1.3544E-03 | 4.0762E+08 | 3.3729E+04 | 8.2870E-05 | 3.0237E+00 | 9.5873E+02 |
| 9100 | 1.3383E-03 | 4.1110E+08 | 3.4807E+04 | 8.3482E-05 | 3.0845E+00 | 1.0412E+03 |
| 9200 | 1.3224E-03 | 4.1474E+08 | 3.6353E+04 | 8.4061E-05 | 3.1514E+00 | 1.1273E+03 |
| 9300 | 1.3068E-03 | 4.1849E+08 | 3.7555E+04 | 8.4603E-05 | 3.2152E+00 | 1.2169E+03 |
| 9400 | 1.2913E-03 | 4.2243E+08 | 3.9355E+04 | 8.5105E-05 | 3.2862E+00 | 1.3098E+03 |
| 9500 | 1.2761E-03 | 4.2653E+08 | 4.1037E+04 | 8.5564E-05 | 3.3596E+00 | 1.4059E+03 |
| 9600 | 1.2610E-03 | 4.3077E+08 | 4.2410E+04 | 8.5976E-05 | 3.4274E+00 | 1.5051E+03 |
| 9700 | 1.2462E-03 | 4.3524E+08 | 4.4668E+04 | 8.6337E-05 | 3.5044E+00 | 1.6070E+03 |

(*continued*)

TABLE A.1. (*Continued*)

| $T$ (K) | Density ($kg/m^3$) | Enthalpy (J/kg) | Sp heat (J/kg·K) | Viscosity (kg/m·s) | Therm cond (W/m·K) | Elec cond (A/V·m) |
|---|---|---|---|---|---|---|
| | | | **HYDROGEN** | | | |
| 9800 | 1.2315E-03 | 4.3986E+08 | 4.6159E+04 | 8.6642E-05 | 3.5741E+00 | 1.7116E+03 |
| 9900 | 1.2169E-03 | 4.4473E+08 | 4.8749E+04 | 8.6889E-05 | 3.6542E+00 | 1.8186E+03 |
| 10000 | 1.2025E-03 | 4.4983E+08 | 5.1015E+04 | 8.7071E-05 | 3.7364E+00 | 1.9278E+03 |
| 10100 | 1.1882E-03 | 4.5510E+08 | 5.2644E+04 | 8.7187E-05 | 3.8078E+00 | 2.0391E+03 |
| 10200 | 1.1740E-03 | 4.6068E+08 | 5.5828E+04 | 8.7230E-05 | 3.8922E+00 | 2.1522E+03 |
| 10300 | 1.1599E-03 | 4.6653E+08 | 5.8490E+04 | 8.7198E-05 | 3.9784E+00 | 2.2669E+03 |
| 10400 | 1.1459E-03 | 4.7255E+08 | 6.0221E+04 | 8.7085E-05 | 4.0499E+00 | 2.3832E+03 |
| 10500 | 1.1321E-03 | 4.7896E+08 | 6.4090E+04 | 8.6890E-05 | 4.1374E+00 | 2.5006E+03 |
| 10600 | 1.1182E-03 | 4.8568E+08 | 6.7175E+04 | 8.6608E-05 | 4.2263E+00 | 2.6191E+03 |
| 10700 | 1.1045E-03 | 4.9272E+08 | 7.0414E+04 | 8.6237E-05 | 4.3165E+00 | 2.7386E+03 |
| 10800 | 1.0908E-03 | 4.9993E+08 | 7.2156E+04 | 8.5773E-05 | 4.3860E+00 | 2.8589E+03 |
| 10900 | 1.0772E-03 | 5.0765E+08 | 7.7139E+04 | 8.5216E-05 | 4.4764E+00 | 2.9798E+03 |
| 11000 | 1.0637E-03 | 5.1573E+08 | 8.0830E+04 | 8.4564E-05 | 4.5678E+00 | 3.1011E+03 |
| 11100 | 1.0501E-03 | 5.2420E+08 | 8.4682E+04 | 8.3816E-05 | 4.6602E+00 | 3.2228E+03 |
| 11200 | 1.0367E-03 | 5.3283E+08 | 8.6290E+04 | 8.2970E-05 | 4.7260E+00 | 3.3448E+03 |
| 11300 | 1.0232E-03 | 5.4209E+08 | 9.2583E+04 | 8.2030E-05 | 4.8180E+00 | 3.4669E+03 |
| 11400 | 1.0098E-03 | 5.5178E+08 | 9.6900E+04 | 8.0997E-05 | 4.9107E+00 | 3.5889E+03 |
| 11500 | 9.9643E-04 | 5.6192E+08 | 1.0138E+05 | 7.9873E-05 | 5.0042E+00 | 3.7109E+03 |
| 11600 | 9.8308E-04 | 5.7252E+08 | 1.0602E+05 | 7.8661E-05 | 5.0985E+00 | 3.8327E+03 |
| 11700 | 9.6975E-04 | 5.8360E+08 | 1.1083E+05 | 7.7365E-05 | 5.1936E+00 | 3.9542E+03 |
| 11800 | 9.5643E-04 | 5.9477E+08 | 1.1169E+05 | 7.5983E-05 | 5.2531E+00 | 4.0757E+03 |
| 11900 | 9.4315E-04 | 6.0682E+08 | 1.2049E+05 | 7.4531E-05 | 5.3476E+00 | 4.1966E+03 |
| 12000 | 9.2990E-04 | 6.1939E+08 | 1.2573E+05 | 7.3010E-05 | 5.4429E+00 | 4.3170E+03 |
| 12100 | 9.1666E-04 | 6.3250E+08 | 1.3112E+05 | 7.1427E-05 | 5.5390E+00 | 4.4370E+03 |
| 12200 | 9.0346E-04 | 6.4617E+08 | 1.3665E+05 | 6.9788E-05 | 5.6357E+00 | 4.5563E+03 |
| 12300 | 8.9028E-04 | 6.6040E+08 | 1.4231E+05 | 6.8100E-05 | 5.7330E+00 | 4.6751E+03 |
| 12400 | 8.7710E-04 | 6.7458E+08 | 1.4180E+05 | 6.6359E-05 | 5.7885E+00 | 4.7937E+03 |
| 12500 | 8.6399E-04 | 6.8993E+08 | 1.5346E+05 | 6.4594E-05 | 5.8852E+00 | 4.9113E+03 |
| 12600 | 8.5092E-04 | 7.0587E+08 | 1.5944E+05 | 6.2803E-05 | 5.9822E+00 | 5.0281E+03 |
| 12700 | 8.3789E-04 | 7.2242E+08 | 1.6552E+05 | 6.0992E-05 | 6.0793E+00 | 5.1442E+03 |
| 12800 | 8.2491E-04 | 7.3959E+08 | 1.7167E+05 | 5.9170E-05 | 6.1764E+00 | 5.2595E+03 |
| 12900 | 8.1198E-04 | 7.5738E+08 | 1.7789E+05 | 5.7343E-05 | 6.2730E+00 | 5.3741E+03 |
| 13000 | 7.9913E-04 | 7.7579E+08 | 1.8415E+05 | 5.5517E-05 | 6.3690E+00 | 5.4878E+03 |
| 13100 | 7.8634E-04 | 7.9484E+08 | 1.9044E+05 | 5.3700E-05 | 6.4638E+00 | 5.6008E+03 |
| 13200 | 7.7363E-04 | 8.1451E+08 | 1.9674E+05 | 5.1897E-05 | 6.5573E+00 | 5.7128E+03 |
| 13300 | 7.6092E-04 | 8.3375E+08 | 1.9240E+05 | 5.0088E-05 | 6.6034E+00 | 5.8248E+03 |
| 13400 | 7.4839E-04 | 8.5461E+08 | 2.0862E+05 | 4.8327E-05 | 6.6923E+00 | 5.9351E+03 |
| 13500 | 7.3595E-04 | 8.7609E+08 | 2.1479E+05 | 4.6596E-05 | 6.7784E+00 | 6.0446E+03 |
| 13600 | 7.2364E-04 | 8.9818E+08 | 2.2088E+05 | 4.4899E-05 | 6.8614E+00 | 6.1531E+03 |
| 13700 | 7.1144E-04 | 9.2087E+08 | 2.2685E+05 | 4.3239E-05 | 6.9406E+00 | 6.2607E+03 |
| 13800 | 6.9937E-04 | 9.4413E+08 | 2.3268E+05 | 4.1619E-05 | 7.0155E+00 | 6.3673E+03 |
| 13900 | 6.8745E-04 | 9.6797E+08 | 2.3833E+05 | 4.0042E-05 | 7.0856E+00 | 6.4730E+03 |
| 14000 | 6.7568E-04 | 9.9234E+08 | 2.4377E+05 | 3.8510E-05 | 7.1503E+00 | 6.5777E+03 |
| 14100 | 6.6406E-04 | 1.0172E+09 | 2.4898E+05 | 3.7026E-05 | 7.2091E+00 | 6.6814E+03 |
| 14200 | 6.5262E-04 | 1.0426E+09 | 2.5392E+05 | 3.5589E-05 | 7.2615E+00 | 6.7841E+03 |
| 14300 | 6.4135E-04 | 1.0685E+09 | 2.5856E+05 | 3.4202E-05 | 7.3069E+00 | 6.8858E+03 |
| 14400 | 6.3028E-04 | 1.0948E+09 | 2.6286E+05 | 3.2865E-05 | 7.3451E+00 | 6.9865E+03 |

**TABLE A.1.** *(Continued)*

| $T$ (K) | Density (kg/m$^3$) | Enthalpy (J/kg) | Sp heat (J/kg · K) | Viscosity (kg/m · s) | Therm cond (W/m · K) | Elec cond (A/V · m) |
|---|---|---|---|---|---|---|
| | | | **HYDROGEN** | | | |
| 14500 | 6.1939E-04 | 1.1215E+09 | 2.6681E+05 | 3.1579E-05 | 7.3755E+00 | 7.0862E+03 |
| 14600 | 6.0872E-04 | 1.1485E+09 | 2.7037E+05 | 3.0343E-05 | 7.3979E+00 | 7.1849E+03 |
| 14700 | 5.9825E-04 | 1.1758E+09 | 2.7351E+05 | 2.9157E-05 | 7.4120E+00 | 7.2826E+03 |
| 14800 | 5.8799E-04 | 1.2035E+09 | 2.7622E+05 | 2.8021E-05 | 7.4174E+00 | 7.3792E+03 |
| 14900 | 5.7797E-04 | 1.2313E+09 | 2.7846E+05 | 2.6935E-05 | 7.4142E+00 | 7.4749E+03 |
| 15000 | 5.6788E-04 | 1.2577E+09 | 2.6355E+05 | 2.5837E-05 | 7.3773E+00 | 7.5715E+03 |
| 15100 | 5.5830E-04 | 1.2858E+09 | 2.8143E+05 | 2.4844E-05 | 7.3583E+00 | 7.6652E+03 |
| 15200 | 5.4895E-04 | 1.3140E+09 | 2.8227E+05 | 2.3899E-05 | 7.3306E+00 | 7.7579E+03 |
| 15300 | 5.3984E-04 | 1.3423E+09 | 2.8260E+05 | 2.2999E-05 | 7.2945E+00 | 7.8497E+03 |
| 15400 | 5.3097E-04 | 1.3705E+09 | 2.8243E+05 | 2.2143E-05 | 7.2495E+00 | 7.9405E+03 |
| 15500 | 5.2235E-04 | 1.3987E+09 | 2.8175E+05 | 2.1331E-05 | 7.1972E+00 | 8.0303E+03 |
| 15600 | 5.1397E-04 | 1.4268E+09 | 2.8056E+05 | 2.0561E-05 | 7.1372E+00 | 8.1192E+03 |
| 15700 | 5.0583E-04 | 1.4547E+09 | 2.7889E+05 | 1.9830E-05 | 7.0701E+00 | 8.2071E+03 |
| 15800 | 4.9793E-04 | 1.4823E+09 | 2.7674E+05 | 1.9139E-05 | 6.9962E+00 | 8.2942E+03 |
| 15900 | 4.9028E-04 | 1.5097E+09 | 2.7414E+05 | 1.8486E-05 | 6.9161E+00 | 8.3803E+03 |
| 16000 | 4.8286E-04 | 1.5369E+09 | 2.7110E+05 | 1.7868E-05 | 6.8305E+00 | 8.4656E+03 |
| 16100 | 4.7568E-04 | 1.5636E+09 | 2.6765E+05 | 1.7285E-05 | 6.7398E+00 | 8.5500E+03 |
| 16200 | 4.6874E-04 | 1.5900E+09 | 2.6381E+05 | 1.6736E-05 | 6.6448E+00 | 8.6336E+03 |
| 16300 | 4.6202E-04 | 1.6160E+09 | 2.5963E+05 | 1.6218E-05 | 6.5459E+00 | 8.7164E+03 |
| 16400 | 4.5552E-04 | 1.6415E+09 | 2.5513E+05 | 1.5731E-05 | 6.4440E+00 | 8.7984E+03 |
| 16500 | 4.4924E-04 | 1.6665E+09 | 2.5034E+05 | 1.5273E-05 | 6.3396E+00 | 8.8797E+03 |
| 16600 | 4.4318E-04 | 1.6910E+09 | 2.4530E+05 | 1.4844E-05 | 6.2334E+00 | 8.9602E+03 |
| 16700 | 4.3732E-04 | 1.7150E+09 | 2.4004E+05 | 1.4440E-05 | 6.1259E+00 | 9.0400E+03 |
| 16800 | 4.3167E-04 | 1.7385E+09 | 2.3461E+05 | 1.4062E-05 | 6.0177E+00 | 9.1192E+03 |
| 16900 | 4.2621E-04 | 1.7614E+09 | 2.2902E+05 | 1.3708E-05 | 5.9095E+00 | 9.1977E+03 |
| 17000 | 4.2094E-04 | 1.7837E+09 | 2.2332E+05 | 1.3377E-05 | 5.8017E+00 | 9.2756E+03 |
| 17100 | 4.1585E-04 | 1.8055E+09 | 2.1754E+05 | 1.3068E-05 | 5.6947E+00 | 9.3529E+03 |
| 17200 | 4.1094E-04 | 1.8267E+09 | 2.1171E+05 | 1.2779E-05 | 5.5892E+00 | 9.4297E+03 |
| 17300 | 4.0620E-04 | 1.8473E+09 | 2.0585E+05 | 1.2510E-05 | 5.4854E+00 | 9.5059E+03 |
| 17400 | 4.0162E-04 | 1.8673E+09 | 2.0000E+05 | 1.2260E-05 | 5.3837E+00 | 9.5816E+03 |
| 17500 | 3.9720E-04 | 1.8867E+09 | 1.9418E+05 | 1.2027E-05 | 5.2845E+00 | 9.6568E+03 |
| 17600 | 3.9293E-04 | 1.9055E+09 | 1.8840E+05 | 1.1812E-05 | 5.1880E+00 | 9.7316E+03 |
| 17700 | 3.8881E-04 | 1.9238E+09 | 1.8270E+05 | 1.1612E-05 | 5.0945E+00 | 9.8059E+03 |
| 17800 | 3.8482E-04 | 1.9415E+09 | 1.7709E+05 | 1.1427E-05 | 5.0041E+00 | 9.8798E+03 |
| 17900 | 3.8096E-04 | 1.9586E+09 | 1.7158E+05 | 1.1256E-05 | 4.9171E+00 | 9.9534E+03 |
| 18000 | 3.7723E-04 | 1.9753E+09 | 1.6618E+05 | 1.1099E-05 | 4.8335E+00 | 1.0027E+04 |
| 18100 | 3.7362E-04 | 1.9914E+09 | 1.6092E+05 | 1.0955E-05 | 4.7535E+00 | 1.0099E+04 |
| 18200 | 3.7070E-04 | 2.0068E+09 | 1.5478E+05 | 1.0920E-05 | 4.6736E+00 | 1.0168E+04 |
| 18300 | 3.6731E-04 | 2.0219E+09 | 1.5019E+05 | 1.0799E-05 | 4.6010E+00 | 1.0241E+04 |
| 18400 | 3.6403E-04 | 2.0364E+09 | 1.4538E+05 | 1.0689E-05 | 4.5321E+00 | 1.0313E+04 |
| 18500 | 3.6084E-04 | 2.0505E+09 | 1.4074E+05 | 1.0590E-05 | 4.4669E+00 | 1.0384E+04 |
| 18600 | 3.5775E-04 | 2.0641E+09 | 1.3625E+05 | 1.0500E-05 | 4.4054E+00 | 1.0456E+04 |
| 18700 | 3.5475E-04 | 2.0773E+09 | 1.3193E+05 | 1.0419E-05 | 4.3475E+00 | 1.0527E+04 |
| 18800 | 3.5184E-04 | 2.0901E+09 | 1.2776E+05 | 1.0347E-05 | 4.2931E+00 | 1.0598E+04 |
| 18900 | 3.4901E-04 | 2.1024E+09 | 1.2376E+05 | 1.0284E-05 | 4.2423E+00 | 1.0669E+04 |
| 19000 | 3.4625E-04 | 2.1144E+09 | 1.1992E+05 | 1.0228E-05 | 4.1949E+00 | 1.0740E+04 |

*(continued)*

**TABLE A.1.** *(Continued)*

| T (K) | Density (kg/m$^3$) | Enthalpy (J/kg) | Sp heat (J/kg · K) | Viscosity (kg/m · s) | Therm cond (W/m · K) | Elec cond (A/V · m) |
|---|---|---|---|---|---|---|
| | | | **HYDROGEN** | | | |
| 19100 | 3.4357E-04 | 2.1261E+09 | 1.1624E+05 | 1.0179E-05 | 4.1509E+00 | 1.0811E+04 |
| 19200 | 3.4096E-04 | 2.1373E+09 | 1.1271E+05 | 1.0137E-05 | 4.1101E+00 | 1.0881E+04 |
| 19300 | 3.3842E-04 | 2.1483E+09 | 1.0934E+05 | 1.0102E-05 | 4.0725E+00 | 1.0951E+04 |
| 19400 | 3.3595E-04 | 2.1589E+09 | 1.0611E+05 | 1.0074E-05 | 4.0379E+00 | 1.1022E+04 |
| 19500 | 3.3353E-04 | 2.1692E+09 | 1.0303E+05 | 1.0051E-05 | 4.0063E+00 | 1.1092E+04 |
| 19600 | 3.3117E-04 | 2.1792E+09 | 1.0009E+05 | 1.0034E-05 | 3.9775E+00 | 1.1162E+04 |
| 19700 | 3.2887E-04 | 2.1889E+09 | 9.7283E+04 | 1.0022E-05 | 3.9514E+00 | 1.1232E+04 |
| 19800 | 3.2662E-04 | 2.1984E+09 | 9.4610E+04 | 1.0015E-05 | 3.9280E+00 | 1.1302E+04 |
| 19900 | 3.2442E-04 | 2.2076E+09 | 9.2064E+04 | 1.0013E-05 | 3.9071E+00 | 1.1372E+04 |
| 20000 | 3.2227E-04 | 2.2165E+09 | 8.9639E+04 | 1.0016E-05 | 3.8886E+00 | 1.1442E+04 |
| 20100 | 3.2017E-04 | 2.2253E+09 | 8.7331E+04 | 1.0023E-05 | 3.8724E+00 | 1.1512E+04 |
| 20200 | 3.1811E-04 | 2.2338E+09 | 8.5136E+04 | 1.0034E-05 | 3.8584E+00 | 1.1582E+04 |
| 20300 | 3.1610E-04 | 2.2421E+09 | 8.3048E+04 | 1.0050E-05 | 3.8465E+00 | 1.1652E+04 |
| 20400 | 3.1412E-04 | 2.2502E+09 | 8.1064E+04 | 1.0068E-05 | 3.8366E+00 | 1.1721E+04 |
| 20500 | 3.1218E-04 | 2.2581E+09 | 7.9177E+04 | 1.0091E-05 | 3.8286E+00 | 1.1791E+04 |
| 20600 | 3.1029E-04 | 2.2659E+09 | 7.7384E+04 | 1.0117E-05 | 3.8225E+00 | 1.1861E+04 |
| 20700 | 3.0842E-04 | 2.2734E+09 | 7.5681E+04 | 1.0146E-05 | 3.8180E+00 | 1.1931E+04 |
| 20800 | 3.0659E-04 | 2.2808E+09 | 7.4066E+04 | 1.0179E-05 | 3.8152E+00 | 1.2001E+04 |
| 20900 | 3.0480E-04 | 2.2881E+09 | 7.2527E+04 | 1.0214E-05 | 3.8140E+00 | 1.2071E+04 |
| 21000 | 3.0303E-04 | 2.2952E+09 | 7.1067E+04 | 1.0252E-05 | 3.8142E+00 | 1.2142E+04 |
| 21100 | 3.0130E-04 | 2.3022E+09 | 6.9681E+04 | 1.0293E-05 | 3.8159E+00 | 1.2212E+04 |
| 21200 | 2.9960E-04 | 2.3090E+09 | 6.8365E+04 | 1.0337E-05 | 3.8189E+00 | 1.2282E+04 |
| 21300 | 2.9792E-04 | 2.3157E+09 | 6.7116E+04 | 1.0383E-05 | 3.8231E+00 | 1.2353E+04 |
| 21400 | 2.9627E-04 | 2.3223E+09 | 6.5929E+04 | 1.0431E-05 | 3.8285E+00 | 1.2423E+04 |
| 21500 | 2.9465E-04 | 2.3288E+09 | 6.4802E+04 | 1.0482E-05 | 3.8351E+00 | 1.2494E+04 |
| 21600 | 2.9305E-04 | 2.3352E+09 | 6.3732E+04 | 1.0535E-05 | 3.8427E+00 | 1.2564E+04 |
| 21700 | 2.9148E-04 | 2.3414E+09 | 6.2715E+04 | 1.0590E-05 | 3.8513E+00 | 1.2635E+04 |
| 21800 | 2.8993E-04 | 2.3476E+09 | 6.1750E+04 | 1.0647E-05 | 3.8608E+00 | 1.2706E+04 |
| 21900 | 2.8841E-04 | 2.3537E+09 | 6.0833E+04 | 1.0706E-05 | 3.8713E+00 | 1.2777E+04 |
| 22000 | 2.8690E-04 | 2.3597E+09 | 5.9962E+04 | 1.0767E-05 | 3.8825E+00 | 1.2849E+04 |
| 22100 | 2.8587E-04 | 2.3643E+09 | 4.6499E+04 | 1.0932E-05 | 3.8807E+00 | 1.2924E+04 |
| 22200 | 2.8440E-04 | 2.3702E+09 | 5.8319E+04 | 1.0996E-05 | 3.8926E+00 | 1.2996E+04 |
| 22300 | 2.8296E-04 | 2.3759E+09 | 5.7573E+04 | 1.1062E-05 | 3.9051E+00 | 1.3068E+04 |
| 22400 | 2.8153E-04 | 2.3816E+09 | 5.6869E+04 | 1.1129E-05 | 3.9183E+00 | 1.3141E+04 |
| 22500 | 2.8012E-04 | 2.3872E+09 | 5.6199E+04 | 1.1198E-05 | 3.9319E+00 | 1.3214E+04 |
| 22600 | 2.7873E-04 | 2.3928E+09 | 5.5562E+04 | 1.1269E-05 | 3.9461E+00 | 1.3287E+04 |
| 22700 | 2.7735E-04 | 2.3983E+09 | 5.4955E+04 | 1.1341E-05 | 3.9608E+00 | 1.3360E+04 |
| 22800 | 2.7600E-04 | 2.4037E+09 | 5.4379E+04 | 1.1415E-05 | 3.9759E+00 | 1.3433E+04 |
| 22900 | 2.7466E-04 | 2.4091E+09 | 5.3830E+04 | 1.1490E-05 | 3.9914E+00 | 1.3507E+04 |
| 23000 | 2.7334E-04 | 2.4144E+09 | 5.3308E+04 | 1.1566E-05 | 4.0073E+00 | 1.3581E+04 |
| 23100 | 2.7203E-04 | 2.4197E+09 | 5.2811E+04 | 1.1644E-05 | 4.0235E+00 | 1.3656E+04 |
| 23200 | 2.7074E-04 | 2.4249E+09 | 5.2338E+04 | 1.1722E-05 | 4.0400E+00 | 1.3730E+04 |
| 23300 | 2.6946E-04 | 2.4301E+09 | 5.1887E+04 | 1.1803E-05 | 4.0567E+00 | 1.3805E+04 |
| 23400 | 2.6820E-04 | 2.4353E+09 | 5.1458E+04 | 1.1884E-05 | 4.0737E+00 | 1.3881E+04 |
| 23500 | 2.6696E-04 | 2.4404E+09 | 5.1049E+04 | 1.1966E-05 | 4.0909E+00 | 1.3956E+04 |
| 23600 | 2.6572E-04 | 2.4454E+09 | 5.0660E+04 | 1.2050E-05 | 4.1083E+00 | 1.4032E+04 |
| 23700 | 2.6451E-04 | 2.4505E+09 | 5.0288E+04 | 1.2135E-05 | 4.1259E+00 | 1.4109E+04 |
| 23800 | 2.6330E-04 | 2.4555E+09 | 4.9934E+04 | 1.2221E-05 | 4.1435E+00 | 1.4186E+04 |
| 23900 | 2.6211E-04 | 2.4604E+09 | 4.9597E+04 | 1.2308E-05 | 4.1613E+00 | 1.4263E+04 |
| 24000 | 2.6093E-04 | 2.4654E+09 | 4.9275E+04 | 1.2396E-05 | 4.1791E+00 | 1.4340E+04 |

## TABLE A.1. (*Continued*)

| $T$ (K) | Density (kg/m$^3$) | Enthalpy (J/kg) | Sp heat (J/kg · K) | Viscosity (kg/m · s) | Therm cond (W/m · K) | Elec cond (A/V · m) |
|---|---|---|---|---|---|---|
| | | | **NITROGEN** | | | |
| 500 | 6.8258E-01 | 2.1308E+05 | 1.0668E+03 | 2.4535E-05 | 3.8606E-02 | 4.7840E-25 |
| 600 | 5.6880E-01 | 3.2206E+05 | 1.0897E+03 | 2.7964E-05 | 4.4988E-02 | 4.5230E-25 |
| 700 | 4.8755E-01 | 4.3263E+05 | 1.1057E+03 | 3.1249E-05 | 5.1294E-02 | 4.3347E-25 |
| 800 | 4.2661E-01 | 5.4479E+05 | 1.1216E+03 | 3.4422E-05 | 5.7542E-02 | 4.1944E-25 |
| 900 | 3.7922E-01 | 6.5852E+05 | 1.1373E+03 | 3.7504E-05 | 6.3739E-02 | 4.0872E-25 |
| 1000 | 3.4131E-01 | 7.7378E+05 | 1.1526E+03 | 4.0508E-05 | 6.9886E-02 | 4.0034E-25 |
| 1100 | 3.1029E-01 | 8.9052E+05 | 1.1675E+03 | 4.3445E-05 | 7.5984E-02 | 3.9365E-25 |
| 1200 | 2.8444E-01 | 1.0087E+06 | 1.1819E+03 | 4.6324E-05 | 8.2030E-02 | 3.8820E-25 |
| 1300 | 2.6256E-01 | 1.1283E+06 | 1.1957E+03 | 4.9150E-05 | 8.8022E-02 | 3.8364E-25 |
| 1400 | 2.4381E-01 | 1.2492E+06 | 1.2090E+03 | 5.1929E-05 | 9.3960E-02 | 1.4059E-22 |
| 1500 | 2.2756E-01 | 1.3713E+06 | 1.2216E+03 | 5.4666E-05 | 9.9840E-02 | 2.7524E-20 |
| 1600 | 2.1334E-01 | 1.4947E+06 | 1.2336E+03 | 5.7365E-05 | 1.0566E-01 | 1.2953E-18 |
| 1700 | 2.0079E-01 | 1.6192E+06 | 1.2449E+03 | 6.0028E-05 | 1.1143E-01 | 3.8527E-17 |
| 1800 | 1.8964E-01 | 1.7447E+06 | 1.2555E+03 | 6.2658E-05 | 1.1713E-01 | 7.8891E-16 |
| 1900 | 1.7966E-01 | 1.8713E+06 | 1.2655E+03 | 6.5259E-05 | 1.2277E-01 | 1.1796E-14 |
| 2000 | 1.7068E-01 | 1.9988E+06 | 1.2747E+03 | 6.7831E-05 | 1.2836E-01 | 1.3498E-13 |
| 2100 | 1.6255E-01 | 2.1271E+06 | 1.2833E+03 | 7.0378E-05 | 1.3388E-01 | 1.2280E-12 |
| 2200 | 1.5517E-01 | 2.2562E+06 | 1.2912E+03 | 7.2900E-05 | 1.3935E-01 | 9.1626E-12 |
| 2300 | 1.4842E-01 | 2.3860E+06 | 1.2984E+03 | 7.5400E-05 | 1.4477E-01 | 5.7535E-11 |
| 2400 | 1.4224E-01 | 2.5165E+06 | 1.3050E+03 | 7.7878E-05 | 1.5012E-01 | 3.1064E-10 |
| 2500 | 1.3655E-01 | 2.6476E+06 | 1.3110E+03 | 8.0336E-05 | 1.5543E-01 | 1.4683E-09 |
| 2600 | 1.3130E-01 | 2.7793E+06 | 1.3163E+03 | 8.2774E-05 | 1.6069E-01 | 6.1693E-09 |
| 2700 | 1.2644E-01 | 2.9114E+06 | 1.3212E+03 | 8.5196E-05 | 1.6591E-01 | 2.3344E-08 |
| 2800 | 1.2192E-01 | 3.0439E+06 | 1.3256E+03 | 8.7600E-05 | 1.7111E-01 | 8.0444E-08 |
| 2900 | 1.1772E-01 | 3.1769E+06 | 1.3296E+03 | 8.9988E-05 | 1.7630E-01 | 2.5488E-07 |
| 3000 | 1.1379E-01 | 3.3103E+06 | 1.3334E+03 | 9.2360E-05 | 1.8152E-01 | 7.4880E-07 |
| 3100 | 1.1012E-01 | 3.4440E+06 | 1.3371E+03 | 9.4718E-05 | 1.8680E-01 | 2.0546E-06 |
| 3200 | 1.0668E-01 | 3.5781E+06 | 1.3409E+03 | 9.7063E-05 | 1.9222E-01 | 5.2989E-06 |
| 3300 | 1.0345E-01 | 3.7126E+06 | 1.3452E+03 | 9.9394E-05 | 1.9787E-01 | 1.2918E-05 |
| 3400 | 1.0040E-01 | 3.8476E+06 | 1.3504E+03 | 1.0171E-04 | 2.0386E-01 | 2.9919E-05 |
| 3500 | 9.7529E-02 | 3.9833E+06 | 1.3570E+03 | 1.0402E-04 | 2.1037E-01 | 6.6122E-05 |
| 3600 | 9.4814E-02 | 4.1199E+06 | 1.3657E+03 | 1.0632E-04 | 2.1761E-01 | 1.4000E-04 |
| 3700 | 9.2243E-02 | 4.2576E+06 | 1.3772E+03 | 1.0861E-04 | 2.2586E-01 | 2.8499E-04 |
| 3800 | 8.9803E-02 | 4.3969E+06 | 1.3927E+03 | 1.1088E-04 | 2.3548E-01 | 5.5961E-04 |
| 3900 | 8.7483E-02 | 4.5382E+06 | 1.4134E+03 | 1.1316E-04 | 2.4689E-01 | 1.0630E-03 |
| 4000 | 8.5273E-02 | 4.6823E+06 | 1.4406E+03 | 1.1542E-04 | 2.6061E-01 | 1.9588E-03 |
| 4100 | 8.3162E-02 | 4.8299E+06 | 1.4761E+03 | 1.1768E-04 | 2.7728E-01 | 3.5096E-03 |
| 4200 | 8.1141E-02 | 4.9821E+06 | 1.5218E+03 | 1.1994E-04 | 2.9762E-01 | 6.1282E-03 |
| 4300 | 7.9201E-02 | 5.1400E+06 | 1.5799E+03 | 1.2220E-04 | 3.2249E-01 | 1.0450E-02 |
| 4400 | 7.7333E-02 | 5.3053E+06 | 1.6528E+03 | 1.2446E-04 | 3.5285E-01 | 1.7435E-02 |
| 4500 | 7.5528E-02 | 5.4797E+06 | 1.7433E+03 | 1.2672E-04 | 3.8980E-01 | 2.8511E-02 |
| 4600 | 7.3779E-02 | 5.6651E+06 | 1.8544E+03 | 1.2900E-04 | 4.3457E-01 | 4.5772E-02 |
| 4700 | 7.2077E-02 | 5.8640E+06 | 1.9894E+03 | 1.3129E-04 | 4.8850E-01 | 7.2240E-02 |
| 4800 | 7.0415E-02 | 6.0792E+06 | 2.1517E+03 | 1.3360E-04 | 5.5303E-01 | 1.1226E-01 |
| 4900 | 6.8783E-02 | 6.3137E+06 | 2.3451E+03 | 1.3593E-04 | 6.2971E-01 | 1.7193E-01 |
| 5000 | 6.7175E-02 | 6.5711E+06 | 2.5737E+03 | 1.3829E-04 | 7.2015E-01 | 2.5990E-01 |

(*continued*)

## TABLE A.1. (*Continued*)

| $T$ (K) | Density (kg/m$^3$) | Enthalpy (J/kg) | Sp heat (J/kg · K) | Viscosity (kg/m · s) | Therm cond (W/m · K) | Elec cond (A/V · m) |
|---|---|---|---|---|---|---|
| | | | **NITROGEN** | | | |
| 5100 | 6.5583E-02 | 6.8552E+06 | 2.8416E+03 | 1.4069E-04 | 8.2600E-01 | 3.8809E-01 |
| 5200 | 6.3999E-02 | 7.1706E+06 | 3.1532E+03 | 1.4313E-04 | 9.4901E-01 | 5.7239E-01 |
| 5300 | 6.2417E-02 | 7.5218E+06 | 3.5129E+03 | 1.4562E-04 | 1.0906E+00 | 8.3606E-01 |
| 5400 | 6.0829E-02 | 7.9144E+06 | 3.9252E+03 | 1.4817E-04 | 1.2525E+00 | 1.2080E+00 |
| 5500 | 5.9231E-02 | 8.3538E+06 | 4.3946E+03 | 1.5078E-04 | 1.4357E+00 | 1.7305E+00 |
| 5600 | 5.7617E-02 | 8.8464E+06 | 4.9253E+03 | 1.5345E-04 | 1.6411E+00 | 2.4577E+00 |
| 5700 | 5.5982E-02 | 9.3984E+06 | 5.5209E+03 | 1.5620E-04 | 1.8693E+00 | 3.4617E+00 |
| 5800 | 5.4325E-02 | 1.0017E+07 | 6.1847E+03 | 1.5902E-04 | 2.1201E+00 | 4.8374E+00 |
| 5900 | 5.2642E-02 | 1.0709E+07 | 6.9186E+03 | 1.6191E-04 | 2.3926E+00 | 6.7085E+00 |
| 6000 | 5.0935E-02 | 1.1481E+07 | 7.7231E+03 | 1.6487E-04 | 2.6848E+00 | 9.2339E+00 |
| 6100 | 4.9205E-02 | 1.2341E+07 | 8.5962E+03 | 1.6789E-04 | 2.9936E+00 | 1.2616E+01 |
| 6200 | 4.7456E-02 | 1.3294E+07 | 9.5333E+03 | 1.7095E-04 | 3.3145E+00 | 1.7105E+01 |
| 6300 | 4.5694E-02 | 1.4347E+07 | 1.0526E+04 | 1.7404E-04 | 3.6415E+00 | 2.3011E+01 |
| 6400 | 4.3927E-02 | 1.5503E+07 | 1.1560E+04 | 1.7713E-04 | 3.9669E+00 | 3.0701E+01 |
| 6500 | 4.2163E-02 | 1.6764E+07 | 1.2616E+04 | 1.8019E-04 | 4.2816E+00 | 4.0600E+01 |
| 6600 | 4.0416E-02 | 1.8131E+07 | 1.3669E+04 | 1.8320E-04 | 4.5747E+00 | 5.3179E+01 |
| 6700 | 3.8698E-02 | 1.9600E+07 | 1.4686E+04 | 1.8611E-04 | 4.8348E+00 | 6.8935E+01 |
| 6800 | 3.7022E-02 | 2.1163E+07 | 1.5628E+04 | 1.8890E-04 | 5.0500E+00 | 8.8365E+01 |
| 6900 | 3.5401E-02 | 2.2808E+07 | 1.6453E+04 | 1.9154E-04 | 5.2094E+00 | 1.1193E+02 |
| 7000 | 3.3848E-02 | 2.4519E+07 | 1.7115E+04 | 1.9400E-04 | 5.3038E+00 | 1.4001E+02 |
| 7100 | 3.2376E-02 | 2.6277E+07 | 1.7574E+04 | 1.9629E-04 | 5.3271E+00 | 1.7290E+02 |
| 7200 | 3.0992E-02 | 2.8056E+07 | 1.7797E+04 | 1.9840E-04 | 5.2773E+00 | 2.1079E+02 |
| 7300 | 2.9704E-02 | 2.9833E+07 | 1.7764E+04 | 2.0034E-04 | 5.1570E+00 | 2.5374E+02 |
| 7400 | 2.8515E-02 | 3.1580E+07 | 1.7474E+04 | 2.0212E-04 | 4.9733E+00 | 3.0174E+02 |
| 7500 | 2.7427E-02 | 3.3274E+07 | 1.6942E+04 | 2.0378E-04 | 4.7373E+00 | 3.5469E+02 |
| 7600 | 2.6436E-02 | 3.4895E+07 | 1.6204E+04 | 2.0535E-04 | 4.4624E+00 | 4.1243E+02 |
| 7700 | 2.5538E-02 | 3.6425E+07 | 1.5302E+04 | 2.0685E-04 | 4.1633E+00 | 4.7474E+02 |
| 7800 | 2.4728E-02 | 3.7854E+07 | 1.4291E+04 | 2.0832E-04 | 3.8541E+00 | 5.4137E+02 |
| 7900 | 2.3997E-02 | 3.9176E+07 | 1.3224E+04 | 2.0977E-04 | 3.5472E+00 | 6.1205E+02 |
| 8000 | 2.3339E-02 | 4.0391E+07 | 1.2149E+04 | 2.1122E-04 | 3.2528E+00 | 6.8649E+02 |
| 8100 | 2.2744E-02 | 4.1502E+07 | 1.1106E+04 | 2.1269E-04 | 2.9782E+00 | 7.6441E+02 |
| 8200 | 2.2206E-02 | 4.2515E+07 | 1.0126E+04 | 2.1418E-04 | 2.7282E+00 | 8.4554E+02 |
| 8300 | 2.1716E-02 | 4.3437E+07 | 9.2288E+03 | 2.1570E-04 | 2.5054E+00 | 9.2962E+02 |
| 8400 | 2.1268E-02 | 4.4280E+07 | 8.4260E+03 | 2.1724E-04 | 2.3104E+00 | 1.0164E+03 |
| 8500 | 2.0857E-02 | 4.5052E+07 | 7.7218E+03 | 2.1880E-04 | 2.1428E+00 | 1.1058E+03 |
| 8600 | 2.0476E-02 | 4.5765E+07 | 7.1232E+03 | 2.2037E-04 | 2.0015E+00 | 1.1974E+03 |
| 8700 | 2.0121E-02 | 4.6425E+07 | 6.6032E+03 | 2.2196E-04 | 1.8837E+00 | 1.2912E+03 |
| 8800 | 1.9790E-02 | 4.7043E+07 | 6.1766E+03 | 2.2355E-04 | 1.7876E+00 | 1.3869E+03 |
| 8900 | 1.9477E-02 | 4.7627E+07 | 5.8422E+03 | 2.2513E-04 | 1.7118E+00 | 1.4845E+03 |
| 9000 | 1.9181E-02 | 4.8182E+07 | 5.5543E+03 | 2.2670E-04 | 1.6525E+00 | 1.5838E+03 |
| 9100 | 1.8898E-02 | 4.8718E+07 | 5.3595E+03 | 2.2825E-04 | 1.6097E+00 | 1.6845E+03 |
| 9200 | 1.8628E-02 | 4.9237E+07 | 5.1867E+03 | 2.2977E-04 | 1.5792E+00 | 1.7867E+03 |
| 9300 | 1.8368E-02 | 4.9747E+07 | 5.1046E+03 | 2.3125E-04 | 1.5621E+00 | 1.8902E+03 |
| 9400 | 1.8117E-02 | 5.0249E+07 | 5.0201E+03 | 2.3269E-04 | 1.5539E+00 | 1.9949E+03 |
| 9500 | 1.7873E-02 | 5.0752E+07 | 5.0296E+03 | 2.3406E-04 | 1.5568E+00 | 2.1005E+03 |
| 9600 | 1.7636E-02 | 5.1253E+07 | 5.0121E+03 | 2.3537E-04 | 1.5660E+00 | 2.2072E+03 |
| 9700 | 1.7405E-02 | 5.1763E+07 | 5.0978E+03 | 2.3661E-04 | 1.5846E+00 | 2.3146E+03 |

**TABLE A.1.** *(Continued)*

| $T$ (K) | Density (kg/m$^3$) | Enthalpy (J/kg) | Sp heat (J/kg · K) | Viscosity (kg/m · s) | Therm cond (W/m · K) | Elec cond (A/V · m) |
|---|---|---|---|---|---|---|
| | | | **NITROGEN** | | | |
| 9800 | 1.7178E-02 | 5.2281E+07 | 5.1810E+03 | 2.3776E-04 | 1.6100E+00 | 2.4228E+03 |
| 9900 | 1.6956E-02 | 5.2805E+07 | 5.2332E+03 | 2.3882E-04 | 1.6386E+00 | 2.5316E+03 |
| 10000 | 1.6737E-02 | 5.3346E+07 | 5.4181E+03 | 2.3977E-04 | 1.6754E+00 | 2.6409E+03 |
| 10100 | 1.6522E-02 | 5.3904E+07 | 5.5773E+03 | 2.4060E-04 | 1.7174E+00 | 2.7507E+03 |
| 10200 | 1.6309E-02 | 5.4472E+07 | 5.6737E+03 | 2.4131E-04 | 1.7600E+00 | 2.8608E+03 |
| 10300 | 1.6099E-02 | 5.5066E+07 | 5.9486E+03 | 2.4188E-04 | 1.8106E+00 | 2.9713E+03 |
| 10400 | 1.5890E-02 | 5.5683E+07 | 6.1709E+03 | 2.4230E-04 | 1.8653E+00 | 3.0819E+03 |
| 10500 | 1.5684E-02 | 5.6313E+07 | 6.2937E+03 | 2.4256E-04 | 1.9184E+00 | 3.1928E+03 |
| 10600 | 1.5478E-02 | 5.6979E+07 | 6.6569E+03 | 2.4266E-04 | 1.9799E+00 | 3.3038E+03 |
| 10700 | 1.5274E-02 | 5.7672E+07 | 6.9354E+03 | 2.4257E-04 | 2.0447E+00 | 3.4148E+03 |
| 10800 | 1.5071E-02 | 5.8395E+07 | 7.2323E+03 | 2.4230E-04 | 2.1127E+00 | 3.5257E+03 |
| 10900 | 1.4869E-02 | 5.9132E+07 | 7.3637E+03 | 2.4182E-04 | 2.1760E+00 | 3.6368E+03 |
| 11000 | 1.4668E-02 | 5.9917E+07 | 7.8554E+03 | 2.4114E-04 | 2.2491E+00 | 3.7477E+03 |
| 11100 | 1.4467E-02 | 6.0738E+07 | 8.2033E+03 | 2.4025E-04 | 2.3247E+00 | 3.8584E+03 |
| 11200 | 1.4266E-02 | 6.1594E+07 | 8.5684E+03 | 2.3914E-04 | 2.4028E+00 | 3.9690E+03 |
| 11300 | 1.4066E-02 | 6.2463E+07 | 8.6839E+03 | 2.3779E-04 | 2.4732E+00 | 4.0795E+03 |
| 11400 | 1.3866E-02 | 6.3394E+07 | 9.3166E+03 | 2.3622E-04 | 2.5548E+00 | 4.1897E+03 |
| 11500 | 1.3667E-02 | 6.4367E+07 | 9.7284E+03 | 2.3441E-04 | 2.6384E+00 | 4.2996E+03 |
| 11600 | 1.3468E-02 | 6.5383E+07 | 1.0156E+04 | 2.3237E-04 | 2.7236E+00 | 4.4092E+03 |
| 11700 | 1.3268E-02 | 6.6443E+07 | 1.0598E+04 | 2.3009E-04 | 2.8104E+00 | 4.5185E+03 |
| 11800 | 1.3070E-02 | 6.7548E+07 | 1.1055E+04 | 2.2758E-04 | 2.8986E+00 | 4.6274E+03 |
| 11900 | 1.2870E-02 | 6.8656E+07 | 1.1080E+04 | 2.2482E-04 | 2.9739E+00 | 4.7363E+03 |
| 12000 | 1.2672E-02 | 6.9853E+07 | 1.1965E+04 | 2.2186E-04 | 3.0631E+00 | 4.8446E+03 |
| 12100 | 1.2474E-02 | 7.1098E+07 | 1.2458E+04 | 2.1867E-04 | 3.1529E+00 | 4.9524E+03 |
| 12200 | 1.2276E-02 | 7.2395E+07 | 1.2962E+04 | 2.1527E-04 | 3.2431E+00 | 5.0597E+03 |
| 12300 | 1.2079E-02 | 7.3742E+07 | 1.3475E+04 | 2.1167E-04 | 3.3333E+00 | 5.1666E+03 |
| 12400 | 1.1882E-02 | 7.5142E+07 | 1.3997E+04 | 2.0788E-04 | 3.4233E+00 | 5.2730E+03 |
| 12500 | 1.1686E-02 | 7.6594E+07 | 1.4526E+04 | 2.0391E-04 | 3.5127E+00 | 5.3789E+03 |
| 12600 | 1.1489E-02 | 7.8029E+07 | 1.4342E+04 | 1.9973E-04 | 3.5841E+00 | 5.4848E+03 |
| 12700 | 1.1295E-02 | 7.9583E+07 | 1.5541E+04 | 1.9544E-04 | 3.6708E+00 | 5.5896E+03 |
| 12800 | 1.1101E-02 | 8.1190E+07 | 1.6077E+04 | 1.9102E-04 | 3.7559E+00 | 5.6938E+03 |
| 12900 | 1.0909E-02 | 8.2852E+07 | 1.6612E+04 | 1.8648E-04 | 3.8391E+00 | 5.7974E+03 |
| 13000 | 1.0718E-02 | 8.4566E+07 | 1.7143E+04 | 1.8185E-04 | 3.9199E+00 | 5.9004E+03 |
| 13100 | 1.0528E-02 | 8.6333E+07 | 1.7669E+04 | 1.7713E-04 | 3.9979E+00 | 6.0027E+03 |
| 13200 | 1.0341E-02 | 8.8151E+07 | 1.8185E+04 | 1.7235E-04 | 4.0727E+00 | 6.1043E+03 |
| 13300 | 1.0155E-02 | 9.0020E+07 | 1.8689E+04 | 1.6753E-04 | 4.1440E+00 | 6.2052E+03 |
| 13400 | 9.9710E-03 | 9.1938E+07 | 1.9179E+04 | 1.6267E-04 | 4.2112E+00 | 6.3054E+03 |
| 13500 | 9.7894E-03 | 9.3903E+07 | 1.9650E+04 | 1.5781E-04 | 4.2741E+00 | 6.4048E+03 |
| 13600 | 9.6103E-03 | 9.5913E+07 | 2.0099E+04 | 1.5295E-04 | 4.3322E+00 | 6.5034E+03 |
| 13700 | 9.4312E-03 | 9.7849E+07 | 1.9358E+04 | 1.4797E-04 | 4.3697E+00 | 6.6025E+03 |
| 13800 | 9.2572E-03 | 9.9937E+07 | 2.0880E+04 | 1.4316E-04 | 4.4177E+00 | 6.6996E+03 |
| 13900 | 9.0862E-03 | 1.0206E+08 | 2.1249E+04 | 1.3840E-04 | 4.4600E+00 | 6.7958E+03 |
| 14000 | 8.9183E-03 | 1.0422E+08 | 2.1585E+04 | 1.3371E-04 | 4.4964E+00 | 6.8913E+03 |
| 14100 | 8.7537E-03 | 1.0641E+08 | 2.1883E+04 | 1.2909E-04 | 4.5267E+00 | 6.9858E+03 |
| 14200 | 8.5925E-03 | 1.0862E+08 | 2.2142E+04 | 1.2456E-04 | 4.5506E+00 | 7.0794E+03 |
| 14300 | 8.4349E-03 | 1.1086E+08 | 2.2360E+04 | 1.2014E-04 | 4.5681E+00 | 7.1722E+03 |
| 14400 | 8.2809E-03 | 1.1311E+08 | 2.2533E+04 | 1.1582E-04 | 4.5792E+00 | 7.2640E+03 |

*(continued)*

**TABLE A.1.** *(Continued)*

| $T$ (K) | Density (kg/m$^3$) | Enthalpy (J/kg) | Sp heat (J/kg · K) | Viscosity (kg/m · s) | Therm cond (W/m · K) | Elec cond (A/V · m) |
|---|---|---|---|---|---|---|
| | | | **NITROGEN** | | | |
| 14500 | 8.1307E-03 | 1.1538E+08 | 2.2660E+04 | 1.1162E-04 | 4.5837E+00 | 7.3550E+03 |
| 14600 | 7.9844E-03 | 1.1765E+08 | 2.2740E+04 | 1.0754E-04 | 4.5818E+00 | 7.4450E+03 |
| 14700 | 7.8420E-03 | 1.1993E+08 | 2.2773E+04 | 1.0360E-04 | 4.5737E+00 | 7.5341E+03 |
| 14800 | 7.7036E-03 | 1.2220E+08 | 2.2756E+04 | 9.9783E-05 | 4.5594E+00 | 7.6223E+03 |
| 14900 | 7.5693E-03 | 1.2447E+08 | 2.2691E+04 | 9.6108E-05 | 4.5393E+00 | 7.7096E+03 |
| 15000 | 7.4390E-03 | 1.2673E+08 | 2.2579E+04 | 9.2572E-05 | 4.5136E+00 | 7.7960E+03 |
| 15100 | 7.3127E-03 | 1.2897E+08 | 2.2419E+04 | 8.9179E-05 | 4.4828E+00 | 7.8815E+03 |
| 15200 | 7.1906E-03 | 1.3120E+08 | 2.2214E+04 | 8.5927E-05 | 4.4472E+00 | 7.9661E+03 |
| 15300 | 7.0724E-03 | 1.3339E+08 | 2.1966E+04 | 8.2819E-05 | 4.4072E+00 | 8.0499E+03 |
| 15400 | 6.9583E-03 | 1.3556E+08 | 2.1678E+04 | 7.9852E-05 | 4.3633E+00 | 8.1328E+03 |
| 15500 | 6.8481E-03 | 1.3769E+08 | 2.1351E+04 | 7.7025E-05 | 4.3160E+00 | 8.2148E+03 |
| 15600 | 6.7418E-03 | 1.3979E+08 | 2.0989E+04 | 7.4337E-05 | 4.2658E+00 | 8.2960E+03 |
| 15700 | 6.6393E-03 | 1.4185E+08 | 2.0595E+04 | 7.1784E-05 | 4.2131E+00 | 8.3765E+03 |
| 15800 | 6.5406E-03 | 1.4387E+08 | 2.0173E+04 | 6.9364E-05 | 4.1586E+00 | 8.4561E+03 |
| 15900 | 6.4455E-03 | 1.4584E+08 | 1.9727E+04 | 6.7074E-05 | 4.1026E+00 | 8.5350E+03 |
| 16000 | 6.3539E-03 | 1.4777E+08 | 1.9260E+04 | 6.4910E-05 | 4.0457E+00 | 8.6131E+03 |
| 16100 | 6.2657E-03 | 1.4965E+08 | 1.8775E+04 | 6.2867E-05 | 3.9881E+00 | 8.6906E+03 |
| 16200 | 6.1809E-03 | 1.5147E+08 | 1.8277E+04 | 6.0943E-05 | 3.9306E+00 | 8.7674E+03 |
| 16300 | 6.0993E-03 | 1.5325E+08 | 1.7769E+04 | 5.9132E-05 | 3.8735E+00 | 8.8435E+03 |
| 16400 | 6.0207E-03 | 1.5498E+08 | 1.7255E+04 | 5.7431E-05 | 3.8171E+00 | 8.9189E+03 |
| 16500 | 5.9452E-03 | 1.5665E+08 | 1.6736E+04 | 5.5835E-05 | 3.7618E+00 | 8.9938E+03 |
| 16600 | 5.8724E-03 | 1.5827E+08 | 1.6217E+04 | 5.4339E-05 | 3.7078E+00 | 9.0681E+03 |
| 16700 | 5.8024E-03 | 1.5984E+08 | 1.5701E+04 | 5.2940E-05 | 3.6554E+00 | 9.1419E+03 |
| 16800 | 5.7350E-03 | 1.6136E+08 | 1.5189E+04 | 5.1632E-05 | 3.6048E+00 | 9.2151E+03 |
| 16900 | 5.6701E-03 | 1.6283E+08 | 1.4684E+04 | 5.0412E-05 | 3.5563E+00 | 9.2878E+03 |
| 17000 | 5.6076E-03 | 1.6425E+08 | 1.4188E+04 | 4.9275E-05 | 3.5100E+00 | 9.3600E+03 |
| 17100 | 5.5473E-03 | 1.6562E+08 | 1.3702E+04 | 4.8218E-05 | 3.4660E+00 | 9.4318E+03 |
| 17200 | 5.4892E-03 | 1.6694E+08 | 1.3229E+04 | 4.7235E-05 | 3.4244E+00 | 9.5032E+03 |
| 17300 | 5.4332E-03 | 1.6822E+08 | 1.2768E+04 | 4.6324E-05 | 3.3854E+00 | 9.5741E+03 |
| 17400 | 5.3790E-03 | 1.6945E+08 | 1.2322E+04 | 4.5481E-05 | 3.3489E+00 | 9.6447E+03 |
| 17500 | 5.3268E-03 | 1.7064E+08 | 1.1892E+04 | 4.4702E-05 | 3.3150E+00 | 9.7148E+03 |
| 17600 | 5.2762E-03 | 1.7179E+08 | 1.1477E+04 | 4.3983E-05 | 3.2837E+00 | 9.7847E+03 |
| 17700 | 5.2274E-03 | 1.7290E+08 | 1.1078E+04 | 4.3322E-05 | 3.2550E+00 | 9.8542E+03 |
| 17800 | 5.1801E-03 | 1.7396E+08 | 1.0695E+04 | 4.2714E-05 | 3.2288E+00 | 9.9233E+03 |
| 17900 | 5.1343E-03 | 1.7500E+08 | 1.0330E+04 | 4.2158E-05 | 3.2052E+00 | 9.9922E+03 |
| 18000 | 5.0899E-03 | 1.7600E+08 | 9.9806E+03 | 4.1650E-05 | 3.1841E+00 | 1.0061E+04 |
| 18100 | 5.0468E-03 | 1.7696E+08 | 9.6482E+03 | 4.1188E-05 | 3.1654E+00 | 1.0129E+04 |
| 18200 | 5.0050E-03 | 1.7789E+08 | 9.3323E+03 | 4.0768E-05 | 3.1491E+00 | 1.0197E+04 |
| 18300 | 4.9645E-03 | 1.7880E+08 | 9.0326E+03 | 4.0389E-05 | 3.1351E+00 | 1.0265E+04 |
| 18400 | 4.9250E-03 | 1.7967E+08 | 8.7490E+03 | 4.0048E-05 | 3.1233E+00 | 1.0332E+04 |
| 18500 | 4.8866E-03 | 1.8052E+08 | 8.4811E+03 | 3.9743E-05 | 3.1137E+00 | 1.0400E+04 |
| 18600 | 4.8493E-03 | 1.8134E+08 | 8.2286E+03 | 3.9472E-05 | 3.1061E+00 | 1.0467E+04 |
| 18700 | 4.8129E-03 | 1.8214E+08 | 7.9911E+03 | 3.9232E-05 | 3.1006E+00 | 1.0534E+04 |
| 18800 | 4.7774E-03 | 1.8292E+08 | 7.7682E+03 | 3.9022E-05 | 3.0970E+00 | 1.0600E+04 |
| 18900 | 4.7428E-03 | 1.8367E+08 | 7.5595E+03 | 3.8840E-05 | 3.0953E+00 | 1.0666E+04 |
| 19000 | 4.7090E-03 | 1.8441E+08 | 7.3646E+03 | 3.8685E-05 | 3.0953E+00 | 1.0733E+04 |

TABLE A.1. (*Continued*)

| $T$ (K) | Density (kg/m$^3$) | Enthalpy (J/kg) | Sp heat (J/kg · K) | Viscosity (kg/m · s) | Therm cond (W/m · K) | Elec cond (A/V · m) |
|---|---|---|---|---|---|---|
| | | | **NITROGEN** | | | |
| 19100 | 4.6760E-03 | 1.8513E+08 | 7.1832E+03 | 3.8554E-05 | 3.0970E+00 | 1.0798E+04 |
| 19200 | 4.6438E-03 | 1.8583E+08 | 7.0149E+03 | 3.8447E-05 | 3.1003E+00 | 1.0864E+04 |
| 19300 | 4.6122E-03 | 1.8652E+08 | 6.8593E+03 | 3.8361E-05 | 3.1052E+00 | 1.0930E+04 |
| 19400 | 4.5891E-03 | 1.8711E+08 | 5.9005E+03 | 3.8804E-05 | 3.1106E+00 | 1.0989E+04 |
| 19500 | 4.5587E-03 | 1.8776E+08 | 6.5703E+03 | 3.8754E-05 | 3.1184E+00 | 1.1054E+04 |
| 19600 | 4.5289E-03 | 1.8841E+08 | 6.4520E+03 | 3.8722E-05 | 3.1276E+00 | 1.1119E+04 |
| 19700 | 4.4997E-03 | 1.8904E+08 | 6.3451E+03 | 3.8707E-05 | 3.1380E+00 | 1.1183E+04 |
| 19800 | 4.4710E-03 | 1.8967E+08 | 6.2494E+03 | 3.8707E-05 | 3.1496E+00 | 1.1247E+04 |
| 19900 | 4.4429E-03 | 1.9029E+08 | 6.1647E+03 | 3.8720E-05 | 3.1624E+00 | 1.1311E+04 |
| 20000 | 4.4152E-03 | 1.9089E+08 | 6.0903E+03 | 3.8747E-05 | 3.1763E+00 | 1.1374E+04 |
| 20100 | 4.3880E-03 | 1.9150E+08 | 6.0267E+03 | 3.8785E-05 | 3.1912E+00 | 1.1437E+04 |
| 20200 | 4.3612E-03 | 1.9209E+08 | 5.9735E+03 | 3.8835E-05 | 3.2071E+00 | 1.1500E+04 |
| 20300 | 4.3348E-03 | 1.9269E+08 | 5.9305E+03 | 3.8894E-05 | 3.2239E+00 | 1.1562E+04 |
| 20400 | 4.3088E-03 | 1.9328E+08 | 5.8978E+03 | 3.8962E-05 | 3.2417E+00 | 1.1624E+04 |
| 20500 | 4.2832E-03 | 1.9386E+08 | 5.8752E+03 | 3.9038E-05 | 3.2602E+00 | 1.1686E+04 |
| 20600 | 4.2580E-03 | 1.9445E+08 | 5.8627E+03 | 3.9121E-05 | 3.2797E+00 | 1.1747E+04 |
| 20700 | 4.2331E-03 | 1.9504E+08 | 5.8603E+03 | 3.9210E-05 | 3.2998E+00 | 1.1808E+04 |
| 20800 | 4.2085E-03 | 1.9562E+08 | 5.8681E+03 | 3.9304E-05 | 3.3207E+00 | 1.1868E+04 |
| 20900 | 4.1841E-03 | 1.9621E+08 | 5.8861E+03 | 3.9402E-05 | 3.3423E+00 | 1.1928E+04 |
| 21000 | 4.1601E-03 | 1.9680E+08 | 5.9143E+03 | 3.9504E-05 | 3.3646E+00 | 1.1987E+04 |
| 21100 | 4.1363E-03 | 1.9740E+08 | 5.9529E+03 | 3.9608E-05 | 3.3875E+00 | 1.2045E+04 |
| 21200 | 4.1128E-03 | 1.9800E+08 | 6.0020E+03 | 3.9713E-05 | 3.4110E+00 | 1.2103E+04 |
| 21300 | 4.0895E-03 | 1.9861E+08 | 6.0616E+03 | 3.9819E-05 | 3.4350E+00 | 1.2160E+04 |
| 21400 | 4.0664E-03 | 1.9922E+08 | 6.1321E+03 | 3.9924E-05 | 3.4596E+00 | 1.2216E+04 |
| 21500 | 4.0436E-03 | 1.9984E+08 | 6.2135E+03 | 4.0029E-05 | 3.4847E+00 | 1.2272E+04 |
| 21600 | 4.0209E-03 | 2.0047E+08 | 6.3061E+03 | 4.0131E-05 | 3.5103E+00 | 1.2327E+04 |
| 21700 | 3.9983E-03 | 2.0111E+08 | 6.4100E+03 | 4.0230E-05 | 3.5364E+00 | 1.2380E+04 |
| 21800 | 3.9760E-03 | 2.0176E+08 | 6.5255E+03 | 4.0326E-05 | 3.5629E+00 | 1.2433E+04 |
| 21900 | 3.9538E-03 | 2.0243E+08 | 6.6528E+03 | 4.0416E-05 | 3.5898E+00 | 1.2485E+04 |
| 22000 | 3.9317E-03 | 2.0311E+08 | 6.7922E+03 | 4.0500E-05 | 3.6171E+00 | 1.2536E+04 |
| 22100 | 3.9097E-03 | 2.0380E+08 | 6.9438E+03 | 4.0578E-05 | 3.6448E+00 | 1.2586E+04 |
| 22200 | 3.8878E-03 | 2.0451E+08 | 7.1080E+03 | 4.0649E-05 | 3.6728E+00 | 1.2635E+04 |
| 22300 | 3.8660E-03 | 2.0524E+08 | 7.2850E+03 | 4.0711E-05 | 3.7011E+00 | 1.2683E+04 |
| 22400 | 3.8443E-03 | 2.0599E+08 | 7.4750E+03 | 4.0763E-05 | 3.7298E+00 | 1.2729E+04 |
| 22500 | 3.8227E-03 | 2.0676E+08 | 7.6783E+03 | 4.0805E-05 | 3.7588E+00 | 1.2774E+04 |
| 22600 | 3.8011E-03 | 2.0755E+08 | 7.8951E+03 | 4.0836E-05 | 3.7880E+00 | 1.2818E+04 |
| 22700 | 3.7796E-03 | 2.0836E+08 | 8.1257E+03 | 4.0855E-05 | 3.8176E+00 | 1.2860E+04 |
| 22800 | 3.7581E-03 | 2.0920E+08 | 8.3701E+03 | 4.0862E-05 | 3.8473E+00 | 1.2901E+04 |
| 22900 | 3.7366E-03 | 2.1006E+08 | 8.6287E+03 | 4.0854E-05 | 3.8773E+00 | 1.2940E+04 |
| 23000 | 3.7202E-03 | 2.1084E+08 | 7.7972E+03 | 4.1275E-05 | 3.9034E+00 | 1.2966E+04 |
| 23100 | 3.6986E-03 | 2.1176E+08 | 9.1868E+03 | 4.1234E-05 | 3.9336E+00 | 1.3002E+04 |
| 23200 | 3.6770E-03 | 2.1271E+08 | 9.4885E+03 | 4.1177E-05 | 3.9640E+00 | 1.3037E+04 |
| 23300 | 3.6554E-03 | 2.1369E+08 | 9.8047E+03 | 4.1104E-05 | 3.9945E+00 | 1.3069E+04 |
| 23400 | 3.6338E-03 | 2.1470E+08 | 1.0136E+04 | 4.1014E-05 | 4.0253E+00 | 1.3100E+04 |
| 23500 | 3.6121E-03 | 2.1575E+08 | 1.0481E+04 | 4.0906E-05 | 4.0561E+00 | 1.3129E+04 |
| 23600 | 3.5904E-03 | 2.1683E+08 | 1.0841E+04 | 4.0781E-05 | 4.0871E+00 | 1.3156E+04 |
| 23700 | 3.5686E-03 | 2.1795E+08 | 1.1216E+04 | 4.0637E-05 | 4.1183E+00 | 1.3182E+04 |
| 23800 | 3.5468E-03 | 2.1911E+08 | 1.1604E+04 | 4.0475E-05 | 4.1495E+00 | 1.3205E+04 |
| 23900 | 3.5249E-03 | 2.2032E+08 | 1.2007E+04 | 4.0294E-05 | 4.1809E+00 | 1.3227E+04 |
| 24000 | 3.5029E-03 | 2.2156E+08 | 1.2424E+04 | 4.0094E-05 | 4.2123E+00 | 1.3247E+04 |

(*continued*)

**TABLE A.1.** *(Continued)*

| $T$ (K) | Density (kg/m$^3$) | Enthalpy (J/kg) | Sp heat (J/kg · K) | Viscosity (kg/m · s) | Therm cond (W/m · K) | Elec cond (A/V · m) |
|---|---|---|---|---|---|---|
| | | | **OXYGEN** | | | |
| 500 | 7.7978E-01 | 1.9310E+05 | 9.7902E+02 | 2.9834E-05 | 4.2106E-02 | 8.2062E-24 |
| 600 | 6.4977E-01 | 2.9295E+05 | 9.9857E+02 | 3.4044E-05 | 4.9132E-02 | 6.9418E-24 |
| 700 | 5.5693E-01 | 3.9479E+05 | 1.0184E+03 | 3.7948E-05 | 5.5894E-02 | 6.0015E-24 |
| 800 | 4.8732E-01 | 4.9848E+05 | 1.0369E+03 | 4.1625E-05 | 6.2459E-02 | 5.2763E-24 |
| 900 | 4.3317E-01 | 6.0390E+05 | 1.0542E+03 | 4.5124E-05 | 6.8866E-02 | 4.7010E-24 |
| 1000 | 3.8986E-01 | 7.1092E+05 | 1.0703E+03 | 4.8480E-05 | 7.5142E-02 | 4.2345E-24 |
| 1100 | 3.5442E-01 | 8.1945E+05 | 1.0853E+03 | 5.1716E-05 | 8.1300E-02 | 1.3791E-21 |
| 1200 | 3.2489E-01 | 9.2939E+05 | 1.0994E+03 | 5.4850E-05 | 8.7354E-02 | 4.0130E-19 |
| 1300 | 2.9990E-01 | 1.0406E+06 | 1.1125E+03 | 5.7894E-05 | 9.3312E-02 | 4.4021E-17 |
| 1400 | 2.7849E-01 | 1.1531E+06 | 1.1249E+03 | 6.0859E-05 | 9.9189E-02 | 2.4276E-15 |
| 1500 | 2.5992E-01 | 1.2668E+06 | 1.1367E+03 | 6.3753E-05 | 1.0501E-01 | 7.7248E-14 |
| 1600 | 2.4368E-01 | 1.3816E+06 | 1.1482E+03 | 6.6583E-05 | 1.1085E-01 | 1.5743E-12 |
| 1700 | 2.2934E-01 | 1.4976E+06 | 1.1600E+03 | 6.9355E-05 | 1.1686E-01 | 2.2266E-11 |
| 1800 | 2.1660E-01 | 1.6150E+06 | 1.1734E+03 | 7.2075E-05 | 1.2330E-01 | 2.3268E-10 |
| 1900 | 2.0518E-01 | 1.7340E+06 | 1.1905E+03 | 7.4746E-05 | 1.3069E-01 | 1.8874E-09 |
| 2000 | 1.9489E-01 | 1.8555E+06 | 1.2146E+03 | 7.7375E-05 | 1.3985E-01 | 1.2360E-08 |
| 2100 | 1.8554E-01 | 1.9806E+06 | 1.2507E+03 | 7.9968E-05 | 1.5202E-01 | 6.7436E-08 |
| 2200 | 1.7699E-01 | 2.1111E+06 | 1.3057E+03 | 8.2533E-05 | 1.6896E-01 | 3.1452E-07 |
| 2300 | 1.6911E-01 | 2.2500E+06 | 1.3887E+03 | 8.5079E-05 | 1.9304E-01 | 1.2806E-06 |
| 2400 | 1.6177E-01 | 2.4011E+06 | 1.5108E+03 | 8.7621E-05 | 2.2728E-01 | 4.6315E-06 |
| 2500 | 1.5485E-01 | 2.5696E+06 | 1.6856E+03 | 9.0176E-05 | 2.7532E-01 | 1.5097E-05 |
| 2600 | 1.4825E-01 | 2.7625E+06 | 1.9285E+03 | 9.2767E-05 | 3.4137E-01 | 4.4896E-05 |
| 2700 | 1.4187E-01 | 2.9882E+06 | 2.2568E+03 | 9.5424E-05 | 4.2999E-01 | 1.2308E-04 |
| 2800 | 1.3560E-01 | 3.2570E+06 | 2.6886E+03 | 9.8178E-05 | 5.4574E-01 | 3.1379E-04 |
| 2900 | 1.2936E-01 | 3.5813E+06 | 3.2425E+03 | 1.0107E-04 | 6.9274E-01 | 7.4961E-04 |
| 3000 | 1.2307E-01 | 3.9749E+06 | 3.9362E+03 | 1.0413E-04 | 8.7388E-01 | 1.6888E-03 |
| 3100 | 1.1668E-01 | 4.4533E+06 | 4.7843E+03 | 1.0739E-04 | 1.0899E+00 | 3.6078E-03 |
| 3200 | 1.1015E-01 | 5.0329E+06 | 5.7952E+03 | 1.1089E-04 | 1.3380E+00 | 7.3423E-03 |
| 3300 | 1.0350E-01 | 5.7294E+06 | 6.9656E+03 | 1.1462E-04 | 1.6105E+00 | 1.4289E-02 |
| 3400 | 9.6761E-02 | 6.5567E+06 | 8.2725E+03 | 1.1856E-04 | 1.8937E+00 | 2.6670E-02 |
| 3500 | 9.0035E-02 | 7.5230E+06 | 9.6634E+03 | 1.2265E-04 | 2.1664E+00 | 4.7853E-02 |
| 3600 | 8.3440E-02 | 8.6276E+06 | 1.1046E+04 | 1.2680E-04 | 2.4014E+00 | 8.2655E-02 |
| 3700 | 7.7124E-02 | 9.8559E+06 | 1.2283E+04 | 1.3090E-04 | 2.5686E+00 | 1.3756E-01 |
| 3800 | 7.1233E-02 | 1.1177E+07 | 1.3209E+04 | 1.3484E-04 | 2.6416E+00 | 2.2070E-01 |
| 3900 | 6.5892E-02 | 1.2543E+07 | 1.3660E+04 | 1.3853E-04 | 2.6063E+00 | 3.4155E-01 |
| 4000 | 6.1181E-02 | 1.3896E+07 | 1.3530E+04 | 1.4193E-04 | 2.4671E+00 | 5.1028E-01 |
| 4100 | 5.7125E-02 | 1.5178E+07 | 1.2822E+04 | 1.4507E-04 | 2.2472E+00 | 7.3739E-01 |
| 4200 | 5.3694E-02 | 1.6343E+07 | 1.1652E+04 | 1.4798E-04 | 1.9813E+00 | 1.0335E+00 |
| 4300 | 5.0822E-02 | 1.7364E+07 | 1.0213E+04 | 1.5074E-04 | 1.7048E+00 | 1.4103E+00 |
| 4400 | 4.8424E-02 | 1.8235E+07 | 8.7059E+03 | 1.5341E-04 | 1.4451E+00 | 1.8652E+00 |
| 4500 | 4.6410E-02 | 1.8964E+07 | 7.2858E+03 | 1.5603E-04 | 1.2182E+00 | 2.4413E+00 |
| 4600 | 4.4702E-02 | 1.9568E+07 | 6.0437E+03 | 1.5863E-04 | 1.0300E+00 | 3.1537E+00 |
| 4700 | 4.3232E-02 | 2.0069E+07 | 5.0124E+03 | 1.6122E-04 | 8.7956E-01 | 4.0382E+00 |
| 4800 | 4.1947E-02 | 2.0488E+07 | 4.1863E+03 | 1.6382E-04 | 7.6234E-01 | 5.1444E+00 |
| 4900 | 4.0805E-02 | 2.0842E+07 | 3.5403E+03 | 1.6642E-04 | 6.7275E-01 | 6.4849E+00 |
| 5000 | 3.9776E-02 | 2.1146E+07 | 3.0426E+03 | 1.6902E-04 | 6.0513E-01 | 8.2090E+00 |

**TABLE A.1.** *(Continued)*

| $T$ (K) | Density (kg/m$^3$) | Enthalpy (J/kg) | Sp heat (J/kg · K) | Viscosity (kg/m · s) | Therm cond (W/m · K) | Elec cond (A/V · m) |
|---|---|---|---|---|---|---|
| | | | **OXYGEN** | | | |
| 5100 | 3.8836E-02 | 2.1412E+07 | 2.6627E+03 | 1.7163E-04 | 5.5469E-01 | 1.0387E+01 |
| 5200 | 3.7968E-02 | 2.1650E+07 | 2.3739E+03 | 1.7423E-04 | 5.1749E-01 | 1.3137E+01 |
| 5300 | 3.7161E-02 | 2.1865E+07 | 2.1547E+03 | 1.7683E-04 | 4.9040E-01 | 1.6597E+01 |
| 5400 | 3.6402E-02 | 2.2064E+07 | 1.9883E+03 | 1.7943E-04 | 4.7105E-01 | 2.0924E+01 |
| 5500 | 3.5686E-02 | 2.2250E+07 | 1.8618E+03 | 1.8203E-04 | 4.5762E-01 | 2.6291E+01 |
| 5600 | 3.5007E-02 | 2.2427E+07 | 1.7655E+03 | 1.8462E-04 | 4.4874E-01 | 3.2889E+01 |
| 5700 | 3.4360E-02 | 2.2596E+07 | 1.6921E+03 | 1.8721E-04 | 4.4341E-01 | 4.0918E+01 |
| 5800 | 3.3742E-02 | 2.2760E+07 | 1.6362E+03 | 1.8978E-04 | 4.4086E-01 | 5.0589E+01 |
| 5900 | 3.3149E-02 | 2.2919E+07 | 1.5936E+03 | 1.9236E-04 | 4.4052E-01 | 6.2121E+01 |
| 6000 | 3.2580E-02 | 2.3075E+07 | 1.5614E+03 | 1.9492E-04 | 4.4198E-01 | 7.5739E+01 |
| 6100 | 3.2033E-02 | 2.3229E+07 | 1.5373E+03 | 1.9748E-04 | 4.4492E-01 | 9.1669E+01 |
| 6200 | 3.1505E-02 | 2.3381E+07 | 1.5195E+03 | 2.0003E-04 | 4.4911E-01 | 1.1014E+02 |
| 6300 | 3.0996E-02 | 2.3532E+07 | 1.5069E+03 | 2.0257E-04 | 4.5440E-01 | 1.3139E+02 |
| 6400 | 3.0503E-02 | 2.3681E+07 | 1.4985E+03 | 2.0511E-04 | 4.6066E-01 | 1.5564E+02 |
| 6500 | 3.0028E-02 | 2.3831E+07 | 1.4937E+03 | 2.0764E-04 | 4.6780E-01 | 1.8311E+02 |
| 6600 | 2.9567E-02 | 2.3980E+07 | 1.4918E+03 | 2.1016E-04 | 4.7576E-01 | 2.1401E+02 |
| 6700 | 2.9120E-02 | 2.4129E+07 | 1.4925E+03 | 2.1267E-04 | 4.8452E-01 | 2.4855E+02 |
| 6800 | 2.8687E-02 | 2.4279E+07 | 1.4956E+03 | 2.1518E-04 | 4.9405E-01 | 2.8691E+02 |
| 6900 | 2.8267E-02 | 2.4429E+07 | 1.5010E+03 | 2.1768E-04 | 5.0435E-01 | 3.2896E+02 |
| 7000 | 2.7859E-02 | 2.4580E+07 | 1.5084E+03 | 2.2016E-04 | 5.1543E-01 | 3.7535E+02 |
| 7100 | 2.7462E-02 | 2.4732E+07 | 1.5179E+03 | 2.2264E-04 | 5.2731E-01 | 4.2591E+02 |
| 7200 | 2.7077E-02 | 2.4884E+07 | 1.5292E+03 | 2.2512E-04 | 5.3999E-01 | 4.8068E+02 |
| 7300 | 2.6701E-02 | 2.5039E+07 | 1.5427E+03 | 2.2758E-04 | 5.5351E-01 | 5.3968E+02 |
| 7400 | 2.6336E-02 | 2.5194E+07 | 1.5578E+03 | 2.3003E-04 | 5.6784E-01 | 6.0284E+02 |
| 7500 | 2.5980E-02 | 2.5352E+07 | 1.5760E+03 | 2.3247E-04 | 5.8312E-01 | 6.7008E+02 |
| 7600 | 2.5633E-02 | 2.5512E+07 | 1.5960E+03 | 2.3491E-04 | 5.9935E-01 | 7.4127E+02 |
| 7700 | 2.5294E-02 | 2.5674E+07 | 1.6185E+03 | 2.3733E-04 | 6.1655E-01 | 8.1625E+02 |
| 7800 | 2.4964E-02 | 2.5838E+07 | 1.6435E+03 | 2.3974E-04 | 6.3476E-01 | 8.9480E+02 |
| 7900 | 2.4641E-02 | 2.6005E+07 | 1.6712E+03 | 2.4213E-04 | 6.5403E-01 | 9.7671E+02 |
| 8000 | 2.4326E-02 | 2.6175E+07 | 1.7017E+03 | 2.4452E-04 | 6.7439E-01 | 1.0617E+03 |
| 8100 | 2.4018E-02 | 2.6349E+07 | 1.7352E+03 | 2.4689E-04 | 6.9588E-01 | 1.1496E+03 |
| 8200 | 2.3716E-02 | 2.6526E+07 | 1.7718E+03 | 2.4924E-04 | 7.1852E-01 | 1.2400E+03 |
| 8300 | 2.3421E-02 | 2.6707E+07 | 1.8149E+03 | 2.5157E-04 | 7.4260E-01 | 1.3328E+03 |
| 8400 | 2.3132E-02 | 2.6893E+07 | 1.8559E+03 | 2.5389E-04 | 7.6770E-01 | 1.4276E+03 |
| 8500 | 2.2849E-02 | 2.7083E+07 | 1.9032E+03 | 2.5618E-04 | 7.9405E-01 | 1.5242E+03 |
| 8600 | 2.2571E-02 | 2.7279E+07 | 1.9596E+03 | 2.5845E-04 | 8.2207E-01 | 1.6224E+03 |
| 8700 | 2.2298E-02 | 2.7480E+07 | 2.0107E+03 | 2.6070E-04 | 8.5106E-01 | 1.7218E+03 |
| 8800 | 2.2030E-02 | 2.7687E+07 | 2.0705E+03 | 2.6291E-04 | 8.8140E-01 | 1.8224E+03 |
| 8900 | 2.1767E-02 | 2.7902E+07 | 2.1429E+03 | 2.6510E-04 | 9.1369E-01 | 1.9239E+03 |
| 9000 | 2.1508E-02 | 2.8122E+07 | 2.2051E+03 | 2.6724E-04 | 9.4684E-01 | 2.0261E+03 |
| 9100 | 2.1254E-02 | 2.8351E+07 | 2.2902E+03 | 2.6935E-04 | 9.8218E-01 | 2.1288E+03 |
| 9200 | 2.1003E-02 | 2.8587E+07 | 2.3600E+03 | 2.7142E-04 | 1.0183E+00 | 2.2320E+03 |
| 9300 | 2.0756E-02 | 2.8833E+07 | 2.4594E+03 | 2.7343E-04 | 1.0568E+00 | 2.3355E+03 |
| 9400 | 2.0512E-02 | 2.9087E+07 | 2.5372E+03 | 2.7540E-04 | 1.0959E+00 | 2.4391E+03 |
| 9500 | 2.0272E-02 | 2.9352E+07 | 2.6525E+03 | 2.7730E-04 | 1.1378E+00 | 2.5428E+03 |
| 9600 | 2.0035E-02 | 2.9628E+07 | 2.7604E+03 | 2.7914E-04 | 1.1814E+00 | 2.6466E+03 |
| 9700 | 1.9800E-02 | 2.9913E+07 | 2.8495E+03 | 2.8091E-04 | 1.2253E+00 | 2.7502E+03 |

*(continued)*

TABLE A.1. *(Continued)*

| $T$ (K) | Density (kg/m$^3$) | Enthalpy (J/kg) | Sp heat (J/kg · K) | Viscosity (kg/m · s) | Therm cond (W/m · K) | Elec cond (A/V · m) |
|---|---|---|---|---|---|---|
| | | | **OXYGEN** | | | |
| 9800 | 1.9568E-02 | 3.0212E+07 | 2.9934E+03 | 2.8260E-04 | 1.2725E+00 | 2.8537E+03 |
| 9900 | 1.9339E-02 | 3.0521E+07 | 3.0904E+03 | 2.8420E-04 | 1.3196E+00 | 2.9571E+03 |
| 10000 | 1.9112E-02 | 3.0847E+07 | 3.2551E+03 | 2.8571E-04 | 1.3704E+00 | 3.0602E+03 |
| 10100 | 1.8887E-02 | 3.1187E+07 | 3.4002E+03 | 2.8712E-04 | 1.4232E+00 | 3.1631E+03 |
| 10200 | 1.8664E-02 | 3.1538E+07 | 3.5074E+03 | 2.8842E-04 | 1.4753E+00 | 3.2657E+03 |
| 10300 | 1.8442E-02 | 3.1909E+07 | 3.7090E+03 | 2.8960E-04 | 1.5318E+00 | 3.3680E+03 |
| 10400 | 1.8222E-02 | 3.2297E+07 | 3.8795E+03 | 2.9066E-04 | 1.5903E+00 | 3.4700E+03 |
| 10500 | 1.8004E-02 | 3.2696E+07 | 3.9953E+03 | 2.9158E-04 | 1.6474E+00 | 3.5717E+03 |
| 10600 | 1.7787E-02 | 3.3120E+07 | 4.2394E+03 | 2.9235E-04 | 1.7098E+00 | 3.6730E+03 |
| 10700 | 1.7571E-02 | 3.3564E+07 | 4.4374E+03 | 2.9297E-04 | 1.7742E+00 | 3.7740E+03 |
| 10800 | 1.7356E-02 | 3.4028E+07 | 4.6454E+03 | 2.9342E-04 | 1.8406E+00 | 3.8747E+03 |
| 10900 | 1.7143E-02 | 3.4505E+07 | 4.7668E+03 | 2.9370E-04 | 1.9043E+00 | 3.9751E+03 |
| 11000 | 1.6930E-02 | 3.5013E+07 | 5.0793E+03 | 2.9379E-04 | 1.9744E+00 | 4.0752E+03 |
| 11100 | 1.6717E-02 | 3.5545E+07 | 5.3170E+03 | 2.9370E-04 | 2.0466E+00 | 4.1749E+03 |
| 11200 | 1.6506E-02 | 3.6101E+07 | 5.5654E+03 | 2.9340E-04 | 2.1208E+00 | 4.2743E+03 |
| 11300 | 1.6295E-02 | 3.6670E+07 | 5.6857E+03 | 2.9289E-04 | 2.1905E+00 | 4.3735E+03 |
| 11400 | 1.6084E-02 | 3.7278E+07 | 6.0782E+03 | 2.9217E-04 | 2.2680E+00 | 4.4723E+03 |
| 11500 | 1.5874E-02 | 3.7913E+07 | 6.3577E+03 | 2.9123E-04 | 2.3474E+00 | 4.5708E+03 |
| 11600 | 1.5664E-02 | 3.8578E+07 | 6.6482E+03 | 2.9005E-04 | 2.4285E+00 | 4.6689E+03 |
| 11700 | 1.5455E-02 | 3.9273E+07 | 6.9497E+03 | 2.8865E-04 | 2.5114E+00 | 4.7668E+03 |
| 11800 | 1.5246E-02 | 3.9999E+07 | 7.2622E+03 | 2.8701E-04 | 2.5960E+00 | 4.8644E+03 |
| 11900 | 1.5037E-02 | 4.0735E+07 | 7.3540E+03 | 2.8511E-04 | 2.6729E+00 | 4.9618E+03 |
| 12000 | 1.4828E-02 | 4.1524E+07 | 7.8961E+03 | 2.8299E-04 | 2.7598E+00 | 5.0588E+03 |
| 12100 | 1.4620E-02 | 4.2348E+07 | 8.2390E+03 | 2.8062E-04 | 2.8479E+00 | 5.1555E+03 |
| 12200 | 1.4412E-02 | 4.3207E+07 | 8.5921E+03 | 2.7801E-04 | 2.9373E+00 | 5.2519E+03 |
| 12300 | 1.4204E-02 | 4.4103E+07 | 8.9552E+03 | 2.7516E-04 | 3.0276E+00 | 5.3480E+03 |
| 12400 | 1.3997E-02 | 4.5036E+07 | 9.3277E+03 | 2.7207E-04 | 3.1189E+00 | 5.4438E+03 |
| 12500 | 1.3790E-02 | 4.6007E+07 | 9.7093E+03 | 2.6876E-04 | 3.2107E+00 | 5.5392E+03 |
| 12600 | 1.3583E-02 | 4.6979E+07 | 9.7217E+03 | 2.6518E-04 | 3.2911E+00 | 5.6347E+03 |
| 12700 | 1.3377E-02 | 4.8025E+07 | 1.0467E+04 | 2.6142E-04 | 3.3832E+00 | 5.7295E+03 |
| 12800 | 1.3172E-02 | 4.9112E+07 | 1.0870E+04 | 2.5745E-04 | 3.4753E+00 | 5.8240E+03 |
| 12900 | 1.2968E-02 | 5.0240E+07 | 1.1280E+04 | 2.5328E-04 | 3.5672E+00 | 5.9182E+03 |
| 13000 | 1.2764E-02 | 5.1410E+07 | 1.1695E+04 | 2.4892E-04 | 3.6587E+00 | 6.0121E+03 |
| 13100 | 1.2561E-02 | 5.2621E+07 | 1.2114E+04 | 2.4439E-04 | 3.7494E+00 | 6.1056E+03 |
| 13200 | 1.2359E-02 | 5.3875E+07 | 1.2536E+04 | 2.3969E-04 | 3.8391E+00 | 6.1987E+03 |
| 13300 | 1.2158E-02 | 5.5171E+07 | 1.2959E+04 | 2.3485E-04 | 3.9275E+00 | 6.2914E+03 |
| 13400 | 1.1959E-02 | 5.6509E+07 | 1.3383E+04 | 2.2988E-04 | 4.0142E+00 | 6.3837E+03 |
| 13500 | 1.1759E-02 | 5.7830E+07 | 1.3204E+04 | 2.2472E-04 | 4.0856E+00 | 6.4761E+03 |
| 13600 | 1.1563E-02 | 5.9249E+07 | 1.4192E+04 | 2.1952E-04 | 4.1679E+00 | 6.5677E+03 |
| 13700 | 1.1368E-02 | 6.0709E+07 | 1.4607E+04 | 2.1424E-04 | 4.2476E+00 | 6.6588E+03 |
| 13800 | 1.1175E-02 | 6.2211E+07 | 1.5015E+04 | 2.0889E-04 | 4.3243E+00 | 6.7495E+03 |
| 13900 | 1.0984E-02 | 6.3752E+07 | 1.5414E+04 | 2.0349E-04 | 4.3977E+00 | 6.8397E+03 |
| 14000 | 1.0795E-02 | 6.5333E+07 | 1.5803E+04 | 1.9806E-04 | 4.4674E+00 | 6.9294E+03 |
| 14100 | 1.0609E-02 | 6.6951E+07 | 1.6179E+04 | 1.9261E-04 | 4.5331E+00 | 7.0187E+03 |
| 14200 | 1.0425E-02 | 6.8605E+07 | 1.6540E+04 | 1.8715E-04 | 4.5944E+00 | 7.1074E+03 |
| 14300 | 1.0243E-02 | 7.0293E+07 | 1.6884E+04 | 1.8171E-04 | 4.6511E+00 | 7.1957E+03 |

TABLE A.1. (*Continued*)

| $T$ (K) | Density (kg/m$^3$) | Enthalpy (J/kg) | Sp heat (J/kg · K) | Viscosity (kg/m · s) | Therm cond (W/m · K) | Elec cond (A/V · m) |
|---|---|---|---|---|---|---|
| | | | **OXYGEN** | | | |
| 14400 | 1.0065E-02 | 7.2014E+07 | 1.7209E+04 | 1.7630E-04 | 4.7029E+00 | 7.2834E+03 |
| 14500 | 9.8894E-03 | 7.3765E+07 | 1.7511E+04 | 1.7093E-04 | 4.7495E+00 | 7.3706E+03 |
| 14600 | 9.7169E-03 | 7.5544E+07 | 1.7791E+04 | 1.6562E-04 | 4.7906E+00 | 7.4573E+03 |
| 14700 | 9.5476E-03 | 7.7348E+07 | 1.8044E+04 | 1.6038E-04 | 4.8260E+00 | 7.5434E+03 |
| 14800 | 9.3816E-03 | 7.9175E+07 | 1.8270E+04 | 1.5522E-04 | 4.8557E+00 | 7.6289E+03 |
| 14900 | 9.2191E-03 | 8.1022E+07 | 1.8467E+04 | 1.5015E-04 | 4.8794E+00 | 7.7139E+03 |
| 15000 | 9.0602E-03 | 8.2885E+07 | 1.8632E+04 | 1.4518E-04 | 4.8972E+00 | 7.7983E+03 |
| 15100 | 8.9049E-03 | 8.4762E+07 | 1.8765E+04 | 1.4032E-04 | 4.9089E+00 | 7.8821E+03 |
| 15200 | 8.7533E-03 | 8.6648E+07 | 1.8865E+04 | 1.3558E-04 | 4.9146E+00 | 7.9654E+03 |
| 15300 | 8.6013E-03 | 8.8461E+07 | 1.8121E+04 | 1.3071E-04 | 4.9098E+00 | 8.0492E+03 |
| 15400 | 8.4572E-03 | 9.0358E+07 | 1.8971E+04 | 1.2622E-04 | 4.9045E+00 | 8.1314E+03 |
| 15500 | 8.3169E-03 | 9.2255E+07 | 1.8970E+04 | 1.2186E-04 | 4.8935E+00 | 8.2129E+03 |
| 15600 | 8.1805E-03 | 9.4148E+07 | 1.8934E+04 | 1.1765E-04 | 4.8771E+00 | 8.2939E+03 |
| 15700 | 8.0481E-03 | 9.6034E+07 | 1.8864E+04 | 1.1357E-04 | 4.8554E+00 | 8.3743E+03 |
| 15800 | 7.9195E-03 | 9.7910E+07 | 1.8759E+04 | 1.0964E-04 | 4.8290E+00 | 8.4541E+03 |
| 15900 | 7.7949E-03 | 9.9773E+07 | 1.8622E+04 | 1.0586E-04 | 4.7979E+00 | 8.5334E+03 |
| 16000 | 7.6742E-03 | 1.0162E+08 | 1.8452E+04 | 1.0223E-04 | 4.7627E+00 | 8.6121E+03 |
| 16100 | 7.5573E-03 | 1.0344E+08 | 1.8252E+04 | 9.8750E-05 | 4.7236E+00 | 8.6903E+03 |
| 16200 | 7.4442E-03 | 1.0525E+08 | 1.8024E+04 | 9.5413E-05 | 4.6813E+00 | 8.7679E+03 |
| 16300 | 7.3349E-03 | 1.0702E+08 | 1.7769E+04 | 9.2223E-05 | 4.6359E+00 | 8.8450E+03 |
| 16400 | 7.2293E-03 | 1.0877E+08 | 1.7489E+04 | 8.9178E-05 | 4.5880E+00 | 8.9217E+03 |
| 16500 | 7.1272E-03 | 1.1049E+08 | 1.7188E+04 | 8.6274E-05 | 4.5380E+00 | 8.9978E+03 |
| 16600 | 7.0287E-03 | 1.1218E+08 | 1.6866E+04 | 8.3511E-05 | 4.4863E+00 | 9.0734E+03 |
| 16700 | 6.9336E-03 | 1.1383E+08 | 1.6527E+04 | 8.0883E-05 | 4.4334E+00 | 9.1486E+03 |
| 16800 | 6.8419E-03 | 1.1545E+08 | 1.6173E+04 | 7.8389E-05 | 4.3796E+00 | 9.2233E+03 |
| 16900 | 6.7533E-03 | 1.1703E+08 | 1.5806E+04 | 7.6024E-05 | 4.3254E+00 | 9.2976E+03 |
| 17000 | 6.6680E-03 | 1.1857E+08 | 1.5430E+04 | 7.3785E-05 | 4.2711E+00 | 9.3715E+03 |
| 17100 | 6.5856E-03 | 1.2007E+08 | 1.5045E+04 | 7.1668E-05 | 4.2170E+00 | 9.4450E+03 |
| 17200 | 6.5062E-03 | 1.2154E+08 | 1.4655E+04 | 6.9669E-05 | 4.1636E+00 | 9.5181E+03 |
| 17300 | 6.4297E-03 | 1.2297E+08 | 1.4262E+04 | 6.7783E-05 | 4.1111E+00 | 9.5909E+03 |
| 17400 | 6.3558E-03 | 1.2435E+08 | 1.3868E+04 | 6.6006E-05 | 4.0597E+00 | 9.6633E+03 |
| 17500 | 6.2845E-03 | 1.2570E+08 | 1.3474E+04 | 6.4335E-05 | 4.0097E+00 | 9.7353E+03 |
| 17600 | 6.2158E-03 | 1.2701E+08 | 1.3082E+04 | 6.2764E-05 | 3.9613E+00 | 9.8071E+03 |
| 17700 | 6.1494E-03 | 1.2828E+08 | 1.2694E+04 | 6.1290E-05 | 3.9147E+00 | 9.8786E+03 |
| 17800 | 6.0853E-03 | 1.2951E+08 | 1.2311E+04 | 5.9908E-05 | 3.8701E+00 | 9.9498E+03 |
| 17900 | 6.0234E-03 | 1.3070E+08 | 1.1935E+04 | 5.8614E-05 | 3.8275E+00 | 1.0021E+04 |
| 18000 | 5.9636E-03 | 1.3186E+08 | 1.1566E+04 | 5.7405E-05 | 3.7871E+00 | 1.0091E+04 |
| 18100 | 5.9126E-03 | 1.3296E+08 | 1.1041E+04 | 5.6732E-05 | 3.7466E+00 | 1.0160E+04 |
| 18200 | 5.8567E-03 | 1.3405E+08 | 1.0822E+04 | 5.5681E-05 | 3.7110E+00 | 1.0230E+04 |
| 18300 | 5.8026E-03 | 1.3509E+08 | 1.0482E+04 | 5.4703E-05 | 3.6777E+00 | 1.0300E+04 |
| 18400 | 5.7501E-03 | 1.3611E+08 | 1.0153E+04 | 5.3794E-05 | 3.6469E+00 | 1.0370E+04 |
| 18500 | 5.6994E-03 | 1.3709E+08 | 9.8342E+03 | 5.2951E-05 | 3.6184E+00 | 1.0440E+04 |
| 18600 | 5.6501E-03 | 1.3805E+08 | 9.5268E+03 | 5.2171E-05 | 3.5924E+00 | 1.0509E+04 |
| 18700 | 5.6024E-03 | 1.3897E+08 | 9.2307E+03 | 5.1449E-05 | 3.5688E+00 | 1.0579E+04 |
| 18800 | 5.5560E-03 | 1.3986E+08 | 8.9459E+03 | 5.0784E-05 | 3.5475E+00 | 1.0648E+04 |
| 18900 | 5.5110E-03 | 1.4073E+08 | 8.6726E+03 | 5.0173E-05 | 3.5285E+00 | 1.0717E+04 |
| 19000 | 5.4673E-03 | 1.4157E+08 | 8.4107E+03 | 4.9612E-05 | 3.5110E+00 | 1.0786E+04 |

(*continued*)

TABLE A.1. (*Continued*)

| $T$ (K) | Density (kg/m$^3$) | Enthalpy (J/kg) | Sp heat (J/kg · K) | Viscosity (kg/m · s) | Therm cond (W/m · K) | Elec cond (A/V · m) |
|---|---|---|---|---|---|---|
| | | | **OXYGEN** | | | |
| 19100 | 5.4248E-03 | 1.4239E+08 | 8.1602E+03 | 4.9100E-05 | 3.4946E+00 | 1.0855E+04 |
| 19200 | 5.3835E-03 | 1.4318E+08 | 7.9204E+03 | 4.8633E-05 | 3.4812E+00 | 1.0924E+04 |
| 19300 | 5.3432E-03 | 1.4395E+08 | 7.6921E+03 | 4.8209E-05 | 3.4705E+00 | 1.0993E+04 |
| 19400 | 5.3040E-03 | 1.4470E+08 | 7.4744E+03 | 4.7827E-05 | 3.4618E+00 | 1.1061E+04 |
| 19500 | 5.2657E-03 | 1.4542E+08 | 7.2672E+03 | 4.7483E-05 | 3.4551E+00 | 1.1128E+04 |
| 19600 | 5.2285E-03 | 1.4613E+08 | 7.0702E+03 | 4.7174E-05 | 3.4501E+00 | 1.1195E+04 |
| 19700 | 5.1921E-03 | 1.4682E+08 | 6.8834E+03 | 4.6901E-05 | 3.4470E+00 | 1.1261E+04 |
| 19800 | 5.1565E-03 | 1.4749E+08 | 6.7063E+03 | 4.6660E-05 | 3.4458E+00 | 1.1327E+04 |
| 19900 | 5.1218E-03 | 1.4814E+08 | 6.5386E+03 | 4.6450E-05 | 3.4462E+00 | 1.1394E+04 |
| 20000 | 5.0879E-03 | 1.4878E+08 | 6.3801E+03 | 4.6270E-05 | 3.4483E+00 | 1.1460E+04 |
| 20100 | 5.0547E-03 | 1.4940E+08 | 6.2305E+03 | 4.6118E-05 | 3.4520E+00 | 1.1526E+04 |
| 20200 | 5.0222E-03 | 1.5001E+08 | 6.0896E+03 | 4.5992E-05 | 3.4572E+00 | 1.1593E+04 |
| 20300 | 4.9903E-03 | 1.5061E+08 | 5.9569E+03 | 4.5892E-05 | 3.4639E+00 | 1.1659E+04 |
| 20400 | 4.9592E-03 | 1.5119E+08 | 5.8324E+03 | 4.5815E-05 | 3.4721E+00 | 1.1725E+04 |
| 20500 | 4.9286E-03 | 1.5176E+08 | 5.7157E+03 | 4.5760E-05 | 3.4815E+00 | 1.1791E+04 |
| 20600 | 4.8986E-03 | 1.5232E+08 | 5.6069E+03 | 4.5727E-05 | 3.4923E+00 | 1.1857E+04 |
| 20700 | 4.8692E-03 | 1.5287E+08 | 5.5050E+03 | 4.5714E-05 | 3.5043E+00 | 1.1923E+04 |
| 20800 | 4.8403E-03 | 1.5342E+08 | 5.4105E+03 | 4.5719E-05 | 3.5176E+00 | 1.1989E+04 |
| 20900 | 4.8119E-03 | 1.5395E+08 | 5.3230E+03 | 4.5743E-05 | 3.5319E+00 | 1.2055E+04 |
| 21000 | 4.7840E-03 | 1.5447E+08 | 5.2422E+03 | 4.5783E-05 | 3.5474E+00 | 1.2120E+04 |
| 21100 | 4.7565E-03 | 1.5499E+08 | 5.1682E+03 | 4.5839E-05 | 3.5639E+00 | 1.2186E+04 |
| 21200 | 4.7295E-03 | 1.5550E+08 | 5.1006E+03 | 4.5910E-05 | 3.5814E+00 | 1.2251E+04 |
| 21300 | 4.7030E-03 | 1.5600E+08 | 5.0395E+03 | 4.5994E-05 | 3.5998E+00 | 1.2317E+04 |
| 21400 | 4.6768E-03 | 1.5650E+08 | 4.9846E+03 | 4.6092E-05 | 3.6192E+00 | 1.2382E+04 |
| 21500 | 4.6510E-03 | 1.5699E+08 | 4.9359E+03 | 4.6202E-05 | 3.6394E+00 | 1.2447E+04 |
| 21600 | 4.6256E-03 | 1.5748E+08 | 4.8933E+03 | 4.6323E-05 | 3.6605E+00 | 1.2512E+04 |
| 21700 | 4.6006E-03 | 1.5797E+08 | 4.8567E+03 | 4.6454E-05 | 3.6824E+00 | 1.2577E+04 |
| 21800 | 4.5759E-03 | 1.5845E+08 | 4.8262E+03 | 4.6596E-05 | 3.7051E+00 | 1.2642E+04 |
| 21900 | 4.5515E-03 | 1.5893E+08 | 4.8017E+03 | 4.6746E-05 | 3.7285E+00 | 1.2706E+04 |
| 22000 | 4.5275E-03 | 1.5941E+08 | 4.7832E+03 | 4.6904E-05 | 3.7527E+00 | 1.2771E+04 |
| 22100 | 4.5084E-03 | 1.5982E+08 | 4.0608E+03 | 4.7501E-05 | 3.7792E+00 | 1.2834E+04 |
| 22200 | 4.4848E-03 | 1.6029E+08 | 4.7655E+03 | 4.7670E-05 | 3.8047E+00 | 1.2897E+04 |
| 22300 | 4.4616E-03 | 1.6077E+08 | 4.7652E+03 | 4.7845E-05 | 3.8308E+00 | 1.2961E+04 |
| 22400 | 4.4385E-03 | 1.6125E+08 | 4.7713E+03 | 4.8025E-05 | 3.8575E+00 | 1.3024E+04 |
| 22500 | 4.4158E-03 | 1.6173E+08 | 4.7835E+03 | 4.8209E-05 | 3.8848E+00 | 1.3087E+04 |
| 22600 | 4.3932E-03 | 1.6221E+08 | 4.8021E+03 | 4.8397E-05 | 3.9126E+00 | 1.3150E+04 |
| 22700 | 4.3709E-03 | 1.6269E+08 | 4.8272E+03 | 4.8588E-05 | 3.9410E+00 | 1.3212E+04 |
| 22800 | 4.3488E-03 | 1.6317E+08 | 4.8588E+03 | 4.8782E-05 | 3.9699E+00 | 1.3274E+04 |
| 22900 | 4.3269E-03 | 1.6366E+08 | 4.8972E+03 | 4.8976E-05 | 3.9993E+00 | 1.3335E+04 |
| 23000 | 4.3052E-03 | 1.6416E+08 | 4.9425E+03 | 4.9172E-05 | 4.0292E+00 | 1.3396E+04 |
| 23100 | 4.2837E-03 | 1.6466E+08 | 4.9948E+03 | 4.9367E-05 | 4.0595E+00 | 1.3457E+04 |
| 23200 | 4.2623E-03 | 1.6516E+08 | 5.0544E+03 | 4.9562E-05 | 4.0903E+00 | 1.3516E+04 |
| 23300 | 4.2411E-03 | 1.6568E+08 | 5.1215E+03 | 4.9754E-05 | 4.1215E+00 | 1.3576E+04 |
| 23400 | 4.2200E-03 | 1.6619E+08 | 5.1962E+03 | 4.9945E-05 | 4.1530E+00 | 1.3634E+04 |
| 23500 | 4.1991E-03 | 1.6672E+08 | 5.2789E+03 | 5.0132E-05 | 4.1850E+00 | 1.3692E+04 |
| 23600 | 4.1783E-03 | 1.6726E+08 | 5.3696E+03 | 5.0314E-05 | 4.2174E+00 | 1.3750E+04 |
| 23700 | 4.1577E-03 | 1.6781E+08 | 5.4688E+03 | 5.0492E-05 | 4.2501E+00 | 1.3806E+04 |
| 23800 | 4.1371E-03 | 1.6836E+08 | 5.5766E+03 | 5.0664E-05 | 4.2831E+00 | 1.3862E+04 |
| 23900 | 4.1166E-03 | 1.6893E+08 | 5.6934E+03 | 5.0829E-05 | 4.3165E+00 | 1.3917E+04 |
| 24000 | 4.0963E-03 | 1.6952E+08 | 5.8192E+03 | 5.0986E-05 | 4.3502E+00 | 1.3971E+04 |

**TABLE A.1.** *(Continued)*

| $T$ (K) | Density (kg/m$^3$) | Enthalpy (J/kg) | Sp heat (J/kg · K) | Viscosity (kg/m · s) | Therm cond (W/m · K) | Elec cond (A/V · m) |
|---|---|---|---|---|---|---|
| | | | **AIR** | | | |
| 500 | 7.0208E-01 | 2.0865E+05 | 1.0473E+03 | 2.7055E-05 | 4.1378E-02 | 0.0000E+00 |
| 600 | 5.8505E-01 | 3.1559E+05 | 1.0694E+03 | 3.0849E-05 | 4.8243E-02 | 0.0000E+00 |
| 700 | 5.0147E-01 | 4.2422E+05 | 1.0863E+03 | 3.4448E-05 | 5.4971E-02 | 0.0000E+00 |
| 800 | 4.3879E-01 | 5.3450E+05 | 1.1028E+03 | 3.7897E-05 | 6.1596E-02 | 0.0000E+00 |
| 900 | 3.9004E-01 | 6.4640E+05 | 1.1190E+03 | 4.1227E-05 | 6.8137E-02 | 0.0000E+00 |
| 1000 | 3.5104E-01 | 7.5990E+05 | 1.1350E+03 | 4.4457E-05 | 7.4598E-02 | 0.0000E+00 |
| 1100 | 3.1913E-01 | 8.7499E+05 | 1.1509E+03 | 4.7602E-05 | 8.1161E-02 | 1.6068E-23 |
| 1200 | 2.9254E-01 | 9.9170E+05 | 1.1670E+03 | 5.0673E-05 | 8.7635E-02 | 8.6758E-21 |
| 1300 | 2.7004E-01 | 1.1101E+06 | 1.1836E+03 | 5.3678E-05 | 9.4153E-02 | 1.8055E-18 |
| 1400 | 2.5076E-01 | 1.2301E+06 | 1.2007E+03 | 5.6624E-05 | 1.0073E-01 | 1.7744E-16 |
| 1500 | 2.3404E-01 | 1.3520E+06 | 1.2185E+03 | 5.9515E-05 | 1.0739E-01 | 9.5610E-15 |
| 1600 | 2.1942E-01 | 1.4757E+06 | 1.2372E+03 | 6.2358E-05 | 1.1419E-01 | 3.1591E-13 |
| 1700 | 2.0651E-01 | 1.6014E+06 | 1.2569E+03 | 6.5154E-05 | 1.2118E-01 | 6.9742E-12 |
| 1800 | 1.9504E-01 | 1.7292E+06 | 1.2780E+03 | 6.7908E-05 | 1.2849E-01 | 1.0997E-10 |
| 1900 | 1.8477E-01 | 1.8593E+06 | 1.3014E+03 | 7.0623E-05 | 1.3633E-01 | 1.3056E-09 |
| 2000 | 1.7552E-01 | 1.9922E+06 | 1.3284E+03 | 7.3301E-05 | 1.4507E-01 | 1.2172E-08 |
| 2100 | 1.6713E-01 | 2.1283E+06 | 1.3611E+03 | 7.5947E-05 | 1.5522E-01 | 9.2195E-08 |
| 2200 | 1.5949E-01 | 2.2685E+06 | 1.4025E+03 | 7.8562E-05 | 1.6752E-01 | 5.8284E-07 |
| 2300 | 1.5248E-01 | 2.4141E+06 | 1.4563E+03 | 8.1152E-05 | 1.8293E-01 | 3.1406E-06 |
| 2400 | 1.4601E-01 | 2.5669E+06 | 1.5272E+03 | 8.3721E-05 | 2.0263E-01 | 1.4640E-05 |
| 2500 | 1.4000E-01 | 2.7289E+06 | 1.6204E+03 | 8.6274E-05 | 2.2793E-01 | 5.9528E-05 |
| 2600 | 1.3437E-01 | 2.9030E+06 | 1.7412E+03 | 8.8819E-05 | 2.6016E-01 | 2.1158E-04 |
| 2700 | 1.2906E-01 | 3.0925E+06 | 1.8944E+03 | 9.1363E-05 | 3.0050E-01 | 6.5681E-04 |
| 2800 | 1.2400E-01 | 3.3008E+06 | 2.0835E+03 | 9.3914E-05 | 3.4962E-01 | 1.7864E-03 |
| 2900 | 1.1915E-01 | 3.5318E+06 | 2.3095E+03 | 9.6481E-05 | 4.0738E-01 | 4.3101E-03 |
| 3000 | 1.1446E-01 | 3.7887E+06 | 2.5694E+03 | 9.9071E-05 | 4.7237E-01 | 9.4093E-03 |
| 3100 | 1.0991E-01 | 4.0742E+06 | 2.8548E+03 | 1.0169E-04 | 5.4154E-01 | 1.8966E-02 |
| 3200 | 1.0548E-01 | 4.3893E+06 | 3.1508E+03 | 1.0434E-04 | 6.1008E-01 | 3.5870E-02 |
| 3300 | 1.0118E-01 | 4.7328E+06 | 3.4351E+03 | 1.0703E-04 | 6.7162E-01 | 6.4378E-02 |
| 3400 | 9.7021E-02 | 5.1007E+06 | 3.6790E+03 | 1.0973E-04 | 7.1911E-01 | 1.1046E-01 |
| 3500 | 9.3032E-02 | 5.4859E+06 | 3.8519E+03 | 1.1245E-04 | 7.4638E-01 | 1.8212E-01 |
| 3600 | 8.9250E-02 | 5.8786E+06 | 3.9273E+03 | 1.1516E-04 | 7.4998E-01 | 2.8954E-01 |
| 3700 | 8.5709E-02 | 6.2677E+06 | 3.8913E+03 | 1.1785E-04 | 7.3068E-01 | 4.4434E-01 |
| 3800 | 8.2435E-02 | 6.6427E+06 | 3.7497E+03 | 1.2051E-04 | 6.9347E-01 | 6.6178E-01 |
| 3900 | 7.9435E-02 | 6.9955E+06 | 3.5278E+03 | 1.2313E-04 | 6.4634E-01 | 9.5778E-01 |
| 4000 | 7.6704E-02 | 7.3218E+06 | 3.2636E+03 | 1.2570E-04 | 5.9775E-01 | 1.3506E+00 |
| 4100 | 7.4219E-02 | 7.6215E+06 | 2.9966E+03 | 1.2822E-04 | 5.5467E-01 | 1.8605E+00 |
| 4200 | 7.1951E-02 | 7.8973E+06 | 2.7579E+03 | 1.3070E-04 | 5.2156E-01 | 2.5101E+00 |
| 4300 | 6.9868E-02 | 8.1540E+06 | 2.5668E+03 | 1.3314E-04 | 5.0064E-01 | 3.3247E+00 |
| 4400 | 6.7941E-02 | 8.3971E+06 | 2.4315E+03 | 1.3555E-04 | 4.9261E-01 | 4.3320E+00 |
| 4500 | 6.6142E-02 | 8.6324E+06 | 2.3532E+03 | 1.3794E-04 | 4.9737E-01 | 5.5628E+00 |
| 4600 | 6.4448E-02 | 8.8654E+06 | 2.3295E+03 | 1.4030E-04 | 5.1468E-01 | 7.0505E+00 |
| 4700 | 6.2841E-02 | 9.1011E+06 | 2.3571E+03 | 1.4264E-04 | 5.4441E-01 | 8.8308E+00 |
| 4800 | 6.1303E-02 | 9.3444E+06 | 2.4331E+03 | 1.4498E-04 | 5.8666E-01 | 1.0942E+01 |
| 4900 | 5.9822E-02 | 9.6000E+06 | 2.5558E+03 | 1.4730E-04 | 6.4182E-01 | 1.3426E+01 |
| 5000 | 5.8385E-02 | 9.8724E+06 | 2.7245E+03 | 1.4962E-04 | 7.1051E-01 | 1.6325E+01 |

*(continued)*

TABLE A.1. *(Continued)*

| $T$ (K) | Density (kg/m³) | Enthalpy (J/kg) | Sp heat (J/kg · K) | Viscosity (kg/m · s) | Therm cond (W/m · K) | Elec cond (A/V · m) |
|---|---|---|---|---|---|---|
| | | | AIR | | | |
| 5100 | 5.6983E-02 | 1.0166E+07 | 2.9398E+03 | 1.5194E-04 | 7.9351E-01 | 1.9684E+01 |
| 5200 | 5.5607E-02 | 1.0487E+07 | 3.2024E+03 | 1.5425E-04 | 8.9169E-01 | 2.3551E+01 |
| 5300 | 5.4249E-02 | 1.0838E+07 | 3.5145E+03 | 1.5657E-04 | 1.0059E+00 | 2.7976E+01 |
| 5400 | 5.2903E-02 | 1.1226E+07 | 3.8780E+03 | 1.5890E-04 | 1.1370E+00 | 3.3011E+01 |
| 5500 | 5.1562E-02 | 1.1655E+07 | 4.2948E+03 | 1.6124E-04 | 1.2856E+00 | 3.8712E+01 |
| 5600 | 5.0222E-02 | 1.2132E+07 | 4.7667E+03 | 1.6359E-04 | 1.4519E+00 | 4.5108E+01 |
| 5700 | 4.8879E-02 | 1.2662E+07 | 5.2948E+03 | 1.6595E-04 | 1.6359E+00 | 5.2304E+01 |
| 5800 | 4.7528E-02 | 1.3249E+07 | 5.8795E+03 | 1.6832E-04 | 1.8367E+00 | 6.0352E+01 |
| 5900 | 4.6170E-02 | 1.3901E+07 | 6.5196E+03 | 1.7069E-04 | 2.0531E+00 | 6.9280E+01 |
| 6000 | 4.4803E-02 | 1.4623E+07 | 7.2121E+03 | 1.7308E-04 | 2.2826E+00 | 7.9181E+01 |
| 6100 | 4.3428E-02 | 1.5418E+07 | 7.9516E+03 | 1.7546E-04 | 2.5219E+00 | 9.0194E+01 |
| 6200 | 4.2047E-02 | 1.6291E+07 | 8.7297E+03 | 1.7783E-04 | 2.7666E+00 | 1.0239E+02 |
| 6300 | 4.0664E-02 | 1.7244E+07 | 9.5342E+03 | 1.8019E-04 | 3.0111E+00 | 1.1590E+02 |
| 6400 | 3.9285E-02 | 1.8279E+07 | 1.0349E+04 | 1.8252E-04 | 3.2484E+00 | 1.3086E+02 |
| 6500 | 3.7916E-02 | 1.9394E+07 | 1.1153E+04 | 1.8480E-04 | 3.4708E+00 | 1.4746E+02 |
| 6600 | 3.6564E-02 | 2.0586E+07 | 1.1921E+04 | 1.8703E-04 | 3.6698E+00 | 1.6591E+02 |
| 6700 | 3.5239E-02 | 2.1849E+07 | 1.2624E+04 | 1.8919E-04 | 3.8367E+00 | 1.8649E+02 |
| 6800 | 3.3949E-02 | 2.3172E+07 | 1.3230E+04 | 1.9127E-04 | 3.9632E+00 | 2.0948E+02 |
| 6900 | 3.2704E-02 | 2.4542E+07 | 1.3707E+04 | 1.9327E-04 | 4.0425E+00 | 2.3523E+02 |
| 7000 | 3.1512E-02 | 2.5945E+07 | 1.4026E+04 | 1.9518E-04 | 4.0697E+00 | 2.6410E+02 |
| 7100 | 3.0381E-02 | 2.7361E+07 | 1.4164E+04 | 1.9701E-04 | 4.0430E+00 | 2.9647E+02 |
| 7200 | 2.9316E-02 | 2.8772E+07 | 1.4108E+04 | 1.9876E-04 | 3.9638E+00 | 3.3269E+02 |
| 7300 | 2.8321E-02 | 3.0158E+07 | 1.3859E+04 | 2.0046E-04 | 3.8368E+00 | 3.7305E+02 |
| 7400 | 2.7399E-02 | 3.1501E+07 | 1.3428E+04 | 2.0211E-04 | 3.6698E+00 | 4.1777E+02 |
| 7500 | 2.6550E-02 | 3.2785E+07 | 1.2842E+04 | 2.0372E-04 | 3.4726E+00 | 4.6700E+02 |
| 7600 | 2.5771E-02 | 3.3999E+07 | 1.2134E+04 | 2.0533E-04 | 3.2561E+00 | 5.2076E+02 |
| 7700 | 2.5058E-02 | 3.5133E+07 | 1.1344E+04 | 2.0694E-04 | 3.0307E+00 | 5.7903E+02 |
| 7800 | 2.4409E-02 | 3.6184E+07 | 1.0513E+04 | 2.0856E-04 | 2.8062E+00 | 6.4167E+02 |
| 7900 | 2.3816E-02 | 3.7152E+07 | 9.6783E+03 | 2.1021E-04 | 2.5906E+00 | 7.0853E+02 |
| 8000 | 2.3274E-02 | 3.8039E+07 | 8.8706E+03 | 2.1188E-04 | 2.3897E+00 | 7.7937E+02 |
| 8100 | 2.2778E-02 | 3.8850E+07 | 8.1133E+03 | 2.1358E-04 | 2.2075E+00 | 8.5399E+02 |
| 8200 | 2.2322E-02 | 3.9593E+07 | 7.4255E+03 | 2.1531E-04 | 2.0464E+00 | 9.3212E+02 |
| 8300 | 2.1901E-02 | 4.0274E+07 | 6.8063E+03 | 2.1707E-04 | 1.9067E+00 | 1.0135E+03 |
| 8400 | 2.1509E-02 | 4.0900E+07 | 6.2682E+03 | 2.1885E-04 | 1.7883E+00 | 1.0980E+03 |
| 8500 | 2.1144E-02 | 4.1481E+07 | 5.8073E+03 | 2.2065E-04 | 1.6902E+00 | 1.1852E+03 |
| 8600 | 2.0802E-02 | 4.2024E+07 | 5.4275E+03 | 2.2246E-04 | 1.6114E+00 | 1.2750E+03 |
| 8700 | 2.0478E-02 | 4.2534E+07 | 5.1025E+03 | 2.2428E-04 | 1.5495E+00 | 1.3673E+03 |
| 8800 | 2.0171E-02 | 4.3020E+07 | 4.8566E+03 | 2.2609E-04 | 1.5037E+00 | 1.4617E+03 |
| 8900 | 1.9879E-02 | 4.3485E+07 | 4.6474E+03 | 2.2790E-04 | 1.4712E+00 | 1.5582E+03 |
| 9000 | 1.9599E-02 | 4.3934E+07 | 4.4973E+03 | 2.2969E-04 | 1.4510E+00 | 1.6565E+03 |
| 9100 | 1.9329E-02 | 4.4375E+07 | 4.4080E+03 | 2.3147E-04 | 1.4428E+00 | 1.7564E+03 |
| 9200 | 1.9069E-02 | 4.4808E+07 | 4.3273E+03 | 2.3321E-04 | 1.4434E+00 | 1.8578E+03 |
| 9300 | 1.8816E-02 | 4.5240E+07 | 4.3184E+03 | 2.3492E-04 | 1.4540E+00 | 1.9605E+03 |
| 9400 | 1.8570E-02 | 4.5670E+07 | 4.3023E+03 | 2.3659E-04 | 1.4713E+00 | 2.0644E+03 |
| 9500 | 1.8331E-02 | 4.6105E+07 | 4.3527E+03 | 2.3821E-04 | 1.4968E+00 | 2.1693E+03 |
| 9600 | 1.8097E-02 | 4.6547E+07 | 4.4180E+03 | 2.3978E-04 | 1.5291E+00 | 2.2752E+03 |
| 9700 | 1.7867E-02 | 4.6993E+07 | 4.4631E+03 | 2.4128E-04 | 1.5652E+00 | 2.3818E+03 |

TABLE A.1. (*Continued*)

| $T$ (K) | Density (kg/m$^3$) | Enthalpy (J/kg) | Sp heat (J/kg · K) | Viscosity (kg/m · s) | Therm cond (W/m · K) | Elec cond (A/V · m) |
|---|---|---|---|---|---|---|
| | | | **AIR** | | | |
| 9800 | 1.7641E-02 | 4.7454E+07 | 4.6054E+03 | 2.4271E-04 | 1.6090E+00 | 2.4891E+03 |
| 9900 | 1.7419E-02 | 4.7923E+07 | 4.6886E+03 | 2.4406E-04 | 1.6555E+00 | 2.5971E+03 |
| 10000 | 1.7200E-02 | 4.8409E+07 | 4.8672E+03 | 2.4532E-04 | 1.7086E+00 | 2.7055E+03 |
| 10100 | 1.6984E-02 | 4.8913E+07 | 5.0401E+03 | 2.4648E-04 | 1.7667E+00 | 2.8142E+03 |
| 10200 | 1.6770E-02 | 4.9429E+07 | 5.1583E+03 | 2.4753E-04 | 1.8258E+00 | 2.9233E+03 |
| 10300 | 1.6558E-02 | 4.9970E+07 | 5.4050E+03 | 2.4847E-04 | 1.8915E+00 | 3.0326E+03 |
| 10400 | 1.6348E-02 | 5.0533E+07 | 5.6319E+03 | 2.4928E-04 | 1.9616E+00 | 3.1421E+03 |
| 10500 | 1.6139E-02 | 5.1120E+07 | 5.8662E+03 | 2.4996E-04 | 2.0356E+00 | 3.2517E+03 |
| 10600 | 1.5932E-02 | 5.1719E+07 | 5.9956E+03 | 2.5049E-04 | 2.1076E+00 | 3.3614E+03 |
| 10700 | 1.5725E-02 | 5.2356E+07 | 6.3662E+03 | 2.5088E-04 | 2.1881E+00 | 3.4710E+03 |
| 10800 | 1.5520E-02 | 5.3020E+07 | 6.6471E+03 | 2.5110E-04 | 2.2721E+00 | 3.5806E+03 |
| 10900 | 1.5315E-02 | 5.3701E+07 | 6.8016E+03 | 2.5115E-04 | 2.3532E+00 | 3.6901E+03 |
| 11000 | 1.5111E-02 | 5.4422E+07 | 7.2153E+03 | 2.5102E-04 | 2.4420E+00 | 3.7995E+03 |
| 11100 | 1.4908E-02 | 5.5178E+07 | 7.5604E+03 | 2.5071E-04 | 2.5347E+00 | 3.9087E+03 |
| 11200 | 1.4705E-02 | 5.5968E+07 | 7.9015E+03 | 2.5021E-04 | 2.6302E+00 | 4.0177E+03 |
| 11300 | 1.4503E-02 | 5.6794E+07 | 8.2576E+03 | 2.4951E-04 | 2.7284E+00 | 4.1264E+03 |
| 11400 | 1.4300E-02 | 5.7634E+07 | 8.3977E+03 | 2.4860E-04 | 2.8204E+00 | 4.2351E+03 |
| 11500 | 1.4098E-02 | 5.8529E+07 | 8.9527E+03 | 2.4749E-04 | 2.9213E+00 | 4.3433E+03 |
| 11600 | 1.3896E-02 | 5.9467E+07 | 9.3811E+03 | 2.4616E-04 | 3.0258E+00 | 4.4513E+03 |
| 11700 | 1.3695E-02 | 6.0446E+07 | 9.7923E+03 | 2.4463E-04 | 3.1321E+00 | 4.5589E+03 |
| 11800 | 1.3493E-02 | 6.1468E+07 | 1.0217E+04 | 2.4288E-04 | 3.2403E+00 | 4.6661E+03 |
| 11900 | 1.3292E-02 | 6.2534E+07 | 1.0654E+04 | 2.4091E-04 | 3.3498E+00 | 4.7730E+03 |
| 12000 | 1.3091E-02 | 6.3606E+07 | 1.0725E+04 | 2.3872E-04 | 3.4487E+00 | 4.8798E+03 |
| 12100 | 1.2890E-02 | 6.4753E+07 | 1.1471E+04 | 2.3633E-04 | 3.5575E+00 | 4.9859E+03 |
| 12200 | 1.2690E-02 | 6.5952E+07 | 1.1991E+04 | 2.3373E-04 | 3.6690E+00 | 5.0916E+03 |
| 12300 | 1.2490E-02 | 6.7199E+07 | 1.2469E+04 | 2.3093E-04 | 3.7807E+00 | 5.1968E+03 |
| 12400 | 1.2291E-02 | 6.8495E+07 | 1.2955E+04 | 2.2794E-04 | 3.8924E+00 | 5.3016E+03 |
| 12500 | 1.2092E-02 | 6.9839E+07 | 1.3448E+04 | 2.2476E-04 | 4.0036E+00 | 5.4058E+03 |
| 12600 | 1.1894E-02 | 7.1234E+07 | 1.3946E+04 | 2.2141E-04 | 4.1140E+00 | 5.5095E+03 |
| 12700 | 1.1696E-02 | 7.2620E+07 | 1.3855E+04 | 2.1784E-04 | 4.2089E+00 | 5.6132E+03 |
| 12800 | 1.1500E-02 | 7.4110E+07 | 1.4910E+04 | 2.1415E-04 | 4.3159E+00 | 5.7159E+03 |
| 12900 | 1.1304E-02 | 7.5642E+07 | 1.5317E+04 | 2.1031E-04 | 4.4176E+00 | 5.8180E+03 |
| 13000 | 1.1110E-02 | 7.7233E+07 | 1.5907E+04 | 2.0634E-04 | 4.5197E+00 | 5.9195E+03 |
| 13100 | 1.0918E-02 | 7.8873E+07 | 1.6405E+04 | 2.0225E-04 | 4.6188E+00 | 6.0204E+03 |
| 13200 | 1.0727E-02 | 8.0563E+07 | 1.6895E+04 | 1.9805E-04 | 4.7142E+00 | 6.1206E+03 |
| 13300 | 1.0538E-02 | 8.2301E+07 | 1.7377E+04 | 1.9376E-04 | 4.8057E+00 | 6.2201E+03 |
| 13400 | 1.0351E-02 | 8.4085E+07 | 1.7847E+04 | 1.8939E-04 | 4.8926E+00 | 6.3190E+03 |
| 13500 | 1.0166E-02 | 8.5916E+07 | 1.8303E+04 | 1.8495E-04 | 4.9744E+00 | 6.4171E+03 |
| 13600 | 9.9834E-03 | 8.7790E+07 | 1.8742E+04 | 1.8046E-04 | 5.0508E+00 | 6.5145E+03 |
| 13700 | 9.8032E-03 | 8.9706E+07 | 1.9160E+04 | 1.7593E-04 | 5.1213E+00 | 6.6112E+03 |
| 13800 | 9.6233E-03 | 9.1569E+07 | 1.8631E+04 | 1.7125E-04 | 5.1737E+00 | 6.7082E+03 |
| 13900 | 9.4484E-03 | 9.3559E+07 | 1.9897E+04 | 1.6668E-04 | 5.2316E+00 | 6.8035E+03 |
| 14000 | 9.2765E-03 | 9.5583E+07 | 2.0241E+04 | 1.6211E-04 | 5.2825E+00 | 6.8979E+03 |
| 14100 | 9.1077E-03 | 9.7638E+07 | 2.0555E+04 | 1.5755E-04 | 5.3262E+00 | 6.9915E+03 |
| 14200 | 8.9421E-03 | 9.9722E+07 | 2.0835E+04 | 1.5302E-04 | 5.3623E+00 | 7.0843E+03 |
| 14300 | 8.7796E-03 | 1.0181E+08 | 2.0908E+04 | 1.4850E-04 | 5.3864E+00 | 7.1763E+03 |

*(continued)*

**TABLE A.1.** *(Continued)*

| $T$ (K) | Density (kg/m$^3$) | Enthalpy (J/kg) | Sp heat (J/kg · K) | Viscosity (kg/m · s) | Therm cond (W/m · K) | Elec cond (A/V · m) |
|---|---|---|---|---|---|---|
| | | | **AIR** | | | |
| 14400 | 8.6210E-03 | 1.0394E+08 | 2.1281E+04 | 1.4405E-04 | 5.4070E+00 | 7.2675E+03 |
| 14500 | 8.4660E-03 | 1.0609E+08 | 2.1448E+04 | 1.3965E-04 | 5.4198E+00 | 7.3577E+03 |
| 14600 | 8.3148E-03 | 1.0824E+08 | 2.1575E+04 | 1.3533E-04 | 5.4246E+00 | 7.4471E+03 |
| 14700 | 8.1673E-03 | 1.1041E+08 | 2.1659E+04 | 1.3108E-04 | 5.4217E+00 | 7.5357E+03 |
| 14800 | 8.0238E-03 | 1.1258E+08 | 2.1700E+04 | 1.2691E-04 | 5.4113E+00 | 7.6234E+03 |
| 14900 | 7.8842E-03 | 1.1475E+08 | 2.1698E+04 | 1.2283E-04 | 5.3934E+00 | 7.7103E+03 |
| 15000 | 7.7485E-03 | 1.1691E+08 | 2.1652E+04 | 1.1885E-04 | 5.3685E+00 | 7.7964E+03 |
| 15100 | 7.6169E-03 | 1.1907E+08 | 2.1563E+04 | 1.1497E-04 | 5.3369E+00 | 7.8816E+03 |
| 15200 | 7.4892E-03 | 1.2121E+08 | 2.1432E+04 | 1.1120E-04 | 5.2989E+00 | 7.9659E+03 |
| 15300 | 7.3656E-03 | 1.2334E+08 | 2.1260E+04 | 1.0754E-04 | 5.2551E+00 | 8.0495E+03 |
| 15400 | 7.2459E-03 | 1.2544E+08 | 2.1049E+04 | 1.0399E-04 | 5.2059E+00 | 8.1322E+03 |
| 15500 | 7.1302E-03 | 1.2752E+08 | 2.0801E+04 | 1.0056E-04 | 5.1518E+00 | 8.2142E+03 |
| 15600 | 7.0184E-03 | 1.2958E+08 | 2.0518E+04 | 9.7246E-05 | 5.0935E+00 | 8.2954E+03 |
| 15700 | 6.9104E-03 | 1.3160E+08 | 2.0202E+04 | 9.4052E-05 | 5.0315E+00 | 8.3758E+03 |
| 15800 | 6.8061E-03 | 1.3358E+08 | 1.9857E+04 | 9.0979E-05 | 4.9660E+00 | 8.4555E+03 |
| 15900 | 6.7056E-03 | 1.3553E+08 | 1.9485E+04 | 8.8027E-05 | 4.8982E+00 | 8.5344E+03 |
| 16000 | 6.6086E-03 | 1.3744E+08 | 1.9089E+04 | 8.5196E-05 | 4.8285E+00 | 8.6127E+03 |
| 16100 | 6.5152E-03 | 1.3931E+08 | 1.8672E+04 | 8.2485E-05 | 4.7572E+00 | 8.6903E+03 |
| 16200 | 6.4251E-03 | 1.4113E+08 | 1.8238E+04 | 7.9893E-05 | 4.6850E+00 | 8.7672E+03 |
| 16300 | 6.3384E-03 | 1.4291E+08 | 1.7790E+04 | 7.7419E-05 | 4.6124E+00 | 8.8435E+03 |
| 16400 | 6.2548E-03 | 1.4464E+08 | 1.7331E+04 | 7.5060E-05 | 4.5399E+00 | 8.9192E+03 |
| 16500 | 6.1743E-03 | 1.4633E+08 | 1.6863E+04 | 7.2815E-05 | 4.4678E+00 | 8.9944E+03 |
| 16600 | 6.0968E-03 | 1.4797E+08 | 1.6391E+04 | 7.0681E-05 | 4.3966E+00 | 9.0689E+03 |
| 16700 | 6.0222E-03 | 1.4956E+08 | 1.5915E+04 | 6.8655E-05 | 4.3266E+00 | 9.1430E+03 |
| 16800 | 5.9502E-03 | 1.5110E+08 | 1.5439E+04 | 6.6735E-05 | 4.2582E+00 | 9.2165E+03 |
| 16900 | 5.8809E-03 | 1.5260E+08 | 1.4966E+04 | 6.4917E-05 | 4.1916E+00 | 9.2896E+03 |
| 17000 | 5.8142E-03 | 1.5405E+08 | 1.4496E+04 | 6.3199E-05 | 4.1271E+00 | 9.3621E+03 |
| 17100 | 5.7498E-03 | 1.5545E+08 | 1.4033E+04 | 6.1576E-05 | 4.0649E+00 | 9.4343E+03 |
| 17200 | 5.6877E-03 | 1.5681E+08 | 1.3578E+04 | 6.0047E-05 | 4.0052E+00 | 9.5060E+03 |
| 17300 | 5.6278E-03 | 1.5812E+08 | 1.3132E+04 | 5.8606E-05 | 3.9482E+00 | 9.5772E+03 |
| 17400 | 5.5700E-03 | 1.5939E+08 | 1.2696E+04 | 5.7252E-05 | 3.8938E+00 | 9.6481E+03 |
| 17500 | 5.5142E-03 | 1.6062E+08 | 1.2272E+04 | 5.5980E-05 | 3.8423E+00 | 9.7187E+03 |
| 17600 | 5.4602E-03 | 1.6181E+08 | 1.1861E+04 | 5.4788E-05 | 3.7936E+00 | 9.7889E+03 |
| 17700 | 5.4081E-03 | 1.6295E+08 | 1.1464E+04 | 5.3671E-05 | 3.7478E+00 | 9.8588E+03 |
| 17800 | 5.3577E-03 | 1.6406E+08 | 1.1080E+04 | 5.2627E-05 | 3.7049E+00 | 9.9284E+03 |
| 17900 | 5.3088E-03 | 1.6513E+08 | 1.0710E+04 | 5.1652E-05 | 3.6650E+00 | 9.9976E+03 |
| 18000 | 5.2616E-03 | 1.6617E+08 | 1.0355E+04 | 5.0743E-05 | 3.6279E+00 | 1.0067E+04 |
| 18100 | 5.2157E-03 | 1.6717E+08 | 1.0015E+04 | 4.9897E-05 | 3.5936E+00 | 1.0135E+04 |
| 18200 | 5.1713E-03 | 1.6814E+08 | 9.6901E+03 | 4.9111E-05 | 3.5622E+00 | 1.0204E+04 |
| 18300 | 5.1282E-03 | 1.6908E+08 | 9.3797E+03 | 4.8382E-05 | 3.5334E+00 | 1.0272E+04 |
| 18400 | 5.0875E-03 | 1.6998E+08 | 9.0167E+03 | 4.7863E-05 | 3.5077E+00 | 1.0340E+04 |
| 18500 | 5.0468E-03 | 1.7086E+08 | 8.7976E+03 | 4.7240E-05 | 3.4842E+00 | 1.0407E+04 |
| 18600 | 5.0072E-03 | 1.7171E+08 | 8.5316E+03 | 4.6666E-05 | 3.4632E+00 | 1.0475E+04 |
| 18700 | 4.9687E-03 | 1.7254E+08 | 8.2799E+03 | 4.6138E-05 | 3.4447E+00 | 1.0542E+04 |
| 18800 | 4.9311E-03 | 1.7334E+08 | 8.0422E+03 | 4.5654E-05 | 3.4285E+00 | 1.0609E+04 |
| 18900 | 4.8946E-03 | 1.7412E+08 | 7.8182E+03 | 4.5211E-05 | 3.4146E+00 | 1.0676E+04 |
| 19000 | 4.8589E-03 | 1.7489E+08 | 7.6075E+03 | 4.4807E-05 | 3.4028E+00 | 1.0743E+04 |

**TABLE A.1.** *(Continued)*

| $T$ (K) | Density (kg/m$^3$) | Enthalpy (J/kg) | Sp heat (J/kg · K) | Viscosity (kg/m · s) | Therm cond (W/m · K) | Elec cond (A/V · m) |
|---|---|---|---|---|---|---|
| | | | **AIR** | | | |
| 19100 | 4.8241E-03 | 1.7563E+08 | 7.4098E+03 | 4.4440E-05 | 3.3929E+00 | 1.0809E+04 |
| 19200 | 4.7901E-03 | 1.7635E+08 | 7.2249E+03 | 4.4109E-05 | 3.3851E+00 | 1.0876E+04 |
| 19300 | 4.7569E-03 | 1.7705E+08 | 7.0523E+03 | 4.3810E-05 | 3.3794E+00 | 1.0942E+04 |
| 19400 | 4.7308E-03 | 1.7768E+08 | 6.2670E+03 | 4.3988E-05 | 3.3763E+00 | 1.1003E+04 |
| 19500 | 4.6989E-03 | 1.7835E+08 | 6.7312E+03 | 4.3748E-05 | 3.3743E+00 | 1.1069E+04 |
| 19600 | 4.6677E-03 | 1.7901E+08 | 6.5949E+03 | 4.3534E-05 | 3.3739E+00 | 1.1134E+04 |
| 19700 | 4.6371E-03 | 1.7966E+08 | 6.4699E+03 | 4.3346E-05 | 3.3752E+00 | 1.1198E+04 |
| 19800 | 4.6071E-03 | 1.8030E+08 | 6.3554E+03 | 4.3183E-05 | 3.3781E+00 | 1.1263E+04 |
| 19900 | 4.5777E-03 | 1.8092E+08 | 6.2517E+03 | 4.3042E-05 | 3.3824E+00 | 1.1327E+04 |
| 20000 | 4.5488E-03 | 1.8154E+08 | 6.1585E+03 | 4.2922E-05 | 3.3881E+00 | 1.1391E+04 |
| 20100 | 4.5204E-03 | 1.8214E+08 | 6.0754E+03 | 4.2823E-05 | 3.3952E+00 | 1.1455E+04 |
| 20200 | 4.4925E-03 | 1.8274E+08 | 6.0024E+03 | 4.2742E-05 | 3.4037E+00 | 1.1518E+04 |
| 20300 | 4.4651E-03 | 1.8334E+08 | 5.9392E+03 | 4.2678E-05 | 3.4133E+00 | 1.1581E+04 |
| 20400 | 4.4381E-03 | 1.8393E+08 | 5.8859E+03 | 4.2630E-05 | 3.4242E+00 | 1.1644E+04 |
| 20500 | 4.4116E-03 | 1.8451E+08 | 5.8422E+03 | 4.2597E-05 | 3.4362E+00 | 1.1707E+04 |
| 20600 | 4.3854E-03 | 1.8509E+08 | 5.8081E+03 | 4.2577E-05 | 3.4492E+00 | 1.1769E+04 |
| 20700 | 4.3596E-03 | 1.8567E+08 | 5.7836E+03 | 4.2570E-05 | 3.4633E+00 | 1.1830E+04 |
| 20800 | 4.3341E-03 | 1.8625E+08 | 5.7685E+03 | 4.2574E-05 | 3.4784E+00 | 1.1892E+04 |
| 20900 | 4.3090E-03 | 1.8682E+08 | 5.7630E+03 | 4.2588E-05 | 3.4944E+00 | 1.1952E+04 |
| 21000 | 4.2842E-03 | 1.8740E+08 | 5.7670E+03 | 4.2611E-05 | 3.5114E+00 | 1.2013E+04 |
| 21100 | 4.2597E-03 | 1.8798E+08 | 5.7806E+03 | 4.2642E-05 | 3.5292E+00 | 1.2073E+04 |
| 21200 | 4.2355E-03 | 1.8856E+08 | 5.8038E+03 | 4.2679E-05 | 3.5478E+00 | 1.2132E+04 |
| 21300 | 4.2115E-03 | 1.8914E+08 | 5.8367E+03 | 4.2723E-05 | 3.5672E+00 | 1.2191E+04 |
| 21400 | 4.1878E-03 | 1.8973E+08 | 5.8794E+03 | 4.2771E-05 | 3.5874E+00 | 1.2249E+04 |
| 21500 | 4.1644E-03 | 1.9032E+08 | 5.9321E+03 | 4.2822E-05 | 3.6082E+00 | 1.2306E+04 |
| 21600 | 4.1411E-03 | 1.9092E+08 | 5.9948E+03 | 4.2876E-05 | 3.6298E+00 | 1.2363E+04 |
| 21700 | 4.1181E-03 | 1.9153E+08 | 6.0678E+03 | 4.2932E-05 | 3.6520E+00 | 1.2419E+04 |
| 21800 | 4.0952E-03 | 1.9215E+08 | 6.1511E+03 | 4.2989E-05 | 3.6748E+00 | 1.2474E+04 |
| 21900 | 4.0725E-03 | 1.9277E+08 | 6.2451E+03 | 4.3045E-05 | 3.6983E+00 | 1.2528E+04 |
| 22000 | 4.0500E-03 | 1.9340E+08 | 6.3498E+03 | 4.3099E-05 | 3.7223E+00 | 1.2582E+04 |
| 22100 | 4.0277E-03 | 1.9405E+08 | 6.4654E+03 | 4.3151E-05 | 3.7468E+00 | 1.2634E+04 |
| 22200 | 4.0062E-03 | 1.9470E+08 | 6.4374E+03 | 4.3350E-05 | 3.7720E+00 | 1.2686E+04 |
| 22300 | 3.9841E-03 | 1.9537E+08 | 6.7311E+03 | 4.3393E-05 | 3.7976E+00 | 1.2736E+04 |
| 22400 | 3.9621E-03 | 1.9606E+08 | 6.8810E+03 | 4.3431E-05 | 3.8237E+00 | 1.2786E+04 |
| 22500 | 3.9402E-03 | 1.9676E+08 | 7.0426E+03 | 4.3462E-05 | 3.8501E+00 | 1.2834E+04 |
| 22600 | 3.9184E-03 | 1.9748E+08 | 7.2163E+03 | 4.3485E-05 | 3.8771E+00 | 1.2881E+04 |
| 22700 | 3.8966E-03 | 1.9822E+08 | 7.4021E+03 | 4.3501E-05 | 3.9044E+00 | 1.2927E+04 |
| 22800 | 3.8749E-03 | 1.9898E+08 | 7.6004E+03 | 4.3507E-05 | 3.9321E+00 | 1.2972E+04 |
| 22900 | 3.8533E-03 | 1.9976E+08 | 7.8113E+03 | 4.3503E-05 | 3.9601E+00 | 1.3015E+04 |
| 23000 | 3.8358E-03 | 2.0048E+08 | 7.1682E+03 | 4.3881E-05 | 3.9860E+00 | 1.3048E+04 |
| 23100 | 3.8142E-03 | 2.0131E+08 | 8.2694E+03 | 4.3850E-05 | 4.0145E+00 | 1.3088E+04 |
| 23200 | 3.7926E-03 | 2.0216E+08 | 8.5189E+03 | 4.3807E-05 | 4.0433E+00 | 1.3127E+04 |
| 23300 | 3.7710E-03 | 2.0304E+08 | 8.7815E+03 | 4.3750E-05 | 4.0725E+00 | 1.3164E+04 |
| 23400 | 3.7493E-03 | 2.0394E+08 | 9.0573E+03 | 4.3679E-05 | 4.1019E+00 | 1.3200E+04 |
| 23500 | 3.7277E-03 | 2.0488E+08 | 9.3464E+03 | 4.3593E-05 | 4.1315E+00 | 1.3234E+04 |
| 23600 | 3.7061E-03 | 2.0584E+08 | 9.6487E+03 | 4.3491E-05 | 4.1614E+00 | 1.3266E+04 |
| 23700 | 3.6844E-03 | 2.0684E+08 | 9.9643E+03 | 4.3374E-05 | 4.1915E+00 | 1.3297E+04 |
| 23800 | 3.6627E-03 | 2.0787E+08 | 1.0293E+04 | 4.3240E-05 | 4.2218E+00 | 1.3326E+04 |
| 23900 | 3.6409E-03 | 2.0893E+08 | 1.0635E+04 | 4.3089E-05 | 4.2524E+00 | 1.3353E+04 |
| 24000 | 3.6191E-03 | 2.1003E+08 | 1.0990E+04 | 4.2921E-05 | 4.2831E+00 | 1.3379E+04 |

## TABLE A.2. Density of Ar/$H_2$ Mixtures (kg/m$^3$)

| | Molar fraction of $H_2$ | | | | | | | | |
|---|---|---|---|---|---|---|---|---|---|
| $T$ (K) | 0.1 | 0.2 | 0.3 | 0.4 | 0.5 | 0.6 | 0.7 | 0.8 | 0.9 |
| 500 | 8.810E-01 | 7.883E-01 | 6.956E-01 | 6.028E-01 | 5.101E-01 | 4.174E-01 | 3.248E-01 | 2.325E-01 | 1.404E-01 |
| 600 | 7.341E-01 | 6.569E-01 | 5.796E-01 | 5.023E-01 | 4.250E-01 | 3.477E-01 | 2.706E-01 | 1.937E-01 | 1.169E-01 |
| 700 | 6.292E-01 | 5.630E-01 | 4.968E-01 | 4.305E-01 | 3.642E-01 | 2.980E-01 | 2.319E-01 | 1.659E-01 | 1.002E-01 |
| 800 | 5.505E-01 | 4.926E-01 | 4.346E-01 | 3.766E-01 | 3.186E-01 | 2.607E-01 | 2.028E-01 | 1.451E-01 | 8.765E-02 |
| 900 | 4.894E-01 | 4.379E-01 | 3.863E-01 | 3.348E-01 | 2.832E-01 | 2.317E-01 | 1.803E-01 | 1.290E-01 | 7.789E-02 |
| 1000 | 4.404E-01 | 3.941E-01 | 3.477E-01 | 3.013E-01 | 2.549E-01 | 2.085E-01 | 1.622E-01 | 1.161E-01 | 7.009E-02 |
| 1100 | 4.004E-01 | 3.582E-01 | 3.161E-01 | 2.739E-01 | 2.317E-01 | 1.895E-01 | 1.474E-01 | 1.055E-01 | 6.370E-02 |
| 1200 | 3.670E-01 | 3.284E-01 | 2.897E-01 | 2.510E-01 | 2.123E-01 | 1.737E-01 | 1.351E-01 | 9.672E-02 | 5.839E-02 |
| 1300 | 3.388E-01 | 3.031E-01 | 2.674E-01 | 2.317E-01 | 1.960E-01 | 1.603E-01 | 1.247E-01 | 8.927E-02 | 5.389E-02 |
| 1400 | 3.146E-01 | 2.815E-01 | 2.483E-01 | 2.152E-01 | 1.820E-01 | 1.489E-01 | 1.158E-01 | 8.289E-02 | 5.003E-02 |
| 1500 | 2.936E-01 | 2.627E-01 | 2.318E-01 | 2.008E-01 | 1.699E-01 | 1.389E-01 | 1.081E-01 | 7.735E-02 | 4.669E-02 |
| 1600 | 2.753E-01 | 2.463E-01 | 2.173E-01 | 1.882E-01 | 1.592E-01 | 1.302E-01 | 1.013E-01 | 7.251E-02 | 4.377E-02 |
| 1700 | 2.591E-01 | 2.318E-01 | 2.045E-01 | 1.772E-01 | 1.498E-01 | 1.226E-01 | 9.539E-02 | 6.824E-02 | 4.119E-02 |
| 1800 | 2.447E-01 | 2.189E-01 | 1.931E-01 | 1.673E-01 | 1.415E-01 | 1.157E-01 | 9.008E-02 | 6.444E-02 | 3.889E-02 |
| 1900 | 2.318E-01 | 2.074E-01 | 1.829E-01 | 1.585E-01 | 1.340E-01 | 1.096E-01 | 8.532E-02 | 6.103E-02 | 3.684E-02 |
| 2000 | 2.201E-01 | 1.969E-01 | 1.737E-01 | 1.505E-01 | 1.273E-01 | 1.041E-01 | 8.102E-02 | 5.796E-02 | 3.498E-02 |
| 2100 | 2.096E-01 | 1.875E-01 | 1.654E-01 | 1.433E-01 | 1.212E-01 | 9.914E-02 | 7.711E-02 | 5.516E-02 | 3.329E-02 |
| 2200 | 2.000E-01 | 1.789E-01 | 1.578E-01 | 1.366E-01 | 1.156E-01 | 9.454E-02 | 7.353E-02 | 5.259E-02 | 3.174E-02 |
| 2300 | 1.912E-01 | 1.710E-01 | 1.507E-01 | 1.305E-01 | 1.104E-01 | 9.029E-02 | 7.022E-02 | 5.022E-02 | 3.030E-02 |
| 2400 | 1.830E-01 | 1.636E-01 | 1.442E-01 | 1.249E-01 | 1.055E-01 | 8.633E-02 | 6.713E-02 | 4.800E-02 | 2.896E-02 |
| 2500 | 1.755E-01 | 1.567E-01 | 1.381E-01 | 1.195E-01 | 1.010E-01 | 8.259E-02 | 6.420E-02 | 4.590E-02 | 2.769E-02 |
| 2600 | 1.684E-01 | 1.503E-01 | 1.323E-01 | 1.145E-01 | 9.673E-02 | 7.903E-02 | 6.141E-02 | 4.389E-02 | 2.646E-02 |
| 2700 | 1.617E-01 | 1.442E-01 | 1.268E-01 | 1.096E-01 | 9.257E-02 | 7.559E-02 | 5.871E-02 | 4.194E-02 | 2.528E-02 |
| 2800 | 1.554E-01 | 1.383E-01 | 1.215E-01 | 1.049E-01 | 8.854E-02 | 7.224E-02 | 5.607E-02 | 4.003E-02 | 2.411E-02 |
| 2900 | 1.494E-01 | 1.327E-01 | 1.164E-01 | 1.004E-01 | 8.458E-02 | 6.894E-02 | 5.346E-02 | 3.813E-02 | 2.295E-02 |
| 3000 | 1.436E-01 | 1.272E-01 | 1.114E-01 | 9.590E-02 | 8.067E-02 | 6.566E-02 | 5.086E-02 | 3.624E-02 | 2.179E-02 |
| 3100 | 1.381E-01 | 1.219E-01 | 1.064E-01 | 9.146E-02 | 7.679E-02 | 6.240E-02 | 4.826E-02 | 3.433E-02 | 2.062E-02 |
| 3200 | 1.328E-01 | 1.168E-01 | 1.016E-01 | 8.708E-02 | 7.294E-02 | 5.915E-02 | 4.565E-02 | 3.243E-02 | 1.944E-02 |
| 3300 | 1.278E-01 | 1.118E-01 | 9.695E-02 | 8.278E-02 | 6.914E-02 | 5.592E-02 | 4.306E-02 | 3.052E-02 | 1.826E-02 |
| 3400 | 1.231E-01 | 1.070E-01 | 9.241E-02 | 7.859E-02 | 6.542E-02 | 5.274E-02 | 4.050E-02 | 2.863E-02 | 1.709E-02 |
| 3500 | 1.187E-01 | 1.025E-01 | 8.805E-02 | 7.456E-02 | 6.181E-02 | 4.966E-02 | 3.801E-02 | 2.679E-02 | 1.595E-02 |
| 3600 | 1.146E-01 | 9.835E-02 | 8.395E-02 | 7.073E-02 | 5.838E-02 | 4.671E-02 | 3.562E-02 | 2.502E-02 | 1.485E-02 |
| 3700 | 1.108E-01 | 9.444E-02 | 8.014E-02 | 6.716E-02 | 5.517E-02 | 4.395E-02 | 3.338E-02 | 2.336E-02 | 1.381E-02 |
| 3800 | 1.073E-01 | 9.088E-02 | 7.665E-02 | 6.389E-02 | 5.222E-02 | 4.142E-02 | 3.132E-02 | 2.183E-02 | 1.286E-02 |
| 3900 | 1.041E-01 | 8.764E-02 | 7.350E-02 | 6.094E-02 | 4.957E-02 | 3.913E-02 | 2.947E-02 | 2.046E-02 | 1.201E-02 |
| 4000 | 1.011E-01 | 8.472E-02 | 7.068E-02 | 5.831E-02 | 4.721E-02 | 3.711E-02 | 2.783E-02 | 1.924E-02 | 1.125E-02 |
| 4100 | 9.846E-02 | 8.209E-02 | 6.817E-02 | 5.599E-02 | 4.514E-02 | 3.534E-02 | 2.640E-02 | 1.819E-02 | 1.060E-02 |
| 4200 | 9.593E-02 | 7.970E-02 | 6.593E-02 | 5.394E-02 | 4.333E-02 | 3.380E-02 | 2.517E-02 | 1.729E-02 | 1.004E-02 |
| 4300 | 9.356E-02 | 7.752E-02 | 6.392E-02 | 5.213E-02 | 4.174E-02 | 3.246E-02 | 2.410E-02 | 1.651E-02 | 9.571E-03 |
| 4400 | 9.134E-02 | 7.551E-02 | 6.211E-02 | 5.052E-02 | 4.035E-02 | 3.130E-02 | 2.318E-02 | 1.584E-02 | 9.165E-03 |
| 4500 | 8.924E-02 | 7.365E-02 | 6.046E-02 | 4.908E-02 | 3.911E-02 | 3.028E-02 | 2.238E-02 | 1.527E-02 | 8.816E-03 |
| 4600 | 8.725E-02 | 7.191E-02 | 5.894E-02 | 4.777E-02 | 3.800E-02 | 2.937E-02 | 2.168E-02 | 1.476E-02 | 8.513E-03 |
| 4700 | 8.535E-02 | 7.028E-02 | 5.754E-02 | 4.657E-02 | 3.700E-02 | 2.856E-02 | 2.105E-02 | 1.432E-02 | 8.248E-03 |
| 4800 | 8.354E-02 | 6.874E-02 | 5.622E-02 | 4.546E-02 | 3.608E-02 | 2.782E-02 | 2.049E-02 | 1.392E-02 | 8.012E-03 |
| 4900 | 8.182E-02 | 6.728E-02 | 5.499E-02 | 4.442E-02 | 3.523E-02 | 2.715E-02 | 1.997E-02 | 1.356E-02 | 7.799E-03 |
| 5000 | 8.017E-02 | 6.590E-02 | 5.382E-02 | 4.346E-02 | 3.444E-02 | 2.652E-02 | 1.950E-02 | 1.324E-02 | 7.606E-03 |

## TABLE A.2. *(Continued)*

| | Molar fraction of $H_2$ | | | | | | | | |
|---|---|---|---|---|---|---|---|---|---|
| $T$ (K) | 0.1 | 0.2 | 0.3 | 0.4 | 0.5 | 0.6 | 0.7 | 0.8 | 0.9 |
| 5100 | 7.858E-02 | 6.457E-02 | 5.272E-02 | 4.254E-02 | 3.370E-02 | 2.594E-02 | 1.906E-02 | 1.293E-02 | 7.428E-03 |
| 5200 | 7.706E-02 | 6.330E-02 | 5.167E-02 | 4.168E-02 | 3.300E-02 | 2.539E-02 | 1.865E-02 | 1.265E-02 | 7.263E-03 |
| 5300 | 7.560E-02 | 6.209E-02 | 5.066E-02 | 4.085E-02 | 3.234E-02 | 2.487E-02 | 1.827E-02 | 1.238E-02 | 7.109E-03 |
| 5400 | 7.419E-02 | 6.093E-02 | 4.970E-02 | 4.007E-02 | 3.171E-02 | 2.438E-02 | 1.790E-02 | 1.213E-02 | 6.963E-03 |
| 5500 | 7.284E-02 | 5.981E-02 | 4.878E-02 | 3.932E-02 | 3.111E-02 | 2.391E-02 | 1.756E-02 | 1.190E-02 | 6.826E-03 |
| 5600 | 7.154E-02 | 5.873E-02 | 4.789E-02 | 3.860E-02 | 3.053E-02 | 2.347E-02 | 1.723E-02 | 1.167E-02 | 6.696E-03 |
| 5700 | 7.028E-02 | 5.769E-02 | 4.704E-02 | 3.790E-02 | 2.998E-02 | 2.304E-02 | 1.691E-02 | 1.146E-02 | 6.572E-03 |
| 5800 | 6.906E-02 | 5.669E-02 | 4.622E-02 | 3.724E-02 | 2.945E-02 | 2.263E-02 | 1.661E-02 | 1.125E-02 | 6.453E-03 |
| 5900 | 6.789E-02 | 5.572E-02 | 4.543E-02 | 3.660E-02 | 2.894E-02 | 2.224E-02 | 1.632E-02 | 1.105E-02 | 6.339E-03 |
| 6000 | 6.676E-02 | 5.479E-02 | 4.466E-02 | 3.598E-02 | 2.845E-02 | 2.186E-02 | 1.604E-02 | 1.086E-02 | 6.230E-03 |
| 6100 | 6.566E-02 | 5.389E-02 | 4.393E-02 | 3.538E-02 | 2.798E-02 | 2.149E-02 | 1.577E-02 | 1.068E-02 | 6.125E-03 |
| 6200 | 6.460E-02 | 5.302E-02 | 4.321E-02 | 3.481E-02 | 2.752E-02 | 2.114E-02 | 1.551E-02 | 1.050E-02 | 6.023E-03 |
| 6300 | 6.357E-02 | 5.217E-02 | 4.252E-02 | 3.425E-02 | 2.708E-02 | 2.080E-02 | 1.526E-02 | 1.033E-02 | 5.926E-03 |
| 6400 | 6.258E-02 | 5.135E-02 | 4.185E-02 | 3.371E-02 | 2.665E-02 | 2.047E-02 | 1.502E-02 | 1.017E-02 | 5.831E-03 |
| 6500 | 6.161E-02 | 5.056E-02 | 4.121E-02 | 3.319E-02 | 2.624E-02 | 2.015E-02 | 1.478E-02 | 1.001E-02 | 5.740E-03 |
| 6600 | 6.068E-02 | 4.979E-02 | 4.058E-02 | 3.268E-02 | 2.584E-02 | 1.985E-02 | 1.456E-02 | 9.860E-03 | 5.652E-03 |
| 6700 | 5.977E-02 | 4.905E-02 | 3.997E-02 | 3.219E-02 | 2.545E-02 | 1.955E-02 | 1.434E-02 | 9.711E-03 | 5.566E-03 |
| 6800 | 5.889E-02 | 4.832E-02 | 3.938E-02 | 3.171E-02 | 2.507E-02 | 1.926E-02 | 1.412E-02 | 9.566E-03 | 5.483E-03 |
| 6900 | 5.803E-02 | 4.762E-02 | 3.881E-02 | 3.125E-02 | 2.470E-02 | 1.897E-02 | 1.392E-02 | 9.426E-03 | 5.403E-03 |
| 7000 | 5.720E-02 | 4.693E-02 | 3.825E-02 | 3.080E-02 | 2.435E-02 | 1.870E-02 | 1.372E-02 | 9.290E-03 | 5.325E-03 |
| 7100 | 5.639E-02 | 4.627E-02 | 3.771E-02 | 3.036E-02 | 2.400E-02 | 1.843E-02 | 1.352E-02 | 9.157E-03 | 5.249E-03 |
| 7200 | 5.560E-02 | 4.562E-02 | 3.718E-02 | 2.994E-02 | 2.367E-02 | 1.818E-02 | 1.333E-02 | 9.029E-03 | 5.175E-03 |
| 7300 | 5.484E-02 | 4.499E-02 | 3.666E-02 | 2.953E-02 | 2.334E-02 | 1.792E-02 | 1.315E-02 | 8.903E-03 | 5.103E-03 |
| 7400 | 5.409E-02 | 4.438E-02 | 3.616E-02 | 2.912E-02 | 2.302E-02 | 1.768E-02 | 1.297E-02 | 8.781E-03 | 5.033E-03 |
| 7500 | 5.336E-02 | 4.378E-02 | 3.568E-02 | 2.873E-02 | 2.271E-02 | 1.744E-02 | 1.279E-02 | 8.663E-03 | 4.965E-03 |
| 7600 | 5.265E-02 | 4.320E-02 | 3.520E-02 | 2.835E-02 | 2.240E-02 | 1.721E-02 | 1.262E-02 | 8.547E-03 | 4.898E-03 |
| 7700 | 5.196E-02 | 4.263E-02 | 3.474E-02 | 2.797E-02 | 2.211E-02 | 1.698E-02 | 1.245E-02 | 8.434E-03 | 4.834E-03 |
| 7800 | 5.128E-02 | 4.207E-02 | 3.428E-02 | 2.761E-02 | 2.182E-02 | 1.676E-02 | 1.229E-02 | 8.323E-03 | 4.770E-03 |
| 7900 | 5.062E-02 | 4.153E-02 | 3.384E-02 | 2.725E-02 | 2.154E-02 | 1.654E-02 | 1.213E-02 | 8.216E-03 | 4.708E-03 |
| 8000 | 4.998E-02 | 4.100E-02 | 3.341E-02 | 2.690E-02 | 2.126E-02 | 1.633E-02 | 1.198E-02 | 8.110E-03 | 4.648E-03 |
| 8100 | 4.935E-02 | 4.048E-02 | 3.299E-02 | 2.656E-02 | 2.099E-02 | 1.612E-02 | 1.182E-02 | 8.007E-03 | 4.589E-03 |
| 8200 | 4.873E-02 | 3.998E-02 | 3.257E-02 | 2.623E-02 | 2.073E-02 | 1.592E-02 | 1.168E-02 | 7.907E-03 | 4.531E-03 |
| 8300 | 4.813E-02 | 3.948E-02 | 3.217E-02 | 2.590E-02 | 2.047E-02 | 1.572E-02 | 1.153E-02 | 7.808E-03 | 4.475E-03 |
| 8400 | 4.753E-02 | 3.899E-02 | 3.177E-02 | 2.558E-02 | 2.022E-02 | 1.553E-02 | 1.139E-02 | 7.711E-03 | 4.419E-03 |
| 8500 | 4.695E-02 | 3.852E-02 | 3.138E-02 | 2.527E-02 | 1.997E-02 | 1.534E-02 | 1.125E-02 | 7.617E-03 | 4.365E-03 |
| 8600 | 4.638E-02 | 3.805E-02 | 3.100E-02 | 2.496E-02 | 1.973E-02 | 1.515E-02 | 1.111E-02 | 7.524E-03 | 4.312E-03 |
| 8700 | 4.582E-02 | 3.759E-02 | 3.063E-02 | 2.466E-02 | 1.949E-02 | 1.497E-02 | 1.098E-02 | 7.433E-03 | 4.260E-03 |
| 8800 | 4.527E-02 | 3.714E-02 | 3.026E-02 | 2.436E-02 | 1.925E-02 | 1.479E-02 | 1.084E-02 | 7.343E-03 | 4.208E-03 |
| 8900 | 4.473E-02 | 3.670E-02 | 2.990E-02 | 2.407E-02 | 1.902E-02 | 1.461E-02 | 1.071E-02 | 7.255E-03 | 4.158E-03 |
| 9000 | 4.420E-02 | 3.626E-02 | 2.954E-02 | 2.378E-02 | 1.880E-02 | 1.443E-02 | 1.059E-02 | 7.168E-03 | 4.108E-03 |
| 9100 | 4.368E-02 | 3.583E-02 | 2.919E-02 | 2.350E-02 | 1.857E-02 | 1.426E-02 | 1.046E-02 | 7.083E-03 | 4.059E-03 |
| 9200 | 4.316E-02 | 3.540E-02 | 2.884E-02 | 2.322E-02 | 1.835E-02 | 1.409E-02 | 1.034E-02 | 6.999E-03 | 4.011E-03 |
| 9300 | 4.265E-02 | 3.499E-02 | 2.850E-02 | 2.295E-02 | 1.814E-02 | 1.393E-02 | 1.021E-02 | 6.916E-03 | 3.964E-03 |
| 9400 | 4.215E-02 | 3.457E-02 | 2.817E-02 | 2.268E-02 | 1.792E-02 | 1.376E-02 | 1.009E-02 | 6.835E-03 | 3.917E-03 |
| 9500 | 4.165E-02 | 3.416E-02 | 2.783E-02 | 2.241E-02 | 1.771E-02 | 1.360E-02 | 9.977E-03 | 6.754E-03 | 3.871E-03 |
| 9600 | 4.116E-02 | 3.376E-02 | 2.751E-02 | 2.215E-02 | 1.750E-02 | 1.344E-02 | 9.860E-03 | 6.674E-03 | 3.825E-03 |

*(continued)*

## TABLE A.2. (*Continued*)

| | Molar fraction of $H_2$ | | | | | | | | |
|---|---|---|---|---|---|---|---|---|---|
| $T$ (K) | 0.1 | 0.2 | 0.3 | 0.4 | 0.5 | 0.6 | 0.7 | 0.8 | 0.9 |
| 9700 | 4.067E-02 | 3.336E-02 | 2.718E-02 | 2.188E-02 | 1.730E-02 | 1.328E-02 | 9.743E-03 | 6.596E-03 | 3.780E-03 |
| 9800 | 4.019E-02 | 3.297E-02 | 2.686E-02 | 2.163E-02 | 1.709E-02 | 1.312E-02 | 9.628E-03 | 6.518E-03 | 3.735E-03 |
| 9900 | 3.971E-02 | 3.257E-02 | 2.654E-02 | 2.137E-02 | 1.689E-02 | 1.297E-02 | 9.514E-03 | 6.441E-03 | 3.691E-03 |
| 10000 | 3.923E-02 | 3.218E-02 | 2.622E-02 | 2.111E-02 | 1.669E-02 | 1.281E-02 | 9.401E-03 | 6.364E-03 | 3.647E-03 |
| 10100 | 3.876E-02 | 3.180E-02 | 2.591E-02 | 2.086E-02 | 1.649E-02 | 1.266E-02 | 9.289E-03 | 6.288E-03 | 3.604E-03 |
| 10200 | 3.829E-02 | 3.141E-02 | 2.560E-02 | 2.061E-02 | 1.629E-02 | 1.251E-02 | 9.178E-03 | 6.213E-03 | 3.561E-03 |
| 10300 | 3.782E-02 | 3.103E-02 | 2.529E-02 | 2.036E-02 | 1.609E-02 | 1.236E-02 | 9.067E-03 | 6.138E-03 | 3.518E-03 |
| 10400 | 3.736E-02 | 3.065E-02 | 2.498E-02 | 2.011E-02 | 1.590E-02 | 1.221E-02 | 8.957E-03 | 6.064E-03 | 3.475E-03 |
| 10500 | 3.689E-02 | 3.027E-02 | 2.467E-02 | 1.986E-02 | 1.570E-02 | 1.206E-02 | 8.848E-03 | 5.990E-03 | 3.433E-03 |
| 10600 | 3.643E-02 | 2.989E-02 | 2.436E-02 | 1.962E-02 | 1.551E-02 | 1.191E-02 | 8.739E-03 | 5.917E-03 | 3.391E-03 |
| 10700 | 3.596E-02 | 2.951E-02 | 2.405E-02 | 1.937E-02 | 1.531E-02 | 1.176E-02 | 8.631E-03 | 5.844E-03 | 3.349E-03 |
| 10800 | 3.550E-02 | 2.914E-02 | 2.375E-02 | 1.913E-02 | 1.512E-02 | 1.161E-02 | 8.524E-03 | 5.771E-03 | 3.308E-03 |
| 10900 | 3.504E-02 | 2.876E-02 | 2.344E-02 | 1.888E-02 | 1.493E-02 | 1.147E-02 | 8.416E-03 | 5.699E-03 | 3.266E-03 |
| 11000 | 3.458E-02 | 2.838E-02 | 2.314E-02 | 1.864E-02 | 1.474E-02 | 1.132E-02 | 8.309E-03 | 5.626E-03 | 3.225E-03 |
| 11100 | 3.411E-02 | 2.801E-02 | 2.283E-02 | 1.840E-02 | 1.455E-02 | 1.117E-02 | 8.202E-03 | 5.554E-03 | 3.184E-03 |
| 11200 | 3.365E-02 | 2.763E-02 | 2.253E-02 | 1.815E-02 | 1.435E-02 | 1.103E-02 | 8.095E-03 | 5.482E-03 | 3.143E-03 |
| 11300 | 3.318E-02 | 2.725E-02 | 2.222E-02 | 1.791E-02 | 1.416E-02 | 1.088E-02 | 7.989E-03 | 5.411E-03 | 3.102E-03 |
| 11400 | 3.272E-02 | 2.687E-02 | 2.192E-02 | 1.767E-02 | 1.397E-02 | 1.074E-02 | 7.883E-03 | 5.339E-03 | 3.061E-03 |
| 11500 | 3.225E-02 | 2.650E-02 | 2.161E-02 | 1.742E-02 | 1.378E-02 | 1.059E-02 | 7.776E-03 | 5.268E-03 | 3.020E-03 |
| 11600 | 3.178E-02 | 2.612E-02 | 2.131E-02 | 1.718E-02 | 1.359E-02 | 1.045E-02 | 7.670E-03 | 5.196E-03 | 2.980E-03 |
| 11700 | 3.132E-02 | 2.574E-02 | 2.100E-02 | 1.693E-02 | 1.340E-02 | 1.030E-02 | 7.564E-03 | 5.125E-03 | 2.939E-03 |
| 11800 | 3.085E-02 | 2.536E-02 | 2.070E-02 | 1.669E-02 | 1.321E-02 | 1.015E-02 | 7.458E-03 | 5.053E-03 | 2.898E-03 |
| 11900 | 3.037E-02 | 2.498E-02 | 2.039E-02 | 1.645E-02 | 1.302E-02 | 1.001E-02 | 7.352E-03 | 4.982E-03 | 2.858E-03 |
| 12000 | 2.990E-02 | 2.459E-02 | 2.008E-02 | 1.620E-02 | 1.283E-02 | 9.867E-03 | 7.246E-03 | 4.911E-03 | 2.817E-03 |
| 12100 | 2.943E-02 | 2.421E-02 | 1.978E-02 | 1.596E-02 | 1.263E-02 | 9.722E-03 | 7.141E-03 | 4.840E-03 | 2.777E-03 |
| 12200 | 2.895E-02 | 2.383E-02 | 1.947E-02 | 1.571E-02 | 1.244E-02 | 9.576E-03 | 7.035E-03 | 4.769E-03 | 2.737E-03 |
| 12300 | 2.848E-02 | 2.345E-02 | 1.916E-02 | 1.547E-02 | 1.225E-02 | 9.431E-03 | 6.929E-03 | 4.698E-03 | 2.696E-03 |
| 12400 | 2.801E-02 | 2.306E-02 | 1.885E-02 | 1.522E-02 | 1.206E-02 | 9.286E-03 | 6.824E-03 | 4.628E-03 | 2.656E-03 |
| 12500 | 2.753E-02 | 2.268E-02 | 1.855E-02 | 1.498E-02 | 1.187E-02 | 9.141E-03 | 6.719E-03 | 4.557E-03 | 2.616E-03 |
| 12600 | 2.706E-02 | 2.230E-02 | 1.824E-02 | 1.474E-02 | 1.168E-02 | 8.997E-03 | 6.614E-03 | 4.487E-03 | 2.576E-03 |
| 12700 | 2.659E-02 | 2.192E-02 | 1.793E-02 | 1.449E-02 | 1.149E-02 | 8.853E-03 | 6.510E-03 | 4.417E-03 | 2.536E-03 |
| 12800 | 2.611E-02 | 2.154E-02 | 1.763E-02 | 1.425E-02 | 1.130E-02 | 8.710E-03 | 6.406E-03 | 4.347E-03 | 2.497E-03 |
| 12900 | 2.564E-02 | 2.116E-02 | 1.732E-02 | 1.401E-02 | 1.111E-02 | 8.567E-03 | 6.302E-03 | 4.277E-03 | 2.457E-03 |
| 13000 | 2.518E-02 | 2.078E-02 | 1.702E-02 | 1.377E-02 | 1.093E-02 | 8.425E-03 | 6.199E-03 | 4.208E-03 | 2.418E-03 |
| 13100 | 2.471E-02 | 2.041E-02 | 1.672E-02 | 1.353E-02 | 1.074E-02 | 8.283E-03 | 6.096E-03 | 4.139E-03 | 2.379E-03 |
| 13200 | 2.425E-02 | 2.004E-02 | 1.642E-02 | 1.329E-02 | 1.056E-02 | 8.143E-03 | 5.994E-03 | 4.071E-03 | 2.340E-03 |
| 13300 | 2.379E-02 | 1.967E-02 | 1.613E-02 | 1.306E-02 | 1.037E-02 | 8.003E-03 | 5.893E-03 | 4.003E-03 | 2.301E-03 |
| 13400 | 2.334E-02 | 1.930E-02 | 1.584E-02 | 1.283E-02 | 1.019E-02 | 7.865E-03 | 5.792E-03 | 3.935E-03 | 2.263E-03 |
| 13500 | 2.289E-02 | 1.894E-02 | 1.555E-02 | 1.260E-02 | 1.001E-02 | 7.728E-03 | 5.693E-03 | 3.869E-03 | 2.225E-03 |
| 13600 | 2.245E-02 | 1.859E-02 | 1.526E-02 | 1.237E-02 | 9.837E-03 | 7.593E-03 | 5.594E-03 | 3.803E-03 | 2.187E-03 |
| 13700 | 2.202E-02 | 1.823E-02 | 1.498E-02 | 1.214E-02 | 9.661E-03 | 7.459E-03 | 5.497E-03 | 3.737E-03 | 2.150E-03 |
| 13800 | 2.159E-02 | 1.789E-02 | 1.470E-02 | 1.192E-02 | 9.487E-03 | 7.327E-03 | 5.401E-03 | 3.673E-03 | 2.113E-03 |
| 13900 | 2.116E-02 | 1.755E-02 | 1.443E-02 | 1.170E-02 | 9.316E-03 | 7.197E-03 | 5.306E-03 | 3.609E-03 | 2.077E-03 |
| 14000 | 2.075E-02 | 1.721E-02 | 1.416E-02 | 1.149E-02 | 9.148E-03 | 7.069E-03 | 5.213E-03 | 3.546E-03 | 2.041E-03 |
| 14100 | 2.035E-02 | 1.688E-02 | 1.389E-02 | 1.128E-02 | 8.983E-03 | 6.943E-03 | 5.121E-03 | 3.484E-03 | 2.006E-03 |
| 14200 | 1.995E-02 | 1.656E-02 | 1.363E-02 | 1.107E-02 | 8.820E-03 | 6.819E-03 | 5.030E-03 | 3.423E-03 | 1.971E-03 |
| 14300 | 1.957E-02 | 1.625E-02 | 1.338E-02 | 1.087E-02 | 8.660E-03 | 6.697E-03 | 4.941E-03 | 3.363E-03 | 1.936E-03 |

**TABLE A.2.** (*Continued*)

| $T$ (K) | Molar fraction of $H_2$ | | | | | | | | |
|---|---|---|---|---|---|---|---|---|---|
| | 0.1 | 0.2 | 0.3 | 0.4 | 0.5 | 0.6 | 0.7 | 0.8 | 0.9 |
| 14400 | 1.919E-02 | 1.594E-02 | 1.313E-02 | 1.067E-02 | 8.504E-03 | 6.577E-03 | 4.854E-03 | 3.304E-03 | 1.903E-03 |
| 14500 | 1.882E-02 | 1.564E-02 | 1.289E-02 | 1.047E-02 | 8.349E-03 | 6.460E-03 | 4.768E-03 | 3.246E-03 | 1.870E-03 |
| 14600 | 1.847E-02 | 1.535E-02 | 1.265E-02 | 1.028E-02 | 8.200E-03 | 6.344E-03 | 4.684E-03 | 3.190E-03 | 1.837E-03 |
| 14700 | 1.812E-02 | 1.507E-02 | 1.242E-02 | 1.010E-02 | 8.054E-03 | 6.232E-03 | 4.601E-03 | 3.134E-03 | 1.805E-03 |
| 14800 | 1.779E-02 | 1.480E-02 | 1.220E-02 | 9.924E-03 | 7.912E-03 | 6.123E-03 | 4.521E-03 | 3.079E-03 | 1.774E-03 |
| 14900 | 1.747E-02 | 1.453E-02 | 1.198E-02 | 9.750E-03 | 7.774E-03 | 6.016E-03 | 4.443E-03 | 3.026E-03 | 1.743E-03 |
| 15000 | 1.716E-02 | 1.428E-02 | 1.177E-02 | 9.580E-03 | 7.640E-03 | 5.913E-03 | 4.366E-03 | 2.974E-03 | 1.713E-03 |
| 15100 | 1.686E-02 | 1.403E-02 | 1.157E-02 | 9.416E-03 | 7.509E-03 | 5.812E-03 | 4.292E-03 | 2.923E-03 | 1.685E-03 |
| 15200 | 1.657E-02 | 1.379E-02 | 1.137E-02 | 9.256E-03 | 7.382E-03 | 5.714E-03 | 4.220E-03 | 2.874E-03 | 1.656E-03 |
| 15300 | 1.629E-02 | 1.356E-02 | 1.118E-02 | 9.102E-03 | 7.259E-03 | 5.619E-03 | 4.150E-03 | 2.827E-03 | 1.629E-03 |
| 15400 | 1.602E-02 | 1.333E-02 | 1.100E-02 | 8.952E-03 | 7.140E-03 | 5.527E-03 | 4.082E-03 | 2.780E-03 | 1.602E-03 |
| 15500 | 1.576E-02 | 1.312E-02 | 1.082E-02 | 8.807E-03 | 7.024E-03 | 5.437E-03 | 4.016E-03 | 2.735E-03 | 1.576E-03 |
| 15600 | 1.551E-02 | 1.291E-02 | 1.065E-02 | 8.668E-03 | 6.913E-03 | 5.351E-03 | 3.951E-03 | 2.691E-03 | 1.551E-03 |
| 15700 | 1.527E-02 | 1.271E-02 | 1.048E-02 | 8.533E-03 | 6.805E-03 | 5.267E-03 | 3.889E-03 | 2.649E-03 | 1.526E-03 |
| 15800 | 1.505E-02 | 1.252E-02 | 1.032E-02 | 8.403E-03 | 6.701E-03 | 5.186E-03 | 3.829E-03 | 2.608E-03 | 1.502E-03 |
| 15900 | 1.483E-02 | 1.234E-02 | 1.017E-02 | 8.277E-03 | 6.600E-03 | 5.107E-03 | 3.771E-03 | 2.568E-03 | 1.479E-03 |
| 16000 | 1.462E-02 | 1.216E-02 | 1.002E-02 | 8.156E-03 | 6.503E-03 | 5.032E-03 | 3.715E-03 | 2.529E-03 | 1.457E-03 |
| 16100 | 1.442E-02 | 1.199E-02 | 9.887E-03 | 8.040E-03 | 6.409E-03 | 4.958E-03 | 3.660E-03 | 2.492E-03 | 1.436E-03 |
| 16200 | 1.422E-02 | 1.183E-02 | 9.750E-03 | 7.928E-03 | 6.318E-03 | 4.888E-03 | 3.608E-03 | 2.456E-03 | 1.415E-03 |
| 16300 | 1.404E-02 | 1.167E-02 | 9.619E-03 | 7.819E-03 | 6.231E-03 | 4.820E-03 | 3.557E-03 | 2.421E-03 | 1.394E-03 |
| 16400 | 1.386E-02 | 1.152E-02 | 9.492E-03 | 7.715E-03 | 6.147E-03 | 4.754E-03 | 3.508E-03 | 2.388E-03 | 1.375E-03 |
| 16500 | 1.369E-02 | 1.137E-02 | 9.371E-03 | 7.615E-03 | 6.066E-03 | 4.691E-03 | 3.461E-03 | 2.355E-03 | 1.356E-03 |
| 16600 | 1.353E-02 | 1.123E-02 | 9.254E-03 | 7.518E-03 | 5.988E-03 | 4.629E-03 | 3.415E-03 | 2.324E-03 | 1.338E-03 |
| 16700 | 1.337E-02 | 1.110E-02 | 9.141E-03 | 7.425E-03 | 5.913E-03 | 4.570E-03 | 3.371E-03 | 2.294E-03 | 1.320E-03 |
| 16800 | 1.322E-02 | 1.097E-02 | 9.033E-03 | 7.335E-03 | 5.840E-03 | 4.514E-03 | 3.329E-03 | 2.265E-03 | 1.303E-03 |
| 16900 | 1.307E-02 | 1.085E-02 | 8.928E-03 | 7.249E-03 | 5.770E-03 | 4.459E-03 | 3.288E-03 | 2.236E-03 | 1.287E-03 |
| 17000 | 1.293E-02 | 1.073E-02 | 8.828E-03 | 7.165E-03 | 5.703E-03 | 4.406E-03 | 3.248E-03 | 2.209E-03 | 1.271E-03 |
| 17100 | 1.280E-02 | 1.061E-02 | 8.731E-03 | 7.085E-03 | 5.638E-03 | 4.355E-03 | 3.210E-03 | 2.183E-03 | 1.256E-03 |
| 17200 | 1.267E-02 | 1.050E-02 | 8.637E-03 | 7.008E-03 | 5.575E-03 | 4.305E-03 | 3.173E-03 | 2.158E-03 | 1.241E-03 |
| 17300 | 1.254E-02 | 1.040E-02 | 8.547E-03 | 6.933E-03 | 5.514E-03 | 4.258E-03 | 3.138E-03 | 2.133E-03 | 1.227E-03 |
| 17400 | 1.242E-02 | 1.029E-02 | 8.460E-03 | 6.861E-03 | 5.456E-03 | 4.212E-03 | 3.104E-03 | 2.110E-03 | 1.213E-03 |
| 17500 | 1.231E-02 | 1.019E-02 | 8.376E-03 | 6.791E-03 | 5.399E-03 | 4.168E-03 | 3.071E-03 | 2.087E-03 | 1.200E-03 |
| 17600 | 1.220E-02 | 1.010E-02 | 8.295E-03 | 6.724E-03 | 5.345E-03 | 4.125E-03 | 3.039E-03 | 2.065E-03 | 1.187E-03 |
| 17700 | 1.209E-02 | 1.000E-02 | 8.216E-03 | 6.658E-03 | 5.292E-03 | 4.084E-03 | 3.008E-03 | 2.044E-03 | 1.175E-03 |
| 17800 | 1.198E-02 | 9.918E-03 | 8.140E-03 | 6.595E-03 | 5.241E-03 | 4.044E-03 | 2.978E-03 | 2.014E-03 | 1.163E-03 |
| 17900 | 1.188E-02 | 9.831E-03 | 8.067E-03 | 6.534E-03 | 5.191E-03 | 4.005E-03 | 2.949E-03 | 2.003E-03 | 1.152E-03 |
| 18000 | 1.178E-02 | 9.746E-03 | 7.995E-03 | 6.475E-03 | 5.144E-03 | 3.967E-03 | 2.921E-03 | 1.984E-03 | 1.140E-03 |
| 18100 | 1.168E-02 | 9.664E-03 | 7.926E-03 | 6.418E-03 | 5.097E-03 | 3.931E-03 | 2.894E-03 | 1.965E-03 | 1.130E-03 |
| 18200 | 1.159E-02 | 9.585E-03 | 7.859E-03 | 6.363E-03 | 5.052E-03 | 3.896E-03 | 2.868E-03 | 1.947E-03 | 1.119E-03 |
| 18300 | 1.150E-02 | 9.507E-03 | 7.794E-03 | 6.309E-03 | 5.009E-03 | 3.862E-03 | 2.842E-03 | 1.933E-03 | 1.109E-03 |
| 18400 | 1.141E-02 | 9.432E-03 | 7.731E-03 | 6.256E-03 | 4.966E-03 | 3.833E-03 | 2.821E-03 | 1.909E-03 | 1.101E-03 |
| 18500 | 1.133E-02 | 9.358E-03 | 7.669E-03 | 6.211E-03 | 4.931E-03 | 3.801E-03 | 2.797E-03 | 1.899E-03 | 1.091E-03 |
| 18600 | 1.124E-02 | 9.287E-03 | 7.614E-03 | 6.161E-03 | 4.891E-03 | 3.771E-03 | 2.774E-03 | 1.883E-03 | 1.082E-03 |
| 18700 | 1.116E-02 | 9.222E-03 | 7.556E-03 | 6.117E-03 | 4.854E-03 | 3.740E-03 | 2.752E-03 | 1.868E-03 | 1.073E-03 |
| 18800 | 1.108E-02 | 9.153E-03 | 7.504E-03 | 6.069E-03 | 4.816E-03 | 3.711E-03 | 2.730E-03 | 1.853E-03 | 1.064E-03 |
| 18900 | 1.100E-02 | 9.095E-03 | 7.448E-03 | 6.024E-03 | 4.779E-03 | 3.682E-03 | 2.708E-03 | 1.838E-03 | 1.056E-03 |
| 19000 | 1.094E-02 | 9.029E-03 | 7.394E-03 | 5.979E-03 | 4.743E-03 | 3.654E-03 | 2.687E-03 | 1.824E-03 | 1.047E-03 |

(*continued*)

## TABLE A.2. (*Continued*)

| $T$ (K) | Molar fraction of $H_2$ | | | | | | | | |
|---|---|---|---|---|---|---|---|---|---|
| | 0.1 | 0.2 | 0.3 | 0.4 | 0.5 | 0.6 | 0.7 | 0.8 | 0.9 |
| 19100 | 1.086E-02 | 8.965E-03 | 7.340E-03 | 5.935E-03 | 4.708E-03 | 3.626E-03 | 2.667E-03 | 1.810E-03 | 1.039E-03 |
| 19200 | 1.079E-02 | 8.903E-03 | 7.288E-03 | 5.892E-03 | 4.673E-03 | 3.600E-03 | 2.647E-03 | 1.796E-03 | 1.032E-03 |
| 19300 | 1.071E-02 | 8.841E-03 | 7.237E-03 | 5.850E-03 | 4.640E-03 | 3.574E-03 | 2.628E-03 | 1.783E-03 | 1.024E-03 |
| 19400 | 1.064E-02 | 8.781E-03 | 7.187E-03 | 5.809E-03 | 4.607E-03 | 3.548E-03 | 2.609E-03 | 1.770E-03 | 1.016E-03 |
| 19500 | 1.057E-02 | 8.721E-03 | 7.138E-03 | 5.769E-03 | 4.575E-03 | 3.523E-03 | 2.590E-03 | 1.757E-03 | 1.009E-03 |
| 19600 | 1.050E-02 | 8.663E-03 | 7.089E-03 | 5.730E-03 | 4.543E-03 | 3.499E-03 | 2.572E-03 | 1.745E-03 | 1.002E-03 |
| 19700 | 1.043E-02 | 8.605E-03 | 7.042E-03 | 5.691E-03 | 4.512E-03 | 3.475E-03 | 2.555E-03 | 1.733E-03 | 9.955E-04 |
| 19800 | 1.036E-02 | 8.549E-03 | 6.995E-03 | 5.653E-03 | 4.482E-03 | 3.451E-03 | 2.537E-03 | 1.721E-03 | 9.887E-04 |
| 19900 | 1.030E-02 | 8.493E-03 | 6.949E-03 | 5.615E-03 | 4.452E-03 | 3.428E-03 | 2.520E-03 | 1.710E-03 | 9.820E-04 |
| 20000 | 1.023E-02 | 8.438E-03 | 6.904E-03 | 5.579E-03 | 4.423E-03 | 3.406E-03 | 2.504E-03 | 1.698E-03 | 9.755E-04 |
| 20100 | 1.016E-02 | 8.383E-03 | 6.859E-03 | 5.543E-03 | 4.394E-03 | 3.384E-03 | 2.487E-03 | 1.687E-03 | 9.692E-04 |
| 20200 | 1.010E-02 | 8.329E-03 | 6.815E-03 | 5.507E-03 | 4.366E-03 | 3.362E-03 | 2.471E-03 | 1.677E-03 | 9.629E-04 |
| 20300 | 1.003E-02 | 8.276E-03 | 6.771E-03 | 5.472E-03 | 4.338E-03 | 3.340E-03 | 2.456E-03 | 1.666E-03 | 9.568E-04 |
| 20400 | 9.975E-03 | 8.223E-03 | 6.728E-03 | 5.437E-03 | 4.311E-03 | 3.319E-03 | 2.440E-03 | 1.655E-03 | 9.508E-04 |
| 20500 | 9.911E-03 | 8.171E-03 | 6.686E-03 | 5.403E-03 | 4.284E-03 | 3.299E-03 | 2.425E-03 | 1.645E-03 | 9.450E-04 |
| 20600 | 9.847E-03 | 8.119E-03 | 6.644E-03 | 5.369E-03 | 4.257E-03 | 3.278E-03 | 2.410E-03 | 1.635E-03 | 9.392E-04 |
| 20700 | 9.784E-03 | 8.068E-03 | 6.602E-03 | 5.335E-03 | 4.231E-03 | 3.258E-03 | 2.395E-03 | 1.625E-03 | 9.335E-04 |
| 20800 | 9.721E-03 | 8.016E-03 | 6.560E-03 | 5.302E-03 | 4.205E-03 | 3.238E-03 | 2.381E-03 | 1.615E-03 | 9.280E-04 |
| 20900 | 9.658E-03 | 7.965E-03 | 6.519E-03 | 5.270E-03 | 4.179E-03 | 3.219E-03 | 2.367E-03 | 1.604E-03 | 9.225E-04 |
| 21000 | 9.596E-03 | 7.915E-03 | 6.478E-03 | 5.237E-03 | 4.153E-03 | 3.199E-03 | 2.353E-03 | 1.596E-03 | 9.171E-04 |
| 21100 | 9.533E-03 | 7.864E-03 | 6.438E-03 | 5.205E-03 | 4.128E-03 | 3.180E-03 | 2.339E-03 | 1.587E-03 | 9.118E-04 |
| 21200 | 9.471E-03 | 7.814E-03 | 6.397E-03 | 5.173E-03 | 4.103E-03 | 3.161E-03 | 2.325E-03 | 1.578E-03 | 9.066E-04 |
| 21300 | 9.408E-03 | 7.764E-03 | 6.357E-03 | 5.141E-03 | 4.079E-03 | 3.142E-03 | 2.311E-03 | 1.569E-03 | 9.014E-04 |
| 21400 | 9.345E-03 | 7.713E-03 | 6.317E-03 | 5.109E-03 | 4.054E-03 | 3.124E-03 | 2.298E-03 | 1.560E-03 | 8.964E-04 |
| 21500 | 9.283E-03 | 7.663E-03 | 6.277E-03 | 5.078E-03 | 4.030E-03 | 3.106E-03 | 2.285E-03 | 1.551E-03 | 8.914E-04 |
| 21600 | 9.220E-03 | 7.613E-03 | 6.238E-03 | 5.047E-03 | 4.005E-03 | 3.087E-03 | 2.272E-03 | 1.542E-03 | 8.865E-04 |
| 21700 | 9.157E-03 | 7.563E-03 | 6.198E-03 | 5.016E-03 | 3.981E-03 | 3.069E-03 | 2.259E-03 | 1.534E-03 | 8.816E-04 |
| 21800 | 9.094E-03 | 7.513E-03 | 6.158E-03 | 4.985E-03 | 3.958E-03 | 3.051E-03 | 2.246E-03 | 1.525E-03 | 8.769E-04 |
| 21900 | 9.030E-03 | 7.463E-03 | 6.119E-03 | 4.954E-03 | 3.934E-03 | 3.034E-03 | 2.233E-03 | 1.517E-03 | 8.721E-04 |
| 22000 | 8.967E-03 | 7.413E-03 | 6.079E-03 | 4.923E-03 | 3.910E-03 | 3.016E-03 | 2.221E-03 | 1.508E-03 | 8.675E-04 |
| 22100 | 8.903E-03 | 7.362E-03 | 6.040E-03 | 4.892E-03 | 3.887E-03 | 2.998E-03 | 2.208E-03 | 1.500E-03 | 8.629E-04 |
| 22200 | 8.839E-03 | 7.312E-03 | 6.000E-03 | 4.862E-03 | 3.863E-03 | 2.981E-03 | 2.196E-03 | 1.494E-03 | 8.596E-04 |
| 22300 | 8.774E-03 | 7.261E-03 | 5.961E-03 | 4.831E-03 | 3.840E-03 | 2.967E-03 | 2.186E-03 | 1.486E-03 | 8.551E-04 |
| 22400 | 8.710E-03 | 7.211E-03 | 5.921E-03 | 4.800E-03 | 3.821E-03 | 2.950E-03 | 2.174E-03 | 1.478E-03 | 8.507E-04 |
| 22500 | 8.645E-03 | 7.160E-03 | 5.882E-03 | 4.774E-03 | 3.799E-03 | 2.934E-03 | 2.162E-03 | 1.470E-03 | 8.463E-04 |
| 22600 | 8.580E-03 | 7.109E-03 | 5.846E-03 | 4.745E-03 | 3.776E-03 | 2.916E-03 | 2.150E-03 | 1.462E-03 | 8.419E-04 |
| 22700 | 8.514E-03 | 7.061E-03 | 5.810E-03 | 4.715E-03 | 3.752E-03 | 2.899E-03 | 2.138E-03 | 1.455E-03 | 8.376E-04 |
| 22800 | 8.450E-03 | 7.010E-03 | 5.770E-03 | 4.684E-03 | 3.729E-03 | 2.882E-03 | 2.126E-03 | 1.447E-03 | 8.333E-04 |
| 22900 | 8.384E-03 | 6.963E-03 | 5.730E-03 | 4.654E-03 | 3.706E-03 | 2.866E-03 | 2.114E-03 | 1.439E-03 | 8.291E-04 |
| 23000 | 8.318E-03 | 6.912E-03 | 5.690E-03 | 4.623E-03 | 3.683E-03 | 2.849E-03 | 2.103E-03 | 1.431E-03 | 8.249E-04 |
| 23100 | 8.258E-03 | 6.860E-03 | 5.650E-03 | 4.593E-03 | 3.660E-03 | 2.832E-03 | 2.091E-03 | 1.424E-03 | 8.208E-04 |
| 23200 | 8.191E-03 | 6.808E-03 | 5.610E-03 | 4.562E-03 | 3.638E-03 | 2.815E-03 | 2.079E-03 | 1.416E-03 | 8.167E-04 |
| 23300 | 8.124E-03 | 6.756E-03 | 5.570E-03 | 4.532E-03 | 3.615E-03 | 2.799E-03 | 2.068E-03 | 1.409E-03 | 8.126E-04 |
| 23400 | 8.057E-03 | 6.704E-03 | 5.530E-03 | 4.501E-03 | 3.592E-03 | 2.782E-03 | 2.056E-03 | 1.402E-03 | 8.086E-04 |
| 23500 | 7.989E-03 | 6.652E-03 | 5.490E-03 | 4.471E-03 | 3.569E-03 | 2.766E-03 | 2.045E-03 | 1.394E-03 | 8.046E-04 |
| 23600 | 7.922E-03 | 6.600E-03 | 5.450E-03 | 4.441E-03 | 3.547E-03 | 2.749E-03 | 2.033E-03 | 1.387E-03 | 8.006E-04 |
| 23700 | 7.854E-03 | 6.548E-03 | 5.410E-03 | 4.410E-03 | 3.524E-03 | 2.733E-03 | 2.022E-03 | 1.380E-03 | 7.967E-04 |
| 23800 | 7.787E-03 | 6.496E-03 | 5.371E-03 | 4.380E-03 | 3.502E-03 | 2.717E-03 | 2.011E-03 | 1.373E-03 | 7.929E-04 |
| 23900 | 7.720E-03 | 6.444E-03 | 5.331E-03 | 4.350E-03 | 3.479E-03 | 2.700E-03 | 2.000E-03 | 1.365E-03 | 7.890E-04 |
| 24000 | 7.652E-03 | 6.392E-03 | 5.291E-03 | 4.320E-03 | 3.457E-03 | 2.684E-03 | 1.989E-03 | 1.358E-03 | 7.852E-04 |

### TABLE A.3. Enthalpy of Ar/$H_2$ Mixtures (J/kg)

| | Molar fraction of $H_2$ | | | | | | | | |
|---|---|---|---|---|---|---|---|---|---|
| $T$ (K) | 0.1 | 0.2 | 0.3 | 0.4 | 0.5 | 0.6 | 0.7 | 0.8 | 0.9 |
| 500 | 1.206E+05 | 1.400E+05 | 1.647E+05 | 1.972E+05 | 2.417E+05 | 3.060E+05 | 4.071E+05 | 5.885E+05 | 1.007E+06 |
| 600 | 1.804E+05 | 2.096E+05 | 2.467E+05 | 2.954E+05 | 3.621E+05 | 4.586E+05 | 6.103E+05 | 8.825E+05 | 1.511E+06 |
| 700 | 2.404E+05 | 2.793E+05 | 3.290E+05 | 3.942E+05 | 4.834E+05 | 6.127E+05 | 8.157E+05 | 1.180E+06 | 2.022E+06 |
| 800 | 3.004E+05 | 3.493E+05 | 4.116E+05 | 4.936E+05 | 6.057E+05 | 7.682E+05 | 1.023E+06 | 1.482E+06 | 2.541E+06 |
| 900 | 3.605E+05 | 4.195E+05 | 4.947E+05 | 5.936E+05 | 7.291E+05 | 9.253E+05 | 1.233E+06 | 1.788E+06 | 3.068E+06 |
| 1000 | 4.207E+05 | 4.899E+05 | 5.781E+05 | 6.943E+05 | 8.534E+05 | 1.084E+06 | 1.446E+06 | 2.098E+06 | 3.604E+06 |
| 1100 | 4.810E+05 | 5.605E+05 | 6.620E+05 | 7.957E+05 | 9.789E+05 | 1.244E+06 | 1.662E+06 | 2.413E+06 | 4.148E+06 |
| 1200 | 5.414E+05 | 6.314E+05 | 7.463E+05 | 8.978E+05 | 1.105E+06 | 1.406E+06 | 1.880E+06 | 2.732E+06 | 4.701E+06 |
| 1300 | 6.019E+05 | 7.025E+05 | 8.311E+05 | 1.000E+06 | 1.233E+06 | 1.570E+06 | 2.101E+06 | 3.056E+06 | 5.263E+06 |
| 1400 | 6.625E+05 | 7.739E+05 | 9.164E+05 | 1.104E+06 | 1.362E+06 | 1.736E+06 | 2.325E+06 | 3.384E+06 | 5.834E+06 |
| 1500 | 7.232E+05 | 8.455E+05 | 1.002E+06 | 1.208E+06 | 1.492E+06 | 1.903E+06 | 2.552E+06 | 3.718E+06 | 6.414E+06 |
| 1600 | 7.841E+05 | 9.175E+05 | 1.088E+06 | 1.314E+06 | 1.623E+06 | 2.073E+06 | 2.782E+06 | 4.056E+06 | 7.004E+06 |
| 1700 | 8.452E+05 | 9.900E+05 | 1.175E+06 | 1.420E+06 | 1.756E+06 | 2.245E+06 | 3.015E+06 | 4.400E+06 | 7.605E+06 |
| 1800 | 9.067E+05 | 1.063E+06 | 1.263E+06 | 1.528E+06 | 1.891E+06 | 2.420E+06 | 3.253E+06 | 4.752E+06 | 8.220E+06 |
| 1900 | 9.686E+05 | 1.137E+06 | 1.352E+06 | 1.637E+06 | 2.029E+06 | 2.599E+06 | 3.497E+06 | 5.112E+06 | 8.851E+06 |
| 2000 | 1.031E+06 | 1.212E+06 | 1.444E+06 | 1.750E+06 | 2.171E+06 | 2.783E+06 | 3.748E+06 | 5.485E+06 | 9.505E+06 |
| 2100 | 1.095E+06 | 1.290E+06 | 1.539E+06 | 1.867E+06 | 2.319E+06 | 2.976E+06 | 4.012E+06 | 5.876E+06 | 1.019E+07 |
| 2200 | 1.162E+06 | 1.371E+06 | 1.638E+06 | 1.991E+06 | 2.476E+06 | 3.181E+06 | 4.292E+06 | 6.293E+06 | 1.092E+07 |
| 2300 | 1.231E+06 | 1.457E+06 | 1.745E+06 | 2.124E+06 | 2.645E+06 | 3.402E+06 | 4.596E+06 | 6.745E+06 | 1.171E+07 |
| 2400 | 1.305E+06 | 1.550E+06 | 1.861E+06 | 2.270E+06 | 2.832E+06 | 3.647E+06 | 4.933E+06 | 7.245E+06 | 1.259E+07 |
| 2500 | 1.385E+06 | 1.653E+06 | 1.991E+06 | 2.434E+06 | 3.042E+06 | 3.923E+06 | 5.312E+06 | 7.810E+06 | 1.359E+07 |
| 2600 | 1.472E+06 | 1.767E+06 | 2.137E+06 | 2.621E+06 | 3.282E+06 | 4.240E+06 | 5.749E+06 | 8.460E+06 | 1.473E+07 |
| 2700 | 1.568E+06 | 1.897E+06 | 2.305E+06 | 2.836E+06 | 3.561E+06 | 4.609E+06 | 6.257E+06 | 9.217E+06 | 1.606E+07 |
| 2800 | 1.675E+06 | 2.045E+06 | 2.499E+06 | 3.088E+06 | 3.888E+06 | 5.042E+06 | 6.855E+06 | 1.010E+07 | 1.762E+07 |
| 2900 | 1.794E+06 | 2.215E+06 | 2.725E+06 | 3.382E+06 | 4.272E+06 | 5.553E+06 | 7.562E+06 | 1.116E+07 | 1.947E+07 |
| 3000 | 1.926E+06 | 2.408E+06 | 2.986E+06 | 3.725E+06 | 4.723E+06 | 6.155E+06 | 8.397E+06 | 1.241E+07 | 2.167E+07 |
| 3100 | 2.071E+06 | 2.627E+06 | 3.286E+06 | 4.124E+06 | 5.249E+06 | 6.862E+06 | 9.380E+06 | 1.388E+07 | 2.426E+07 |
| 3200 | 2.227E+06 | 2.871E+06 | 3.627E+06 | 4.581E+06 | 5.858E+06 | 7.683E+06 | 1.052E+07 | 1.560E+07 | 2.731E+07 |
| 3300 | 2.391E+06 | 3.137E+06 | 4.006E+06 | 5.097E+06 | 6.552E+06 | 8.625E+06 | 1.185E+07 | 1.760E+07 | 3.085E+07 |
| 3400 | 2.558E+06 | 3.421E+06 | 4.419E+06 | 5.668E+06 | 7.328E+06 | 9.688E+06 | 1.335E+07 | 1.989E+07 | 3.492E+07 |
| 3500 | 2.724E+06 | 3.715E+06 | 4.859E+06 | 6.285E+06 | 8.177E+06 | 1.086E+07 | 1.503E+07 | 2.245E+07 | 3.950E+07 |
| 3600 | 2.884E+06 | 4.010E+06 | 5.311E+06 | 6.932E+06 | 9.080E+06 | 1.212E+07 | 1.684E+07 | 2.525E+07 | 4.456E+07 |
| 3700 | 3.033E+06 | 4.296E+06 | 5.762E+06 | 7.590E+06 | 1.001E+07 | 1.344E+07 | 1.876E+07 | 2.823E+07 | 4.998E+07 |
| 3800 | 3.168E+06 | 4.564E+06 | 6.196E+06 | 8.237E+06 | 1.094E+07 | 1.478E+07 | 2.072E+07 | 3.130E+07 | 5.561E+07 |
| 3900 | 3.291E+06 | 4.808E+06 | 6.601E+06 | 8.850E+06 | 1.183E+07 | 1.608E+07 | 2.266E+07 | 3.436E+07 | 6.125E+07 |
| 4000 | 3.400E+06 | 5.027E+06 | 6.907E+06 | 9.414E+06 | 1.267E+07 | 1.730E+07 | 2.449E+07 | 3.729E+07 | 6.670E+07 |
| 4100 | 3.499E+06 | 5.219E+06 | 7.291E+06 | 9.918E+06 | 1.342E+07 | 1.842E+07 | 2.618E+07 | 4.001E+07 | 7.179E+07 |
| 4200 | 3.590E+06 | 5.387E+06 | 7.574E+06 | 1.036E+07 | 1.409E+07 | 1.941E+07 | 2.769E+07 | 4.244E+07 | 7.638E+07 |
| 4300 | 3.673E+06 | 5.536E+06 | 7.819E+06 | 1.074E+07 | 1.466E+07 | 2.027E+07 | 2.901E+07 | 4.458E+07 | 8.042E+07 |
| 4400 | 3.751E+06 | 5.668E+06 | 8.032E+06 | 1.107E+07 | 1.516E+07 | 2.101E+07 | 3.014E+07 | 4.642E+07 | 8.391E+07 |
| 4500 | 3.826E+06 | 5.787E+06 | 8.219E+06 | 1.135E+07 | 1.558E+07 | 2.164E+07 | 3.110E+07 | 4.799E+07 | 8.688E+07 |
| 4600 | 3.897E+06 | 5.896E+06 | 8.385E+06 | 1.160E+07 | 1.595E+07 | 2.219E+07 | 3.193E+07 | 4.933E+07 | 8.941E+07 |
| 4700 | 3.967E+06 | 5.997E+06 | 8.535E+06 | 1.182E+07 | 1.627E+07 | 2.266E+07 | 3.264E+07 | 5.047E+07 | 9.156E+07 |
| 4800 | 4.035E+06 | 6.093E+06 | 8.672E+06 | 1.201E+07 | 1.655E+07 | 2.307E+07 | 3.325E+07 | 5.146E+07 | 9.341E+07 |
| 4900 | 4.102E+06 | 6.185E+06 | 8.800E+06 | 1.219E+07 | 1.680E+07 | 2.343E+07 | 3.379E+07 | 5.232E+07 | 9.501E+07 |
| 5000 | 4.168E+06 | 6.273E+06 | 8.920E+06 | 1.236E+07 | 1.703E+07 | 2.375E+07 | 3.427E+07 | 5.308E+07 | 9.642E+07 |

*(continued)*

**TABLE A.3.** *(Continued)*

| | Molar fraction of $H_2$ | | | | | | | | |
|---|---|---|---|---|---|---|---|---|---|
| $T$ (K) | 0.1 | 0.2 | 0.3 | 0.4 | 0.5 | 0.6 | 0.7 | 0.8 | 0.9 |
| 5100 | 4.233E+06 | 6.359E+06 | 9.035E+06 | 1.251E+07 | 1.725E+07 | 2.405E+07 | 3.470E+07 | 5.376E+07 | 9.767E+07 |
| 5200 | 4.298E+06 | 6.443E+06 | 9.145E+06 | 1.266E+07 | 1.744E+07 | 2.433E+07 | 3.510E+07 | 5.438E+07 | 9.881E+07 |
| 5300 | 4.363E+06 | 6.525E+06 | 9.253E+06 | 1.280E+07 | 1.763E+07 | 2.459E+07 | 3.548E+07 | 5.495E+07 | 9.986E+07 |
| 5400 | 4.427E+06 | 6.607E+06 | 9.357E+06 | 1.294E+07 | 1.781E+07 | 2.484E+07 | 3.583E+07 | 5.549E+07 | 1.008E+08 |
| 5500 | 4.491E+06 | 6.687E+06 | 9.460E+06 | 1.307E+07 | 1.799E+07 | 2.507E+07 | 3.616E+07 | 5.600E+07 | 1.017E+08 |
| 5600 | 4.556E+06 | 6.767E+06 | 9.561E+06 | 1.320E+07 | 1.816E+07 | 2.530E+07 | 3.648E+07 | 5.649E+07 | 1.026E+08 |
| 5700 | 4.619E+06 | 6.847E+06 | 9.661E+06 | 1.333E+07 | 1.833E+07 | 2.552E+07 | 3.679E+07 | 5.696E+07 | 1.034E+08 |
| 5800 | 4.683E+06 | 6.926E+06 | 9.760E+06 | 1.346E+07 | 1.849E+07 | 2.574E+07 | 3.709E+07 | 5.741E+07 | 1.042E+08 |
| 5900 | 4.747E+06 | 7.005E+06 | 9.859E+06 | 1.358E+07 | 1.865E+07 | 2.595E+07 | 3.739E+07 | 5.786E+07 | 1.050E+08 |
| 6000 | 4.811E+06 | 7.084E+06 | 9.957E+06 | 1.370E+07 | 1.881E+07 | 2.616E+07 | 3.768E+07 | 5.829E+07 | 1.058E+08 |
| 6100 | 4.875E+06 | 7.163E+06 | 1.005E+07 | 1.383E+07 | 1.897E+07 | 2.637E+07 | 3.797E+07 | 5.872E+07 | 1.065E+08 |
| 6200 | 4.939E+06 | 7.241E+06 | 1.015E+07 | 1.395E+07 | 1.912E+07 | 2.658E+07 | 3.825E+07 | 5.915E+07 | 1.073E+08 |
| 6300 | 5.003E+06 | 7.320E+06 | 1.025E+07 | 1.407E+07 | 1.928E+07 | 2.678E+07 | 3.853E+07 | 5.957E+07 | 1.080E+08 |
| 6400 | 5.067E+06 | 7.399E+06 | 1.034E+07 | 1.419E+07 | 1.943E+07 | 2.699E+07 | 3.882E+07 | 5.999E+07 | 1.088E+08 |
| 6500 | 5.132E+06 | 7.478E+06 | 1.044E+07 | 1.432E+07 | 1.959E+07 | 2.719E+07 | 3.910E+07 | 6.040E+07 | 1.095E+08 |
| 6600 | 5.197E+06 | 7.557E+06 | 1.054E+07 | 1.444E+07 | 1.975E+07 | 2.740E+07 | 3.938E+07 | 6.082E+07 | 1.102E+08 |
| 6700 | 5.261E+06 | 7.636E+06 | 1.064E+07 | 1.456E+07 | 1.990E+07 | 2.760E+07 | 3.966E+07 | 6.124E+07 | 1.110E+08 |
| 6800 | 5.327E+06 | 7.716E+06 | 1.074E+07 | 1.468E+07 | 2.006E+07 | 2.780E+07 | 3.994E+07 | 6.165E+07 | 1.117E+08 |
| 6900 | 5.392E+06 | 7.797E+06 | 1.083E+07 | 1.481E+07 | 2.022E+07 | 2.801E+07 | 4.022E+07 | 6.207E+07 | 1.124E+08 |
| 7000 | 5.458E+06 | 7.877E+06 | 1.093E+07 | 1.493E+07 | 2.037E+07 | 2.822E+07 | 4.050E+07 | 6.249E+07 | 1.132E+08 |
| 7100 | 5.525E+06 | 7.959E+06 | 1.103E+07 | 1.506E+07 | 2.053E+07 | 2.842E+07 | 4.078E+07 | 6.291E+07 | 1.139E+08 |
| 7200 | 5.592E+06 | 8.041E+06 | 1.114E+07 | 1.518E+07 | 2.069E+07 | 2.863E+07 | 4.107E+07 | 6.333E+07 | 1.146E+08 |
| 7300 | 5.660E+06 | 8.124E+06 | 1.124E+07 | 1.531E+07 | 2.085E+07 | 2.885E+07 | 4.136E+07 | 6.376E+07 | 1.154E+08 |
| 7400 | 5.728E+06 | 8.208E+06 | 1.134E+07 | 1.544E+07 | 2.102E+07 | 2.906E+07 | 4.165E+07 | 6.420E+07 | 1.161E+08 |
| 7500 | 5.798E+06 | 8.293E+06 | 1.145E+07 | 1.557E+07 | 2.118E+07 | 2.927E+07 | 4.195E+07 | 6.463E+07 | 1.169E+08 |
| 7600 | 5.868E+06 | 8.379E+06 | 1.155E+07 | 1.570E+07 | 2.135E+07 | 2.949E+07 | 4.225E+07 | 6.508E+07 | 1.177E+08 |
| 7700 | 5.939E+06 | 8.466E+06 | 1.166E+07 | 1.584E+07 | 2.152E+07 | 2.972E+07 | 4.255E+07 | 6.553E+07 | 1.185E+08 |
| 7800 | 6.012E+06 | 8.555E+06 | 1.177E+07 | 1.597E+07 | 2.170E+07 | 2.994E+07 | 4.286E+07 | 6.599E+07 | 1.193E+08 |
| 7900 | 6.086E+06 | 8.646E+06 | 1.188E+07 | 1.611E+07 | 2.187E+07 | 3.017E+07 | 4.318E+07 | 6.645E+07 | 1.201E+08 |
| 8000 | 6.162E+06 | 8.739E+06 | 1.200E+07 | 1.625E+07 | 2.205E+07 | 3.041E+07 | 4.350E+07 | 6.693E+07 | 1.209E+08 |
| 8100 | 6.240E+06 | 8.834E+06 | 1.211E+07 | 1.640E+07 | 2.224E+07 | 3.065E+07 | 4.383E+07 | 6.742E+07 | 1.218E+08 |
| 8200 | 6.319E+06 | 8.931E+06 | 1.223E+07 | 1.655E+07 | 2.243E+07 | 3.090E+07 | 4.416E+07 | 6.791E+07 | 1.226E+08 |
| 8300 | 6.401E+06 | 9.031E+06 | 1.235E+07 | 1.670E+07 | 2.262E+07 | 3.115E+07 | 4.451E+07 | 6.842E+07 | 1.235E+08 |
| 8400 | 6.485E+06 | 9.134E+06 | 1.248E+07 | 1.686E+07 | 2.282E+07 | 3.141E+07 | 4.486E+07 | 6.895E+07 | 1.244E+08 |
| 8500 | 6.572E+06 | 9.240E+06 | 1.261E+07 | 1.702E+07 | 2.302E+07 | 3.168E+07 | 4.523E+07 | 6.949E+07 | 1.254E+08 |
| 8600 | 6.662E+06 | 9.349E+06 | 1.275E+07 | 1.719E+07 | 2.324E+07 | 3.195E+07 | 4.560E+07 | 7.004E+07 | 1.264E+08 |
| 8700 | 6.756E+06 | 9.463E+06 | 1.289E+07 | 1.736E+07 | 2.345E+07 | 3.224E+07 | 4.599E+07 | 7.062E+07 | 1.274E+08 |
| 8800 | 6.852E+06 | 9.581E+06 | 1.303E+07 | 1.754E+07 | 2.368E+07 | 3.253E+07 | 4.639E+07 | 7.121E+07 | 1.284E+08 |
| 8900 | 6.953E+06 | 9.703E+06 | 1.318E+07 | 1.773E+07 | 2.392E+07 | 3.284E+07 | 4.681E+07 | 7.182E+07 | 1.295E+08 |
| 9000 | 7.058E+06 | 9.831E+06 | 1.334E+07 | 1.792E+07 | 2.416E+07 | 3.315E+07 | 4.724E+07 | 7.246E+07 | 1.306E+08 |
| 9100 | 7.168E+06 | 9.964E+06 | 1.350E+07 | 1.812E+07 | 2.441E+07 | 3.348E+07 | 4.769E+07 | 7.312E+07 | 1.317E+08 |
| 9200 | 7.283E+06 | 1.010E+07 | 1.367E+07 | 1.833E+07 | 2.468E+07 | 3.383E+07 | 4.816E+07 | 7.381E+07 | 1.329E+08 |
| 9300 | 7.403E+06 | 1.024E+07 | 1.385E+07 | 1.855E+07 | 2.495E+07 | 3.419E+07 | 4.865E+07 | 7.452E+07 | 1.342E+08 |
| 9400 | 7.529E+06 | 1.040E+07 | 1.403E+07 | 1.878E+07 | 2.524E+07 | 3.456E+07 | 4.915E+07 | 7.527E+07 | 1.355E+08 |
| 9500 | 7.662E+06 | 1.056E+07 | 1.423E+07 | 1.902E+07 | 2.554E+07 | 3.495E+07 | 4.968E+07 | 7.605E+07 | 1.368E+08 |
| 9600 | 7.802E+06 | 1.072E+07 | 1.443E+07 | 1.927E+07 | 2.586E+07 | 3.535E+07 | 5.023E+07 | 7.686E+07 | 1.382E+08 |
| 9700 | 7.948E+06 | 1.090E+07 | 1.464E+07 | 1.953E+07 | 2.619E+07 | 3.578E+07 | 5.081E+07 | 7.771E+07 | 1.397E+08 |
| 9800 | 8.104E+06 | 1.109E+07 | 1.487E+07 | 1.981E+07 | 2.653E+07 | 3.623E+07 | 5.141E+07 | 7.860E+07 | 1.412E+08 |

TABLE A.3. *(Continued)*

| $T$ (K) | Molar fraction of $H_2$ | | | | | | | | |
|---|---|---|---|---|---|---|---|---|---|
| | 0.1 | 0.2 | 0.3 | 0.4 | 0.5 | 0.6 | 0.7 | 0.8 | 0.9 |
| 9900 | 8.267E+06 | 1.128E+07 | 1.511E+07 | 2.010E+07 | 2.690E+07 | 3.670E+07 | 5.205E+07 | 7.953E+07 | 1.429E+08 |
| 10000 | 8.440E+06 | 1.149E+07 | 1.536E+07 | 2.041E+07 | 2.728E+07 | 3.719E+07 | 5.272E+07 | 8.051E+07 | 1.446E+08 |
| 10100 | 8.622E+06 | 1.171E+07 | 1.562E+07 | 2.073E+07 | 2.768E+07 | 3.771E+07 | 5.341E+07 | 8.152E+07 | 1.463E+08 |
| 10200 | 8.816E+06 | 1.194E+07 | 1.590E+07 | 2.107E+07 | 2.811E+07 | 3.825E+07 | 5.415E+07 | 8.260E+07 | 1.482E+08 |
| 10300 | 9.021E+06 | 1.218E+07 | 1.619E+07 | 2.143E+07 | 2.855E+07 | 3.883E+07 | 5.492E+07 | 8.372E+07 | 1.501E+08 |
| 10400 | 9.235E+06 | 1.244E+07 | 1.650E+07 | 2.180E+07 | 2.902E+07 | 3.942E+07 | 5.572E+07 | 8.489E+07 | 1.521E+08 |
| 10500 | 9.465E+06 | 1.271E+07 | 1.682E+07 | 2.220E+07 | 2.951E+07 | 4.005E+07 | 5.657E+07 | 8.613E+07 | 1.542E+08 |
| 10600 | 9.708E+06 | 1.300E+07 | 1.717E+07 | 2.262E+07 | 3.003E+07 | 4.072E+07 | 5.746E+07 | 8.743E+07 | 1.565E+08 |
| 10700 | 9.962E+06 | 1.330E+07 | 1.754E+07 | 2.306E+07 | 3.058E+07 | 4.142E+07 | 5.840E+07 | 8.880E+07 | 1.588E+08 |
| 10800 | 1.023E+07 | 1.362E+07 | 1.791E+07 | 2.352E+07 | 3.115E+07 | 4.215E+07 | 5.937E+07 | 9.020E+07 | 1.613E+08 |
| 10900 | 1.052E+07 | 1.396E+07 | 1.832E+07 | 2.401E+07 | 3.176E+07 | 4.292E+07 | 6.041E+07 | 9.171E+07 | 1.638E+08 |
| 11000 | 1.082E+07 | 1.432E+07 | 1.875E+07 | 2.453E+07 | 3.240E+07 | 4.374E+07 | 6.150E+07 | 9.329E+07 | 1.665E+08 |
| 11100 | 1.114E+07 | 1.470E+07 | 1.919E+07 | 2.507E+07 | 3.307E+07 | 4.459E+07 | 6.264E+07 | 9.494E+07 | 1.694E+08 |
| 11200 | 1.148E+07 | 1.510E+07 | 1.967E+07 | 2.564E+07 | 3.376E+07 | 4.547E+07 | 6.381E+07 | 9.664E+07 | 1.723E+08 |
| 11300 | 1.184E+07 | 1.552E+07 | 2.017E+07 | 2.624E+07 | 3.451E+07 | 4.642E+07 | 6.507E+07 | 9.845E+07 | 1.754E+08 |
| 11400 | 1.222E+07 | 1.597E+07 | 2.070E+07 | 2.688E+07 | 3.529E+07 | 4.741E+07 | 6.639E+07 | 1.003E+08 | 1.787E+08 |
| 11500 | 1.262E+07 | 1.644E+07 | 2.126E+07 | 2.755E+07 | 3.611E+07 | 4.845E+07 | 6.777E+07 | 1.023E+08 | 1.821E+08 |
| 11600 | 1.304E+07 | 1.692E+07 | 2.183E+07 | 2.823E+07 | 3.697E+07 | 4.954E+07 | 6.923E+07 | 1.044E+08 | 1.856E+08 |
| 11700 | 1.349E+07 | 1.744E+07 | 2.244E+07 | 2.897E+07 | 3.786E+07 | 5.066E+07 | 7.070E+07 | 1.065E+08 | 1.892E+08 |
| 11800 | 1.396E+07 | 1.799E+07 | 2.309E+07 | 2.975E+07 | 3.881E+07 | 5.186E+07 | 7.229E+07 | 1.088E+08 | 1.931E+08 |
| 11900 | 1.445E+07 | 1.857E+07 | 2.377E+07 | 3.056E+07 | 3.980E+07 | 5.311E+07 | 7.395E+07 | 1.112E+08 | 1.972E+08 |
| 12000 | 1.498E+07 | 1.918E+07 | 2.449E+07 | 3.142E+07 | 4.084E+07 | 5.442E+07 | 7.569E+07 | 1.137E+08 | 2.015E+08 |
| 12100 | 1.551E+07 | 1.979E+07 | 2.523E+07 | 3.231E+07 | 4.193E+07 | 5.580E+07 | 7.751E+07 | 1.163E+08 | 2.059E+08 |
| 12200 | 1.608E+07 | 2.046E+07 | 2.599E+07 | 3.321E+07 | 4.306E+07 | 5.724E+07 | 7.941E+07 | 1.190E+08 | 2.105E+08 |
| 12300 | 1.668E+07 | 2.116E+07 | 2.681E+07 | 3.419E+07 | 4.422E+07 | 5.868E+07 | 8.131E+07 | 1.219E+08 | 2.153E+08 |
| 12400 | 1.731E+07 | 2.189E+07 | 2.767E+07 | 3.521E+07 | 4.546E+07 | 6.024E+07 | 8.337E+07 | 1.247E+08 | 2.201E+08 |
| 12500 | 1.797E+07 | 2.265E+07 | 2.856E+07 | 3.627E+07 | 4.676E+07 | 6.186E+07 | 8.551E+07 | 1.278E+08 | 2.253E+08 |
| 12600 | 1.866E+07 | 2.345E+07 | 2.949E+07 | 3.738E+07 | 4.810E+07 | 6.355E+07 | 8.774E+07 | 1.310E+08 | 2.307E+08 |
| 12700 | 1.938E+07 | 2.428E+07 | 3.046E+07 | 3.853E+07 | 4.950E+07 | 6.531E+07 | 9.005E+07 | 1.343E+08 | 2.363E+08 |
| 12800 | 2.010E+07 | 2.512E+07 | 3.147E+07 | 3.973E+07 | 5.096E+07 | 6.713E+07 | 9.245E+07 | 1.377E+08 | 2.422E+08 |
| 12900 | 2.088E+07 | 2.600E+07 | 3.248E+07 | 4.097E+07 | 5.247E+07 | 6.903E+07 | 9.495E+07 | 1.413E+08 | 2.482E+08 |
| 13000 | 2.168E+07 | 2.693E+07 | 3.356E+07 | 4.221E+07 | 5.401E+07 | 7.099E+07 | 9.753E+07 | 1.450E+08 | 2.544E+08 |
| 13100 | 2.252E+07 | 2.790E+07 | 3.469E+07 | 4.354E+07 | 5.558E+07 | 7.293E+07 | 1.001E+08 | 1.488E+08 | 2.609E+08 |
| 13200 | 2.338E+07 | 2.889E+07 | 3.585E+07 | 4.492E+07 | 5.725E+07 | 7.502E+07 | 1.028E+08 | 1.525E+08 | 2.676E+08 |
| 13300 | 2.428E+07 | 2.992E+07 | 3.705E+07 | 4.634E+07 | 5.897E+07 | 7.717E+07 | 1.056E+08 | 1.566E+08 | 2.741E+08 |
| 13400 | 2.520E+07 | 3.098E+07 | 3.828E+07 | 4.780E+07 | 6.075E+07 | 7.939E+07 | 1.085E+08 | 1.608E+08 | 2.812E+08 |
| 13500 | 2.615E+07 | 3.207E+07 | 3.955E+07 | 4.931E+07 | 6.257E+07 | 8.168E+07 | 1.115E+08 | 1.651E+08 | 2.885E+08 |
| 13600 | 2.712E+07 | 3.319E+07 | 4.085E+07 | 5.085E+07 | 6.445E+07 | 8.403E+07 | 1.146E+08 | 1.695E+08 | 2.959E+08 |
| 13700 | 2.812E+07 | 3.434E+07 | 4.219E+07 | 5.243E+07 | 6.637E+07 | 8.643E+07 | 1.178E+08 | 1.740E+08 | 3.036E+08 |
| 13800 | 2.911E+07 | 3.551E+07 | 4.356E+07 | 5.405E+07 | 6.833E+07 | 8.889E+07 | 1.211E+08 | 1.787E+08 | 3.115E+08 |
| 13900 | 3.014E+07 | 3.671E+07 | 4.495E+07 | 5.570E+07 | 7.034E+07 | 9.141E+07 | 1.244E+08 | 1.834E+08 | 3.196E+08 |
| 14000 | 3.119E+07 | 3.787E+07 | 4.637E+07 | 5.739E+07 | 7.238E+07 | 9.398E+07 | 1.278E+08 | 1.883E+08 | 3.278E+08 |
| 14100 | 3.227E+07 | 3.910E+07 | 4.777E+07 | 5.910E+07 | 7.446E+07 | 9.659E+07 | 1.312E+08 | 1.932E+08 | 3.362E+08 |
| 14200 | 3.335E+07 | 4.035E+07 | 4.920E+07 | 6.080E+07 | 7.657E+07 | 9.925E+07 | 1.347E+08 | 1.983E+08 | 3.448E+08 |
| 14300 | 3.445E+07 | 4.162E+07 | 5.068E+07 | 6.251E+07 | 7.868E+07 | 1.019E+08 | 1.383E+08 | 2.034E+08 | 3.535E+08 |
| 14400 | 3.555E+07 | 4.289E+07 | 5.217E+07 | 6.428E+07 | 8.085E+07 | 1.046E+08 | 1.419E+08 | 2.086E+08 | 3.624E+08 |
| 14500 | 3.666E+07 | 4.417E+07 | 5.367E+07 | 6.607E+07 | 8.296E+07 | 1.074E+08 | 1.456E+08 | 2.139E+08 | 3.714E+08 |
| 14600 | 3.777E+07 | 4.546E+07 | 5.518E+07 | 6.788E+07 | 8.516E+07 | 1.100E+08 | 1.492E+08 | 2.192E+08 | 3.804E+08 |

*(continued)*

**TABLE A.3.** *(Continued)*

| | Molar fraction of $H_2$ | | | | | | | | |
|---|---|---|---|---|---|---|---|---|---|
| $T$ (K) | 0.1 | 0.2 | 0.3 | 0.4 | 0.5 | 0.6 | 0.7 | 0.8 | 0.9 |
| 14700 | 3.888E+07 | 4.674E+07 | 5.669E+07 | 6.968E+07 | 8.737E+07 | 1.128E+08 | 1.528E+08 | 2.245E+08 | 3.896E+08 |
| 14800 | 3.999E+07 | 4.803E+07 | 5.820E+07 | 7.150E+07 | 8.959E+07 | 1.156E+08 | 1.565E+08 | 2.297E+08 | 3.989E+08 |
| 14900 | 4.109E+07 | 4.931E+07 | 5.971E+07 | 7.331E+07 | 9.182E+07 | 1.185E+08 | 1.603E+08 | 2.351E+08 | 4.077E+08 |
| 15000 | 4.218E+07 | 5.058E+07 | 6.122E+07 | 7.511E+07 | 9.404E+07 | 1.213E+08 | 1.640E+08 | 2.406E+08 | 4.171E+08 |
| 15100 | 4.325E+07 | 5.184E+07 | 6.271E+07 | 7.691E+07 | 9.626E+07 | 1.241E+08 | 1.678E+08 | 2.461E+08 | 4.265E+08 |
| 15200 | 4.432E+07 | 5.308E+07 | 6.419E+07 | 7.870E+07 | 9.847E+07 | 1.269E+08 | 1.716E+08 | 2.515E+08 | 4.359E+08 |
| 15300 | 4.536E+07 | 5.431E+07 | 6.565E+07 | 8.047E+07 | 1.006E+08 | 1.297E+08 | 1.754E+08 | 2.570E+08 | 4.453E+08 |
| 15400 | 4.639E+07 | 5.552E+07 | 6.710E+07 | 8.223E+07 | 1.028E+08 | 1.325E+08 | 1.791E+08 | 2.625E+08 | 4.548E+08 |
| 15500 | 4.739E+07 | 5.671E+07 | 6.852E+07 | 8.396E+07 | 1.050E+08 | 1.353E+08 | 1.828E+08 | 2.679E+08 | 4.641E+08 |
| 15600 | 4.837E+07 | 5.788E+07 | 6.992E+07 | 8.567E+07 | 1.071E+08 | 1.380E+08 | 1.865E+08 | 2.733E+08 | 4.735E+08 |
| 15700 | 4.933E+07 | 5.902E+07 | 7.129E+07 | 8.735E+07 | 1.092E+08 | 1.407E+08 | 1.901E+08 | 2.786E+08 | 4.827E+08 |
| 15800 | 5.026E+07 | 6.013E+07 | 7.264E+07 | 8.900E+07 | 1.112E+08 | 1.434E+08 | 1.938E+08 | 2.839E+08 | 4.919E+08 |
| 15900 | 5.116E+07 | 6.122E+07 | 7.396E+07 | 9.062E+07 | 1.133E+08 | 1.460E+08 | 1.973E+08 | 2.892E+08 | 5.010E+08 |
| 16000 | 5.204E+07 | 6.227E+07 | 7.524E+07 | 9.220E+07 | 1.153E+08 | 1.486E+08 | 2.008E+08 | 2.943E+08 | 5.099E+08 |
| 16100 | 5.289E+07 | 6.330E+07 | 7.650E+07 | 9.375E+07 | 1.172E+08 | 1.511E+08 | 2.043E+08 | 2.994E+08 | 5.188E+08 |
| 16200 | 5.371E+07 | 6.430E+07 | 7.772E+07 | 9.526E+07 | 1.191E+08 | 1.536E+08 | 2.077E+08 | 3.044E+08 | 5.275E+08 |
| 16300 | 5.450E+07 | 6.526E+07 | 7.890E+07 | 9.674E+07 | 1.210E+08 | 1.561E+08 | 2.110E+08 | 3.093E+08 | 5.361E+08 |
| 16400 | 5.526E+07 | 6.620E+07 | 8.006E+07 | 9.818E+07 | 1.228E+08 | 1.584E+08 | 2.142E+08 | 3.141E+08 | 5.445E+08 |
| 16500 | 5.600E+07 | 6.710E+07 | 8.117E+07 | 9.957E+07 | 1.246E+08 | 1.608E+08 | 2.174E+08 | 3.188E+08 | 5.527E+08 |
| 16600 | 5.671E+07 | 6.798E+07 | 8.226E+07 | 1.009E+08 | 1.263E+08 | 1.630E+08 | 2.205E+08 | 3.234E+08 | 5.608E+08 |
| 16700 | 5.739E+07 | 6.882E+07 | 8.331E+07 | 1.022E+08 | 1.280E+08 | 1.652E+08 | 2.235E+08 | 3.279E+08 | 5.686E+08 |
| 16800 | 5.805E+07 | 6.964E+07 | 8.432E+07 | 1.035E+08 | 1.296E+08 | 1.674E+08 | 2.265E+08 | 3.323E+08 | 5.763E+08 |
| 16900 | 5.868E+07 | 7.042E+07 | 8.531E+07 | 1.047E+08 | 1.312E+08 | 1.695E+08 | 2.294E+08 | 3.366E+08 | 5.839E+08 |
| 17000 | 5.928E+07 | 7.118E+07 | 8.626E+07 | 1.059E+08 | 1.328E+08 | 1.715E+08 | 2.322E+08 | 3.407E+08 | 5.912E+08 |
| 17100 | 5.986E+07 | 7.191E+07 | 8.717E+07 | 1.071E+08 | 1.343E+08 | 1.735E+08 | 2.349E+08 | 3.448E+08 | 5.983E+08 |
| 17200 | 6.042E+07 | 7.261E+07 | 8.806E+07 | 1.082E+08 | 1.357E+08 | 1.754E+08 | 2.375E+08 | 3.487E+08 | 6.052E+08 |
| 17300 | 6.096E+07 | 7.329E+07 | 8.892E+07 | 1.093E+08 | 1.371E+08 | 1.772E+08 | 2.401E+08 | 3.526E+08 | 6.120E+08 |
| 17400 | 6.148E+07 | 7.395E+07 | 8.974E+07 | 1.103E+08 | 1.385E+08 | 1.790E+08 | 2.425E+08 | 3.563E+08 | 6.185E+08 |
| 17500 | 6.197E+07 | 7.458E+07 | 9.054E+07 | 1.114E+08 | 1.398E+08 | 1.808E+08 | 2.449E+08 | 3.599E+08 | 6.248E+08 |
| 17600 | 6.245E+07 | 7.518E+07 | 9.131E+07 | 1.123E+08 | 1.410E+08 | 1.824E+08 | 2.473E+08 | 3.633E+08 | 6.310E+08 |
| 17700 | 6.291E+07 | 7.577E+07 | 9.205E+07 | 1.133E+08 | 1.423E+08 | 1.841E+08 | 2.495E+08 | 3.667E+08 | 6.370E+08 |
| 17800 | 6.336E+07 | 7.633E+07 | 9.277E+07 | 1.142E+08 | 1.435E+08 | 1.856E+08 | 2.517E+08 | 3.726E+08 | 6.427E+08 |
| 17900 | 6.379E+07 | 7.688E+07 | 9.346E+07 | 1.151E+08 | 1.446E+08 | 1.872E+08 | 2.538E+08 | 3.732E+08 | 6.483E+08 |
| 18000 | 6.420E+07 | 7.741E+07 | 9.414E+07 | 1.160E+08 | 1.457E+08 | 1.886E+08 | 2.559E+08 | 3.762E+08 | 6.538E+08 |
| 18100 | 6.460E+07 | 7.792E+07 | 9.479E+07 | 1.168E+08 | 1.468E+08 | 1.901E+08 | 2.579E+08 | 3.792E+08 | 6.590E+08 |
| 18200 | 6.499E+07 | 7.841E+07 | 9.542E+07 | 1.176E+08 | 1.478E+08 | 1.915E+08 | 2.598E+08 | 3.821E+08 | 6.641E+08 |
| 18300 | 6.537E+07 | 7.890E+07 | 9.603E+07 | 1.184E+08 | 1.489E+08 | 1.928E+08 | 2.616E+08 | 3.848E+08 | 6.706E+08 |
| 18400 | 6.573E+07 | 7.936E+07 | 9.662E+07 | 1.191E+08 | 1.498E+08 | 1.941E+08 | 2.634E+08 | 3.894E+08 | 6.737E+08 |
| 18500 | 6.609E+07 | 7.981E+07 | 9.719E+07 | 1.198E+08 | 1.508E+08 | 1.953E+08 | 2.652E+08 | 3.901E+08 | 6.783E+08 |
| 18600 | 6.644E+07 | 8.026E+07 | 9.774E+07 | 1.206E+08 | 1.517E+08 | 1.965E+08 | 2.668E+08 | 3.926E+08 | 6.827E+08 |
| 18700 | 6.678E+07 | 8.068E+07 | 9.829E+07 | 1.212E+08 | 1.526E+08 | 1.977E+08 | 2.685E+08 | 3.951E+08 | 6.870E+08 |
| 18800 | 6.711E+07 | 8.110E+07 | 9.880E+07 | 1.219E+08 | 1.534E+08 | 1.989E+08 | 2.700E+08 | 3.975E+08 | 6.912E+08 |
| 18900 | 6.743E+07 | 8.150E+07 | 9.932E+07 | 1.226E+08 | 1.543E+08 | 2.000E+08 | 2.716E+08 | 3.998E+08 | 6.953E+08 |
| 19000 | 6.774E+07 | 8.190E+07 | 9.983E+07 | 1.232E+08 | 1.551E+08 | 2.011E+08 | 2.731E+08 | 4.020E+08 | 6.992E+08 |
| 19100 | 6.806E+07 | 8.230E+07 | 1.003E+08 | 1.238E+08 | 1.559E+08 | 2.021E+08 | 2.745E+08 | 4.042E+08 | 7.030E+08 |
| 19200 | 6.837E+07 | 8.269E+07 | 1.008E+08 | 1.245E+08 | 1.567E+08 | 2.032E+08 | 2.760E+08 | 4.063E+08 | 7.067E+08 |
| 19300 | 6.868E+07 | 8.308E+07 | 1.013E+08 | 1.251E+08 | 1.575E+08 | 2.042E+08 | 2.773E+08 | 4.083E+08 | 7.103E+08 |
| 19400 | 6.899E+07 | 8.346E+07 | 1.017E+08 | 1.256E+08 | 1.582E+08 | 2.052E+08 | 2.787E+08 | 4.103E+08 | 7.138E+08 |

## TABLE A.3. (*Continued*)

| $T$ (K) | Molar fraction of $H_2$ | | | | | | | | |
|---|---|---|---|---|---|---|---|---|---|
| | 0.1 | 0.2 | 0.3 | 0.4 | 0.5 | 0.6 | 0.7 | 0.8 | 0.9 |
| 19500 | 6.930E+07 | 8.384E+07 | 1.022E+08 | 1.262E+08 | 1.590E+08 | 2.061E+08 | 2.800E+08 | 4.123E+08 | 7.172E+08 |
| 19600 | 6.962E+07 | 8.422E+07 | 1.027E+08 | 1.268E+08 | 1.597E+08 | 2.071E+08 | 2.813E+08 | 4.142E+08 | 7.205E+08 |
| 19700 | 6.993E+07 | 8.460E+07 | 1.031E+08 | 1.274E+08 | 1.604E+08 | 2.080E+08 | 2.826E+08 | 4.160E+08 | 7.237E+08 |
| 19800 | 7.024E+07 | 8.498E+07 | 1.036E+08 | 1.279E+08 | 1.611E+08 | 2.089E+08 | 2.838E+08 | 4.178E+08 | 7.269E+08 |
| 19900 | 7.056E+07 | 8.535E+07 | 1.040E+08 | 1.285E+08 | 1.618E+08 | 2.098E+08 | 2.850E+08 | 4.196E+08 | 7.299E+08 |
| 20000 | 7.088E+07 | 8.574E+07 | 1.045E+08 | 1.291E+08 | 1.625E+08 | 2.107E+08 | 2.862E+08 | 4.213E+08 | 7.329E+08 |
| 20100 | 7.120E+07 | 8.612E+07 | 1.050E+08 | 1.296E+08 | 1.632E+08 | 2.116E+08 | 2.874E+08 | 4.230E+08 | 7.358E+08 |
| 20200 | 7.154E+07 | 8.651E+07 | 1.054E+08 | 1.302E+08 | 1.639E+08 | 2.124E+08 | 2.885E+08 | 4.247E+08 | 7.387E+08 |
| 20300 | 7.187E+07 | 8.690E+07 | 1.059E+08 | 1.307E+08 | 1.646E+08 | 2.133E+08 | 2.897E+08 | 4.264E+08 | 7.415E+08 |
| 20400 | 7.222E+07 | 8.730E+07 | 1.063E+08 | 1.313E+08 | 1.652E+08 | 2.142E+08 | 2.908E+08 | 4.280E+08 | 7.443E+08 |
| 20500 | 7.257E+07 | 8.771E+07 | 1.068E+08 | 1.318E+08 | 1.659E+08 | 2.150E+08 | 2.919E+08 | 4.296E+08 | 7.469E+08 |
| 20600 | 7.293E+07 | 8.812E+07 | 1.073E+08 | 1.324E+08 | 1.666E+08 | 2.159E+08 | 2.930E+08 | 4.311E+08 | 7.496E+08 |
| 20700 | 7.330E+07 | 8.854E+07 | 1.078E+08 | 1.330E+08 | 1.673E+08 | 2.167E+08 | 2.941E+08 | 4.327E+08 | 7.522E+08 |
| 20800 | 7.369E+07 | 8.898E+07 | 1.083E+08 | 1.336E+08 | 1.680E+08 | 2.175E+08 | 2.952E+08 | 4.342E+08 | 7.547E+08 |
| 20900 | 7.408E+07 | 8.942E+07 | 1.088E+08 | 1.341E+08 | 1.686E+08 | 2.184E+08 | 2.963E+08 | 4.363E+08 | 7.573E+08 |
| 21000 | 7.449E+07 | 8.987E+07 | 1.093E+08 | 1.347E+08 | 1.693E+08 | 2.192E+08 | 2.974E+08 | 4.372E+08 | 7.597E+08 |
| 21100 | 7.491E+07 | 9.034E+07 | 1.098E+08 | 1.353E+08 | 1.701E+08 | 2.201E+08 | 2.985E+08 | 4.387E+08 | 7.622E+08 |
| 21200 | 7.535E+07 | 9.083E+07 | 1.104E+08 | 1.359E+08 | 1.708E+08 | 2.210E+08 | 2.996E+08 | 4.402E+08 | 7.646E+08 |
| 21300 | 7.580E+07 | 9.132E+07 | 1.109E+08 | 1.366E+08 | 1.715E+08 | 2.218E+08 | 3.006E+08 | 4.417E+08 | 7.670E+08 |
| 21400 | 7.627E+07 | 9.184E+07 | 1.115E+08 | 1.372E+08 | 1.722E+08 | 2.227E+08 | 3.017E+08 | 4.432E+08 | 7.694E+08 |
| 21500 | 7.675E+07 | 9.237E+07 | 1.121E+08 | 1.379E+08 | 1.730E+08 | 2.236E+08 | 3.028E+08 | 4.447E+08 | 7.717E+08 |
| 21600 | 7.726E+07 | 9.291E+07 | 1.127E+08 | 1.385E+08 | 1.737E+08 | 2.245E+08 | 3.039E+08 | 4.461E+08 | 7.740E+08 |
| 21700 | 7.778E+07 | 9.348E+07 | 1.133E+08 | 1.392E+08 | 1.745E+08 | 2.254E+08 | 3.050E+08 | 4.476E+08 | 7.763E+08 |
| 21800 | 7.833E+07 | 9.407E+07 | 1.139E+08 | 1.399E+08 | 1.753E+08 | 2.263E+08 | 3.062E+08 | 4.491E+08 | 7.786E+08 |
| 21900 | 7.889E+07 | 9.468E+07 | 1.146E+08 | 1.407E+08 | 1.761E+08 | 2.273E+08 | 3.073E+08 | 4.506E+08 | 7.809E+08 |
| 22000 | 7.948E+07 | 9.530E+07 | 1.153E+08 | 1.414E+08 | 1.770E+08 | 2.282E+08 | 3.085E+08 | 4.521E+08 | 7.832E+08 |
| 22100 | 8.009E+07 | 9.596E+07 | 1.160E+08 | 1.422E+08 | 1.778E+08 | 2.292E+08 | 3.096E+08 | 4.536E+08 | 7.855E+08 |
| 22200 | 8.072E+07 | 9.663E+07 | 1.167E+08 | 1.430E+08 | 1.787E+08 | 2.302E+08 | 3.108E+08 | 4.549E+08 | 7.873E+08 |
| 22300 | 8.138E+07 | 9.733E+07 | 1.175E+08 | 1.438E+08 | 1.796E+08 | 2.311E+08 | 3.119E+08 | 4.563E+08 | 7.895E+08 |
| 22400 | 8.206E+07 | 9.805E+07 | 1.182E+08 | 1.446E+08 | 1.804E+08 | 2.321E+08 | 3.130E+08 | 4.579E+08 | 7.918E+08 |
| 22500 | 8.277E+07 | 9.880E+07 | 1.190E+08 | 1.454E+08 | 1.813E+08 | 2.331E+08 | 3.143E+08 | 4.594E+08 | 7.940E+08 |
| 22600 | 8.350E+07 | 9.957E+07 | 1.198E+08 | 1.463E+08 | 1.823E+08 | 2.342E+08 | 3.155E+08 | 4.610E+08 | 7.963E+08 |
| 22700 | 8.426E+07 | 1.003E+08 | 1.206E+08 | 1.472E+08 | 1.833E+08 | 2.353E+08 | 3.168E+08 | 4.625E+08 | 7.985E+08 |
| 22800 | 8.504E+07 | 1.011E+08 | 1.215E+08 | 1.481E+08 | 1.843E+08 | 2.364E+08 | 3.181E+08 | 4.641E+08 | 8.008E+08 |
| 22900 | 8.586E+07 | 1.020E+08 | 1.224E+08 | 1.491E+08 | 1.853E+08 | 2.376E+08 | 3.194E+08 | 4.657E+08 | 8.030E+08 |
| 23000 | 8.670E+07 | 1.028E+08 | 1.233E+08 | 1.500E+08 | 1.864E+08 | 2.387E+08 | 3.207E+08 | 4.673E+08 | 8.053E+08 |
| 23100 | 8.752E+07 | 1.037E+08 | 1.243E+08 | 1.511E+08 | 1.875E+08 | 2.399E+08 | 3.220E+08 | 4.689E+08 | 8.076E+08 |
| 23200 | 8.842E+07 | 1.047E+08 | 1.253E+08 | 1.521E+08 | 1.886E+08 | 2.411E+08 | 3.234E+08 | 4.706E+08 | 8.099E+08 |
| 23300 | 8.933E+07 | 1.056E+08 | 1.263E+08 | 1.532E+08 | 1.897E+08 | 2.424E+08 | 3.248E+08 | 4.723E+08 | 8.121E+08 |
| 23400 | 9.028E+07 | 1.066E+08 | 1.273E+08 | 1.542E+08 | 1.909E+08 | 2.436E+08 | 3.262E+08 | 4.739E+08 | 8.144E+08 |
| 23500 | 9.125E+07 | 1.076E+08 | 1.284E+08 | 1.554E+08 | 1.921E+08 | 2.449E+08 | 3.277E+08 | 4.756E+08 | 8.167E+08 |
| 23600 | 9.225E+07 | 1.087E+08 | 1.294E+08 | 1.565E+08 | 1.933E+08 | 2.462E+08 | 3.291E+08 | 4.774E+08 | 8.190E+08 |
| 23700 | 9.327E+07 | 1.097E+08 | 1.305E+08 | 1.577E+08 | 1.945E+08 | 2.476E+08 | 3.306E+08 | 4.791E+08 | 8.213E+08 |
| 23800 | 9.431E+07 | 1.108E+08 | 1.317E+08 | 1.589E+08 | 1.958E+08 | 2.489E+08 | 3.321E+08 | 4.808E+08 | 8.237E+08 |
| 23900 | 9.538E+07 | 1.119E+08 | 1.328E+08 | 1.601E+08 | 1.971E+08 | 2.503E+08 | 3.336E+08 | 4.826E+08 | 8.260E+08 |
| 24000 | 9.647E+07 | 1.131E+08 | 1.340E+08 | 1.613E+08 | 1.984E+08 | 2.517E+08 | 3.352E+08 | 4.844E+08 | 8.283E+08 |

# TABLE A.4. Specific Heat of Ar/$H_2$ Mixtures (J/kg · K)

| | Molar fraction of $H_2$ | | | | | | | | |
|---|---|---|---|---|---|---|---|---|---|
| $T$ (K) | 0.1 | 0.2 | 0.3 | 0.4 | 0.5 | 0.6 | 0.7 | 0.8 | 0.9 |
| 500 | 5.980E+02 | 6.938E+02 | 8.152E+02 | 9.737E+02 | 1.189E+03 | 1.500E+03 | 1.987E+03 | 2.859E+03 | 4.870E+03 |
| 600 | 5.986E+02 | 6.957E+02 | 8.193E+02 | 9.818E+02 | 1.204E+03 | 1.526E+03 | 2.031E+03 | 2.939E+03 | 5.037E+03 |
| 700 | 5.993E+02 | 6.975E+02 | 8.228E+02 | 9.875E+02 | 1.213E+03 | 1.540E+03 | 2.054E+03 | 2.977E+03 | 5.111E+03 |
| 800 | 6.001E+02 | 6.995E+02 | 8.265E+02 | 9.937E+02 | 1.223E+03 | 1.555E+03 | 2.078E+03 | 3.017E+03 | 5.189E+03 |
| 900 | 6.010E+02 | 7.017E+02 | 8.304E+02 | 1.000E+03 | 1.233E+03 | 1.571E+03 | 2.103E+03 | 3.059E+03 | 5.271E+03 |
| 1000 | 6.019E+02 | 7.039E+02 | 8.346E+02 | 1.007E+03 | 1.243E+03 | 1.587E+03 | 2.129E+03 | 3.102E+03 | 5.355E+03 |
| 1100 | 6.029E+02 | 7.063E+02 | 8.388E+02 | 1.014E+03 | 1.254E+03 | 1.604E+03 | 2.155E+03 | 3.147E+03 | 5.441E+03 |
| 1200 | 6.039E+02 | 7.087E+02 | 8.432E+02 | 1.021E+03 | 1.265E+03 | 1.621E+03 | 2.182E+03 | 3.192E+03 | 5.529E+03 |
| 1300 | 6.050E+02 | 7.112E+02 | 8.478E+02 | 1.028E+03 | 1.277E+03 | 1.639E+03 | 2.210E+03 | 3.238E+03 | 5.618E+03 |
| 1400 | 6.061E+02 | 7.138E+02 | 8.524E+02 | 1.036E+03 | 1.288E+03 | 1.657E+03 | 2.238E+03 | 3.285E+03 | 5.709E+03 |
| 1500 | 6.073E+02 | 7.166E+02 | 8.574E+02 | 1.043E+03 | 1.301E+03 | 1.675E+03 | 2.267E+03 | 3.333E+03 | 5.802E+03 |
| 1600 | 6.088E+02 | 7.199E+02 | 8.630E+02 | 1.052E+03 | 1.314E+03 | 1.696E+03 | 2.298E+03 | 3.385E+03 | 5.901E+03 |
| 1700 | 6.110E+02 | 7.242E+02 | 8.699E+02 | 1.063E+03 | 1.330E+03 | 1.719E+03 | 2.334E+03 | 3.443E+03 | 6.012E+03 |
| 1800 | 6.144E+02 | 7.303E+02 | 8.795E+02 | 1.077E+03 | 1.350E+03 | 1.748E+03 | 2.378E+03 | 3.513E+03 | 6.143E+03 |
| 1900 | 6.198E+02 | 7.398E+02 | 8.937E+02 | 1.097E+03 | 1.378E+03 | 1.788E+03 | 2.436E+03 | 3.605E+03 | 6.313E+03 |
| 2000 | 6.287E+02 | 7.546E+02 | 9.152E+02 | 1.127E+03 | 1.419E+03 | 1.845E+03 | 2.518E+03 | 3.731E+03 | 6.542E+03 |
| 2100 | 6.426E+02 | 7.774E+02 | 9.478E+02 | 1.172E+03 | 1.480E+03 | 1.928E+03 | 2.635E+03 | 3.910E+03 | 6.863E+03 |
| 2200 | 6.638E+02 | 8.117E+02 | 9.961E+02 | 1.237E+03 | 1.567E+03 | 2.047E+03 | 2.803E+03 | 4.163E+03 | 7.313E+03 |
| 2300 | 6.946E+02 | 8.612E+02 | 1.065E+03 | 1.331E+03 | 1.692E+03 | 2.216E+03 | 3.039E+03 | 4.518E+03 | 7.939E+03 |
| 2400 | 7.376E+02 | 9.305E+02 | 1.162E+03 | 1.461E+03 | 1.865E+03 | 2.448E+03 | 3.362E+03 | 5.003E+03 | 8.792E+03 |
| 2500 | 7.954E+02 | 1.023E+03 | 1.293E+03 | 1.637E+03 | 2.098E+03 | 2.761E+03 | 3.796E+03 | 5.651E+03 | 9.931E+03 |
| 2600 | 8.698E+02 | 1.145E+03 | 1.464E+03 | 1.866E+03 | 2.402E+03 | 3.169E+03 | 4.363E+03 | 6.497E+03 | 1.141E+04 |
| 2700 | 9.618E+02 | 1.297E+03 | 1.680E+03 | 2.156E+03 | 2.789E+03 | 3.688E+03 | 5.085E+03 | 7.574E+03 | 1.330E+04 |
| 2800 | 1.070E+03 | 1.481E+03 | 1.944E+03 | 2.514E+03 | 3.267E+03 | 4.331E+03 | 5.981E+03 | 8.914E+03 | 1.565E+04 |
| 2900 | 1.192E+03 | 1.696E+03 | 2.256E+03 | 2.941E+03 | 3.841E+03 | 5.108E+03 | 7.067E+03 | 1.054E+04 | 1.852E+04 |
| 3000 | 1.321E+03 | 1.935E+03 | 2.611E+03 | 3.434E+03 | 4.510E+03 | 6.021E+03 | 8.351E+03 | 1.248E+04 | 2.194E+04 |
| 3100 | 1.448E+03 | 2.188E+03 | 3.000E+03 | 3.984E+03 | 5.266E+03 | 7.063E+03 | 9.827E+03 | 1.472E+04 | 2.593E+04 |
| 3200 | 1.558E+03 | 2.438E+03 | 3.402E+03 | 4.569E+03 | 6.087E+03 | 8.210E+03 | 1.147E+04 | 1.724E+04 | 3.046E+04 |
| 3300 | 1.638E+03 | 2.663E+03 | 3.793E+03 | 5.160E+03 | 6.936E+03 | 9.422E+03 | 1.324E+04 | 1.998E+04 | 3.543E+04 |
| 3400 | 1.675E+03 | 2.839E+03 | 4.135E+03 | 5.711E+03 | 7.761E+03 | 1.063E+04 | 1.503E+04 | 2.283E+04 | 4.066E+04 |
| 3500 | 1.660E+03 | 2.939E+03 | 4.391E+03 | 6.168E+03 | 8.488E+03 | 1.174E+04 | 1.674E+04 | 2.559E+04 | 4.584E+04 |
| 3600 | 1.594E+03 | 2.947E+03 | 4.524E+03 | 6.474E+03 | 9.035E+03 | 1.263E+04 | 1.818E+04 | 2.801E+04 | 5.053E+04 |
| 3700 | 1.487E+03 | 2.857E+03 | 4.508E+03 | 6.581E+03 | 9.324E+03 | 1.320E+04 | 1.919E+04 | 2.982E+04 | 5.419E+04 |
| 3800 | 1.357E+03 | 2.681E+03 | 4.341E+03 | 6.464E+03 | 9.302E+03 | 1.333E+04 | 1.959E+04 | 3.072E+04 | 5.629E+04 |
| 3900 | 1.222E+03 | 2.445E+03 | 4.045E+03 | 6.134E+03 | 8.960E+03 | 1.300E+04 | 1.931E+04 | 3.056E+04 | 5.643E+04 |
| 4000 | 1.097E+03 | 2.182E+03 | 3.661E+03 | 5.636E+03 | 8.342E+03 | 1.224E+04 | 1.836E+04 | 2.931E+04 | 5.453E+04 |
| 4100 | 9.892E+02 | 1.921E+03 | 3.241E+03 | 5.039E+03 | 7.535E+03 | 1.116E+04 | 1.688E+04 | 2.715E+04 | 5.085E+04 |
| 4200 | 9.020E+02 | 1.685E+03 | 2.827E+03 | 4.414E+03 | 6.642E+03 | 9.910E+03 | 1.508E+04 | 2.438E+04 | 4.592E+04 |
| 4300 | 8.342E+02 | 1.483E+03 | 2.452E+03 | 3.819E+03 | 5.757E+03 | 8.619E+03 | 1.316E+04 | 2.137E+04 | 4.039E+04 |
| 4400 | 7.829E+02 | 1.319E+03 | 2.132E+03 | 3.290E+03 | 4.946E+03 | 7.404E+03 | 1.132E+04 | 1.841E+04 | 3.486E+04 |
| 4500 | 7.446E+02 | 1.190E+03 | 1.869E+03 | 2.844E+03 | 4.244E+03 | 6.330E+03 | 9.665E+03 | 1.571E+04 | 2.974E+04 |
| 4600 | 7.164E+02 | 1.091E+03 | 1.660E+03 | 2.480E+03 | 3.661E+03 | 5.424E+03 | 8.248E+03 | 1.337E+04 | 2.527E+04 |
| 4700 | 6.956E+02 | 1.015E+03 | 1.498E+03 | 2.191E+03 | 3.191E+03 | 4.684E+03 | 7.078E+03 | 1.142E+04 | 2.152E+04 |
| 4800 | 6.803E+02 | 9.583E+02 | 1.372E+03 | 1.965E+03 | 2.819E+03 | 4.093E+03 | 6.136E+03 | 9.844E+03 | 1.846E+04 |
| 4900 | 6.690E+02 | 9.151E+02 | 1.276E+03 | 1.790E+03 | 2.528E+03 | 3.628E+03 | 5.390E+03 | 8.586E+03 | 1.601E+04 |
| 5000 | 6.606E+02 | 8.825E+02 | 1.203E+03 | 1.655E+03 | 2.302E+03 | 3.265E+03 | 4.805E+03 | 7.596E+03 | 1.408E+04 |
| 5100 | 6.543E+02 | 8.579E+02 | 1.147E+03 | 1.552E+03 | 2.128E+03 | 2.983E+03 | 4.348E+03 | 6.821E+03 | 1.256E+04 |
| 5200 | 6.497E+02 | 8.393E+02 | 1.104E+03 | 1.472E+03 | 1.994E+03 | 2.765E+03 | 3.994E+03 | 6.217E+03 | 1.137E+04 |
| 5300 | 6.462E+02 | 8.252E+02 | 1.071E+03 | 1.411E+03 | 1.890E+03 | 2.595E+03 | 3.718E+03 | 5.747E+03 | 1.045E+04 |

**TABLE A.4.** *(Continued)*

| | Molar fraction of $H_2$ | | | | | | | | |
|---|---|---|---|---|---|---|---|---|---|
| $T$ (K) | 0.1 | 0.2 | 0.3 | 0.4 | 0.5 | 0.6 | 0.7 | 0.8 | 0.9 |
| 5400 | 6.436E+02 | 8.144E+02 | 1.046E+03 | 1.364E+03 | 1.809E+03 | 2.464E+03 | 3.503E+03 | 5.380E+03 | 9.729E+03 |
| 5500 | 6.418E+02 | 8.062E+02 | 1.027E+03 | 1.327E+03 | 1.746E+03 | 2.362E+03 | 3.336E+03 | 5.093E+03 | 9.163E+03 |
| 5600 | 6.404E+02 | 8.000E+02 | 1.012E+03 | 1.299E+03 | 1.698E+03 | 2.282E+03 | 3.205E+03 | 4.869E+03 | 8.720E+03 |
| 5700 | 6.395E+02 | 7.953E+02 | 1.001E+03 | 1.277E+03 | 1.660E+03 | 2.219E+03 | 3.103E+03 | 4.693E+03 | 8.372E+03 |
| 5800 | 6.390E+02 | 7.919E+02 | 9.923E+02 | 1.260E+03 | 1.631E+03 | 2.171E+03 | 3.023E+03 | 4.555E+03 | 8.098E+03 |
| 5900 | 6.388E+02 | 7.894E+02 | 9.857E+02 | 1.247E+03 | 1.608E+03 | 2.133E+03 | 2.960E+03 | 4.447E+03 | 7.883E+03 |
| 6000 | 6.389E+02 | 7.878E+02 | 9.810E+02 | 1.237E+03 | 1.590E+03 | 2.104E+03 | 2.911E+03 | 4.362E+03 | 7.715E+03 |
| 6100 | 6.393E+02 | 7.870E+02 | 9.777E+02 | 1.230E+03 | 1.577E+03 | 2.081E+03 | 2.874E+03 | 4.297E+03 | 7.584E+03 |
| 6200 | 6.401E+02 | 7.868E+02 | 9.757E+02 | 1.225E+03 | 1.568E+03 | 2.064E+03 | 2.845E+03 | 4.247E+03 | 7.483E+03 |
| 6300 | 6.411E+02 | 7.872E+02 | 9.748E+02 | 1.222E+03 | 1.561E+03 | 2.053E+03 | 2.825E+03 | 4.209E+03 | 7.407E+03 |
| 6400 | 6.424E+02 | 7.882E+02 | 9.749E+02 | 1.221E+03 | 1.557E+03 | 2.045E+03 | 2.810E+03 | 4.183E+03 | 7.351E+03 |
| 6500 | 6.442E+02 | 7.898E+02 | 9.760E+02 | 1.221E+03 | 1.556E+03 | 2.041E+03 | 2.801E+03 | 4.166E+03 | 7.314E+03 |
| 6600 | 6.463E+02 | 7.921E+02 | 9.781E+02 | 1.222E+03 | 1.556E+03 | 2.040E+03 | 2.798E+03 | 4.157E+03 | 7.292E+03 |
| 6700 | 6.489E+02 | 7.949E+02 | 9.811E+02 | 1.225E+03 | 1.559E+03 | 2.042E+03 | 2.799E+03 | 4.155E+03 | 7.285E+03 |
| 6800 | 6.520E+02 | 7.985E+02 | 9.851E+02 | 1.229E+03 | 1.564E+03 | 2.046E+03 | 2.804E+03 | 4.160E+03 | 7.290E+03 |
| 6900 | 6.557E+02 | 8.029E+02 | 9.901E+02 | 1.235E+03 | 1.570E+03 | 2.054E+03 | 2.813E+03 | 4.172E+03 | 7.308E+03 |
| 7000 | 6.600E+02 | 8.081E+02 | 9.963E+02 | 1.242E+03 | 1.579E+03 | 2.065E+03 | 2.826E+03 | 4.191E+03 | 7.338E+03 |
| 7100 | 6.650E+02 | 8.142E+02 | 1.003E+03 | 1.251E+03 | 1.590E+03 | 2.078E+03 | 2.844E+03 | 4.216E+03 | 7.379E+03 |
| 7200 | 6.709E+02 | 8.213E+02 | 1.012E+03 | 1.262E+03 | 1.603E+03 | 2.094E+03 | 2.865E+03 | 4.246E+03 | 7.429E+03 |
| 7300 | 6.776E+02 | 8.295E+02 | 1.022E+03 | 1.274E+03 | 1.618E+03 | 2.114E+03 | 2.890E+03 | 4.282E+03 | 7.495E+03 |
| 7400 | 6.852E+02 | 8.385E+02 | 1.033E+03 | 1.288E+03 | 1.635E+03 | 2.136E+03 | 2.921E+03 | 4.328E+03 | 7.571E+03 |
| 7500 | 6.940E+02 | 8.494E+02 | 1.046E+03 | 1.304E+03 | 1.655E+03 | 2.162E+03 | 2.957E+03 | 4.379E+03 | 7.660E+03 |
| 7600 | 7.037E+02 | 8.615E+02 | 1.061E+03 | 1.322E+03 | 1.678E+03 | 2.192E+03 | 2.997E+03 | 4.438E+03 | 7.758E+03 |
| 7700 | 7.152E+02 | 8.747E+02 | 1.078E+03 | 1.342E+03 | 1.704E+03 | 2.226E+03 | 3.043E+03 | 4.505E+03 | 7.878E+03 |
| 7800 | 7.279E+02 | 8.905E+02 | 1.096E+03 | 1.366E+03 | 1.733E+03 | 2.263E+03 | 3.094E+03 | 4.581E+03 | 8.009E+03 |
| 7900 | 7.422E+02 | 9.078E+02 | 1.117E+03 | 1.392E+03 | 1.766E+03 | 2.306E+03 | 3.151E+03 | 4.665E+03 | 8.163E+03 |
| 8000 | 7.582E+02 | 9.270E+02 | 1.141E+03 | 1.421E+03 | 1.802E+03 | 2.353E+03 | 3.215E+03 | 4.759E+03 | 8.319E+03 |
| 8100 | 7.760E+02 | 9.485E+02 | 1.167E+03 | 1.454E+03 | 1.843E+03 | 2.405E+03 | 3.286E+03 | 4.862E+03 | 8.499E+03 |
| 8200 | 7.959E+02 | 9.724E+02 | 1.196E+03 | 1.489E+03 | 1.890E+03 | 2.462E+03 | 3.363E+03 | 4.977E+03 | 8.697E+03 |
| 8300 | 8.179E+02 | 9.988E+02 | 1.228E+03 | 1.528E+03 | 1.937E+03 | 2.530E+03 | 3.449E+03 | 5.102E+03 | 8.914E+03 |
| 8400 | 8.423E+02 | 1.029E+03 | 1.263E+03 | 1.571E+03 | 1.991E+03 | 2.596E+03 | 3.550E+03 | 5.251E+03 | 9.173E+03 |
| 8500 | 8.693E+02 | 1.060E+03 | 1.304E+03 | 1.618E+03 | 2.049E+03 | 2.671E+03 | 3.646E+03 | 5.390E+03 | 9.414E+03 |
| 8600 | 8.989E+02 | 1.095E+03 | 1.344E+03 | 1.674E+03 | 2.120E+03 | 2.755E+03 | 3.757E+03 | 5.553E+03 | 9.695E+03 |
| 8700 | 9.336E+02 | 1.134E+03 | 1.391E+03 | 1.727E+03 | 2.185E+03 | 2.852E+03 | 3.891E+03 | 5.750E+03 | 1.003E+04 |
| 8800 | 9.674E+02 | 1.179E+03 | 1.446E+03 | 1.788E+03 | 2.261E+03 | 2.943E+03 | 4.011E+03 | 5.924E+03 | 1.033E+04 |
| 8900 | 1.006E+03 | 1.222E+03 | 1.497E+03 | 1.862E+03 | 2.353E+03 | 3.061E+03 | 4.171E+03 | 6.132E+03 | 1.069E+04 |
| 9000 | 1.052E+03 | 1.277E+03 | 1.557E+03 | 1.928E+03 | 2.434E+03 | 3.164E+03 | 4.308E+03 | 6.382E+03 | 1.112E+04 |
| 9100 | 1.095E+03 | 1.327E+03 | 1.629E+03 | 2.016E+03 | 2.544E+03 | 3.305E+03 | 4.476E+03 | 6.596E+03 | 1.149E+04 |
| 9200 | 1.150E+03 | 1.386E+03 | 1.692E+03 | 2.092E+03 | 2.637E+03 | 3.422E+03 | 4.675E+03 | 6.894E+03 | 1.200E+04 |
| 9300 | 1.200E+03 | 1.458E+03 | 1.778E+03 | 2.196E+03 | 2.766E+03 | 3.588E+03 | 4.876E+03 | 7.132E+03 | 1.241E+04 |
| 9400 | 1.264E+03 | 1.519E+03 | 1.850E+03 | 2.282E+03 | 2.871E+03 | 3.720E+03 | 5.050E+03 | 7.481E+03 | 1.301E+04 |
| 9500 | 1.324E+03 | 1.604E+03 | 1.951E+03 | 2.405E+03 | 3.023E+03 | 3.914E+03 | 5.311E+03 | 7.811E+03 | 1.357E+04 |
| 9600 | 1.400E+03 | 1.674E+03 | 2.033E+03 | 2.502E+03 | 3.140E+03 | 4.061E+03 | 5.504E+03 | 8.087E+03 | 1.404E+04 |
| 9700 | 1.465E+03 | 1.773E+03 | 2.151E+03 | 2.645E+03 | 3.318E+03 | 4.287E+03 | 5.806E+03 | 8.526E+03 | 1.479E+04 |
| 9800 | 1.556E+03 | 1.868E+03 | 2.263E+03 | 2.762E+03 | 3.448E+03 | 4.450E+03 | 6.019E+03 | 8.828E+03 | 1.530E+04 |
| 9900 | 1.628E+03 | 1.951E+03 | 2.359E+03 | 2.910E+03 | 3.653E+03 | 4.711E+03 | 6.368E+03 | 9.334E+03 | 1.617E+04 |
| 10000 | 1.734E+03 | 2.076E+03 | 2.508E+03 | 3.072E+03 | 3.840E+03 | 4.947E+03 | 6.681E+03 | 9.784E+03 | 1.693E+04 |

*(continued)*

## TABLE A.4. (*Continued*)

| | Molar fraction of $H_2$ | | | | | | | | |
|---|---|---|---|---|---|---|---|---|---|
| $T$ (K) | 0.1 | 0.2 | 0.3 | 0.4 | 0.5 | 0.6 | 0.7 | 0.8 | 0.9 |
| 10100 | 1.815E+03 | 2.168E+03 | 2.614E+03 | 3.196E+03 | 3.989E+03 | 5.130E+03 | 6.919E+03 | 1.011E+04 | 1.749E+04 |
| 10200 | 1.938E+03 | 2.312E+03 | 2.786E+03 | 3.403E+03 | 4.244E+03 | 5.454E+03 | 7.350E+03 | 1.074E+04 | 1.856E+04 |
| 10300 | 2.052E+03 | 2.444E+03 | 2.940E+03 | 3.588E+03 | 4.468E+03 | 5.736E+03 | 7.722E+03 | 1.127E+04 | 1.947E+04 |
| 10400 | 2.144E+03 | 2.549E+03 | 3.060E+03 | 3.726E+03 | 4.632E+03 | 5.938E+03 | 7.981E+03 | 1.163E+04 | 2.007E+04 |
| 10500 | 2.297E+03 | 2.728E+03 | 3.272E+03 | 3.981E+03 | 4.946E+03 | 6.335E+03 | 8.510E+03 | 1.240E+04 | 2.137E+04 |
| 10600 | 2.433E+03 | 2.885E+03 | 3.456E+03 | 4.199E+03 | 5.210E+03 | 6.666E+03 | 8.945E+03 | 1.302E+04 | 2.242E+04 |
| 10700 | 2.537E+03 | 3.020E+03 | 3.650E+03 | 4.430E+03 | 5.490E+03 | 7.015E+03 | 9.403E+03 | 1.367E+04 | 2.352E+04 |
| 10800 | 2.724E+03 | 3.202E+03 | 3.783E+03 | 4.582E+03 | 5.668E+03 | 7.231E+03 | 9.677E+03 | 1.405E+04 | 2.413E+04 |
| 10900 | 2.884E+03 | 3.405E+03 | 4.061E+03 | 4.916E+03 | 6.077E+03 | 7.748E+03 | 1.036E+04 | 1.503E+04 | 2.581E+04 |
| 11000 | 3.053E+03 | 3.599E+03 | 4.287E+03 | 5.183E+03 | 6.400E+03 | 8.150E+03 | 1.088E+04 | 1.578E+04 | 2.707E+04 |
| 11100 | 3.168E+03 | 3.726E+03 | 4.429E+03 | 5.464E+03 | 6.739E+03 | 8.572E+03 | 1.144E+04 | 1.657E+04 | 2.839E+04 |
| 11200 | 3.410E+03 | 4.008E+03 | 4.762E+03 | 5.622E+03 | 6.921E+03 | 8.789E+03 | 1.171E+04 | 1.693E+04 | 2.897E+04 |
| 11300 | 3.605E+03 | 4.232E+03 | 5.022E+03 | 6.048E+03 | 7.442E+03 | 9.446E+03 | 1.258E+04 | 1.818E+04 | 3.109E+04 |
| 11400 | 3.810E+03 | 4.467E+03 | 5.293E+03 | 6.368E+03 | 7.826E+03 | 9.923E+03 | 1.320E+04 | 1.906E+04 | 3.257E+04 |
| 11500 | 3.947E+03 | 4.712E+03 | 5.577E+03 | 6.701E+03 | 8.227E+03 | 1.042E+04 | 1.385E+04 | 1.998E+04 | 3.411E+04 |
| 11600 | 4.219E+03 | 4.838E+03 | 5.715E+03 | 6.854E+03 | 8.560E+03 | 1.093E+04 | 1.452E+04 | 2.093E+04 | 3.570E+04 |
| 11700 | 4.467E+03 | 5.217E+03 | 6.159E+03 | 7.384E+03 | 8.885E+03 | 1.111E+04 | 1.474E+04 | 2.121E+04 | 3.612E+04 |
| 11800 | 4.707E+03 | 5.491E+03 | 6.475E+03 | 7.755E+03 | 9.491E+03 | 1.198E+04 | 1.588E+04 | 2.286E+04 | 3.892E+04 |
| 11900 | 4.956E+03 | 5.774E+03 | 6.802E+03 | 8.138E+03 | 9.950E+03 | 1.255E+04 | 1.662E+04 | 2.390E+04 | 4.067E+04 |
| 12000 | 5.214E+03 | 6.067E+03 | 7.140E+03 | 8.533E+03 | 1.042E+04 | 1.314E+04 | 1.738E+04 | 2.498E+04 | 4.247E+04 |
| 12100 | 5.317E+03 | 6.175E+03 | 7.487E+03 | 8.940E+03 | 1.091E+04 | 1.374E+04 | 1.817E+04 | 2.608E+04 | 4.432E+04 |
| 12200 | 5.736E+03 | 6.659E+03 | 7.584E+03 | 9.039E+03 | 1.127E+04 | 1.436E+04 | 1.897E+04 | 2.722E+04 | 4.621E+04 |
| 12300 | 6.015E+03 | 6.976E+03 | 8.184E+03 | 9.753E+03 | 1.162E+04 | 1.443E+04 | 1.903E+04 | 2.824E+04 | 4.816E+04 |
| 12400 | 6.300E+03 | 7.299E+03 | 8.555E+03 | 1.018E+04 | 1.240E+04 | 1.558E+04 | 2.056E+04 | 2.848E+04 | 4.805E+04 |
| 12500 | 6.591E+03 | 7.628E+03 | 8.933E+03 | 1.063E+04 | 1.293E+04 | 1.624E+04 | 2.141E+04 | 3.066E+04 | 5.198E+04 |
| 12600 | 6.886E+03 | 7.962E+03 | 9.316E+03 | 1.107E+04 | 1.346E+04 | 1.690E+04 | 2.227E+04 | 3.188E+04 | 5.403E+04 |
| 12700 | 7.184E+03 | 8.300E+03 | 9.704E+03 | 1.153E+04 | 1.401E+04 | 1.757E+04 | 2.315E+04 | 3.312E+04 | 5.610E+04 |
| 12800 | 7.217E+03 | 8.420E+03 | 1.009E+04 | 1.198E+04 | 1.455E+04 | 1.825E+04 | 2.403E+04 | 3.437E+04 | 5.820E+04 |
| 12900 | 7.764E+03 | 8.856E+03 | 1.007E+04 | 1.244E+04 | 1.510E+04 | 1.893E+04 | 2.492E+04 | 3.563E+04 | 6.032E+04 |
| 13000 | 8.064E+03 | 9.293E+03 | 1.084E+04 | 1.236E+04 | 1.541E+04 | 1.961E+04 | 2.581E+04 | 3.690E+04 | 6.246E+04 |
| 13100 | 8.360E+03 | 9.629E+03 | 1.122E+04 | 1.331E+04 | 1.571E+04 | 1.937E+04 | 2.646E+04 | 3.816E+04 | 6.459E+04 |
| 13200 | 8.653E+03 | 9.960E+03 | 1.161E+04 | 1.376E+04 | 1.668E+04 | 2.090E+04 | 2.649E+04 | 3.750E+04 | 6.654E+04 |
| 13300 | 8.939E+03 | 1.028E+04 | 1.198E+04 | 1.420E+04 | 1.721E+04 | 2.156E+04 | 2.836E+04 | 4.055E+04 | 6.550E+04 |
| 13400 | 9.217E+03 | 1.059E+04 | 1.234E+04 | 1.463E+04 | 1.773E+04 | 2.221E+04 | 2.922E+04 | 4.178E+04 | 7.073E+04 |
| 13500 | 9.483E+03 | 1.090E+04 | 1.269E+04 | 1.504E+04 | 1.824E+04 | 2.285E+04 | 3.007E+04 | 4.299E+04 | 7.281E+04 |
| 13600 | 9.737E+03 | 1.119E+04 | 1.303E+04 | 1.544E+04 | 1.873E+04 | 2.346E+04 | 3.089E+04 | 4.418E+04 | 7.484E+04 |
| 13700 | 9.976E+03 | 1.146E+04 | 1.335E+04 | 1.582E+04 | 1.919E+04 | 2.406E+04 | 3.168E+04 | 4.534E+04 | 7.683E+04 |
| 13800 | 9.852E+03 | 1.172E+04 | 1.365E+04 | 1.618E+04 | 1.964E+04 | 2.463E+04 | 3.244E+04 | 4.645E+04 | 7.877E+04 |
| 13900 | 1.031E+04 | 1.195E+04 | 1.393E+04 | 1.652E+04 | 2.006E+04 | 2.516E+04 | 3.317E+04 | 4.752E+04 | 8.063E+04 |
| 14000 | 1.056E+04 | 1.163E+04 | 1.418E+04 | 1.683E+04 | 2.045E+04 | 2.567E+04 | 3.386E+04 | 4.854E+04 | 8.242E+04 |
| 14100 | 1.072E+04 | 1.234E+04 | 1.405E+04 | 1.711E+04 | 2.080E+04 | 2.614E+04 | 3.451E+04 | 4.951E+04 | 8.412E+04 |
| 14200 | 1.085E+04 | 1.250E+04 | 1.429E+04 | 1.701E+04 | 2.113E+04 | 2.657E+04 | 3.510E+04 | 5.041E+04 | 8.572E+04 |
| 14300 | 1.096E+04 | 1.263E+04 | 1.476E+04 | 1.708E+04 | 2.107E+04 | 2.695E+04 | 3.565E+04 | 5.123E+04 | 8.721E+04 |
| 14400 | 1.104E+04 | 1.274E+04 | 1.490E+04 | 1.774E+04 | 2.166E+04 | 2.729E+04 | 3.614E+04 | 5.199E+04 | 8.858E+04 |
| 14500 | 1.109E+04 | 1.281E+04 | 1.501E+04 | 1.789E+04 | 2.109E+04 | 2.725E+04 | 3.656E+04 | 5.266E+04 | 8.982E+04 |
| 14600 | 1.111E+04 | 1.285E+04 | 1.508E+04 | 1.801E+04 | 2.201E+04 | 2.668E+04 | 3.661E+04 | 5.325E+04 | 9.092E+04 |
| 14700 | 1.110E+04 | 1.286E+04 | 1.511E+04 | 1.808E+04 | 2.213E+04 | 2.800E+04 | 3.550E+04 | 5.346E+04 | 9.188E+04 |
| 14800 | 1.106E+04 | 1.284E+04 | 1.512E+04 | 1.811E+04 | 2.221E+04 | 2.814E+04 | 3.744E+04 | 5.138E+04 | 9.245E+04 |

**TABLE A.4.** *(Continued)*

| | Molar fraction of $H_2$ | | | | | | | | |
|---|---|---|---|---|---|---|---|---|---|
| $T$ (K) | 0.1 | 0.2 | 0.3 | 0.4 | 0.5 | 0.6 | 0.7 | 0.8 | 0.9 |
| 14900 | 1.099E+04 | 1.279E+04 | 1.509E+04 | 1.811E+04 | 2.224E+04 | 2.822E+04 | 3.761E+04 | 5.444E+04 | 8.814E+04 |
| 15000 | 1.089E+04 | 1.270E+04 | 1.502E+04 | 1.807E+04 | 2.223E+04 | 2.826E+04 | 3.771E+04 | 5.466E+04 | 9.378E+04 |
| 15100 | 1.077E+04 | 1.259E+04 | 1.492E+04 | 1.799E+04 | 2.218E+04 | 2.823E+04 | 3.774E+04 | 5.479E+04 | 9.411E+04 |
| 15200 | 1.062E+04 | 1.245E+04 | 1.479E+04 | 1.787E+04 | 2.208E+04 | 2.816E+04 | 3.770E+04 | 5.481E+04 | 9.428E+04 |
| 15300 | 1.044E+04 | 1.228E+04 | 1.463E+04 | 1.772E+04 | 2.194E+04 | 2.803E+04 | 3.760E+04 | 5.473E+04 | 9.428E+04 |
| 15400 | 1.025E+04 | 1.209E+04 | 1.445E+04 | 1.753E+04 | 2.175E+04 | 2.785E+04 | 3.742E+04 | 5.456E+04 | 9.410E+04 |
| 15500 | 1.004E+04 | 1.188E+04 | 1.423E+04 | 1.732E+04 | 2.153E+04 | 2.762E+04 | 3.717E+04 | 5.429E+04 | 9.376E+04 |
| 15600 | 9.809E+03 | 1.165E+04 | 1.399E+04 | 1.707E+04 | 2.128E+04 | 2.735E+04 | 3.687E+04 | 5.392E+04 | 9.326E+04 |
| 15700 | 9.563E+03 | 1.140E+04 | 1.373E+04 | 1.680E+04 | 2.098E+04 | 2.702E+04 | 3.650E+04 | 5.346E+04 | 9.259E+04 |
| 15800 | 9.306E+03 | 1.113E+04 | 1.346E+04 | 1.650E+04 | 2.066E+04 | 2.666E+04 | 3.607E+04 | 5.291E+04 | 9.176E+04 |
| 15900 | 9.039E+03 | 1.085E+04 | 1.316E+04 | 1.619E+04 | 2.031E+04 | 2.626E+04 | 3.558E+04 | 5.228E+04 | 9.079E+04 |
| 16000 | 8.765E+03 | 1.056E+04 | 1.285E+04 | 1.585E+04 | 1.993E+04 | 2.582E+04 | 3.505E+04 | 5.158E+04 | 8.968E+04 |
| 16100 | 8.486E+03 | 1.027E+04 | 1.253E+04 | 1.549E+04 | 1.953E+04 | 2.535E+04 | 3.447E+04 | 5.080E+04 | 8.844E+04 |
| 16200 | 8.204E+03 | 9.969E+03 | 1.220E+04 | 1.513E+04 | 1.911E+04 | 2.485E+04 | 3.385E+04 | 4.995E+04 | 8.708E+04 |
| 16300 | 7.921E+03 | 9.663E+03 | 1.187E+04 | 1.475E+04 | 1.867E+04 | 2.433E+04 | 3.319E+04 | 4.905E+04 | 8.561E+04 |
| 16400 | 7.640E+03 | 9.356E+03 | 1.152E+04 | 1.436E+04 | 1.822E+04 | 2.379E+04 | 3.250E+04 | 4.809E+04 | 8.404E+04 |
| 16500 | 7.361E+03 | 9.050E+03 | 1.118E+04 | 1.397E+04 | 1.776E+04 | 2.323E+04 | 3.178E+04 | 4.709E+04 | 8.238E+04 |
| 16600 | 7.087E+03 | 8.745E+03 | 1.084E+04 | 1.357E+04 | 1.729E+04 | 2.265E+04 | 3.104E+04 | 4.605E+04 | 8.064E+04 |
| 16700 | 6.818E+03 | 8.444E+03 | 1.049E+04 | 1.317E+04 | 1.682E+04 | 2.206E+04 | 3.028E+04 | 4.497E+04 | 7.884E+04 |
| 16800 | 6.556E+03 | 8.148E+03 | 1.016E+04 | 1.278E+04 | 1.634E+04 | 2.147E+04 | 2.950E+04 | 4.387E+04 | 7.699E+04 |
| 16900 | 6.301E+03 | 7.859E+03 | 9.824E+03 | 1.238E+04 | 1.586E+04 | 2.088E+04 | 2.872E+04 | 4.275E+04 | 7.510E+04 |
| 17000 | 6.055E+03 | 7.576E+03 | 9.495E+03 | 1.199E+04 | 1.539E+04 | 2.028E+04 | 2.793E+04 | 4.162E+04 | 7.317E+04 |
| 17100 | 5.818E+03 | 7.302E+03 | 9.173E+03 | 1.161E+04 | 1.492E+04 | 1.968E+04 | 2.714E+04 | 4.048E+04 | 7.123E+04 |
| 17200 | 5.590E+03 | 7.036E+03 | 8.859E+03 | 1.123E+04 | 1.445E+04 | 1.909E+04 | 2.635E+04 | 3.934E+04 | 6.927E+04 |
| 17300 | 5.373E+03 | 6.780E+03 | 8.553E+03 | 1.086E+04 | 1.400E+04 | 1.851E+04 | 2.557E+04 | 3.821E+04 | 6.732E+04 |
| 17400 | 5.165E+03 | 6.534E+03 | 8.258E+03 | 1.050E+04 | 1.355E+04 | 1.794E+04 | 2.480E+04 | 3.708E+04 | 6.537E+04 |
| 17500 | 4.968E+03 | 6.298E+03 | 7.973E+03 | 1.015E+04 | 1.311E+04 | 1.737E+04 | 2.404E+04 | 3.596E+04 | 6.343E+04 |
| 17600 | 4.782E+03 | 6.072E+03 | 7.698E+03 | 9.814E+03 | 1.268E+04 | 1.682E+04 | 2.329E+04 | 3.486E+04 | 6.152E+04 |
| 17700 | 4.606E+03 | 5.858E+03 | 7.435E+03 | 9.487E+03 | 1.227E+04 | 1.628E+04 | 2.256E+04 | 3.378E+04 | 5.964E+04 |
| 17800 | 4.440E+03 | 5.654E+03 | 7.183E+03 | 9.173E+03 | 1.187E+04 | 1.576E+04 | 2.184E+04 | 5.914E+04 | 5.779E+04 |
| 17900 | 4.285E+03 | 5.462E+03 | 6.943E+03 | 8.871E+03 | 1.149E+04 | 1.525E+04 | 2.115E+04 | 5.275E+03 | 5.598E+04 |
| 18000 | 4.140E+03 | 5.280E+03 | 6.715E+03 | 8.583E+03 | 1.112E+04 | 1.476E+04 | 2.047E+04 | 3.068E+04 | 5.421E+04 |
| 18100 | 4.006E+03 | 5.109E+03 | 6.499E+03 | 8.307E+03 | 1.076E+04 | 1.429E+04 | 1.982E+04 | 2.971E+04 | 5.249E+04 |
| 18200 | 3.881E+03 | 4.950E+03 | 6.295E+03 | 8.045E+03 | 1.042E+04 | 1.384E+04 | 1.919E+04 | 2.876E+04 | 5.082E+04 |
| 18300 | 3.767E+03 | 4.801E+03 | 6.102E+03 | 7.797E+03 | 1.009E+04 | 1.341E+04 | 1.859E+04 | 2.754E+04 | 6.514E+04 |
| 18400 | 3.662E+03 | 4.662E+03 | 5.922E+03 | 7.562E+03 | 9.790E+03 | 1.280E+04 | 1.774E+04 | 4.590E+04 | 3.098E+04 |
| 18500 | 3.567E+03 | 4.535E+03 | 5.754E+03 | 7.227E+03 | 9.335E+03 | 1.255E+04 | 1.739E+04 | 6.879E+03 | 4.585E+04 |
| 18600 | 3.481E+03 | 4.417E+03 | 5.508E+03 | 7.116E+03 | 9.194E+03 | 1.203E+04 | 1.671E+04 | 2.522E+04 | 4.450E+04 |
| 18700 | 3.404E+03 | 4.250E+03 | 5.442E+03 | 6.749E+03 | 8.767E+03 | 1.182E+04 | 1.635E+04 | 2.444E+04 | 4.311E+04 |
| 18800 | 3.306E+03 | 4.208E+03 | 5.124E+03 | 6.736E+03 | 8.683E+03 | 1.148E+04 | 1.586E+04 | 2.371E+04 | 4.177E+04 |
| 18900 | 3.276E+03 | 3.926E+03 | 5.182E+03 | 6.570E+03 | 8.455E+03 | 1.116E+04 | 1.541E+04 | 2.300E+04 | 4.050E+04 |
| 19000 | 3.019E+03 | 4.041E+03 | 5.073E+03 | 6.416E+03 | 8.242E+03 | 1.087E+04 | 1.498E+04 | 2.233E+04 | 3.928E+04 |
| 19100 | 3.182E+03 | 3.975E+03 | 4.974E+03 | 6.276E+03 | 8.044E+03 | 1.059E+04 | 1.457E+04 | 2.169E+04 | 3.811E+04 |
| 19200 | 3.150E+03 | 3.918E+03 | 4.887E+03 | 6.148E+03 | 7.862E+03 | 1.032E+04 | 1.418E+04 | 2.109E+04 | 3.700E+04 |
| 19300 | 3.127E+03 | 3.872E+03 | 4.810E+03 | 6.033E+03 | 7.694E+03 | 1.008E+04 | 1.382E+04 | 2.051E+04 | 3.594E+04 |
| 19400 | 3.112E+03 | 3.835E+03 | 4.745E+03 | 5.930E+03 | 7.541E+03 | 9.860E+03 | 1.348E+04 | 1.997E+04 | 3.493E+04 |
| 19500 | 3.106E+03 | 3.807E+03 | 4.690E+03 | 5.840E+03 | 7.403E+03 | 9.653E+03 | 1.317E+04 | 1.946E+04 | 3.397E+04 |

*(continued)*

TABLE A.4. (*Continued*)

| | Molar fraction of $H_2$ | | | | | | | | |
|---|---|---|---|---|---|---|---|---|---|
| $T$ (K) | 0.1 | 0.2 | 0.3 | 0.4 | 0.5 | 0.6 | 0.7 | 0.8 | 0.9 |
| 19600 | 3.109E+03 | 3.789E+03 | 4.646E+03 | 5.762E+03 | 7.279E+03 | 9.462E+03 | 1.287E+04 | 1.899E+04 | 3.307E+04 |
| 19700 | 3.120E+03 | 3.781E+03 | 4.613E+03 | 5.697E+03 | 7.170E+03 | 9.289E+03 | 1.260E+04 | 1.854E+04 | 3.221E+04 |
| 19800 | 3.140E+03 | 3.782E+03 | 4.590E+03 | 5.643E+03 | 7.074E+03 | 9.133E+03 | 1.235E+04 | 1.811E+04 | 3.140E+04 |
| 19900 | 3.169E+03 | 3.792E+03 | 4.578E+03 | 5.602E+03 | 6.992E+03 | 8.994E+03 | 1.212E+04 | 1.772E+04 | 3.063E+04 |
| 20000 | 3.206E+03 | 3.812E+03 | 4.577E+03 | 5.572E+03 | 6.924E+03 | 8.871E+03 | 1.191E+04 | 1.736E+04 | 2.991E+04 |
| 20100 | 3.252E+03 | 3.842E+03 | 4.586E+03 | 5.555E+03 | 6.870E+03 | 8.764E+03 | 1.172E+04 | 1.702E+04 | 2.923E+04 |
| 20200 | 3.307E+03 | 3.882E+03 | 4.606E+03 | 5.549E+03 | 6.830E+03 | 8.672E+03 | 1.155E+04 | 1.671E+04 | 2.859E+04 |
| 20300 | 3.371E+03 | 3.931E+03 | 4.637E+03 | 5.555E+03 | 6.803E+03 | 8.597E+03 | 1.140E+04 | 1.642E+04 | 2.799E+04 |
| 20400 | 3.444E+03 | 3.990E+03 | 4.678E+03 | 5.573E+03 | 6.789E+03 | 8.537E+03 | 1.127E+04 | 1.616E+04 | 2.743E+04 |
| 20500 | 3.526E+03 | 4.059E+03 | 4.730E+03 | 5.604E+03 | 6.789E+03 | 8.493E+03 | 1.115E+04 | 1.592E+04 | 2.691E+04 |
| 20600 | 3.617E+03 | 4.138E+03 | 4.793E+03 | 5.645E+03 | 6.802E+03 | 8.464E+03 | 1.106E+04 | 1.571E+04 | 2.642E+04 |
| 20700 | 3.718E+03 | 4.227E+03 | 4.867E+03 | 5.699E+03 | 6.828E+03 | 8.450E+03 | 1.098E+04 | 1.552E+04 | 2.597E+04 |
| 20800 | 3.828E+03 | 4.326E+03 | 4.952E+03 | 5.765E+03 | 6.867E+03 | 8.451E+03 | 1.092E+04 | 1.535E+04 | 2.555E+04 |
| 20900 | 3.947E+03 | 4.435E+03 | 5.047E+03 | 5.842E+03 | 6.919E+03 | 8.466E+03 | 1.088E+04 | 2.117E+04 | 2.516E+04 |
| 21000 | 4.076E+03 | 4.554E+03 | 5.154E+03 | 5.931E+03 | 6.984E+03 | 8.496E+03 | 1.086E+04 | 9.114E+03 | 2.480E+04 |
| 21100 | 4.214E+03 | 4.684E+03 | 5.271E+03 | 6.032E+03 | 7.062E+03 | 8.541E+03 | 1.085E+04 | 1.497E+04 | 2.447E+04 |
| 21200 | 4.362E+03 | 4.823E+03 | 5.399E+03 | 6.145E+03 | 7.153E+03 | 8.600E+03 | 1.085E+04 | 1.489E+04 | 2.418E+04 |
| 21300 | 4.520E+03 | 4.973E+03 | 5.538E+03 | 6.269E+03 | 7.256E+03 | 8.672E+03 | 1.088E+04 | 1.482E+04 | 2.391E+04 |
| 21400 | 4.687E+03 | 5.132E+03 | 5.687E+03 | 6.404E+03 | 7.372E+03 | 8.759E+03 | 1.092E+04 | 1.478E+04 | 2.366E+04 |
| 21500 | 4.863E+03 | 5.302E+03 | 5.847E+03 | 6.551E+03 | 7.500E+03 | 8.859E+03 | 1.097E+04 | 1.475E+04 | 2.345E+04 |
| 21600 | 5.048E+03 | 5.481E+03 | 6.018E+03 | 6.709E+03 | 7.640E+03 | 8.972E+03 | 1.104E+04 | 1.474E+04 | 2.326E+04 |
| 21700 | 5.242E+03 | 5.669E+03 | 6.198E+03 | 6.878E+03 | 7.792E+03 | 9.098E+03 | 1.113E+04 | 1.475E+04 | 2.309E+04 |
| 21800 | 5.445E+03 | 5.867E+03 | 6.388E+03 | 7.057E+03 | 7.955E+03 | 9.237E+03 | 1.123E+04 | 1.478E+04 | 2.294E+04 |
| 21900 | 5.656E+03 | 6.074E+03 | 6.588E+03 | 7.247E+03 | 8.129E+03 | 9.387E+03 | 1.134E+04 | 1.482E+04 | 2.282E+04 |
| 22000 | 5.876E+03 | 6.289E+03 | 6.797E+03 | 7.446E+03 | 8.314E+03 | 9.550E+03 | 1.146E+04 | 1.488E+04 | 2.272E+04 |
| 22100 | 6.102E+03 | 6.513E+03 | 7.015E+03 | 7.655E+03 | 8.509E+03 | 9.723E+03 | 1.160E+04 | 1.495E+04 | 2.263E+04 |
| 22200 | 6.336E+03 | 6.744E+03 | 7.241E+03 | 7.872E+03 | 8.713E+03 | 9.907E+03 | 1.175E+04 | 1.296E+04 | 1.865E+04 |
| 22300 | 6.577E+03 | 6.982E+03 | 7.474E+03 | 8.098E+03 | 8.927E+03 | 9.259E+03 | 1.063E+04 | 1.482E+04 | 2.226E+04 |
| 22400 | 6.824E+03 | 7.227E+03 | 7.715E+03 | 8.331E+03 | 8.587E+03 | 1.030E+04 | 1.174E+04 | 1.524E+04 | 2.249E+04 |
| 22500 | 7.076E+03 | 7.477E+03 | 7.962E+03 | 8.201E+03 | 9.017E+03 | 1.016E+04 | 1.226E+04 | 1.537E+04 | 2.248E+04 |
| 22600 | 7.332E+03 | 7.733E+03 | 7.982E+03 | 8.450E+03 | 9.611E+03 | 1.073E+04 | 1.245E+04 | 1.550E+04 | 2.248E+04 |
| 22700 | 7.593E+03 | 7.861E+03 | 8.099E+03 | 9.066E+03 | 9.853E+03 | 1.095E+04 | 1.265E+04 | 1.565E+04 | 2.250E+04 |
| 22800 | 7.801E+03 | 8.257E+03 | 8.728E+03 | 9.322E+03 | 1.010E+04 | 1.118E+04 | 1.285E+04 | 1.580E+04 | 2.253E+04 |
| 22900 | 8.123E+03 | 8.142E+03 | 8.992E+03 | 9.581E+03 | 1.035E+04 | 1.142E+04 | 1.307E+04 | 1.596E+04 | 2.257E+04 |
| 23000 | 8.390E+03 | 8.785E+03 | 9.258E+03 | 9.842E+03 | 1.060E+04 | 1.166E+04 | 1.328E+04 | 1.613E+04 | 2.261E+04 |
| 23100 | 8.272E+03 | 9.053E+03 | 9.524E+03 | 1.010E+04 | 1.085E+04 | 1.190E+04 | 1.350E+04 | 1.630E+04 | 2.267E+04 |
| 23200 | 8.920E+03 | 9.321E+03 | 9.790E+03 | 1.036E+04 | 1.111E+04 | 1.214E+04 | 1.372E+04 | 1.647E+04 | 2.274E+04 |
| 23300 | 9.186E+03 | 9.586E+03 | 1.005E+04 | 1.062E+04 | 1.136E+04 | 1.239E+04 | 1.394E+04 | 1.665E+04 | 2.281E+04 |
| 23400 | 9.449E+03 | 9.850E+03 | 1.031E+04 | 1.088E+04 | 1.162E+04 | 1.263E+04 | 1.416E+04 | 1.683E+04 | 2.288E+04 |
| 23500 | 9.708E+03 | 1.011E+04 | 1.057E+04 | 1.114E+04 | 1.187E+04 | 1.287E+04 | 1.438E+04 | 1.701E+04 | 2.296E+04 |
| 23600 | 9.963E+03 | 1.036E+04 | 1.083E+04 | 1.139E+04 | 1.211E+04 | 1.310E+04 | 1.459E+04 | 1.718E+04 | 2.304E+04 |
| 23700 | 1.021E+04 | 1.061E+04 | 1.107E+04 | 1.164E+04 | 1.235E+04 | 1.333E+04 | 1.480E+04 | 1.735E+04 | 2.312E+04 |
| 23800 | 1.045E+04 | 1.085E+04 | 1.131E+04 | 1.187E+04 | 1.258E+04 | 1.355E+04 | 1.500E+04 | 1.752E+04 | 2.320E+04 |
| 23900 | 1.068E+04 | 1.109E+04 | 1.155E+04 | 1.210E+04 | 1.281E+04 | 1.377E+04 | 1.520E+04 | 1.768E+04 | 2.328E+04 |
| 24000 | 1.091E+04 | 1.131E+04 | 1.177E+04 | 1.232E+04 | 1.302E+04 | 1.397E+04 | 1.539E+04 | 1.784E+04 | 2.336E+04 |

**TABLE A.5.** Viscosity of Ar/$H_2$ Mixtures (kg/m · s)

| | Molar fraction of $H_2$ | | | | | | | | |
|---|---|---|---|---|---|---|---|---|---|
| $T$ (K) | 0.1 | 0.2 | 0.3 | 0.4 | 0.5 | 0.6 | 0.7 | 0.8 | 0.9 |
| 500 | 3.344E-05 | 3.252E-05 | 3.145E-05 | 3.018E-05 | 2.864E-05 | 2.676E-05 | 2.441E-05 | 2.140E-05 | 1.744E-05 |
| 600 | 3.838E-05 | 3.737E-05 | 3.617E-05 | 3.474E-05 | 3.301E-05 | 3.087E-05 | 2.817E-05 | 2.470E-05 | 2.010E-05 |
| 700 | 4.385E-05 | 4.268E-05 | 4.129E-05 | 3.964E-05 | 3.763E-05 | 3.515E-05 | 3.203E-05 | 2.802E-05 | 2.271E-05 |
| 800 | 4.865E-05 | 4.737E-05 | 4.586E-05 | 4.404E-05 | 4.183E-05 | 3.909E-05 | 3.563E-05 | 3.116E-05 | 2.522E-05 |
| 900 | 5.279E-05 | 5.146E-05 | 4.988E-05 | 4.797E-05 | 4.562E-05 | 4.270E-05 | 3.898E-05 | 3.414E-05 | 2.764E-05 |
| 1000 | 5.650E-05 | 5.515E-05 | 5.353E-05 | 5.157E-05 | 4.914E-05 | 4.608E-05 | 4.216E-05 | 3.699E-05 | 2.999E-05 |
| 1100 | 5.995E-05 | 5.860E-05 | 5.697E-05 | 5.497E-05 | 5.247E-05 | 4.931E-05 | 4.521E-05 | 3.976E-05 | 3.229E-05 |
| 1200 | 6.326E-05 | 6.191E-05 | 6.027E-05 | 5.825E-05 | 5.570E-05 | 5.244E-05 | 4.818E-05 | 4.246E-05 | 3.453E-05 |
| 1300 | 6.649E-05 | 6.516E-05 | 6.351E-05 | 6.146E-05 | 5.886E-05 | 5.551E-05 | 5.110E-05 | 4.512E-05 | 3.675E-05 |
| 1400 | 6.969E-05 | 6.836E-05 | 6.671E-05 | 6.463E-05 | 6.198E-05 | 5.854E-05 | 5.397E-05 | 4.773E-05 | 3.893E-05 |
| 1500 | 7.287E-05 | 7.154E-05 | 6.988E-05 | 6.777E-05 | 6.507E-05 | 6.154E-05 | 5.681E-05 | 5.032E-05 | 4.108E-05 |
| 1600 | 7.604E-05 | 7.471E-05 | 7.303E-05 | 7.090E-05 | 6.814E-05 | 6.451E-05 | 5.963E-05 | 5.288E-05 | 4.321E-05 |
| 1700 | 7.921E-05 | 7.787E-05 | 7.617E-05 | 7.401E-05 | 7.119E-05 | 6.746E-05 | 6.242E-05 | 5.541E-05 | 4.532E-05 |
| 1800 | 8.236E-05 | 8.101E-05 | 7.930E-05 | 7.710E-05 | 7.422E-05 | 7.039E-05 | 6.519E-05 | 5.792E-05 | 4.740E-05 |
| 1900 | 8.550E-05 | 8.415E-05 | 8.241E-05 | 8.017E-05 | 7.723E-05 | 7.330E-05 | 6.794E-05 | 6.041E-05 | 4.947E-05 |
| 2000 | 8.863E-05 | 8.727E-05 | 8.551E-05 | 8.323E-05 | 8.022E-05 | 7.619E-05 | 7.067E-05 | 6.288E-05 | 5.152E-05 |
| 2100 | 9.175E-05 | 9.037E-05 | 8.859E-05 | 8.627E-05 | 8.320E-05 | 7.906E-05 | 7.337E-05 | 6.533E-05 | 5.355E-05 |
| 2200 | 9.485E-05 | 9.346E-05 | 9.165E-05 | 8.929E-05 | 8.615E-05 | 8.191E-05 | 7.606E-05 | 6.776E-05 | 5.557E-05 |
| 2300 | 9.793E-05 | 9.652E-05 | 9.469E-05 | 9.229E-05 | 8.908E-05 | 8.474E-05 | 7.873E-05 | 7.018E-05 | 5.757E-05 |
| 2400 | 1.009E-04 | 9.957E-05 | 9.771E-05 | 9.527E-05 | 9.200E-05 | 8.755E-05 | 8.138E-05 | 7.257E-05 | 5.956E-05 |
| 2500 | 1.040E-04 | 1.026E-04 | 1.007E-04 | 9.823E-05 | 9.489E-05 | 9.034E-05 | 8.400E-05 | 7.495E-05 | 6.154E-05 |
| 2600 | 1.070E-04 | 1.056E-04 | 1.037E-04 | 1.011E-04 | 9.776E-05 | 9.310E-05 | 8.661E-05 | 7.731E-05 | 6.351E-05 |
| 2700 | 1.100E-04 | 1.085E-04 | 1.066E-04 | 1.040E-04 | 1.006E-04 | 9.585E-05 | 8.920E-05 | 7.965E-05 | 6.546E-05 |
| 2800 | 1.129E-04 | 1.115E-04 | 1.095E-04 | 1.069E-04 | 1.034E-04 | 9.857E-05 | 9.176E-05 | 8.197E-05 | 6.740E-05 |
| 2900 | 1.159E-04 | 1.144E-04 | 1.124E-04 | 1.098E-04 | 1.062E-04 | 1.012E-04 | 9.429E-05 | 8.426E-05 | 6.931E-05 |
| 3000 | 1.188E-04 | 1.173E-04 | 1.153E-04 | 1.126E-04 | 1.089E-04 | 1.039E-04 | 9.677E-05 | 8.649E-05 | 7.118E-05 |
| 3100 | 1.217E-04 | 1.202E-04 | 1.182E-04 | 1.154E-04 | 1.117E-04 | 1.065E-04 | 9.920E-05 | 8.866E-05 | 7.297E-05 |
| 3200 | 1.245E-04 | 1.231E-04 | 1.210E-04 | 1.182E-04 | 1.143E-04 | 1.090E-04 | 1.015E-04 | 9.073E-05 | 7.465E-05 |
| 3300 | 1.274E-04 | 1.259E-04 | 1.238E-04 | 1.209E-04 | 1.169E-04 | 1.114E-04 | 1.037E-04 | 9.265E-05 | 7.617E-05 |
| 3400 | 1.302E-04 | 1.287E-04 | 1.265E-04 | 1.235E-04 | 1.194E-04 | 1.138E-04 | 1.058E-04 | 9.440E-05 | 7.749E-05 |
| 3500 | 1.330E-04 | 1.314E-04 | 1.292E-04 | 1.261E-04 | 1.219E-04 | 1.160E-04 | 1.077E-04 | 9.593E-05 | 7.854E-05 |
| 3600 | 1.357E-04 | 1.341E-04 | 1.318E-04 | 1.286E-04 | 1.242E-04 | 1.180E-04 | 1.094E-04 | 9.722E-05 | 7.932E-05 |
| 3700 | 1.384E-04 | 1.368E-04 | 1.344E-04 | 1.311E-04 | 1.264E-04 | 1.199E-04 | 1.110E-04 | 9.828E-05 | 7.983E-05 |
| 3800 | 1.411E-04 | 1.395E-04 | 1.370E-04 | 1.334E-04 | 1.285E-04 | 1.217E-04 | 1.123E-04 | 9.915E-05 | 8.010E-05 |
| 3900 | 1.437E-04 | 1.421E-04 | 1.395E-04 | 1.358E-04 | 1.306E-04 | 1.235E-04 | 1.136E-04 | 9.989E-05 | 8.021E-05 |
| 4000 | 1.464E-04 | 1.447E-04 | 1.420E-04 | 1.381E-04 | 1.326E-04 | 1.251E-04 | 1.148E-04 | 1.005E-04 | 8.027E-05 |
| 4100 | 1.490E-04 | 1.472E-04 | 1.444E-04 | 1.404E-04 | 1.347E-04 | 1.268E-04 | 1.161E-04 | 1.012E-04 | 8.034E-05 |
| 4200 | 1.515E-04 | 1.498E-04 | 1.469E-04 | 1.427E-04 | 1.367E-04 | 1.286E-04 | 1.174E-04 | 1.020E-04 | 8.052E-05 |
| 4300 | 1.541E-04 | 1.523E-04 | 1.494E-04 | 1.450E-04 | 1.388E-04 | 1.304E-04 | 1.188E-04 | 1.029E-04 | 8.084E-05 |
| 4400 | 1.566E-04 | 1.548E-04 | 1.518E-04 | 1.473E-04 | 1.410E-04 | 1.322E-04 | 1.203E-04 | 1.040E-04 | 8.132E-05 |
| 4500 | 1.591E-04 | 1.573E-04 | 1.543E-04 | 1.497E-04 | 1.431E-04 | 1.342E-04 | 1.219E-04 | 1.051E-04 | 8.196E-05 |
| 4600 | 1.616E-04 | 1.598E-04 | 1.567E-04 | 1.520E-04 | 1.453E-04 | 1.361E-04 | 1.236E-04 | 1.064E-04 | 8.275E-05 |
| 4700 | 1.640E-04 | 1.623E-04 | 1.592E-04 | 1.544E-04 | 1.476E-04 | 1.382E-04 | 1.253E-04 | 1.078E-04 | 8.367E-05 |
| 4800 | 1.664E-04 | 1.647E-04 | 1.616E-04 | 1.567E-04 | 1.498E-04 | 1.402E-04 | 1.271E-04 | 1.093E-04 | 8.470E-05 |
| 4900 | 1.689E-04 | 1.671E-04 | 1.640E-04 | 1.591E-04 | 1.521E-04 | 1.423E-04 | 1.290E-04 | 1.108E-04 | 8.581E-05 |
| 5000 | 1.712E-04 | 1.696E-04 | 1.664E-04 | 1.614E-04 | 1.543E-04 | 1.444E-04 | 1.309E-04 | 1.125E-04 | 8.699E-05 |
| 5100 | 1.736E-04 | 1.719E-04 | 1.688E-04 | 1.638E-04 | 1.566E-04 | 1.466E-04 | 1.329E-04 | 1.141E-04 | 8.822E-05 |
| 5200 | 1.760E-04 | 1.743E-04 | 1.711E-04 | 1.661E-04 | 1.588E-04 | 1.487E-04 | 1.348E-04 | 1.158E-04 | 8.950E-05 |

(*continued*)

## TABLE A.5. (*Continued*)

| | Molar fraction of $H_2$ | | | | | | | | |
|---|---|---|---|---|---|---|---|---|---|
| $T$ (K) | 0.1 | 0.2 | 0.3 | 0.4 | 0.5 | 0.6 | 0.7 | 0.8 | 0.9 |
| 5300 | 1.783E-04 | 1.767E-04 | 1.735E-04 | 1.684E-04 | 1.611E-04 | 1.509E-04 | 1.368E-04 | 1.175E-04 | 9.081E-05 |
| 5400 | 1.806E-04 | 1.790E-04 | 1.758E-04 | 1.708E-04 | 1.634E-04 | 1.530E-04 | 1.388E-04 | 1.192E-04 | 9.215E-05 |
| 5500 | 1.829E-04 | 1.813E-04 | 1.782E-04 | 1.731E-04 | 1.656E-04 | 1.551E-04 | 1.407E-04 | 1.209E-04 | 9.351E-05 |
| 5600 | 1.852E-04 | 1.836E-04 | 1.805E-04 | 1.754E-04 | 1.679E-04 | 1.573E-04 | 1.427E-04 | 1.227E-04 | 9.489E-05 |
| 5700 | 1.874E-04 | 1.859E-04 | 1.828E-04 | 1.777E-04 | 1.701E-04 | 1.594E-04 | 1.447E-04 | 1.244E-04 | 9.628E-05 |
| 5800 | 1.897E-04 | 1.882E-04 | 1.851E-04 | 1.799E-04 | 1.723E-04 | 1.616E-04 | 1.467E-04 | 1.262E-04 | 9.767E-05 |
| 5900 | 1.919E-04 | 1.905E-04 | 1.873E-04 | 1.822E-04 | 1.745E-04 | 1.637E-04 | 1.487E-04 | 1.280E-04 | 9.908E-05 |
| 6000 | 1.941E-04 | 1.927E-04 | 1.896E-04 | 1.845E-04 | 1.768E-04 | 1.659E-04 | 1.507E-04 | 1.297E-04 | 1.005E-04 |
| 6100 | 1.964E-04 | 1.950E-04 | 1.919E-04 | 1.867E-04 | 1.790E-04 | 1.680E-04 | 1.527E-04 | 1.315E-04 | 1.019E-04 |
| 6200 | 1.985E-04 | 1.972E-04 | 1.941E-04 | 1.889E-04 | 1.812E-04 | 1.701E-04 | 1.547E-04 | 1.333E-04 | 1.033E-04 |
| 6300 | 2.007E-04 | 1.994E-04 | 1.963E-04 | 1.912E-04 | 1.834E-04 | 1.722E-04 | 1.567E-04 | 1.350E-04 | 1.047E-04 |
| 6400 | 2.029E-04 | 2.016E-04 | 1.986E-04 | 1.934E-04 | 1.855E-04 | 1.743E-04 | 1.586E-04 | 1.368E-04 | 1.061E-04 |
| 6500 | 2.051E-04 | 2.038E-04 | 2.008E-04 | 1.956E-04 | 1.877E-04 | 1.764E-04 | 1.606E-04 | 1.386E-04 | 1.076E-04 |
| 6600 | 2.072E-04 | 2.060E-04 | 2.030E-04 | 1.978E-04 | 1.899E-04 | 1.785E-04 | 1.626E-04 | 1.403E-04 | 1.090E-04 |
| 6700 | 2.093E-04 | 2.082E-04 | 2.052E-04 | 2.000E-04 | 1.920E-04 | 1.806E-04 | 1.646E-04 | 1.421E-04 | 1.104E-04 |
| 6800 | 2.115E-04 | 2.103E-04 | 2.073E-04 | 2.021E-04 | 1.942E-04 | 1.827E-04 | 1.665E-04 | 1.439E-04 | 1.118E-04 |
| 6900 | 2.136E-04 | 2.125E-04 | 2.095E-04 | 2.043E-04 | 1.963E-04 | 1.848E-04 | 1.685E-04 | 1.456E-04 | 1.132E-04 |
| 7000 | 2.157E-04 | 2.146E-04 | 2.117E-04 | 2.065E-04 | 1.985E-04 | 1.868E-04 | 1.704E-04 | 1.474E-04 | 1.147E-04 |
| 7100 | 2.178E-04 | 2.167E-04 | 2.138E-04 | 2.086E-04 | 2.006E-04 | 1.889E-04 | 1.723E-04 | 1.491E-04 | 1.161E-04 |
| 7200 | 2.198E-04 | 2.189E-04 | 2.160E-04 | 2.108E-04 | 2.027E-04 | 1.909E-04 | 1.743E-04 | 1.508E-04 | 1.175E-04 |
| 7300 | 2.219E-04 | 2.210E-04 | 2.181E-04 | 2.129E-04 | 2.048E-04 | 1.930E-04 | 1.762E-04 | 1.525E-04 | 1.189E-04 |
| 7400 | 2.240E-04 | 2.231E-04 | 2.202E-04 | 2.150E-04 | 2.069E-04 | 1.950E-04 | 1.781E-04 | 1.543E-04 | 1.203E-04 |
| 7500 | 2.260E-04 | 2.252E-04 | 2.223E-04 | 2.171E-04 | 2.090E-04 | 1.970E-04 | 1.800E-04 | 1.560E-04 | 1.217E-04 |
| 7600 | 2.281E-04 | 2.272E-04 | 2.244E-04 | 2.192E-04 | 2.110E-04 | 1.990E-04 | 1.819E-04 | 1.577E-04 | 1.230E-04 |
| 7700 | 2.301E-04 | 2.293E-04 | 2.265E-04 | 2.213E-04 | 2.131E-04 | 2.010E-04 | 1.838E-04 | 1.594E-04 | 1.244E-04 |
| 7800 | 2.321E-04 | 2.314E-04 | 2.286E-04 | 2.234E-04 | 2.151E-04 | 2.030E-04 | 1.856E-04 | 1.610E-04 | 1.258E-04 |
| 7900 | 2.341E-04 | 2.334E-04 | 2.306E-04 | 2.254E-04 | 2.172E-04 | 2.050E-04 | 1.875E-04 | 1.627E-04 | 1.271E-04 |
| 8000 | 2.361E-04 | 2.354E-04 | 2.327E-04 | 2.275E-04 | 2.192E-04 | 2.069E-04 | 1.893E-04 | 1.643E-04 | 1.284E-04 |
| 8100 | 2.381E-04 | 2.375E-04 | 2.347E-04 | 2.295E-04 | 2.212E-04 | 2.088E-04 | 1.911E-04 | 1.659E-04 | 1.298E-04 |
| 8200 | 2.401E-04 | 2.395E-04 | 2.367E-04 | 2.315E-04 | 2.231E-04 | 2.107E-04 | 1.929E-04 | 1.675E-04 | 1.311E-04 |
| 8300 | 2.421E-04 | 2.414E-04 | 2.387E-04 | 2.335E-04 | 2.251E-04 | 2.126E-04 | 1.947E-04 | 1.691E-04 | 1.323E-04 |
| 8400 | 2.440E-04 | 2.434E-04 | 2.407E-04 | 2.354E-04 | 2.270E-04 | 2.144E-04 | 1.964E-04 | 1.707E-04 | 1.336E-04 |
| 8500 | 2.459E-04 | 2.454E-04 | 2.426E-04 | 2.374E-04 | 2.289E-04 | 2.163E-04 | 1.981E-04 | 1.722E-04 | 1.348E-04 |
| 8600 | 2.479E-04 | 2.473E-04 | 2.446E-04 | 2.393E-04 | 2.307E-04 | 2.180E-04 | 1.998E-04 | 1.737E-04 | 1.360E-04 |
| 8700 | 2.498E-04 | 2.492E-04 | 2.465E-04 | 2.411E-04 | 2.326E-04 | 2.198E-04 | 2.014E-04 | 1.751E-04 | 1.372E-04 |
| 8800 | 2.516E-04 | 2.510E-04 | 2.483E-04 | 2.430E-04 | 2.343E-04 | 2.215E-04 | 2.030E-04 | 1.765E-04 | 1.384E-04 |
| 8900 | 2.535E-04 | 2.529E-04 | 2.501E-04 | 2.448E-04 | 2.361E-04 | 2.232E-04 | 2.045E-04 | 1.779E-04 | 1.395E-04 |
| 9000 | 2.553E-04 | 2.547E-04 | 2.519E-04 | 2.465E-04 | 2.378E-04 | 2.248E-04 | 2.060E-04 | 1.792E-04 | 1.405E-04 |
| 9100 | 2.570E-04 | 2.565E-04 | 2.537E-04 | 2.482E-04 | 2.394E-04 | 2.263E-04 | 2.074E-04 | 1.805E-04 | 1.415E-04 |
| 9200 | 2.588E-04 | 2.582E-04 | 2.554E-04 | 2.499E-04 | 2.410E-04 | 2.278E-04 | 2.088E-04 | 1.817E-04 | 1.425E-04 |
| 9300 | 2.605E-04 | 2.599E-04 | 2.570E-04 | 2.514E-04 | 2.425E-04 | 2.292E-04 | 2.101E-04 | 1.828E-04 | 1.434E-04 |
| 9400 | 2.621E-04 | 2.615E-04 | 2.586E-04 | 2.529E-04 | 2.439E-04 | 2.306E-04 | 2.113E-04 | 1.839E-04 | 1.442E-04 |
| 9500 | 2.637E-04 | 2.630E-04 | 2.601E-04 | 2.544E-04 | 2.453E-04 | 2.318E-04 | 2.124E-04 | 1.849E-04 | 1.450E-04 |
| 9600 | 2.652E-04 | 2.645E-04 | 2.615E-04 | 2.557E-04 | 2.465E-04 | 2.330E-04 | 2.135E-04 | 1.857E-04 | 1.457E-04 |
| 9700 | 2.667E-04 | 2.659E-04 | 2.628E-04 | 2.570E-04 | 2.477E-04 | 2.340E-04 | 2.144E-04 | 1.865E-04 | 1.464E-04 |
| 9800 | 2.681E-04 | 2.672E-04 | 2.641E-04 | 2.581E-04 | 2.487E-04 | 2.349E-04 | 2.152E-04 | 1.872E-04 | 1.469E-04 |
| 9900 | 2.694E-04 | 2.684E-04 | 2.652E-04 | 2.591E-04 | 2.497E-04 | 2.358E-04 | 2.159E-04 | 1.878E-04 | 1.474E-04 |
| 10000 | 2.706E-04 | 2.695E-04 | 2.662E-04 | 2.600E-04 | 2.505E-04 | 2.364E-04 | 2.165E-04 | 1.883E-04 | 1.477E-04 |

**TABLE A.5.** *(Continued)*

| $T$ (K) | Molar fraction of $H_2$ | | | | | | | | |
|---|---|---|---|---|---|---|---|---|---|
| | 0.1 | 0.2 | 0.3 | 0.4 | 0.5 | 0.6 | 0.7 | 0.8 | 0.9 |
| 10100 | 2.717E-04 | 2.705E-04 | 2.671E-04 | 2.608E-04 | 2.511E-04 | 2.370E-04 | 2.169E-04 | 1.886E-04 | 1.480E-04 |
| 10200 | 2.726E-04 | 2.714E-04 | 2.678E-04 | 2.614E-04 | 2.516E-04 | 2.373E-04 | 2.172E-04 | 1.888E-04 | 1.481E-04 |
| 10300 | 2.735E-04 | 2.721E-04 | 2.684E-04 | 2.619E-04 | 2.519E-04 | 2.375E-04 | 2.173E-04 | 1.889E-04 | 1.481E-04 |
| 10400 | 2.742E-04 | 2.726E-04 | 2.688E-04 | 2.621E-04 | 2.520E-04 | 2.376E-04 | 2.173E-04 | 1.888E-04 | 1.480E-04 |
| 10500 | 2.747E-04 | 2.730E-04 | 2.690E-04 | 2.622E-04 | 2.520E-04 | 2.374E-04 | 2.170E-04 | 1.885E-04 | 1.478E-04 |
| 10600 | 2.750E-04 | 2.732E-04 | 2.690E-04 | 2.621E-04 | 2.517E-04 | 2.370E-04 | 2.166E-04 | 1.881E-04 | 1.475E-04 |
| 10700 | 2.752E-04 | 2.732E-04 | 2.688E-04 | 2.617E-04 | 2.512E-04 | 2.364E-04 | 2.160E-04 | 1.875E-04 | 1.470E-04 |
| 10800 | 2.751E-04 | 2.729E-04 | 2.684E-04 | 2.611E-04 | 2.505E-04 | 2.356E-04 | 2.151E-04 | 1.867E-04 | 1.463E-04 |
| 10900 | 2.749E-04 | 2.725E-04 | 2.677E-04 | 2.603E-04 | 2.495E-04 | 2.346E-04 | 2.141E-04 | 1.857E-04 | 1.456E-04 |
| 11000 | 2.743E-04 | 2.717E-04 | 2.668E-04 | 2.592E-04 | 2.483E-04 | 2.333E-04 | 2.128E-04 | 1.846E-04 | 1.446E-04 |
| 11100 | 2.736E-04 | 2.708E-04 | 2.656E-04 | 2.579E-04 | 2.468E-04 | 2.318E-04 | 2.113E-04 | 1.832E-04 | 1.435E-04 |
| 11200 | 2.725E-04 | 2.695E-04 | 2.642E-04 | 2.562E-04 | 2.451E-04 | 2.300E-04 | 2.095E-04 | 1.817E-04 | 1.423E-04 |
| 11300 | 2.712E-04 | 2.679E-04 | 2.624E-04 | 2.543E-04 | 2.431E-04 | 2.279E-04 | 2.076E-04 | 1.799E-04 | 1.409E-04 |
| 11400 | 2.696E-04 | 2.661E-04 | 2.604E-04 | 2.521E-04 | 2.407E-04 | 2.256E-04 | 2.054E-04 | 1.779E-04 | 1.394E-04 |
| 11500 | 2.677E-04 | 2.639E-04 | 2.580E-04 | 2.496E-04 | 2.382E-04 | 2.230E-04 | 2.029E-04 | 1.758E-04 | 1.377E-04 |
| 11600 | 2.655E-04 | 2.615E-04 | 2.553E-04 | 2.468E-04 | 2.353E-04 | 2.202E-04 | 2.003E-04 | 1.734E-04 | 1.358E-04 |
| 11700 | 2.629E-04 | 2.587E-04 | 2.524E-04 | 2.437E-04 | 2.321E-04 | 2.171E-04 | 1.973E-04 | 1.709E-04 | 1.338E-04 |
| 11800 | 2.601E-04 | 2.556E-04 | 2.491E-04 | 2.403E-04 | 2.287E-04 | 2.137E-04 | 1.942E-04 | 1.682E-04 | 1.317E-04 |
| 11900 | 2.569E-04 | 2.522E-04 | 2.455E-04 | 2.366E-04 | 2.250E-04 | 2.102E-04 | 1.909E-04 | 1.653E-04 | 1.294E-04 |
| 12000 | 2.534E-04 | 2.485E-04 | 2.416E-04 | 2.326E-04 | 2.211E-04 | 2.063E-04 | 1.874E-04 | 1.622E-04 | 1.270E-04 |
| 12100 | 2.495E-04 | 2.444E-04 | 2.375E-04 | 2.284E-04 | 2.169E-04 | 2.023E-04 | 1.836E-04 | 1.590E-04 | 1.245E-04 |
| 12200 | 2.454E-04 | 2.401E-04 | 2.330E-04 | 2.239E-04 | 2.125E-04 | 1.981E-04 | 1.797E-04 | 1.556E-04 | 1.219E-04 |
| 12300 | 2.410E-04 | 2.355E-04 | 2.283E-04 | 2.192E-04 | 2.078E-04 | 1.936E-04 | 1.757E-04 | 1.521E-04 | 1.192E-04 |
| 12400 | 2.363E-04 | 2.307E-04 | 2.234E-04 | 2.143E-04 | 2.030E-04 | 1.891E-04 | 1.714E-04 | 1.484E-04 | 1.164E-04 |
| 12500 | 2.313E-04 | 2.256E-04 | 2.183E-04 | 2.092E-04 | 1.980E-04 | 1.843E-04 | 1.671E-04 | 1.447E-04 | 1.135E-04 |
| 12600 | 2.261E-04 | 2.203E-04 | 2.129E-04 | 2.039E-04 | 1.929E-04 | 1.795E-04 | 1.627E-04 | 1.409E-04 | 1.105E-04 |
| 12700 | 2.207E-04 | 2.148E-04 | 2.074E-04 | 1.984E-04 | 1.876E-04 | 1.745E-04 | 1.581E-04 | 1.370E-04 | 1.075E-04 |
| 12800 | 2.151E-04 | 2.091E-04 | 2.018E-04 | 1.929E-04 | 1.823E-04 | 1.694E-04 | 1.535E-04 | 1.330E-04 | 1.045E-04 |
| 12900 | 2.093E-04 | 2.033E-04 | 1.959E-04 | 1.872E-04 | 1.768E-04 | 1.643E-04 | 1.489E-04 | 1.290E-04 | 1.014E-04 |
| 13000 | 2.034E-04 | 1.974E-04 | 1.901E-04 | 1.815E-04 | 1.713E-04 | 1.591E-04 | 1.442E-04 | 1.250E-04 | 9.834E-05 |
| 13100 | 1.973E-04 | 1.914E-04 | 1.842E-04 | 1.757E-04 | 1.657E-04 | 1.539E-04 | 1.394E-04 | 1.210E-04 | 9.523E-05 |
| 13200 | 1.913E-04 | 1.853E-04 | 1.782E-04 | 1.699E-04 | 1.602E-04 | 1.487E-04 | 1.347E-04 | 1.169E-04 | 9.209E-05 |
| 13300 | 1.851E-04 | 1.792E-04 | 1.722E-04 | 1.641E-04 | 1.547E-04 | 1.435E-04 | 1.300E-04 | 1.128E-04 | 8.897E-05 |
| 13400 | 1.790E-04 | 1.731E-04 | 1.663E-04 | 1.584E-04 | 1.492E-04 | 1.384E-04 | 1.254E-04 | 1.088E-04 | 8.589E-05 |
| 13500 | 1.728E-04 | 1.671E-04 | 1.604E-04 | 1.526E-04 | 1.437E-04 | 1.333E-04 | 1.208E-04 | 1.049E-04 | 8.284E-05 |
| 13600 | 1.667E-04 | 1.611E-04 | 1.545E-04 | 1.470E-04 | 1.384E-04 | 1.283E-04 | 1.163E-04 | 1.010E-04 | 7.983E-05 |
| 13700 | 1.607E-04 | 1.552E-04 | 1.488E-04 | 1.415E-04 | 1.331E-04 | 1.234E-04 | 1.118E-04 | 9.719E-05 | 7.686E-05 |
| 13800 | 1.546E-04 | 1.494E-04 | 1.431E-04 | 1.360E-04 | 1.279E-04 | 1.186E-04 | 1.074E-04 | 9.343E-05 | 7.395E-05 |
| 13900 | 1.487E-04 | 1.437E-04 | 1.376E-04 | 1.307E-04 | 1.229E-04 | 1.139E-04 | 1.032E-04 | 8.976E-05 | 7.111E-05 |
| 14000 | 1.430E-04 | 1.379E-04 | 1.322E-04 | 1.256E-04 | 1.180E-04 | 1.093E-04 | 9.907E-05 | 8.617E-05 | 6.832E-05 |
| 14100 | 1.374E-04 | 1.325E-04 | 1.268E-04 | 1.205E-04 | 1.132E-04 | 1.049E-04 | 9.503E-05 | 8.269E-05 | 6.562E-05 |
| 14200 | 1.320E-04 | 1.273E-04 | 1.217E-04 | 1.155E-04 | 1.086E-04 | 1.006E-04 | 9.113E-05 | 7.930E-05 | 6.298E-05 |
| 14300 | 1.268E-04 | 1.222E-04 | 1.168E-04 | 1.108E-04 | 1.040E-04 | 9.645E-05 | 8.734E-05 | 7.603E-05 | 6.043E-05 |
| 14400 | 1.218E-04 | 1.173E-04 | 1.121E-04 | 1.062E-04 | 9.975E-05 | 9.244E-05 | 8.370E-05 | 7.286E-05 | 5.796E-05 |
| 14500 | 1.169E-04 | 1.126E-04 | 1.075E-04 | 1.019E-04 | 9.558E-05 | 8.844E-05 | 8.019E-05 | 6.981E-05 | 5.558E-05 |
| 14600 | 1.123E-04 | 1.080E-04 | 1.032E-04 | 9.773E-05 | 9.161E-05 | 8.470E-05 | 7.669E-05 | 6.688E-05 | 5.328E-05 |
| 14700 | 1.078E-04 | 1.037E-04 | 9.904E-05 | 9.373E-05 | 8.781E-05 | 8.114E-05 | 7.341E-05 | 6.396E-05 | 5.107E-05 |
| 14800 | 1.036E-04 | 9.965E-05 | 9.506E-05 | 8.991E-05 | 8.418E-05 | 7.775E-05 | 7.031E-05 | 6.121E-05 | 4.888E-05 |

*(continued)*

**TABLE A.5.** *(Continued)*

| | Molar fraction of $H_2$ | | | | | | | | |
|---|---|---|---|---|---|---|---|---|---|
| $T$ (K) | 0.1 | 0.2 | 0.3 | 0.4 | 0.5 | 0.6 | 0.7 | 0.8 | 0.9 |
| 14900 | 9.958E-05 | 9.574E-05 | 9.127E-05 | 8.627E-05 | 8.072E-05 | 7.451E-05 | 6.735E-05 | 5.862E-05 | 4.679E-05 |
| 15000 | 9.576E-05 | 9.202E-05 | 8.767E-05 | 8.281E-05 | 7.743E-05 | 7.142E-05 | 6.452E-05 | 5.615E-05 | 4.484E-05 |
| 15100 | 9.215E-05 | 8.850E-05 | 8.425E-05 | 7.952E-05 | 7.430E-05 | 6.848E-05 | 6.183E-05 | 5.379E-05 | 4.298E-05 |
| 15200 | 8.874E-05 | 8.517E-05 | 8.102E-05 | 7.641E-05 | 7.133E-05 | 6.569E-05 | 5.927E-05 | 5.155E-05 | 4.120E-05 |
| 15300 | 8.553E-05 | 8.203E-05 | 7.797E-05 | 7.347E-05 | 6.852E-05 | 6.305E-05 | 5.685E-05 | 4.942E-05 | 3.951E-05 |
| 15400 | 8.251E-05 | 7.907E-05 | 7.509E-05 | 7.069E-05 | 6.587E-05 | 6.055E-05 | 5.455E-05 | 4.739E-05 | 3.790E-05 |
| 15500 | 7.968E-05 | 7.629E-05 | 7.239E-05 | 6.807E-05 | 6.336E-05 | 5.819E-05 | 5.237E-05 | 4.547E-05 | 3.638E-05 |
| 15600 | 7.703E-05 | 7.369E-05 | 6.984E-05 | 6.560E-05 | 6.100E-05 | 5.596E-05 | 5.031E-05 | 4.366E-05 | 3.493E-05 |
| 15700 | 7.455E-05 | 7.125E-05 | 6.745E-05 | 6.329E-05 | 5.877E-05 | 5.385E-05 | 4.837E-05 | 4.194E-05 | 3.356E-05 |
| 15800 | 7.224E-05 | 6.897E-05 | 6.522E-05 | 6.111E-05 | 5.668E-05 | 5.187E-05 | 4.654E-05 | 4.032E-05 | 3.226E-05 |
| 15900 | 7.010E-05 | 6.684E-05 | 6.313E-05 | 5.908E-05 | 5.472E-05 | 5.001E-05 | 4.481E-05 | 3.879E-05 | 3.103E-05 |
| 16000 | 6.810E-05 | 6.486E-05 | 6.117E-05 | 5.717E-05 | 5.288E-05 | 4.826E-05 | 4.319E-05 | 3.734E-05 | 2.987E-05 |
| 16100 | 6.625E-05 | 6.301E-05 | 5.935E-05 | 5.539E-05 | 5.116E-05 | 4.662E-05 | 4.167E-05 | 3.599E-05 | 2.878E-05 |
| 16200 | 6.455E-05 | 6.130E-05 | 5.765E-05 | 5.372E-05 | 4.954E-05 | 4.509E-05 | 4.024E-05 | 3.471E-05 | 2.775E-05 |
| 16300 | 6.297E-05 | 5.971E-05 | 5.608E-05 | 5.217E-05 | 4.804E-05 | 4.365E-05 | 3.890E-05 | 3.351E-05 | 2.679E-05 |
| 16400 | 6.151E-05 | 5.825E-05 | 5.461E-05 | 5.073E-05 | 4.663E-05 | 4.230E-05 | 3.764E-05 | 3.239E-05 | 2.588E-05 |
| 16500 | 6.018E-05 | 5.689E-05 | 5.325E-05 | 4.939E-05 | 4.533E-05 | 4.105E-05 | 3.647E-05 | 3.134E-05 | 2.502E-05 |
| 16600 | 5.895E-05 | 5.564E-05 | 5.200E-05 | 4.814E-05 | 4.411E-05 | 3.988E-05 | 3.537E-05 | 3.035E-05 | 2.422E-05 |
| 16700 | 5.783E-05 | 5.449E-05 | 5.084E-05 | 4.699E-05 | 4.298E-05 | 3.879E-05 | 3.434E-05 | 2.943E-05 | 2.347E-05 |
| 16800 | 5.681E-05 | 5.343E-05 | 4.977E-05 | 4.592E-05 | 4.192E-05 | 3.777E-05 | 3.338E-05 | 2.856E-05 | 2.277E-05 |
| 16900 | 5.588E-05 | 5.247E-05 | 4.878E-05 | 4.493E-05 | 4.095E-05 | 3.683E-05 | 3.249E-05 | 2.776E-05 | 2.211E-05 |
| 17000 | 5.503E-05 | 5.158E-05 | 4.788E-05 | 4.402E-05 | 4.005E-05 | 3.595E-05 | 3.166E-05 | 2.700E-05 | 2.149E-05 |
| 17100 | 5.427E-05 | 5.078E-05 | 4.705E-05 | 4.318E-05 | 3.921E-05 | 3.514E-05 | 3.089E-05 | 2.630E-05 | 2.092E-05 |
| 17200 | 5.358E-05 | 5.004E-05 | 4.629E-05 | 4.241E-05 | 3.844E-05 | 3.438E-05 | 3.017E-05 | 2.565E-05 | 2.038E-05 |
| 17300 | 5.296E-05 | 4.938E-05 | 4.559E-05 | 4.170E-05 | 3.773E-05 | 3.369E-05 | 2.951E-05 | 2.504E-05 | 1.988E-05 |
| 17400 | 5.241E-05 | 4.878E-05 | 4.496E-05 | 4.105E-05 | 3.708E-05 | 3.304E-05 | 2.889E-05 | 2.448E-05 | 1.942E-05 |
| 17500 | 5.192E-05 | 4.824E-05 | 4.439E-05 | 4.046E-05 | 3.648E-05 | 3.245E-05 | 2.832E-05 | 2.396E-05 | 1.898E-05 |
| 17600 | 5.148E-05 | 4.776E-05 | 4.387E-05 | 3.992E-05 | 3.593E-05 | 3.190E-05 | 2.779E-05 | 2.347E-05 | 1.858E-05 |
| 17700 | 5.110E-05 | 4.733E-05 | 4.340E-05 | 3.943E-05 | 3.543E-05 | 3.140E-05 | 2.731E-05 | 2.302E-05 | 1.821E-05 |
| 17800 | 5.077E-05 | 4.694E-05 | 4.299E-05 | 3.898E-05 | 3.497E-05 | 3.094E-05 | 2.686E-05 | 1.752E-05 | 1.786E-05 |
| 17900 | 5.049E-05 | 4.661E-05 | 4.261E-05 | 3.858E-05 | 3.455E-05 | 3.052E-05 | 2.645E-05 | 2.223E-05 | 1.754E-05 |
| 18000 | 5.025E-05 | 4.631E-05 | 4.228E-05 | 3.822E-05 | 3.417E-05 | 3.014E-05 | 2.607E-05 | 2.187E-05 | 1.725E-05 |
| 18100 | 5.004E-05 | 4.606E-05 | 4.198E-05 | 3.790E-05 | 3.383E-05 | 2.979E-05 | 2.573E-05 | 2.155E-05 | 1.698E-05 |
| 18200 | 4.988E-05 | 4.584E-05 | 4.172E-05 | 3.761E-05 | 3.353E-05 | 2.947E-05 | 2.541E-05 | 2.125E-05 | 1.672E-05 |
| 18300 | 4.975E-05 | 4.565E-05 | 4.150E-05 | 3.735E-05 | 3.325E-05 | 2.918E-05 | 2.512E-05 | 2.108E-05 | 1.392E-05 |
| 18400 | 4.964E-05 | 4.550E-05 | 4.130E-05 | 3.713E-05 | 3.300E-05 | 2.903E-05 | 2.497E-05 | 1.707E-05 | 1.638E-05 |
| 18500 | 4.957E-05 | 4.537E-05 | 4.114E-05 | 3.702E-05 | 3.288E-05 | 2.879E-05 | 2.473E-05 | 2.078E-05 | 1.631E-05 |
| 18600 | 4.952E-05 | 4.527E-05 | 4.108E-05 | 3.685E-05 | 3.269E-05 | 2.883E-05 | 2.473E-05 | 2.058E-05 | 1.613E-05 |
| 18700 | 4.949E-05 | 4.526E-05 | 4.096E-05 | 3.701E-05 | 3.279E-05 | 2.865E-05 | 2.454E-05 | 2.039E-05 | 1.597E-05 |
| 18800 | 4.953E-05 | 4.521E-05 | 4.120E-05 | 3.689E-05 | 3.265E-05 | 2.848E-05 | 2.437E-05 | 2.022E-05 | 1.583E-05 |
| 18900 | 4.954E-05 | 4.553E-05 | 4.113E-05 | 3.678E-05 | 3.252E-05 | 2.834E-05 | 2.421E-05 | 2.007E-05 | 1.570E-05 |
| 19000 | 4.996E-05 | 4.551E-05 | 4.107E-05 | 3.670E-05 | 3.241E-05 | 2.821E-05 | 2.408E-05 | 1.993E-05 | 1.558E-05 |
| 19100 | 5.000E-05 | 4.551E-05 | 4.103E-05 | 3.663E-05 | 3.232E-05 | 2.810E-05 | 2.396E-05 | 1.981E-05 | 1.547E-05 |
| 19200 | 5.005E-05 | 4.552E-05 | 4.101E-05 | 3.658E-05 | 3.224E-05 | 2.801E-05 | 2.385E-05 | 1.970E-05 | 1.538E-05 |
| 19300 | 5.011E-05 | 4.554E-05 | 4.100E-05 | 3.654E-05 | 3.218E-05 | 2.793E-05 | 2.376E-05 | 1.961E-05 | 1.529E-05 |
| 19400 | 5.018E-05 | 4.557E-05 | 4.100E-05 | 3.651E-05 | 3.213E-05 | 2.787E-05 | 2.368E-05 | 1.953E-05 | 1.522E-05 |
| 19500 | 5.025E-05 | 4.561E-05 | 4.101E-05 | 3.650E-05 | 3.210E-05 | 2.781E-05 | 2.362E-05 | 1.946E-05 | 1.516E-05 |
| 19600 | 5.033E-05 | 4.565E-05 | 4.102E-05 | 3.649E-05 | 3.207E-05 | 2.777E-05 | 2.357E-05 | 1.940E-05 | 1.510E-05 |

## TABLE A.5. (*Continued*)

| $T$ (K) | Molar fraction of $H_2$ | | | | | | | | |
|---|---|---|---|---|---|---|---|---|---|
| | 0.1 | 0.2 | 0.3 | 0.4 | 0.5 | 0.6 | 0.7 | 0.8 | 0.9 |
| 19700 | 5.040E-05 | 4.570E-05 | 4.105E-05 | 3.649E-05 | 3.206E-05 | 2.774E-05 | 2.352E-05 | 1.935E-05 | 1.506E-05 |
| 19800 | 5.048E-05 | 4.575E-05 | 4.108E-05 | 3.650E-05 | 3.205E-05 | 2.772E-05 | 2.349E-05 | 1.931E-05 | 1.502E-05 |
| 19900 | 5.055E-05 | 4.580E-05 | 4.111E-05 | 3.652E-05 | 3.205E-05 | 2.770E-05 | 2.346E-05 | 1.927E-05 | 1.499E-05 |
| 20000 | 5.062E-05 | 4.585E-05 | 4.114E-05 | 3.654E-05 | 3.205E-05 | 2.770E-05 | 2.344E-05 | 1.925E-05 | 1.497E-05 |
| 20100 | 5.068E-05 | 4.590E-05 | 4.117E-05 | 3.656E-05 | 3.206E-05 | 2.770E-05 | 2.343E-05 | 1.923E-05 | 1.495E-05 |
| 20200 | 5.073E-05 | 4.594E-05 | 4.121E-05 | 3.658E-05 | 3.208E-05 | 2.770E-05 | 2.343E-05 | 1.922E-05 | 1.494E-05 |
| 20300 | 5.077E-05 | 4.597E-05 | 4.124E-05 | 3.660E-05 | 3.209E-05 | 2.771E-05 | 2.343E-05 | 1.922E-05 | 1.494E-05 |
| 20400 | 5.080E-05 | 4.600E-05 | 4.126E-05 | 3.663E-05 | 3.211E-05 | 2.772E-05 | 2.344E-05 | 1.922E-05 | 1.494E-05 |
| 20500 | 5.081E-05 | 4.602E-05 | 4.128E-05 | 3.665E-05 | 3.213E-05 | 2.773E-05 | 2.345E-05 | 1.922E-05 | 1.494E-05 |
| 20600 | 5.081E-05 | 4.602E-05 | 4.129E-05 | 3.666E-05 | 3.215E-05 | 2.775E-05 | 2.346E-05 | 1.923E-05 | 1.495E-05 |
| 20700 | 5.079E-05 | 4.602E-05 | 4.130E-05 | 3.667E-05 | 3.216E-05 | 2.777E-05 | 2.348E-05 | 1.925E-05 | 1.497E-05 |
| 20800 | 5.075E-05 | 4.599E-05 | 4.129E-05 | 3.668E-05 | 3.218E-05 | 2.779E-05 | 2.349E-05 | 1.926E-05 | 1.499E-05 |
| 20900 | 5.068E-05 | 4.596E-05 | 4.128E-05 | 3.668E-05 | 3.219E-05 | 2.780E-05 | 2.351E-05 | 1.771E-05 | 1.501E-05 |
| 21000 | 5.060E-05 | 4.590E-05 | 4.125E-05 | 3.667E-05 | 3.219E-05 | 2.782E-05 | 2.353E-05 | 1.931E-05 | 1.503E-05 |
| 21100 | 5.049E-05 | 4.583E-05 | 4.120E-05 | 3.665E-05 | 3.219E-05 | 2.783E-05 | 2.355E-05 | 1.933E-05 | 1.506E-05 |
| 21200 | 5.035E-05 | 4.573E-05 | 4.115E-05 | 3.662E-05 | 3.219E-05 | 2.784E-05 | 2.358E-05 | 1.936E-05 | 1.509E-05 |
| 21300 | 5.019E-05 | 4.562E-05 | 4.107E-05 | 3.658E-05 | 3.217E-05 | 2.785E-05 | 2.359E-05 | 1.939E-05 | 1.513E-05 |
| 21400 | 5.000E-05 | 4.548E-05 | 4.098E-05 | 3.653E-05 | 3.215E-05 | 2.785E-05 | 2.361E-05 | 1.942E-05 | 1.516E-05 |
| 21500 | 4.978E-05 | 4.532E-05 | 4.087E-05 | 3.647E-05 | 3.212E-05 | 2.785E-05 | 2.363E-05 | 1.944E-05 | 1.520E-05 |
| 21600 | 4.953E-05 | 4.514E-05 | 4.074E-05 | 3.639E-05 | 3.208E-05 | 2.784E-05 | 2.364E-05 | 1.947E-05 | 1.524E-05 |
| 21700 | 4.925E-05 | 4.493E-05 | 4.060E-05 | 3.629E-05 | 3.203E-05 | 2.782E-05 | 2.365E-05 | 1.950E-05 | 1.528E-05 |
| 21800 | 4.894E-05 | 4.469E-05 | 4.043E-05 | 3.618E-05 | 3.197E-05 | 2.780E-05 | 2.366E-05 | 1.953E-05 | 1.532E-05 |
| 21900 | 4.860E-05 | 4.443E-05 | 4.024E-05 | 3.606E-05 | 3.190E-05 | 2.777E-05 | 2.366E-05 | 1.956E-05 | 1.537E-05 |
| 22000 | 4.823E-05 | 4.415E-05 | 4.004E-05 | 3.592E-05 | 3.182E-05 | 2.773E-05 | 2.366E-05 | 1.958E-05 | 1.541E-05 |
| 22100 | 4.783E-05 | 4.384E-05 | 3.981E-05 | 3.576E-05 | 3.172E-05 | 2.769E-05 | 2.365E-05 | 1.960E-05 | 1.545E-05 |
| 22200 | 4.740E-05 | 4.350E-05 | 3.956E-05 | 3.559E-05 | 3.161E-05 | 2.763E-05 | 2.364E-05 | 1.973E-05 | 1.560E-05 |
| 22300 | 4.694E-05 | 4.314E-05 | 3.928E-05 | 3.539E-05 | 3.149E-05 | 2.767E-05 | 2.373E-05 | 1.992E-05 | 1.577E-05 |
| 22400 | 4.645E-05 | 4.275E-05 | 3.899E-05 | 3.519E-05 | 3.145E-05 | 2.760E-05 | 2.391E-05 | 1.994E-05 | 1.582E-05 |
| 22500 | 4.594E-05 | 4.234E-05 | 3.868E-05 | 3.505E-05 | 3.154E-05 | 2.774E-05 | 2.388E-05 | 1.995E-05 | 1.586E-05 |
| 22600 | 4.540E-05 | 4.191E-05 | 3.841E-05 | 3.507E-05 | 3.138E-05 | 2.764E-05 | 2.384E-05 | 1.996E-05 | 1.590E-05 |
| 22700 | 4.483E-05 | 4.151E-05 | 3.834E-05 | 3.480E-05 | 3.120E-05 | 2.753E-05 | 2.380E-05 | 1.996E-05 | 1.594E-05 |
| 22800 | 4.428E-05 | 4.103E-05 | 3.797E-05 | 3.452E-05 | 3.101E-05 | 2.742E-05 | 2.374E-05 | 1.996E-05 | 1.598E-05 |
| 22900 | 4.367E-05 | 4.083E-05 | 3.757E-05 | 3.423E-05 | 3.080E-05 | 2.729E-05 | 2.369E-05 | 1.996E-05 | 1.602E-05 |
| 23000 | 4.305E-05 | 4.031E-05 | 3.716E-05 | 3.392E-05 | 3.058E-05 | 2.715E-05 | 2.362E-05 | 1.995E-05 | 1.606E-05 |
| 23100 | 4.271E-05 | 3.978E-05 | 3.673E-05 | 3.359E-05 | 3.035E-05 | 2.701E-05 | 2.355E-05 | 1.994E-05 | 1.610E-05 |
| 23200 | 4.204E-05 | 3.923E-05 | 3.629E-05 | 3.325E-05 | 3.011E-05 | 2.685E-05 | 2.347E-05 | 1.993E-05 | 1.614E-05 |
| 23300 | 4.137E-05 | 3.866E-05 | 3.584E-05 | 3.290E-05 | 2.986E-05 | 2.669E-05 | 2.338E-05 | 1.991E-05 | 1.617E-05 |
| 23400 | 4.068E-05 | 3.809E-05 | 3.537E-05 | 3.254E-05 | 2.959E-05 | 2.652E-05 | 2.329E-05 | 1.988E-05 | 1.621E-05 |
| 23500 | 3.999E-05 | 3.750E-05 | 3.490E-05 | 3.217E-05 | 2.932E-05 | 2.634E-05 | 2.319E-05 | 1.986E-05 | 1.624E-05 |
| 23600 | 3.928E-05 | 3.691E-05 | 3.441E-05 | 3.179E-05 | 2.904E-05 | 2.615E-05 | 2.309E-05 | 1.983E-05 | 1.627E-05 |
| 23700 | 3.858E-05 | 3.631E-05 | 3.392E-05 | 3.140E-05 | 2.875E-05 | 2.595E-05 | 2.298E-05 | 1.979E-05 | 1.631E-05 |
| 23800 | 3.787E-05 | 3.571E-05 | 3.342E-05 | 3.100E-05 | 2.845E-05 | 2.575E-05 | 2.287E-05 | 1.976E-05 | 1.634E-05 |
| 23900 | 3.716E-05 | 3.510E-05 | 3.291E-05 | 3.060E-05 | 2.815E-05 | 2.554E-05 | 2.275E-05 | 1.972E-05 | 1.636E-05 |
| 24000 | 3.645E-05 | 3.449E-05 | 3.241E-05 | 3.020E-05 | 2.784E-05 | 2.533E-05 | 2.262E-05 | 1.967E-05 | 1.639E-05 |

**TABLE A.6.** Thermal Conductivity of $Ar/H_2$ Mixtures (W/m · K)

| | Molar fraction of $H_2$ | | | | | | | | |
|---|---|---|---|---|---|---|---|---|---|
| $T$ (K) | 0.1 | 0.2 | 0.3 | 0.4 | 0.5 | 0.6 | 0.7 | 0.8 | 0.9 |
| 500 | 3.685E-02 | 4.841E-02 | 6.168E-02 | 7.703E-02 | 9.495E-02 | 1.160E-01 | 1.413E-01 | 1.720E-01 | 2.098E-01 |
| 600 | 4.265E-02 | 5.630E-02 | 7.192E-02 | 8.991E-02 | 1.108E-01 | 1.354E-01 | 1.646E-01 | 1.999E-01 | 2.431E-01 |
| 700 | 4.882E-02 | 6.447E-02 | 8.234E-02 | 1.028E-01 | 1.267E-01 | 1.546E-01 | 1.877E-01 | 2.275E-01 | 2.761E-01 |
| 800 | 5.450E-02 | 7.223E-02 | 9.242E-02 | 1.155E-01 | 1.423E-01 | 1.736E-01 | 2.106E-01 | 2.549E-01 | 3.088E-01 |
| 900 | 5.971E-02 | 7.959E-02 | 1.021E-01 | 1.280E-01 | 1.578E-01 | 1.925E-01 | 2.335E-01 | 2.823E-01 | 3.415E-01 |
| 1000 | 6.461E-02 | 8.671E-02 | 1.117E-01 | 1.403E-01 | 1.732E-01 | 2.114E-01 | 2.563E-01 | 3.096E-01 | 3.741E-01 |
| 1100 | 6.935E-02 | 9.370E-02 | 1.212E-01 | 1.526E-01 | 1.886E-01 | 2.303E-01 | 2.791E-01 | 3.371E-01 | 4.069E-01 |
| 1200 | 7.400E-02 | 1.006E-01 | 1.307E-01 | 1.648E-01 | 2.040E-01 | 2.492E-01 | 3.021E-01 | 3.647E-01 | 4.398E-01 |
| 1300 | 7.863E-02 | 1.075E-01 | 1.402E-01 | 1.772E-01 | 2.195E-01 | 2.683E-01 | 3.253E-01 | 3.925E-01 | 4.730E-01 |
| 1400 | 8.329E-02 | 1.146E-01 | 1.498E-01 | 1.897E-01 | 2.352E-01 | 2.877E-01 | 3.487E-01 | 4.207E-01 | 5.066E-01 |
| 1500 | 8.808E-02 | 1.218E-01 | 1.597E-01 | 2.025E-01 | 2.513E-01 | 3.075E-01 | 3.728E-01 | 4.496E-01 | 5.411E-01 |
| 1600 | 9.319E-02 | 1.295E-01 | 1.702E-01 | 2.161E-01 | 2.684E-01 | 3.285E-01 | 3.981E-01 | 4.799E-01 | 5.772E-01 |
| 1700 | 9.897E-02 | 1.382E-01 | 1.820E-01 | 2.314E-01 | 2.874E-01 | 3.517E-01 | 4.260E-01 | 5.132E-01 | 6.166E-01 |
| 1800 | 1.060E-01 | 1.488E-01 | 1.964E-01 | 2.497E-01 | 3.100E-01 | 3.790E-01 | 4.588E-01 | 5.520E-01 | 6.624E-01 |
| 1900 | 1.154E-01 | 1.629E-01 | 2.151E-01 | 2.733E-01 | 3.390E-01 | 4.138E-01 | 5.000E-01 | 6.005E-01 | 7.192E-01 |
| 2000 | 1.285E-01 | 1.825E-01 | 2.410E-01 | 3.057E-01 | 3.783E-01 | 4.606E-01 | 5.551E-01 | 6.649E-01 | 7.941E-01 |
| 2100 | 1.473E-01 | 2.107E-01 | 2.780E-01 | 3.516E-01 | 4.336E-01 | 5.260E-01 | 6.316E-01 | 7.537E-01 | 8.968E-01 |
| 2200 | 1.744E-01 | 2.513E-01 | 3.311E-01 | 4.172E-01 | 5.122E-01 | 6.185E-01 | 7.393E-01 | 8.782E-01 | 1.040E+00 |
| 2300 | 2.128E-01 | 3.091E-01 | 4.064E-01 | 5.100E-01 | 6.230E-01 | 7.486E-01 | 8.903E-01 | 1.052E+00 | 1.240E+00 |
| 2400 | 2.658E-01 | 3.891E-01 | 5.109E-01 | 6.387E-01 | 7.766E-01 | 9.287E-01 | 1.099E+00 | 1.292E+00 | 1.515E+00 |
| 2500 | 3.365E-01 | 4.970E-01 | 6.521E-01 | 8.127E-01 | 9.844E-01 | 1.172E+00 | 1.381E+00 | 1.617E+00 | 1.887E+00 |
| 2600 | 4.271E-01 | 6.373E-01 | 8.370E-01 | 1.041E+00 | 1.257E+00 | 1.492E+00 | 1.752E+00 | 2.045E+00 | 2.378E+00 |
| 2700 | 5.383E-01 | 8.134E-01 | 1.071E+00 | 1.332E+00 | 1.606E+00 | 1.903E+00 | 2.229E+00 | 2.594E+00 | 3.009E+00 |
| 2800 | 6.678E-01 | 1.025E+00 | 1.356E+00 | 1.689E+00 | 2.038E+00 | 2.412E+00 | 2.823E+00 | 3.280E+00 | 3.797E+00 |
| 2900 | 8.098E-01 | 1.268E+00 | 1.691E+00 | 2.113E+00 | 2.554E+00 | 3.025E+00 | 3.539E+00 | 4.110E+00 | 4.755E+00 |
| 3000 | 9.541E-01 | 1.533E+00 | 2.065E+00 | 2.596E+00 | 3.147E+00 | 3.734E+00 | 4.374E+00 | 5.083E+00 | 5.880E+00 |
| 3100 | 1.086E+00 | 1.802E+00 | 2.462E+00 | 3.118E+00 | 3.799E+00 | 4.523E+00 | 5.310E+00 | 6.180E+00 | 7.158E+00 |
| 3200 | 1.189E+00 | 2.053E+00 | 2.854E+00 | 3.652E+00 | 4.479E+00 | 5.358E+00 | 6.313E+00 | 7.366E+00 | 8.546E+00 |
| 3300 | 1.249E+00 | 2.258E+00 | 3.206E+00 | 4.155E+00 | 5.141E+00 | 6.189E+00 | 7.325E+00 | 8.578E+00 | 9.979E+00 |
| 3400 | 1.254E+00 | 2.389E+00 | 3.480E+00 | 4.581E+00 | 5.727E+00 | 6.948E+00 | 8.272E+00 | 9.730E+00 | 1.136E+01 |
| 3500 | 1.205E+00 | 2.429E+00 | 3.641E+00 | 4.878E+00 | 6.175E+00 | 7.560E+00 | 9.063E+00 | 1.071E+01 | 1.256E+01 |
| 3600 | 1.112E+00 | 2.371E+00 | 3.665E+00 | 5.008E+00 | 6.427E+00 | 7.950E+00 | 9.606E+00 | 1.143E+01 | 1.346E+01 |
| 3700 | 9.910E-01 | 2.226E+00 | 3.549E+00 | 4.951E+00 | 6.448E+00 | 8.064E+00 | 9.827E+00 | 1.177E+01 | 1.394E+01 |
| 3800 | 8.618E-01 | 2.018E+00 | 3.313E+00 | 4.716E+00 | 6.233E+00 | 7.884E+00 | 9.692E+00 | 1.169E+01 | 1.392E+01 |
| 3900 | 7.401E-01 | 1.779E+00 | 2.993E+00 | 4.340E+00 | 5.817E+00 | 7.437E+00 | 9.220E+00 | 1.119E+01 | 1.340E+01 |
| 4000 | 6.350E-01 | 1.538E+00 | 2.635E+00 | 3.879E+00 | 5.261E+00 | 6.790E+00 | 8.482E+00 | 1.036E+01 | 1.246E+01 |
| 4100 | 5.500E-01 | 1.318E+00 | 2.279E+00 | 3.390E+00 | 4.639E+00 | 6.031E+00 | 7.580E+00 | 9.305E+00 | 1.123E+01 |
| 4200 | 4.844E-01 | 1.130E+00 | 1.954E+00 | 2.920E+00 | 4.017E+00 | 5.248E+00 | 6.622E+00 | 8.155E+00 | 9.872E+00 |
| 4300 | 4.355E-01 | 9.774E-01 | 1.674E+00 | 2.500E+00 | 3.443E+00 | 4.506E+00 | 5.696E+00 | 7.026E+00 | 8.514E+00 |
| 4400 | 3.999E-01 | 8.573E-01 | 1.445E+00 | 2.143E+00 | 2.944E+00 | 3.848E+00 | 4.861E+00 | 5.993E+00 | 7.258E+00 |
| 4500 | 3.747E-01 | 7.656E-01 | 1.263E+00 | 1.853E+00 | 2.529E+00 | 3.291E+00 | 4.144E+00 | 5.096E+00 | 6.159E+00 |
| 4600 | 3.573E-01 | 6.970E-01 | 1.122E+00 | 1.622E+00 | 2.193E+00 | 2.835E+00 | 3.551E+00 | 4.348E+00 | 5.235E+00 |
| 4700 | 3.456E-01 | 6.465E-01 | 1.015E+00 | 1.443E+00 | 1.928E+00 | 2.471E+00 | 3.073E+00 | 3.741E+00 | 4.481E+00 |
| 4800 | 3.381E-01 | 6.100E-01 | 9.347E-01 | 1.306E+00 | 1.723E+00 | 2.185E+00 | 2.695E+00 | 3.257E+00 | 3.876E+00 |
| 4900 | 3.338E-01 | 5.842E-01 | 8.752E-01 | 1.202E+00 | 1.565E+00 | 1.964E+00 | 2.400E+00 | 2.876E+00 | 3.399E+00 |
| 5000 | 3.318E-01 | 5.663E-01 | 8.318E-01 | 1.125E+00 | 1.446E+00 | 1.794E+00 | 2.171E+00 | 2.581E+00 | 3.026E+00 |

**TABLE A.6.** *(Continued)*

| | Molar fraction of $H_2$ | | | | | | | | |
|---|---|---|---|---|---|---|---|---|---|
| $T$ (K) | 0.1 | 0.2 | 0.3 | 0.4 | 0.5 | 0.6 | 0.7 | 0.8 | 0.9 |
| 5100 | 3.315E-01 | 5.545E-01 | 8.008E-01 | 1.068E+00 | 1.356E+00 | 1.665E+00 | 1.996E+00 | 2.352E+00 | 2.736E+00 |
| 5200 | 3.325E-01 | 5.473E-01 | 7.792E-01 | 1.026E+00 | 1.289E+00 | 1.568E+00 | 1.863E+00 | 2.177E+00 | 2.513E+00 |
| 5300 | 3.345E-01 | 5.436E-01 | 7.648E-01 | 9.972E-01 | 1.240E+00 | 1.495E+00 | 1.763E+00 | 2.044E+00 | 2.342E+00 |
| 5400 | 3.372E-01 | 5.425E-01 | 7.560E-01 | 9.771E-01 | 1.205E+00 | 1.442E+00 | 1.688E+00 | 1.943E+00 | 2.211E+00 |
| 5500 | 3.405E-01 | 5.435E-01 | 7.515E-01 | 9.641E-01 | 1.181E+00 | 1.404E+00 | 1.632E+00 | 1.868E+00 | 2.113E+00 |
| 5600 | 3.443E-01 | 5.460E-01 | 7.503E-01 | 9.568E-01 | 1.165E+00 | 1.377E+00 | 1.593E+00 | 1.813E+00 | 2.039E+00 |
| 5700 | 3.485E-01 | 5.499E-01 | 7.517E-01 | 9.539E-01 | 1.156E+00 | 1.360E+00 | 1.565E+00 | 1.773E+00 | 1.985E+00 |
| 5800 | 3.530E-01 | 5.548E-01 | 7.553E-01 | 9.544E-01 | 1.152E+00 | 1.350E+00 | 1.547E+00 | 1.746E+00 | 1.946E+00 |
| 5900 | 3.579E-01 | 5.606E-01 | 7.605E-01 | 9.577E-01 | 1.152E+00 | 1.345E+00 | 1.537E+00 | 1.728E+00 | 1.920E+00 |
| 6000 | 3.630E-01 | 5.670E-01 | 7.671E-01 | 9.633E-01 | 1.156E+00 | 1.346E+00 | 1.533E+00 | 1.719E+00 | 1.904E+00 |
| 6100 | 3.685E-01 | 5.742E-01 | 7.748E-01 | 9.707E-01 | 1.162E+00 | 1.350E+00 | 1.534E+00 | 1.716E+00 | 1.895E+00 |
| 6200 | 3.742E-01 | 5.819E-01 | 7.836E-01 | 9.797E-01 | 1.170E+00 | 1.357E+00 | 1.539E+00 | 1.717E+00 | 1.893E+00 |
| 6300 | 3.803E-01 | 5.901E-01 | 7.932E-01 | 9.901E-01 | 1.181E+00 | 1.366E+00 | 1.547E+00 | 1.724E+00 | 1.896E+00 |
| 6400 | 3.867E-01 | 5.989E-01 | 8.037E-01 | 1.001E+00 | 1.193E+00 | 1.378E+00 | 1.558E+00 | 1.733E+00 | 1.904E+00 |
| 6500 | 3.935E-01 | 6.082E-01 | 8.149E-01 | 1.014E+00 | 1.206E+00 | 1.392E+00 | 1.571E+00 | 1.746E+00 | 1.915E+00 |
| 6600 | 4.007E-01 | 6.181E-01 | 8.269E-01 | 1.027E+00 | 1.221E+00 | 1.407E+00 | 1.587E+00 | 1.761E+00 | 1.929E+00 |
| 6700 | 4.083E-01 | 6.285E-01 | 8.396E-01 | 1.042E+00 | 1.236E+00 | 1.424E+00 | 1.604E+00 | 1.778E+00 | 1.946E+00 |
| 6800 | 4.164E-01 | 6.396E-01 | 8.530E-01 | 1.057E+00 | 1.253E+00 | 1.442E+00 | 1.623E+00 | 1.797E+00 | 1.965E+00 |
| 6900 | 4.249E-01 | 6.513E-01 | 8.673E-01 | 1.074E+00 | 1.272E+00 | 1.461E+00 | 1.644E+00 | 1.819E+00 | 1.987E+00 |
| 7000 | 4.341E-01 | 6.636E-01 | 8.823E-01 | 1.091E+00 | 1.291E+00 | 1.482E+00 | 1.666E+00 | 1.841E+00 | 2.010E+00 |
| 7100 | 4.438E-01 | 6.767E-01 | 8.982E-01 | 1.109E+00 | 1.311E+00 | 1.504E+00 | 1.689E+00 | 1.866E+00 | 2.035E+00 |
| 7200 | 4.541E-01 | 6.905E-01 | 9.150E-01 | 1.128E+00 | 1.332E+00 | 1.527E+00 | 1.714E+00 | 1.892E+00 | 2.062E+00 |
| 7300 | 4.650E-01 | 7.052E-01 | 9.326E-01 | 1.149E+00 | 1.355E+00 | 1.552E+00 | 1.740E+00 | 1.919E+00 | 2.091E+00 |
| 7400 | 4.767E-01 | 7.206E-01 | 9.513E-01 | 1.170E+00 | 1.379E+00 | 1.577E+00 | 1.767E+00 | 1.948E+00 | 2.121E+00 |
| 7500 | 4.891E-01 | 7.370E-01 | 9.708E-01 | 1.192E+00 | 1.403E+00 | 1.604E+00 | 1.796E+00 | 1.979E+00 | 2.152E+00 |
| 7600 | 5.021E-01 | 7.543E-01 | 9.915E-01 | 1.216E+00 | 1.430E+00 | 1.633E+00 | 1.826E+00 | 2.011E+00 | 2.185E+00 |
| 7700 | 5.161E-01 | 7.725E-01 | 1.013E+00 | 1.241E+00 | 1.457E+00 | 1.663E+00 | 1.858E+00 | 2.044E+00 | 2.220E+00 |
| 7800 | 5.308E-01 | 7.918E-01 | 1.036E+00 | 1.267E+00 | 1.486E+00 | 1.694E+00 | 1.892E+00 | 2.080E+00 | 2.257E+00 |
| 7900 | 5.465E-01 | 8.123E-01 | 1.060E+00 | 1.294E+00 | 1.516E+00 | 1.727E+00 | 1.927E+00 | 2.116E+00 | 2.296E+00 |
| 8000 | 5.631E-01 | 8.339E-01 | 1.086E+00 | 1.323E+00 | 1.548E+00 | 1.761E+00 | 1.963E+00 | 2.155E+00 | 2.336E+00 |
| 8100 | 5.806E-01 | 8.566E-01 | 1.113E+00 | 1.354E+00 | 1.581E+00 | 1.797E+00 | 2.002E+00 | 2.195E+00 | 2.378E+00 |
| 8200 | 5.991E-01 | 8.806E-01 | 1.141E+00 | 1.386E+00 | 1.617E+00 | 1.835E+00 | 2.042E+00 | 2.237E+00 | 2.422E+00 |
| 8300 | 6.187E-01 | 9.058E-01 | 1.171E+00 | 1.420E+00 | 1.654E+00 | 1.875E+00 | 2.084E+00 | 2.281E+00 | 2.467E+00 |
| 8400 | 6.392E-01 | 9.328E-01 | 1.202E+00 | 1.455E+00 | 1.692E+00 | 1.917E+00 | 2.128E+00 | 2.329E+00 | 2.516E+00 |
| 8500 | 6.609E-01 | 9.606E-01 | 1.236E+00 | 1.491E+00 | 1.732E+00 | 1.960E+00 | 2.174E+00 | 2.376E+00 | 2.566E+00 |
| 8600 | 6.835E-01 | 9.898E-01 | 1.270E+00 | 1.531E+00 | 1.776E+00 | 2.005E+00 | 2.222E+00 | 2.426E+00 | 2.617E+00 |
| 8700 | 7.078E-01 | 1.020E+00 | 1.306E+00 | 1.571E+00 | 1.819E+00 | 2.053E+00 | 2.273E+00 | 2.480E+00 | 2.673E+00 |
| 8800 | 7.327E-01 | 1.052E+00 | 1.344E+00 | 1.613E+00 | 1.865E+00 | 2.102E+00 | 2.324E+00 | 2.533E+00 | 2.728E+00 |
| 8900 | 7.587E-01 | 1.085E+00 | 1.383E+00 | 1.658E+00 | 1.914E+00 | 2.155E+00 | 2.380E+00 | 2.589E+00 | 2.785E+00 |
| 9000 | 7.865E-01 | 1.121E+00 | 1.423E+00 | 1.703E+00 | 1.963E+00 | 2.207E+00 | 2.436E+00 | 2.649E+00 | 2.847E+00 |
| 9100 | 8.147E-01 | 1.157E+00 | 1.466E+00 | 1.751E+00 | 2.016E+00 | 2.264E+00 | 2.493E+00 | 2.709E+00 | 2.908E+00 |
| 9200 | 8.450E-01 | 1.194E+00 | 1.510E+00 | 1.800E+00 | 2.069E+00 | 2.320E+00 | 2.555E+00 | 2.774E+00 | 2.975E+00 |
| 9300 | 8.754E-01 | 1.234E+00 | 1.556E+00 | 1.852E+00 | 2.127E+00 | 2.382E+00 | 2.621E+00 | 2.837E+00 | 3.039E+00 |
| 9400 | 9.072E-01 | 1.273E+00 | 1.602E+00 | 1.904E+00 | 2.182E+00 | 2.442E+00 | 2.683E+00 | 2.907E+00 | 3.111E+00 |
| 9500 | 9.406E-01 | 1.316E+00 | 1.652E+00 | 1.960E+00 | 2.244E+00 | 2.508E+00 | 2.753E+00 | 2.980E+00 | 3.185E+00 |
| 9600 | 9.757E-01 | 1.357E+00 | 1.701E+00 | 2.014E+00 | 2.303E+00 | 2.571E+00 | 2.819E+00 | 3.048E+00 | 3.254E+00 |
| 9700 | 1.010E+00 | 1.402E+00 | 1.754E+00 | 2.074E+00 | 2.368E+00 | 2.641E+00 | 2.893E+00 | 3.125E+00 | 3.332E+00 |
| 9800 | 1.047E+00 | 1.449E+00 | 1.809E+00 | 2.131E+00 | 2.430E+00 | 2.707E+00 | 2.962E+00 | 3.196E+00 | 3.403E+00 |

*(continued)*

## TABLE A.6. (*Continued*)

| | Molar fraction of $H_2$ | | | | | | | | |
|---|---|---|---|---|---|---|---|---|---|
| $T$ (K) | 0.1 | 0.2 | 0.3 | 0.4 | 0.5 | 0.6 | 0.7 | 0.8 | 0.9 |
| 9900 | 1.084E+00 | 1.494E+00 | 1.860E+00 | 2.193E+00 | 2.499E+00 | 2.781E+00 | 3.040E+00 | 3.277E+00 | 3.485E+00 |
| 10000 | 1.124E+00 | 1.543E+00 | 1.918E+00 | 2.259E+00 | 2.570E+00 | 2.857E+00 | 3.121E+00 | 3.360E+00 | 3.569E+00 |
| 10100 | 1.162E+00 | 1.589E+00 | 1.972E+00 | 2.318E+00 | 2.635E+00 | 2.927E+00 | 3.193E+00 | 3.434E+00 | 3.644E+00 |
| 10200 | 1.204E+00 | 1.641E+00 | 2.032E+00 | 2.386E+00 | 2.710E+00 | 3.007E+00 | 3.278E+00 | 3.521E+00 | 3.731E+00 |
| 10300 | 1.247E+00 | 1.694E+00 | 2.094E+00 | 2.456E+00 | 2.787E+00 | 3.089E+00 | 3.364E+00 | 3.610E+00 | 3.820E+00 |
| 10400 | 1.288E+00 | 1.742E+00 | 2.149E+00 | 2.518E+00 | 2.854E+00 | 3.161E+00 | 3.440E+00 | 3.687E+00 | 3.896E+00 |
| 10500 | 1.333E+00 | 1.797E+00 | 2.213E+00 | 2.590E+00 | 2.934E+00 | 3.247E+00 | 3.529E+00 | 3.779E+00 | 3.988E+00 |
| 10600 | 1.380E+00 | 1.853E+00 | 2.279E+00 | 2.664E+00 | 3.015E+00 | 3.334E+00 | 3.622E+00 | 3.873E+00 | 4.081E+00 |
| 10700 | 1.422E+00 | 1.909E+00 | 2.346E+00 | 2.740E+00 | 3.099E+00 | 3.424E+00 | 3.716E+00 | 3.970E+00 | 4.177E+00 |
| 10800 | 1.471E+00 | 1.961E+00 | 2.403E+00 | 2.804E+00 | 3.169E+00 | 3.499E+00 | 3.794E+00 | 4.048E+00 | 4.253E+00 |
| 10900 | 1.521E+00 | 2.020E+00 | 2.472E+00 | 2.882E+00 | 3.255E+00 | 3.591E+00 | 3.891E+00 | 4.147E+00 | 4.350E+00 |
| 11000 | 1.571E+00 | 2.081E+00 | 2.542E+00 | 2.961E+00 | 3.342E+00 | 3.686E+00 | 3.990E+00 | 4.248E+00 | 4.448E+00 |
| 11100 | 1.617E+00 | 2.132E+00 | 2.600E+00 | 3.042E+00 | 3.432E+00 | 3.783E+00 | 4.092E+00 | 4.351E+00 | 4.548E+00 |
| 11200 | 1.669E+00 | 2.194E+00 | 2.671E+00 | 3.107E+00 | 3.504E+00 | 3.860E+00 | 4.172E+00 | 4.431E+00 | 4.623E+00 |
| 11300 | 1.723E+00 | 2.257E+00 | 2.744E+00 | 3.190E+00 | 3.595E+00 | 3.959E+00 | 4.276E+00 | 4.536E+00 | 4.725E+00 |
| 11400 | 1.777E+00 | 2.321E+00 | 2.818E+00 | 3.274E+00 | 3.689E+00 | 4.061E+00 | 4.383E+00 | 4.644E+00 | 4.828E+00 |
| 11500 | 1.829E+00 | 2.386E+00 | 2.893E+00 | 3.359E+00 | 3.785E+00 | 4.165E+00 | 4.492E+00 | 4.753E+00 | 4.932E+00 |
| 11600 | 1.879E+00 | 2.437E+00 | 2.950E+00 | 3.424E+00 | 3.879E+00 | 4.271E+00 | 4.603E+00 | 4.865E+00 | 5.038E+00 |
| 11700 | 1.936E+00 | 2.502E+00 | 3.026E+00 | 3.511E+00 | 3.954E+00 | 4.349E+00 | 4.685E+00 | 4.946E+00 | 5.112E+00 |
| 11800 | 1.993E+00 | 2.569E+00 | 3.103E+00 | 3.598E+00 | 4.053E+00 | 4.457E+00 | 4.799E+00 | 5.060E+00 | 5.220E+00 |
| 11900 | 2.051E+00 | 2.636E+00 | 3.180E+00 | 3.687E+00 | 4.153E+00 | 4.567E+00 | 4.914E+00 | 5.176E+00 | 5.329E+00 |
| 12000 | 2.110E+00 | 2.703E+00 | 3.258E+00 | 3.777E+00 | 4.254E+00 | 4.678E+00 | 5.032E+00 | 5.294E+00 | 5.439E+00 |
| 12100 | 2.157E+00 | 2.753E+00 | 3.337E+00 | 3.867E+00 | 4.356E+00 | 4.791E+00 | 5.151E+00 | 5.414E+00 | 5.551E+00 |
| 12200 | 2.215E+00 | 2.820E+00 | 3.391E+00 | 3.929E+00 | 4.456E+00 | 4.905E+00 | 5.273E+00 | 5.536E+00 | 5.665E+00 |
| 12300 | 2.274E+00 | 2.887E+00 | 3.469E+00 | 4.019E+00 | 4.529E+00 | 4.983E+00 | 5.356E+00 | 5.656E+00 | 5.780E+00 |
| 12400 | 2.333E+00 | 2.955E+00 | 3.547E+00 | 4.109E+00 | 4.632E+00 | 5.097E+00 | 5.478E+00 | 5.741E+00 | 5.853E+00 |
| 12500 | 2.392E+00 | 3.022E+00 | 3.624E+00 | 4.199E+00 | 4.735E+00 | 5.212E+00 | 5.601E+00 | 5.865E+00 | 5.968E+00 |
| 12600 | 2.450E+00 | 3.089E+00 | 3.701E+00 | 4.288E+00 | 4.837E+00 | 5.327E+00 | 5.724E+00 | 5.989E+00 | 6.084E+00 |
| 12700 | 2.508E+00 | 3.155E+00 | 3.778E+00 | 4.376E+00 | 4.939E+00 | 5.441E+00 | 5.846E+00 | 6.113E+00 | 6.200E+00 |
| 12800 | 2.549E+00 | 3.215E+00 | 3.853E+00 | 4.464E+00 | 5.040E+00 | 5.554E+00 | 5.969E+00 | 6.238E+00 | 6.316E+00 |
| 12900 | 2.605E+00 | 3.259E+00 | 3.896E+00 | 4.550E+00 | 5.139E+00 | 5.666E+00 | 6.090E+00 | 6.361E+00 | 6.432E+00 |
| 13000 | 2.660E+00 | 3.321E+00 | 3.967E+00 | 4.598E+00 | 5.231E+00 | 5.776E+00 | 6.210E+00 | 6.483E+00 | 6.547E+00 |
| 13100 | 2.713E+00 | 3.381E+00 | 4.037E+00 | 4.679E+00 | 5.291E+00 | 5.841E+00 | 6.323E+00 | 6.604E+00 | 6.660E+00 |
| 13200 | 2.765E+00 | 3.439E+00 | 4.104E+00 | 4.757E+00 | 5.382E+00 | 5.945E+00 | 6.397E+00 | 6.677E+00 | 6.769E+00 |
| 13300 | 2.815E+00 | 3.495E+00 | 4.168E+00 | 4.832E+00 | 5.469E+00 | 6.045E+00 | 6.507E+00 | 6.791E+00 | 6.834E+00 |
| 13400 | 2.863E+00 | 3.549E+00 | 4.230E+00 | 4.903E+00 | 5.553E+00 | 6.141E+00 | 6.613E+00 | 6.901E+00 | 6.940E+00 |
| 13500 | 2.908E+00 | 3.600E+00 | 4.288E+00 | 4.971E+00 | 5.632E+00 | 6.232E+00 | 6.715E+00 | 7.008E+00 | 7.042E+00 |
| 13600 | 2.951E+00 | 3.647E+00 | 4.342E+00 | 5.035E+00 | 5.706E+00 | 6.318E+00 | 6.811E+00 | 7.109E+00 | 7.139E+00 |
| 13700 | 2.991E+00 | 3.692E+00 | 4.393E+00 | 5.094E+00 | 5.776E+00 | 6.399E+00 | 6.901E+00 | 7.204E+00 | 7.232E+00 |
| 13800 | 3.024E+00 | 3.733E+00 | 4.439E+00 | 5.148E+00 | 5.839E+00 | 6.472E+00 | 6.984E+00 | 7.293E+00 | 7.319E+00 |
| 13900 | 3.043E+00 | 3.770E+00 | 4.481E+00 | 5.196E+00 | 5.896E+00 | 6.539E+00 | 7.060E+00 | 7.375E+00 | 7.400E+00 |
| 14000 | 3.074E+00 | 3.775E+00 | 4.518E+00 | 5.239E+00 | 5.947E+00 | 6.599E+00 | 7.128E+00 | 7.449E+00 | 7.475E+00 |
| 14100 | 3.101E+00 | 3.804E+00 | 4.547E+00 | 5.277E+00 | 5.991E+00 | 6.650E+00 | 7.188E+00 | 7.515E+00 | 7.541E+00 |
| 14200 | 3.125E+00 | 3.829E+00 | 4.545E+00 | 5.304E+00 | 6.027E+00 | 6.694E+00 | 7.239E+00 | 7.572E+00 | 7.600E+00 |
| 14300 | 3.145E+00 | 3.850E+00 | 4.567E+00 | 5.297E+00 | 6.054E+00 | 6.729E+00 | 7.281E+00 | 7.619E+00 | 7.650E+00 |
| 14400 | 3.161E+00 | 3.866E+00 | 4.584E+00 | 5.317E+00 | 6.076E+00 | 6.756E+00 | 7.313E+00 | 7.657E+00 | 7.692E+00 |
| 14500 | 3.174E+00 | 3.878E+00 | 4.597E+00 | 5.330E+00 | 6.058E+00 | 6.771E+00 | 7.335E+00 | 7.685E+00 | 7.723E+00 |
| 14600 | 3.184E+00 | 3.887E+00 | 4.604E+00 | 5.337E+00 | 6.066E+00 | 6.749E+00 | 7.347E+00 | 7.703E+00 | 7.746E+00 |

## TABLE A.6. (*Continued*)

| T (K) | Molar fraction of $H_2$ | | | | | | | | |
|---|---|---|---|---|---|---|---|---|---|
| | 0.1 | 0.2 | 0.3 | 0.4 | 0.5 | 0.6 | 0.7 | 0.8 | 0.9 |
| 14700 | 3.190E+00 | 3.891E+00 | 4.607E+00 | 5.338E+00 | 6.067E+00 | 6.751E+00 | 7.320E+00 | 7.710E+00 | 7.758E+00 |
| 14800 | 3.194E+00 | 3.892E+00 | 4.604E+00 | 5.334E+00 | 6.060E+00 | 6.744E+00 | 7.315E+00 | 7.679E+00 | 7.760E+00 |
| 14900 | 3.194E+00 | 3.889E+00 | 4.598E+00 | 5.323E+00 | 6.047E+00 | 6.728E+00 | 7.299E+00 | 7.667E+00 | 7.726E+00 |
| 15000 | 3.192E+00 | 3.882E+00 | 4.587E+00 | 5.308E+00 | 6.027E+00 | 6.704E+00 | 7.275E+00 | 7.645E+00 | 7.710E+00 |
| 15100 | 3.187E+00 | 3.873E+00 | 4.572E+00 | 5.287E+00 | 6.000E+00 | 6.673E+00 | 7.240E+00 | 7.613E+00 | 7.684E+00 |
| 15200 | 3.180E+00 | 3.861E+00 | 4.554E+00 | 5.262E+00 | 5.967E+00 | 6.634E+00 | 7.198E+00 | 7.571E+00 | 7.649E+00 |
| 15300 | 3.171E+00 | 3.846E+00 | 4.532E+00 | 5.232E+00 | 5.929E+00 | 6.587E+00 | 7.146E+00 | 7.520E+00 | 7.603E+00 |
| 15400 | 3.160E+00 | 3.829E+00 | 4.507E+00 | 5.198E+00 | 5.885E+00 | 6.535E+00 | 7.087E+00 | 7.459E+00 | 7.549E+00 |
| 15500 | 3.148E+00 | 3.810E+00 | 4.480E+00 | 5.161E+00 | 5.837E+00 | 6.476E+00 | 7.020E+00 | 7.391E+00 | 7.486E+00 |
| 15600 | 3.135E+00 | 3.790E+00 | 4.450E+00 | 5.120E+00 | 5.784E+00 | 6.411E+00 | 6.947E+00 | 7.314E+00 | 7.414E+00 |
| 15700 | 3.121E+00 | 3.768E+00 | 4.419E+00 | 5.077E+00 | 5.728E+00 | 6.341E+00 | 6.867E+00 | 7.231E+00 | 7.335E+00 |
| 15800 | 3.107E+00 | 3.745E+00 | 4.386E+00 | 5.031E+00 | 5.668E+00 | 6.267E+00 | 6.782E+00 | 7.140E+00 | 7.249E+00 |
| 15900 | 3.093E+00 | 3.722E+00 | 4.351E+00 | 4.983E+00 | 5.605E+00 | 6.189E+00 | 6.691E+00 | 7.044E+00 | 7.157E+00 |
| 16000 | 3.078E+00 | 3.698E+00 | 4.316E+00 | 4.933E+00 | 5.539E+00 | 6.108E+00 | 6.597E+00 | 6.943E+00 | 7.058E+00 |
| 16100 | 3.064E+00 | 3.674E+00 | 4.279E+00 | 4.882E+00 | 5.471E+00 | 6.023E+00 | 6.498E+00 | 6.837E+00 | 6.955E+00 |
| 16200 | 3.050E+00 | 3.650E+00 | 4.243E+00 | 4.830E+00 | 5.402E+00 | 5.937E+00 | 6.397E+00 | 6.727E+00 | 6.847E+00 |
| 16300 | 3.036E+00 | 3.626E+00 | 4.206E+00 | 4.777E+00 | 5.332E+00 | 5.849E+00 | 6.293E+00 | 6.615E+00 | 6.736E+00 |
| 16400 | 3.023E+00 | 3.603E+00 | 4.169E+00 | 4.724E+00 | 5.261E+00 | 5.759E+00 | 6.188E+00 | 6.500E+00 | 6.622E+00 |
| 16500 | 3.011E+00 | 3.580E+00 | 4.132E+00 | 4.671E+00 | 5.189E+00 | 5.669E+00 | 6.081E+00 | 6.383E+00 | 6.505E+00 |
| 16600 | 3.000E+00 | 3.558E+00 | 4.096E+00 | 4.618E+00 | 5.117E+00 | 5.578E+00 | 5.974E+00 | 6.266E+00 | 6.388E+00 |
| 16700 | 2.990E+00 | 3.536E+00 | 4.060E+00 | 4.565E+00 | 5.045E+00 | 5.487E+00 | 5.867E+00 | 6.148E+00 | 6.269E+00 |
| 16800 | 2.981E+00 | 3.516E+00 | 4.025E+00 | 4.513E+00 | 4.974E+00 | 5.397E+00 | 5.760E+00 | 6.030E+00 | 6.150E+00 |
| 16900 | 2.973E+00 | 3.496E+00 | 3.990E+00 | 4.461E+00 | 4.904E+00 | 5.308E+00 | 5.654E+00 | 5.913E+00 | 6.031E+00 |
| 17000 | 2.966E+00 | 3.477E+00 | 3.957E+00 | 4.410E+00 | 4.834E+00 | 5.219E+00 | 5.550E+00 | 5.797E+00 | 5.914E+00 |
| 17100 | 2.960E+00 | 3.459E+00 | 3.924E+00 | 4.360E+00 | 4.766E+00 | 5.133E+00 | 5.447E+00 | 5.683E+00 | 5.797E+00 |
| 17200 | 2.955E+00 | 3.442E+00 | 3.892E+00 | 4.311E+00 | 4.699E+00 | 5.048E+00 | 5.346E+00 | 5.571E+00 | 5.683E+00 |
| 17300 | 2.952E+00 | 3.426E+00 | 3.861E+00 | 4.264E+00 | 4.633E+00 | 4.964E+00 | 5.247E+00 | 5.462E+00 | 5.570E+00 |
| 17400 | 2.949E+00 | 3.412E+00 | 3.832E+00 | 4.217E+00 | 4.569E+00 | 4.883E+00 | 5.151E+00 | 5.355E+00 | 5.461E+00 |
| 17500 | 2.948E+00 | 3.398E+00 | 3.803E+00 | 4.172E+00 | 4.507E+00 | 4.804E+00 | 5.058E+00 | 5.251E+00 | 5.354E+00 |
| 17600 | 2.947E+00 | 3.385E+00 | 3.776E+00 | 4.129E+00 | 4.447E+00 | 4.728E+00 | 4.967E+00 | 5.151E+00 | 5.250E+00 |
| 17700 | 2.948E+00 | 3.373E+00 | 3.750E+00 | 4.087E+00 | 4.388E+00 | 4.654E+00 | 4.880E+00 | 5.054E+00 | 5.150E+00 |
| 17800 | 2.950E+00 | 3.363E+00 | 3.724E+00 | 4.046E+00 | 4.332E+00 | 4.583E+00 | 4.796E+00 | 4.560E+00 | 5.054E+00 |
| 17900 | 2.953E+00 | 3.353E+00 | 3.701E+00 | 4.007E+00 | 4.278E+00 | 4.515E+00 | 4.715E+00 | 4.871E+00 | 4.961E+00 |
| 18000 | 2.956E+00 | 3.344E+00 | 3.678E+00 | 3.970E+00 | 4.226E+00 | 4.449E+00 | 4.638E+00 | 4.785E+00 | 4.872E+00 |
| 18100 | 2.961E+00 | 3.337E+00 | 3.657E+00 | 3.935E+00 | 4.177E+00 | 4.387E+00 | 4.564E+00 | 4.703E+00 | 4.787E+00 |
| 18200 | 2.966E+00 | 3.330E+00 | 3.637E+00 | 3.901E+00 | 4.130E+00 | 4.327E+00 | 4.494E+00 | 4.625E+00 | 4.705E+00 |
| 18300 | 2.973E+00 | 3.324E+00 | 3.618E+00 | 3.869E+00 | 4.085E+00 | 4.270E+00 | 4.427E+00 | 4.547E+00 | 4.439E+00 |
| 18400 | 2.980E+00 | 3.320E+00 | 3.600E+00 | 3.838E+00 | 4.042E+00 | 4.213E+00 | 4.360E+00 | 4.232E+00 | 4.551E+00 |
| 18500 | 2.988E+00 | 3.316E+00 | 3.584E+00 | 3.809E+00 | 3.999E+00 | 4.162E+00 | 4.300E+00 | 4.411E+00 | 4.483E+00 |
| 18600 | 2.997E+00 | 3.313E+00 | 3.570E+00 | 3.781E+00 | 3.961E+00 | 4.116E+00 | 4.245E+00 | 4.349E+00 | 4.417E+00 |
| 18700 | 3.006E+00 | 3.313E+00 | 3.555E+00 | 3.758E+00 | 3.927E+00 | 4.071E+00 | 4.192E+00 | 4.290E+00 | 4.356E+00 |
| 18800 | 3.019E+00 | 3.311E+00 | 3.544E+00 | 3.734E+00 | 3.893E+00 | 4.028E+00 | 4.142E+00 | 4.235E+00 | 4.299E+00 |
| 18900 | 3.030E+00 | 3.312E+00 | 3.532E+00 | 3.712E+00 | 3.862E+00 | 3.988E+00 | 4.096E+00 | 4.183E+00 | 4.245E+00 |
| 19000 | 3.041E+00 | 3.312E+00 | 3.522E+00 | 3.692E+00 | 3.833E+00 | 3.952E+00 | 4.052E+00 | 4.135E+00 | 4.195E+00 |
| 19100 | 3.053E+00 | 3.313E+00 | 3.513E+00 | 3.673E+00 | 3.806E+00 | 3.917E+00 | 4.012E+00 | 4.091E+00 | 4.148E+00 |
| 19200 | 3.066E+00 | 3.315E+00 | 3.505E+00 | 3.656E+00 | 3.781E+00 | 3.886E+00 | 3.975E+00 | 4.050E+00 | 4.105E+00 |
| 19300 | 3.079E+00 | 3.318E+00 | 3.498E+00 | 3.641E+00 | 3.758E+00 | 3.857E+00 | 3.941E+00 | 4.012E+00 | 4.065E+00 |
| 19400 | 3.092E+00 | 3.322E+00 | 3.493E+00 | 3.628E+00 | 3.738E+00 | 3.830E+00 | 3.909E+00 | 3.977E+00 | 4.029E+00 |

(*continued*)

**TABLE A.6.** *(Continued)*

| T (K) | Molar fraction of $H_2$ | | | | | | | | |
|---|---|---|---|---|---|---|---|---|---|
| | 0.1 | 0.2 | 0.3 | 0.4 | 0.5 | 0.6 | 0.7 | 0.8 | 0.9 |
| 19500 | 3.107E+00 | 3.326E+00 | 3.489E+00 | 3.616E+00 | 3.719E+00 | 3.806E+00 | 3.881E+00 | 3.945E+00 | 3.995E+00 |
| 19600 | 3.121E+00 | 3.332E+00 | 3.486E+00 | 3.606E+00 | 3.703E+00 | 3.784E+00 | 3.855E+00 | 3.915E+00 | 3.965E+00 |
| 19700 | 3.137E+00 | 3.338E+00 | 3.484E+00 | 3.597E+00 | 3.688E+00 | 3.765E+00 | 3.831E+00 | 3.889E+00 | 3.937E+00 |
| 19800 | 3.152E+00 | 3.345E+00 | 3.484E+00 | 3.590E+00 | 3.676E+00 | 3.748E+00 | 3.810E+00 | 3.865E+00 | 3.912E+00 |
| 19900 | 3.169E+00 | 3.353E+00 | 3.484E+00 | 3.584E+00 | 3.665E+00 | 3.732E+00 | 3.791E+00 | 3.844E+00 | 3.889E+00 |
| 20000 | 3.186E+00 | 3.362E+00 | 3.486E+00 | 3.580E+00 | 3.656E+00 | 3.719E+00 | 3.775E+00 | 3.825E+00 | 3.870E+00 |
| 20100 | 3.203E+00 | 3.371E+00 | 3.488E+00 | 3.577E+00 | 3.648E+00 | 3.708E+00 | 3.761E+00 | 3.809E+00 | 3.852E+00 |
| 20200 | 3.221E+00 | 3.381E+00 | 3.492E+00 | 3.576E+00 | 3.642E+00 | 3.698E+00 | 3.748E+00 | 3.795E+00 | 3.837E+00 |
| 20300 | 3.239E+00 | 3.392E+00 | 3.497E+00 | 3.575E+00 | 3.638E+00 | 3.691E+00 | 3.738E+00 | 3.783E+00 | 3.824E+00 |
| 20400 | 3.258E+00 | 3.404E+00 | 3.503E+00 | 3.576E+00 | 3.635E+00 | 3.685E+00 | 3.730E+00 | 3.772E+00 | 3.813E+00 |
| 20500 | 3.277E+00 | 3.416E+00 | 3.509E+00 | 3.579E+00 | 3.634E+00 | 3.681E+00 | 3.723E+00 | 3.764E+00 | 3.804E+00 |
| 20600 | 3.296E+00 | 3.429E+00 | 3.517E+00 | 3.582E+00 | 3.634E+00 | 3.678E+00 | 3.719E+00 | 3.758E+00 | 3.797E+00 |
| 20700 | 3.316E+00 | 3.442E+00 | 3.526E+00 | 3.587E+00 | 3.635E+00 | 3.677E+00 | 3.716E+00 | 3.753E+00 | 3.791E+00 |
| 20800 | 3.336E+00 | 3.456E+00 | 3.535E+00 | 3.593E+00 | 3.638E+00 | 3.677E+00 | 3.714E+00 | 3.750E+00 | 3.788E+00 |
| 20900 | 3.357E+00 | 3.471E+00 | 3.545E+00 | 3.599E+00 | 3.642E+00 | 3.679E+00 | 3.714E+00 | 3.706E+00 | 3.786E+00 |
| 21000 | 3.378E+00 | 3.486E+00 | 3.556E+00 | 3.607E+00 | 3.647E+00 | 3.682E+00 | 3.715E+00 | 3.749E+00 | 3.785E+00 |
| 21100 | 3.399E+00 | 3.502E+00 | 3.568E+00 | 3.616E+00 | 3.653E+00 | 3.686E+00 | 3.718E+00 | 3.751E+00 | 3.786E+00 |
| 21200 | 3.420E+00 | 3.519E+00 | 3.581E+00 | 3.625E+00 | 3.661E+00 | 3.692E+00 | 3.722E+00 | 3.753E+00 | 3.788E+00 |
| 21300 | 3.442E+00 | 3.536E+00 | 3.594E+00 | 3.636E+00 | 3.669E+00 | 3.698E+00 | 3.727E+00 | 3.757E+00 | 3.792E+00 |
| 21400 | 3.464E+00 | 3.553E+00 | 3.608E+00 | 3.647E+00 | 3.678E+00 | 3.706E+00 | 3.733E+00 | 3.763E+00 | 3.796E+00 |
| 21500 | 3.487E+00 | 3.571E+00 | 3.623E+00 | 3.659E+00 | 3.689E+00 | 3.715E+00 | 3.741E+00 | 3.769E+00 | 3.802E+00 |
| 21600 | 3.509E+00 | 3.589E+00 | 3.638E+00 | 3.672E+00 | 3.700E+00 | 3.724E+00 | 3.749E+00 | 3.777E+00 | 3.809E+00 |
| 21700 | 3.532E+00 | 3.608E+00 | 3.654E+00 | 3.686E+00 | 3.711E+00 | 3.735E+00 | 3.758E+00 | 3.785E+00 | 3.817E+00 |
| 21800 | 3.555E+00 | 3.627E+00 | 3.670E+00 | 3.700E+00 | 3.724E+00 | 3.746E+00 | 3.769E+00 | 3.794E+00 | 3.826E+00 |
| 21900 | 3.578E+00 | 3.647E+00 | 3.687E+00 | 3.715E+00 | 3.737E+00 | 3.758E+00 | 3.780E+00 | 3.805E+00 | 3.836E+00 |
| 22000 | 3.601E+00 | 3.667E+00 | 3.705E+00 | 3.731E+00 | 3.751E+00 | 3.771E+00 | 3.791E+00 | 3.816E+00 | 3.847E+00 |
| 22100 | 3.625E+00 | 3.687E+00 | 3.722E+00 | 3.747E+00 | 3.766E+00 | 3.784E+00 | 3.804E+00 | 3.828E+00 | 3.858E+00 |
| 22200 | 3.649E+00 | 3.707E+00 | 3.741E+00 | 3.763E+00 | 3.781E+00 | 3.799E+00 | 3.817E+00 | 3.827E+00 | 3.856E+00 |
| 22300 | 3.672E+00 | 3.728E+00 | 3.759E+00 | 3.780E+00 | 3.797E+00 | 3.801E+00 | 3.818E+00 | 3.840E+00 | 3.869E+00 |
| 22400 | 3.696E+00 | 3.749E+00 | 3.778E+00 | 3.798E+00 | 3.802E+00 | 3.816E+00 | 3.833E+00 | 3.853E+00 | 3.882E+00 |
| 22500 | 3.720E+00 | 3.770E+00 | 3.798E+00 | 3.805E+00 | 3.819E+00 | 3.832E+00 | 3.847E+00 | 3.867E+00 | 3.895E+00 |
| 22600 | 3.744E+00 | 3.792E+00 | 3.808E+00 | 3.824E+00 | 3.836E+00 | 3.848E+00 | 3.862E+00 | 3.881E+00 | 3.909E+00 |
| 22700 | 3.769E+00 | 3.806E+00 | 3.829E+00 | 3.842E+00 | 3.853E+00 | 3.864E+00 | 3.877E+00 | 3.896E+00 | 3.923E+00 |
| 22800 | 3.788E+00 | 3.828E+00 | 3.849E+00 | 3.861E+00 | 3.870E+00 | 3.880E+00 | 3.893E+00 | 3.911E+00 | 3.937E+00 |
| 22900 | 3.813E+00 | 3.851E+00 | 3.869E+00 | 3.880E+00 | 3.888E+00 | 3.897E+00 | 3.909E+00 | 3.926E+00 | 3.952E+00 |
| 23000 | 3.837E+00 | 3.873E+00 | 3.889E+00 | 3.899E+00 | 3.906E+00 | 3.914E+00 | 3.925E+00 | 3.942E+00 | 3.968E+00 |
| 23100 | 3.862E+00 | 3.895E+00 | 3.909E+00 | 3.918E+00 | 3.924E+00 | 3.932E+00 | 3.942E+00 | 3.958E+00 | 3.983E+00 |
| 23200 | 3.887E+00 | 3.917E+00 | 3.930E+00 | 3.937E+00 | 3.943E+00 | 3.949E+00 | 3.959E+00 | 3.974E+00 | 3.999E+00 |
| 23300 | 3.911E+00 | 3.939E+00 | 3.951E+00 | 3.957E+00 | 3.961E+00 | 3.967E+00 | 3.976E+00 | 3.990E+00 | 4.015E+00 |
| 23400 | 3.935E+00 | 3.961E+00 | 3.971E+00 | 3.976E+00 | 3.980E+00 | 3.985E+00 | 3.993E+00 | 4.007E+00 | 4.032E+00 |
| 23500 | 3.960E+00 | 3.983E+00 | 3.992E+00 | 3.996E+00 | 3.999E+00 | 4.003E+00 | 4.011E+00 | 4.024E+00 | 4.048E+00 |
| 23600 | 3.984E+00 | 4.005E+00 | 4.013E+00 | 4.016E+00 | 4.018E+00 | 4.021E+00 | 4.028E+00 | 4.041E+00 | 4.065E+00 |
| 23700 | 4.008E+00 | 4.028E+00 | 4.034E+00 | 4.036E+00 | 4.037E+00 | 4.040E+00 | 4.046E+00 | 4.058E+00 | 4.082E+00 |
| 23800 | 4.033E+00 | 4.050E+00 | 4.055E+00 | 4.055E+00 | 4.056E+00 | 4.058E+00 | 4.063E+00 | 4.075E+00 | 4.099E+00 |
| 23900 | 4.057E+00 | 4.072E+00 | 4.076E+00 | 4.075E+00 | 4.075E+00 | 4.076E+00 | 4.081E+00 | 4.093E+00 | 4.116E+00 |
| 24000 | 4.081E+00 | 4.094E+00 | 4.096E+00 | 4.095E+00 | 4.094E+00 | 4.094E+00 | 4.099E+00 | 4.110E+00 | 4.133E+00 |

**TABLE A.7.** Electrical Conductivity of $Ar/H_2$ Mixtures (A/V · m)

| | Molar fraction of $H_2$ | | | | | | | | |
|---|---|---|---|---|---|---|---|---|---|
| $T$ (K) | 0.1 | 0.2 | 0.3 | 0.4 | 0.5 | 0.6 | 0.7 | 0.8 | 0.9 |
| 500 | 0.000E+00 | 0.000E+00 | 0.000E+00 | 0.000E+00 | 0.000E+00 | 0.000E+00 | 0.000E+00 | 0.000E+00 | 0.000E+00 |
| 600 | 0.000E+00 | 0.000E+00 | 0.000E+00 | 0.000E+00 | 0.000E+00 | 0.000E+00 | 0.000E+00 | 0.000E+00 | 0.000E+00 |
| 700 | 0.000E+00 | 0.000E+00 | 0.000E+00 | 0.000E+00 | 0.000E+00 | 0.000E+00 | 0.000E+00 | 0.000E+00 | 0.000E+00 |
| 800 | 0.000E+00 | 0.000E+00 | 0.000E+00 | 0.000E+00 | 0.000E+00 | 0.000E+00 | 0.000E+00 | 0.000E+00 | 0.000E+00 |
| 900 | 0.000E+00 | 0.000E+00 | 0.000E+00 | 0.000E+00 | 0.000E+00 | 0.000E+00 | 0.000E+00 | 0.000E+00 | 0.000E+00 |
| 1000 | 0.000E+00 | 0.000E+00 | 0.000E+00 | 0.000E+00 | 0.000E+00 | 0.000E+00 | 0.000E+00 | 0.000E+00 | 0.000E+00 |
| 1100 | 0.000E+00 | 0.000E+00 | 0.000E+00 | 0.000E+00 | 0.000E+00 | 0.000E+00 | 0.000E+00 | 0.000E+00 | 0.000E+00 |
| 1200 | 0.000E+00 | 0.000E+00 | 0.000E+00 | 0.000E+00 | 0.000E+00 | 0.000E+00 | 0.000E+00 | 0.000E+00 | 0.000E+00 |
| 1300 | 0.000E+00 | 0.000E+00 | 0.000E+00 | 0.000E+00 | 2.939E-24 | 2.680E-24 | 2.474E-24 | 2.303E-24 | 2.158E-24 |
| 1400 | 1.345E-20 | 8.815E-21 | 6.730E-21 | 5.515E-21 | 4.710E-21 | 4.131E-21 | 3.693E-21 | 3.347E-21 | 3.066E-21 |
| 1500 | 1.188E-18 | 7.712E-19 | 5.860E-19 | 4.789E-19 | 4.082E-19 | 3.575E-19 | 3.192E-19 | 2.890E-19 | 2.646E-19 |
| 1600 | 5.813E-17 | 3.758E-17 | 2.852E-17 | 2.329E-17 | 1.984E-17 | 1.737E-17 | 1.550E-17 | 1.404E-17 | 1.285E-17 |
| 1700 | 1.801E-15 | 1.160E-15 | 8.792E-16 | 7.177E-16 | 6.113E-16 | 5.352E-16 | 4.777E-16 | 4.325E-16 | 3.959E-16 |
| 1800 | 3.807E-14 | 2.445E-14 | 1.852E-14 | 1.512E-14 | 1.288E-14 | 1.127E-14 | 1.006E-14 | 9.115E-15 | 8.344E-15 |
| 1900 | 5.828E-13 | 3.738E-13 | 2.832E-13 | 2.312E-13 | 1.970E-13 | 1.725E-13 | 1.540E-13 | 1.395E-13 | 1.277E-13 |
| 2000 | 6.769E-12 | 4.342E-12 | 3.293E-12 | 2.691E-12 | 2.294E-12 | 2.010E-12 | 1.795E-12 | 1.626E-12 | 1.489E-12 |
| 2100 | 6.193E-11 | 3.983E-11 | 3.026E-11 | 2.476E-11 | 2.112E-11 | 1.852E-11 | 1.655E-11 | 1.500E-11 | 1.374E-11 |
| 2200 | 4.600E-10 | 2.974E-10 | 2.266E-10 | 1.858E-10 | 1.587E-10 | 1.393E-10 | 1.246E-10 | 1.130E-10 | 1.035E-10 |
| 2300 | 2.844E-09 | 1.853E-09 | 1.418E-09 | 1.165E-09 | 9.979E-10 | 8.771E-10 | 7.853E-10 | 7.128E-10 | 6.539E-10 |
| 2400 | 1.492E-08 | 9.837E-09 | 7.571E-09 | 6.244E-09 | 5.358E-09 | 4.718E-09 | 4.231E-09 | 3.845E-09 | 3.531E-09 |
| 2500 | 6.767E-08 | 4.522E-08 | 3.505E-08 | 2.904E-08 | 2.500E-08 | 2.207E-08 | 1.983E-08 | 1.805E-08 | 1.660E-08 |
| 2600 | 2.687E-07 | 1.826E-07 | 1.428E-07 | 1.189E-07 | 1.028E-07 | 9.105E-08 | 8.201E-08 | 7.481E-08 | 6.892E-08 |
| 2700 | 9.476E-07 | 6.561E-07 | 5.181E-07 | 4.344E-07 | 3.773E-07 | 3.354E-07 | 3.030E-07 | 2.771E-07 | 2.559E-07 |
| 2800 | 3.001E-06 | 2.119E-06 | 1.692E-06 | 1.429E-06 | 1.248E-06 | 1.115E-06 | 1.011E-06 | 9.277E-07 | 8.589E-07 |
| 2900 | 8.638E-06 | 6.223E-06 | 5.026E-06 | 4.281E-06 | 3.763E-06 | 3.376E-06 | 3.075E-06 | 2.831E-06 | 2.630E-06 |
| 3000 | 2.284E-05 | 1.676E-05 | 1.370E-05 | 1.177E-05 | 1.041E-05 | 9.396E-06 | 8.596E-06 | 7.948E-06 | 7.408E-06 |
| 3100 | 5.604E-05 | 4.184E-05 | 3.458E-05 | 2.996E-05 | 2.669E-05 | 2.421E-05 | 2.226E-05 | 2.067E-05 | 1.934E-05 |
| 3200 | 1.288E-04 | 9.759E-05 | 8.150E-05 | 7.119E-05 | 6.384E-05 | 5.826E-05 | 5.383E-05 | 5.020E-05 | 4.716E-05 |
| 3300 | 2.797E-04 | 2.144E-04 | 1.807E-04 | 1.590E-04 | 1.435E-04 | 1.317E-04 | 1.223E-04 | 1.145E-04 | 1.080E-04 |
| 3400 | 5.782E-04 | 4.471E-04 | 3.798E-04 | 3.365E-04 | 3.054E-04 | 2.817E-04 | 2.628E-04 | 2.473E-04 | 2.342E-04 |
| 3500 | 1.145E-03 | 8.908E-04 | 7.615E-04 | 6.785E-04 | 6.191E-04 | 5.737E-04 | 5.375E-04 | 5.077E-04 | 4.826E-04 |
| 3600 | 2.183E-03 | 1.704E-03 | 1.464E-03 | 1.311E-03 | 1.201E-03 | 1.118E-03 | 1.051E-03 | 9.971E-04 | 9.511E-04 |
| 3700 | 4.026E-03 | 3.149E-03 | 2.714E-03 | 2.439E-03 | 2.244E-03 | 2.096E-03 | 1.978E-03 | 1.881E-03 | 1.800E-03 |
| 3800 | 7.203E-03 | 5.637E-03 | 4.870E-03 | 4.389E-03 | 4.050E-03 | 3.794E-03 | 3.591E-03 | 3.424E-03 | 3.285E-03 |
| 3900 | 1.253E-02 | 9.806E-03 | 8.486E-03 | 7.664E-03 | 7.089E-03 | 6.656E-03 | 6.315E-03 | 6.038E-03 | 5.805E-03 |
| 4000 | 2.128E-02 | 1.662E-02 | 1.439E-02 | 1.302E-02 | 1.206E-02 | 1.135E-02 | 1.079E-02 | 1.033E-02 | 9.960E-03 |
| 4100 | 3.528E-02 | 2.751E-02 | 2.383E-02 | 2.158E-02 | 2.002E-02 | 1.887E-02 | 1.796E-02 | 1.723E-02 | 1.663E-02 |
| 4200 | 5.722E-02 | 4.453E-02 | 3.857E-02 | 3.496E-02 | 3.247E-02 | 3.063E-02 | 2.920E-02 | 2.805E-02 | 2.710E-02 |
| 4300 | 9.093E-02 | 7.060E-02 | 6.114E-02 | 5.544E-02 | 5.152E-02 | 4.865E-02 | 4.642E-02 | 4.464E-02 | 4.317E-02 |
| 4400 | 1.417E-01 | 1.098E-01 | 9.506E-02 | 8.620E-02 | 8.016E-02 | 7.573E-02 | 7.231E-02 | 6.958E-02 | 6.734E-02 |
| 4500 | 2.170E-01 | 1.676E-01 | 1.450E-01 | 1.315E-01 | 1.223E-01 | 1.156E-01 | 1.105E-01 | 1.063E-01 | 1.030E-01 |
| 4600 | 3.267E-01 | 2.517E-01 | 2.176E-01 | 1.973E-01 | 1.835E-01 | 1.735E-01 | 1.658E-01 | 1.597E-01 | 1.547E-01 |
| 4700 | 4.841E-01 | 3.718E-01 | 3.212E-01 | 2.911E-01 | 2.709E-01 | 2.561E-01 | 2.447E-01 | 2.357E-01 | 2.284E-01 |
| 4800 | 7.067E-01 | 5.411E-01 | 4.664E-01 | 4.226E-01 | 3.932E-01 | 3.718E-01 | 3.555E-01 | 3.426E-01 | 3.321E-01 |
| 4900 | 1.017E+00 | 7.749E-01 | 6.681E-01 | 6.052E-01 | 5.630E-01 | 5.324E-01 | 5.090E-01 | 4.906E-01 | 4.756E-01 |
| 5000 | 1.440E+00 | 1.096E+00 | 9.441E-01 | 8.548E-01 | 7.950E-01 | 7.517E-01 | 7.188E-01 | 6.928E-01 | 6.717E-01 |
| 5100 | 2.019E+00 | 1.531E+00 | 1.317E+00 | 1.191E+00 | 1.108E+00 | 1.047E+00 | 1.001E+00 | 9.655E-01 | 9.362E-01 |
| 5200 | 2.796E+00 | 2.113E+00 | 1.815E+00 | 1.641E+00 | 1.525E+00 | 1.442E+00 | 1.379E+00 | 1.329E+00 | 1.288E+00 |

(*continued*)

## TABLE A.7. (*Continued*)

| $T$ (K) | Molar fraction of $H_2$ | | | | | | | | |
|---|---|---|---|---|---|---|---|---|---|
| | 0.1 | 0.2 | 0.3 | 0.4 | 0.5 | 0.6 | 0.7 | 0.8 | 0.9 |
| 5300 | 3.818E+00 | 2.883E+00 | 2.473E+00 | 2.234E+00 | 2.076E+00 | 1.962E+00 | 1.876E+00 | 1.808E+00 | 1.753E+00 |
| 5400 | 5.165E+00 | 3.885E+00 | 3.332E+00 | 3.009E+00 | 2.795E+00 | 2.641E+00 | 2.524E+00 | 2.432E+00 | 2.358E+00 |
| 5500 | 6.914E+00 | 5.187E+00 | 4.440E+00 | 4.010E+00 | 3.723E+00 | 3.516E+00 | 3.360E+00 | 3.238E+00 | 3.139E+00 |
| 5600 | 9.160E+00 | 6.855E+00 | 5.861E+00 | 5.287E+00 | 4.907E+00 | 4.636E+00 | 4.430E+00 | 4.268E+00 | 4.137E+00 |
| 5700 | 1.201E+01 | 8.973E+00 | 7.663E+00 | 6.908E+00 | 6.409E+00 | 6.052E+00 | 5.784E+00 | 5.571E+00 | 5.400E+00 |
| 5800 | 1.561E+01 | 1.163E+01 | 9.928E+00 | 8.944E+00 | 8.295E+00 | 7.831E+00 | 7.483E+00 | 7.207E+00 | 6.984E+00 |
| 5900 | 2.008E+01 | 1.495E+01 | 1.275E+01 | 1.148E+01 | 1.064E+01 | 1.004E+01 | 9.593E+00 | 9.242E+00 | 8.955E+00 |
| 6000 | 2.561E+01 | 1.906E+01 | 1.623E+01 | 1.461E+01 | 1.354E+01 | 1.277E+01 | 1.220E+01 | 1.174E+01 | 1.138E+01 |
| 6100 | 3.237E+01 | 2.408E+01 | 2.050E+01 | 1.844E+01 | 1.708E+01 | 1.612E+01 | 1.539E+01 | 1.482E+01 | 1.436E+01 |
| 6200 | 4.055E+01 | 3.019E+01 | 2.569E+01 | 2.310E+01 | 2.140E+01 | 2.018E+01 | 1.926E+01 | 1.855E+01 | 1.797E+01 |
| 6300 | 5.037E+01 | 3.755E+01 | 3.195E+01 | 2.872E+01 | 2.660E+01 | 2.508E+01 | 2.394E+01 | 2.304E+01 | 2.233E+01 |
| 6400 | 6.206E+01 | 4.635E+01 | 3.944E+01 | 3.545E+01 | 3.282E+01 | 3.095E+01 | 2.953E+01 | 2.843E+01 | 2.754E+01 |
| 6500 | 7.583E+01 | 5.679E+01 | 4.833E+01 | 4.345E+01 | 4.022E+01 | 3.792E+01 | 3.619E+01 | 3.483E+01 | 3.374E+01 |
| 6600 | 9.194E+01 | 6.907E+01 | 5.883E+01 | 5.288E+01 | 4.896E+01 | 4.616E+01 | 4.404E+01 | 4.239E+01 | 4.106E+01 |
| 6700 | 1.106E+02 | 8.343E+01 | 7.112E+01 | 6.395E+01 | 5.921E+01 | 5.582E+01 | 5.326E+01 | 5.126E+01 | 4.966E+01 |
| 6800 | 1.321E+02 | 1.001E+02 | 8.542E+01 | 7.684E+01 | 7.115E+01 | 6.708E+01 | 6.401E+01 | 6.161E+01 | 5.967E+01 |
| 6900 | 1.567E+02 | 1.193E+02 | 1.019E+02 | 9.175E+01 | 8.497E+01 | 8.012E+01 | 7.646E+01 | 7.359E+01 | 7.129E+01 |
| 7000 | 1.845E+02 | 1.412E+02 | 1.209E+02 | 1.089E+02 | 1.008E+02 | 9.514E+01 | 9.080E+01 | 8.740E+01 | 8.466E+01 |
| 7100 | 2.159E+02 | 1.662E+02 | 1.425E+02 | 1.285E+02 | 1.190E+02 | 1.123E+02 | 1.072E+02 | 1.032E+02 | 9.999E+01 |
| 7200 | 2.511E+02 | 1.945E+02 | 1.671E+02 | 1.507E+02 | 1.397E+02 | 1.319E+02 | 1.259E+02 | 1.212E+02 | 1.174E+02 |
| 7300 | 2.902E+02 | 2.262E+02 | 1.948E+02 | 1.759E+02 | 1.632E+02 | 1.540E+02 | 1.470E+02 | 1.416E+02 | 1.372E+02 |
| 7400 | 3.334E+02 | 2.616E+02 | 2.258E+02 | 2.041E+02 | 1.895E+02 | 1.789E+02 | 1.709E+02 | 1.646E+02 | 1.595E+02 |
| 7500 | 3.809E+02 | 3.009E+02 | 2.604E+02 | 2.357E+02 | 2.190E+02 | 2.068E+02 | 1.976E+02 | 1.903E+02 | 1.845E+02 |
| 7600 | 4.328E+02 | 3.443E+02 | 2.987E+02 | 2.708E+02 | 2.517E+02 | 2.379E+02 | 2.274E+02 | 2.190E+02 | 2.123E+02 |
| 7700 | 4.891E+02 | 3.918E+02 | 3.410E+02 | 3.095E+02 | 2.880E+02 | 2.723E+02 | 2.604E+02 | 2.509E+02 | 2.433E+02 |
| 7800 | 5.499E+02 | 4.437E+02 | 3.873E+02 | 3.521E+02 | 3.280E+02 | 3.103E+02 | 2.968E+02 | 2.861E+02 | 2.775E+02 |
| 7900 | 6.152E+02 | 5.000E+02 | 4.379E+02 | 3.987E+02 | 3.717E+02 | 3.520E+02 | 3.368E+02 | 3.248E+02 | 3.150E+02 |
| 8000 | 6.850E+02 | 5.607E+02 | 4.927E+02 | 4.495E+02 | 4.195E+02 | 3.974E+02 | 3.805E+02 | 3.671E+02 | 3.562E+02 |
| 8100 | 7.591E+02 | 6.260E+02 | 5.518E+02 | 5.044E+02 | 4.713E+02 | 4.469E+02 | 4.281E+02 | 4.131E+02 | 4.009E+02 |
| 8200 | 8.374E+02 | 6.956E+02 | 6.154E+02 | 5.636E+02 | 5.273E+02 | 5.003E+02 | 4.796E+02 | 4.630E+02 | 4.495E+02 |
| 8300 | 9.198E+02 | 7.696E+02 | 6.833E+02 | 6.271E+02 | 5.874E+02 | 5.579E+02 | 5.351E+02 | 5.169E+02 | 5.020E+02 |
| 8400 | 1.006E+03 | 8.479E+02 | 7.556E+02 | 6.949E+02 | 6.518E+02 | 6.197E+02 | 5.947E+02 | 5.747E+02 | 5.584E+02 |
| 8500 | 1.095E+03 | 9.303E+02 | 8.321E+02 | 7.669E+02 | 7.204E+02 | 6.856E+02 | 6.584E+02 | 6.366E+02 | 6.188E+02 |
| 8600 | 1.189E+03 | 1.016E+03 | 9.127E+02 | 8.432E+02 | 7.932E+02 | 7.556E+02 | 7.262E+02 | 7.026E+02 | 6.832E+02 |
| 8700 | 1.285E+03 | 1.106E+03 | 9.974E+02 | 9.235E+02 | 8.702E+02 | 8.297E+02 | 7.981E+02 | 7.726E+02 | 7.516E+02 |
| 8800 | 1.384E+03 | 1.200E+03 | 1.085E+03 | 1.007E+03 | 9.510E+02 | 9.079E+02 | 8.740E+02 | 8.466E+02 | 8.240E+02 |
| 8900 | 1.486E+03 | 1.297E+03 | 1.178E+03 | 1.096E+03 | 1.035E+03 | 9.900E+02 | 9.538E+02 | 9.245E+02 | 9.003E+02 |
| 9000 | 1.590E+03 | 1.397E+03 | 1.273E+03 | 1.187E+03 | 1.124E+03 | 1.076E+03 | 1.037E+03 | 1.006E+03 | 9.804E+02 |
| 9100 | 1.696E+03 | 1.499E+03 | 1.372E+03 | 1.283E+03 | 1.216E+03 | 1.165E+03 | 1.124E+03 | 1.091E+03 | 1.064E+03 |
| 9200 | 1.804E+03 | 1.605E+03 | 1.474E+03 | 1.381E+03 | 1.312E+03 | 1.258E+03 | 1.215E+03 | 1.180E+03 | 1.151E+03 |
| 9300 | 1.913E+03 | 1.712E+03 | 1.578E+03 | 1.483E+03 | 1.411E+03 | 1.355E+03 | 1.310E+03 | 1.273E+03 | 1.242E+03 |
| 9400 | 2.024E+03 | 1.822E+03 | 1.685E+03 | 1.587E+03 | 1.512E+03 | 1.454E+03 | 1.407E+03 | 1.369E+03 | 1.337E+03 |
| 9500 | 2.135E+03 | 1.933E+03 | 1.795E+03 | 1.694E+03 | 1.617E+03 | 1.557E+03 | 1.508E+03 | 1.468E+03 | 1.434E+03 |
| 9600 | 2.248E+03 | 2.046E+03 | 1.906E+03 | 1.803E+03 | 1.724E+03 | 1.662E+03 | 1.611E+03 | 1.570E+03 | 1.535E+03 |
| 9700 | 2.361E+03 | 2.160E+03 | 2.019E+03 | 1.914E+03 | 1.833E+03 | 1.769E+03 | 1.717E+03 | 1.674E+03 | 1.638E+03 |
| 9800 | 2.475E+03 | 2.275E+03 | 2.133E+03 | 2.027E+03 | 1.945E+03 | 1.879E+03 | 1.826E+03 | 1.781E+03 | 1.743E+03 |
| 9900 | 2.589E+03 | 2.392E+03 | 2.249E+03 | 2.142E+03 | 2.058E+03 | 1.991E+03 | 1.936E+03 | 1.890E+03 | 1.852E+03 |
| 10000 | 2.704E+03 | 2.508E+03 | 2.366E+03 | 2.258E+03 | 2.173E+03 | 2.105E+03 | 2.049E+03 | 2.002E+03 | 1.962E+03 |

**TABLE A.7.** *(Continued)*

| $T$ (K) | Molar fraction of $H_2$ | | | | | | | | |
|---|---|---|---|---|---|---|---|---|---|
| | 0.1 | 0.2 | 0.3 | 0.4 | 0.5 | 0.6 | 0.7 | 0.8 | 0.9 |
| 10100 | 2.819E+03 | 2.626E+03 | 2.484E+03 | 2.376E+03 | 2.290E+03 | 2.221E+03 | 2.163E+03 | 2.115E+03 | 2.074E+03 |
| 10200 | 2.934E+03 | 2.744E+03 | 2.603E+03 | 2.494E+03 | 2.408E+03 | 2.337E+03 | 2.279E+03 | 2.230E+03 | 2.188E+03 |
| 10300 | 3.048E+03 | 2.862E+03 | 2.722E+03 | 2.613E+03 | 2.527E+03 | 2.456E+03 | 2.396E+03 | 2.346E+03 | 2.304E+03 |
| 10400 | 3.163E+03 | 2.980E+03 | 2.842E+03 | 2.733E+03 | 2.646E+03 | 2.575E+03 | 2.515E+03 | 2.464E+03 | 2.421E+03 |
| 10500 | 3.278E+03 | 3.099E+03 | 2.962E+03 | 2.854E+03 | 2.767E+03 | 2.695E+03 | 2.634E+03 | 2.583E+03 | 2.539E+03 |
| 10600 | 3.392E+03 | 3.217E+03 | 3.082E+03 | 2.975E+03 | 2.888E+03 | 2.815E+03 | 2.755E+03 | 2.703E+03 | 2.658E+03 |
| 10700 | 3.507E+03 | 3.335E+03 | 3.202E+03 | 3.096E+03 | 3.009E+03 | 2.937E+03 | 2.875E+03 | 2.823E+03 | 2.778E+03 |
| 10800 | 3.621E+03 | 3.453E+03 | 3.322E+03 | 3.217E+03 | 3.131E+03 | 3.058E+03 | 2.997E+03 | 2.944E+03 | 2.899E+03 |
| 10900 | 3.734E+03 | 3.571E+03 | 3.442E+03 | 3.338E+03 | 3.252E+03 | 3.180E+03 | 3.119E+03 | 3.066E+03 | 3.020E+03 |
| 11000 | 3.848E+03 | 3.689E+03 | 3.562E+03 | 3.459E+03 | 3.374E+03 | 3.302E+03 | 3.241E+03 | 3.188E+03 | 3.141E+03 |
| 11100 | 3.961E+03 | 3.806E+03 | 3.682E+03 | 3.580E+03 | 3.496E+03 | 3.424E+03 | 3.363E+03 | 3.310E+03 | 3.263E+03 |
| 11200 | 4.073E+03 | 3.922E+03 | 3.801E+03 | 3.701E+03 | 3.617E+03 | 3.546E+03 | 3.485E+03 | 3.432E+03 | 3.386E+03 |
| 11300 | 4.185E+03 | 4.039E+03 | 3.920E+03 | 3.821E+03 | 3.739E+03 | 3.668E+03 | 3.607E+03 | 3.554E+03 | 3.508E+03 |
| 11400 | 4.297E+03 | 4.154E+03 | 4.038E+03 | 3.941E+03 | 3.860E+03 | 3.790E+03 | 3.729E+03 | 3.676E+03 | 3.630E+03 |
| 11500 | 4.409E+03 | 4.270E+03 | 4.156E+03 | 4.061E+03 | 3.980E+03 | 3.911E+03 | 3.851E+03 | 3.798E+03 | 3.752E+03 |
| 11600 | 4.520E+03 | 4.385E+03 | 4.273E+03 | 4.180E+03 | 4.100E+03 | 4.032E+03 | 3.972E+03 | 3.920E+03 | 3.874E+03 |
| 11700 | 4.630E+03 | 4.499E+03 | 4.390E+03 | 4.298E+03 | 4.220E+03 | 4.152E+03 | 4.093E+03 | 4.041E+03 | 3.995E+03 |
| 11800 | 4.740E+03 | 4.612E+03 | 4.506E+03 | 4.416E+03 | 4.339E+03 | 4.272E+03 | 4.214E+03 | 4.162E+03 | 4.116E+03 |
| 11900 | 4.849E+03 | 4.725E+03 | 4.622E+03 | 4.533E+03 | 4.458E+03 | 4.392E+03 | 4.334E+03 | 4.283E+03 | 4.237E+03 |
| 12000 | 4.958E+03 | 4.838E+03 | 4.736E+03 | 4.650E+03 | 4.576E+03 | 4.511E+03 | 4.453E+03 | 4.403E+03 | 4.357E+03 |
| 12100 | 5.067E+03 | 4.950E+03 | 4.851E+03 | 4.766E+03 | 4.693E+03 | 4.629E+03 | 4.572E+03 | 4.522E+03 | 4.477E+03 |
| 12200 | 5.175E+03 | 5.061E+03 | 4.964E+03 | 4.881E+03 | 4.809E+03 | 4.746E+03 | 4.690E+03 | 4.641E+03 | 4.596E+03 |
| 12300 | 5.282E+03 | 5.171E+03 | 5.077E+03 | 4.996E+03 | 4.925E+03 | 4.863E+03 | 4.808E+03 | 4.759E+03 | 4.715E+03 |
| 12400 | 5.389E+03 | 5.281E+03 | 5.189E+03 | 5.110E+03 | 5.040E+03 | 4.979E+03 | 4.925E+03 | 4.876E+03 | 4.833E+03 |
| 12500 | 5.495E+03 | 5.390E+03 | 5.301E+03 | 5.223E+03 | 5.155E+03 | 5.094E+03 | 5.041E+03 | 4.993E+03 | 4.950E+03 |
| 12600 | 5.601E+03 | 5.499E+03 | 5.411E+03 | 5.335E+03 | 5.268E+03 | 5.209E+03 | 5.156E+03 | 5.109E+03 | 5.066E+03 |
| 12700 | 5.706E+03 | 5.606E+03 | 5.521E+03 | 5.446E+03 | 5.381E+03 | 5.323E+03 | 5.271E+03 | 5.224E+03 | 5.182E+03 |
| 12800 | 5.810E+03 | 5.714E+03 | 5.630E+03 | 5.557E+03 | 5.492E+03 | 5.435E+03 | 5.384E+03 | 5.338E+03 | 5.297E+03 |
| 12900 | 5.914E+03 | 5.820E+03 | 5.738E+03 | 5.666E+03 | 5.603E+03 | 5.547E+03 | 5.497E+03 | 5.452E+03 | 5.411E+03 |
| 13000 | 6.017E+03 | 5.925E+03 | 5.845E+03 | 5.776E+03 | 5.713E+03 | 5.658E+03 | 5.609E+03 | 5.564E+03 | 5.524E+03 |
| 13100 | 6.119E+03 | 6.030E+03 | 5.952E+03 | 5.883E+03 | 5.823E+03 | 5.769E+03 | 5.720E+03 | 5.676E+03 | 5.636E+03 |
| 13200 | 6.220E+03 | 6.133E+03 | 6.057E+03 | 5.990E+03 | 5.931E+03 | 5.878E+03 | 5.831E+03 | 5.788E+03 | 5.748E+03 |
| 13300 | 6.321E+03 | 6.236E+03 | 6.162E+03 | 6.097E+03 | 6.038E+03 | 5.987E+03 | 5.940E+03 | 5.898E+03 | 5.859E+03 |
| 13400 | 6.421E+03 | 6.338E+03 | 6.266E+03 | 6.202E+03 | 6.145E+03 | 6.094E+03 | 6.048E+03 | 6.007E+03 | 5.969E+03 |
| 13500 | 6.520E+03 | 6.439E+03 | 6.368E+03 | 6.306E+03 | 6.250E+03 | 6.200E+03 | 6.155E+03 | 6.115E+03 | 6.078E+03 |
| 13600 | 6.618E+03 | 6.539E+03 | 6.470E+03 | 6.409E+03 | 6.355E+03 | 6.306E+03 | 6.262E+03 | 6.222E+03 | 6.186E+03 |
| 13700 | 6.715E+03 | 6.638E+03 | 6.571E+03 | 6.511E+03 | 6.458E+03 | 6.410E+03 | 6.367E+03 | 6.328E+03 | 6.293E+03 |
| 13800 | 6.812E+03 | 6.736E+03 | 6.671E+03 | 6.612E+03 | 6.560E+03 | 6.514E+03 | 6.472E+03 | 6.433E+03 | 6.399E+03 |
| 13900 | 6.908E+03 | 6.834E+03 | 6.769E+03 | 6.712E+03 | 6.662E+03 | 6.616E+03 | 6.575E+03 | 6.538E+03 | 6.504E+03 |
| 14000 | 7.002E+03 | 6.931E+03 | 6.867E+03 | 6.811E+03 | 6.762E+03 | 6.717E+03 | 6.677E+03 | 6.641E+03 | 6.608E+03 |
| 14100 | 7.096E+03 | 7.026E+03 | 6.965E+03 | 6.910E+03 | 6.861E+03 | 6.818E+03 | 6.778E+03 | 6.743E+03 | 6.711E+03 |
| 14200 | 7.188E+03 | 7.120E+03 | 7.061E+03 | 7.007E+03 | 6.959E+03 | 6.917E+03 | 6.879E+03 | 6.844E+03 | 6.813E+03 |
| 14300 | 7.280E+03 | 7.214E+03 | 7.156E+03 | 7.104E+03 | 7.057E+03 | 7.015E+03 | 6.978E+03 | 6.944E+03 | 6.914E+03 |
| 14400 | 7.370E+03 | 7.306E+03 | 7.249E+03 | 7.199E+03 | 7.153E+03 | 7.113E+03 | 7.076E+03 | 7.043E+03 | 7.013E+03 |
| 14500 | 7.460E+03 | 7.397E+03 | 7.342E+03 | 7.293E+03 | 7.249E+03 | 7.209E+03 | 7.174E+03 | 7.141E+03 | 7.112E+03 |
| 14600 | 7.549E+03 | 7.487E+03 | 7.434E+03 | 7.386E+03 | 7.344E+03 | 7.306E+03 | 7.270E+03 | 7.239E+03 | 7.210E+03 |
| 14700 | 7.636E+03 | 7.577E+03 | 7.524E+03 | 7.478E+03 | 7.437E+03 | 7.400E+03 | 7.367E+03 | 7.335E+03 | 7.307E+03 |
| 14800 | 7.723E+03 | 7.665E+03 | 7.614E+03 | 7.569E+03 | 7.529E+03 | 7.493E+03 | 7.461E+03 | 7.431E+03 | 7.403E+03 |

*(continued)*

## TABLE A.7. *(Continued)*

| T (K) | Molar fraction of $H_2$ | | | | | | | | |
|---|---|---|---|---|---|---|---|---|---|
| | 0.1 | 0.2 | 0.3 | 0.4 | 0.5 | 0.6 | 0.7 | 0.8 | 0.9 |
| 14900 | 7.808E+03 | 7.752E+03 | 7.703E+03 | 7.659E+03 | 7.620E+03 | 7.585E+03 | 7.554E+03 | 7.526E+03 | 7.500E+03 |
| 15000 | 7.893E+03 | 7.838E+03 | 7.790E+03 | 7.748E+03 | 7.710E+03 | 7.677E+03 | 7.646E+03 | 7.619E+03 | 7.594E+03 |
| 15100 | 7.977E+03 | 7.924E+03 | 7.877E+03 | 7.836E+03 | 7.800E+03 | 7.767E+03 | 7.737E+03 | 7.711E+03 | 7.687E+03 |
| 15200 | 8.059E+03 | 8.008E+03 | 7.963E+03 | 7.923E+03 | 7.888E+03 | 7.856E+03 | 7.828E+03 | 7.802E+03 | 7.779E+03 |
| 15300 | 8.141E+03 | 8.091E+03 | 8.048E+03 | 8.009E+03 | 7.975E+03 | 7.944E+03 | 7.917E+03 | 7.892E+03 | 7.870E+03 |
| 15400 | 8.222E+03 | 8.174E+03 | 8.132E+03 | 8.094E+03 | 8.061E+03 | 8.032E+03 | 8.005E+03 | 7.981E+03 | 7.960E+03 |
| 15500 | 8.302E+03 | 8.255E+03 | 8.215E+03 | 8.179E+03 | 8.147E+03 | 8.118E+03 | 8.093E+03 | 8.070E+03 | 8.049E+03 |
| 15600 | 8.381E+03 | 8.336E+03 | 8.297E+03 | 8.262E+03 | 8.231E+03 | 8.204E+03 | 8.179E+03 | 8.157E+03 | 8.137E+03 |
| 15700 | 8.460E+03 | 8.416E+03 | 8.378E+03 | 8.345E+03 | 8.315E+03 | 8.289E+03 | 8.265E+03 | 8.244E+03 | 8.224E+03 |
| 15800 | 8.537E+03 | 8.495E+03 | 8.458E+03 | 8.426E+03 | 8.398E+03 | 8.372E+03 | 8.350E+03 | 8.329E+03 | 8.311E+03 |
| 15900 | 8.614E+03 | 8.574E+03 | 8.538E+03 | 8.507E+03 | 8.480E+03 | 8.455E+03 | 8.433E+03 | 8.414E+03 | 8.396E+03 |
| 16000 | 8.690E+03 | 8.651E+03 | 8.617E+03 | 8.587E+03 | 8.561E+03 | 8.537E+03 | 8.517E+03 | 8.498E+03 | 8.481E+03 |
| 16100 | 8.766E+03 | 8.728E+03 | 8.695E+03 | 8.667E+03 | 8.641E+03 | 8.619E+03 | 8.599E+03 | 8.581E+03 | 8.564E+03 |
| 16200 | 8.840E+03 | 8.804E+03 | 8.773E+03 | 8.745E+03 | 8.721E+03 | 8.699E+03 | 8.680E+03 | 8.663E+03 | 8.647E+03 |
| 16300 | 8.914E+03 | 8.880E+03 | 8.849E+03 | 8.823E+03 | 8.800E+03 | 8.779E+03 | 8.761E+03 | 8.744E+03 | 8.729E+03 |
| 16400 | 8.988E+03 | 8.954E+03 | 8.926E+03 | 8.900E+03 | 8.878E+03 | 8.858E+03 | 8.841E+03 | 8.825E+03 | 8.811E+03 |
| 16500 | 9.061E+03 | 9.029E+03 | 9.001E+03 | 8.977E+03 | 8.956E+03 | 8.937E+03 | 8.920E+03 | 8.905E+03 | 8.892E+03 |
| 16600 | 9.133E+03 | 9.102E+03 | 9.076E+03 | 9.053E+03 | 9.033E+03 | 9.015E+03 | 8.999E+03 | 8.984E+03 | 8.971E+03 |
| 16700 | 9.205E+03 | 9.175E+03 | 9.150E+03 | 9.128E+03 | 9.109E+03 | 9.092E+03 | 9.077E+03 | 9.063E+03 | 9.051E+03 |
| 16800 | 9.276E+03 | 9.248E+03 | 9.224E+03 | 9.203E+03 | 9.185E+03 | 9.168E+03 | 9.154E+03 | 9.141E+03 | 9.129E+03 |
| 16900 | 9.347E+03 | 9.320E+03 | 9.297E+03 | 9.277E+03 | 9.260E+03 | 9.244E+03 | 9.231E+03 | 9.218E+03 | 9.207E+03 |
| 17000 | 9.417E+03 | 9.392E+03 | 9.370E+03 | 9.351E+03 | 9.334E+03 | 9.320E+03 | 9.307E+03 | 9.295E+03 | 9.285E+03 |
| 17100 | 9.487E+03 | 9.463E+03 | 9.442E+03 | 9.424E+03 | 9.408E+03 | 9.395E+03 | 9.382E+03 | 9.371E+03 | 9.361E+03 |
| 17200 | 9.556E+03 | 9.533E+03 | 9.514E+03 | 9.497E+03 | 9.482E+03 | 9.469E+03 | 9.457E+03 | 9.447E+03 | 9.438E+03 |
| 17300 | 9.625E+03 | 9.603E+03 | 9.585E+03 | 9.569E+03 | 9.555E+03 | 9.543E+03 | 9.532E+03 | 9.522E+03 | 9.513E+03 |
| 17400 | 9.693E+03 | 9.673E+03 | 9.656E+03 | 9.641E+03 | 9.628E+03 | 9.616E+03 | 9.606E+03 | 9.597E+03 | 9.589E+03 |
| 17500 | 9.761E+03 | 9.742E+03 | 9.726E+03 | 9.712E+03 | 9.700E+03 | 9.689E+03 | 9.679E+03 | 9.671E+03 | 9.663E+03 |
| 17600 | 9.829E+03 | 9.811E+03 | 9.796E+03 | 9.783E+03 | 9.772E+03 | 9.761E+03 | 9.753E+03 | 9.745E+03 | 9.738E+03 |
| 17700 | 9.896E+03 | 9.880E+03 | 9.866E+03 | 9.853E+03 | 9.843E+03 | 9.834E+03 | 9.825E+03 | 9.818E+03 | 9.811E+03 |
| 17800 | 9.963E+03 | 9.948E+03 | 9.935E+03 | 9.923E+03 | 9.914E+03 | 9.905E+03 | 9.898E+03 | 9.902E+03 | 9.885E+03 |
| 17900 | 1.003E+04 | 1.001E+04 | 1.000E+04 | 9.993E+03 | 9.984E+03 | 9.976E+03 | 9.970E+03 | 9.963E+03 | 9.958E+03 |
| 18000 | 1.009E+04 | 1.008E+04 | 1.007E+04 | 1.006E+04 | 1.005E+04 | 1.004E+04 | 1.004E+04 | 1.003E+04 | 1.003E+04 |
| 18100 | 1.016E+04 | 1.015E+04 | 1.014E+04 | 1.013E+04 | 1.012E+04 | 1.011E+04 | 1.011E+04 | 1.010E+04 | 1.010E+04 |
| 18200 | 1.022E+04 | 1.021E+04 | 1.020E+04 | 1.020E+04 | 1.019E+04 | 1.018E+04 | 1.018E+04 | 1.017E+04 | 1.017E+04 |
| 18300 | 1.029E+04 | 1.028E+04 | 1.027E+04 | 1.026E+04 | 1.026E+04 | 1.025E+04 | 1.025E+04 | 1.024E+04 | 1.024E+04 |
| 18400 | 1.035E+04 | 1.034E+04 | 1.034E+04 | 1.033E+04 | 1.033E+04 | 1.032E+04 | 1.032E+04 | 1.032E+04 | 1.031E+04 |
| 18500 | 1.042E+04 | 1.041E+04 | 1.040E+04 | 1.040E+04 | 1.039E+04 | 1.039E+04 | 1.039E+04 | 1.038E+04 | 1.038E+04 |
| 18600 | 1.048E+04 | 1.047E+04 | 1.047E+04 | 1.046E+04 | 1.046E+04 | 1.046E+04 | 1.046E+04 | 1.045E+04 | 1.045E+04 |
| 18700 | 1.054E+04 | 1.054E+04 | 1.053E+04 | 1.053E+04 | 1.053E+04 | 1.053E+04 | 1.053E+04 | 1.052E+04 | 1.052E+04 |
| 18800 | 1.060E+04 | 1.060E+04 | 1.060E+04 | 1.060E+04 | 1.060E+04 | 1.060E+04 | 1.059E+04 | 1.059E+04 | 1.059E+04 |
| 18900 | 1.067E+04 | 1.066E+04 | 1.066E+04 | 1.066E+04 | 1.066E+04 | 1.066E+04 | 1.066E+04 | 1.066E+04 | 1.066E+04 |
| 19000 | 1.072E+04 | 1.073E+04 | 1.073E+04 | 1.073E+04 | 1.073E+04 | 1.073E+04 | 1.073E+04 | 1.073E+04 | 1.073E+04 |
| 19100 | 1.079E+04 | 1.079E+04 | 1.079E+04 | 1.079E+04 | 1.080E+04 | 1.080E+04 | 1.080E+04 | 1.080E+04 | 1.080E+04 |
| 19200 | 1.085E+04 | 1.085E+04 | 1.085E+04 | 1.086E+04 | 1.086E+04 | 1.087E+04 | 1.087E+04 | 1.087E+04 | 1.087E+04 |
| 19300 | 1.090E+04 | 1.091E+04 | 1.092E+04 | 1.092E+04 | 1.093E+04 | 1.093E+04 | 1.094E+04 | 1.094E+04 | 1.094E+04 |
| 19400 | 1.096E+04 | 1.097E+04 | 1.098E+04 | 1.099E+04 | 1.099E+04 | 1.100E+04 | 1.100E+04 | 1.101E+04 | 1.101E+04 |
| 19500 | 1.102E+04 | 1.103E+04 | 1.104E+04 | 1.105E+04 | 1.106E+04 | 1.107E+04 | 1.107E+04 | 1.108E+04 | 1.108E+04 |
| 19600 | 1.108E+04 | 1.109E+04 | 1.110E+04 | 1.111E+04 | 1.112E+04 | 1.113E+04 | 1.114E+04 | 1.115E+04 | 1.115E+04 |

**TABLE A.7.** *(Continued)*

| T (K) | Molar fraction of $H_2$ | | | | | | | | |
|---|---|---|---|---|---|---|---|---|---|
| | 0.1 | 0.2 | 0.3 | 0.4 | 0.5 | 0.6 | 0.7 | 0.8 | 0.9 |
| 19700 | 1.113E+04 | 1.115E+04 | 1.116E+04 | 1.118E+04 | 1.119E+04 | 1.120E+04 | 1.121E+04 | 1.121E+04 | 1.122E+04 |
| 19800 | 1.119E+04 | 1.121E+04 | 1.122E+04 | 1.124E+04 | 1.125E+04 | 1.126E+04 | 1.127E+04 | 1.128E+04 | 1.129E+04 |
| 19900 | 1.124E+04 | 1.126E+04 | 1.128E+04 | 1.130E+04 | 1.131E+04 | 1.133E+04 | 1.134E+04 | 1.135E+04 | 1.136E+04 |
| 20000 | 1.130E+04 | 1.132E+04 | 1.134E+04 | 1.136E+04 | 1.138E+04 | 1.139E+04 | 1.140E+04 | 1.142E+04 | 1.143E+04 |
| 20100 | 1.135E+04 | 1.138E+04 | 1.140E+04 | 1.142E+04 | 1.144E+04 | 1.145E+04 | 1.147E+04 | 1.148E+04 | 1.150E+04 |
| 20200 | 1.140E+04 | 1.143E+04 | 1.146E+04 | 1.148E+04 | 1.150E+04 | 1.152E+04 | 1.154E+04 | 1.155E+04 | 1.156E+04 |
| 20300 | 1.145E+04 | 1.148E+04 | 1.151E+04 | 1.154E+04 | 1.156E+04 | 1.158E+04 | 1.160E+04 | 1.162E+04 | 1.163E+04 |
| 20400 | 1.150E+04 | 1.153E+04 | 1.157E+04 | 1.160E+04 | 1.162E+04 | 1.164E+04 | 1.167E+04 | 1.168E+04 | 1.170E+04 |
| 20500 | 1.154E+04 | 1.159E+04 | 1.162E+04 | 1.165E+04 | 1.168E+04 | 1.171E+04 | 1.173E+04 | 1.175E+04 | 1.177E+04 |
| 20600 | 1.159E+04 | 1.163E+04 | 1.167E+04 | 1.171E+04 | 1.174E+04 | 1.177E+04 | 1.179E+04 | 1.182E+04 | 1.184E+04 |
| 20700 | 1.163E+04 | 1.168E+04 | 1.173E+04 | 1.176E+04 | 1.180E+04 | 1.183E+04 | 1.186E+04 | 1.188E+04 | 1.191E+04 |
| 20800 | 1.168E+04 | 1.173E+04 | 1.178E+04 | 1.182E+04 | 1.186E+04 | 1.189E+04 | 1.192E+04 | 1.195E+04 | 1.197E+04 |
| 20900 | 1.172E+04 | 1.177E+04 | 1.183E+04 | 1.187E+04 | 1.191E+04 | 1.195E+04 | 1.198E+04 | 1.202E+04 | 1.204E+04 |
| 21000 | 1.175E+04 | 1.182E+04 | 1.187E+04 | 1.192E+04 | 1.197E+04 | 1.201E+04 | 1.205E+04 | 1.208E+04 | 1.211E+04 |
| 21100 | 1.179E+04 | 1.186E+04 | 1.192E+04 | 1.197E+04 | 1.202E+04 | 1.207E+04 | 1.211E+04 | 1.214E+04 | 1.218E+04 |
| 21200 | 1.183E+04 | 1.190E+04 | 1.197E+04 | 1.202E+04 | 1.208E+04 | 1.213E+04 | 1.217E+04 | 1.221E+04 | 1.224E+04 |
| 21300 | 1.186E+04 | 1.194E+04 | 1.201E+04 | 1.207E+04 | 1.213E+04 | 1.218E+04 | 1.223E+04 | 1.227E+04 | 1.231E+04 |
| 21400 | 1.189E+04 | 1.198E+04 | 1.205E+04 | 1.212E+04 | 1.218E+04 | 1.224E+04 | 1.229E+04 | 1.234E+04 | 1.238E+04 |
| 21500 | 1.192E+04 | 1.201E+04 | 1.209E+04 | 1.217E+04 | 1.223E+04 | 1.229E+04 | 1.235E+04 | 1.240E+04 | 1.245E+04 |
| 21600 | 1.195E+04 | 1.204E+04 | 1.213E+04 | 1.221E+04 | 1.228E+04 | 1.235E+04 | 1.241E+04 | 1.246E+04 | 1.251E+04 |
| 21700 | 1.197E+04 | 1.208E+04 | 1.217E+04 | 1.226E+04 | 1.233E+04 | 1.240E+04 | 1.247E+04 | 1.253E+04 | 1.258E+04 |
| 21800 | 1.199E+04 | 1.210E+04 | 1.221E+04 | 1.230E+04 | 1.238E+04 | 1.246E+04 | 1.253E+04 | 1.259E+04 | 1.265E+04 |
| 21900 | 1.201E+04 | 1.213E+04 | 1.224E+04 | 1.234E+04 | 1.243E+04 | 1.251E+04 | 1.258E+04 | 1.265E+04 | 1.271E+04 |
| 22000 | 1.203E+04 | 1.216E+04 | 1.227E+04 | 1.238E+04 | 1.247E+04 | 1.256E+04 | 1.264E+04 | 1.271E+04 | 1.278E+04 |
| 22100 | 1.205E+04 | 1.218E+04 | 1.230E+04 | 1.241E+04 | 1.252E+04 | 1.261E+04 | 1.269E+04 | 1.277E+04 | 1.285E+04 |
| 22200 | 1.206E+04 | 1.220E+04 | 1.233E+04 | 1.245E+04 | 1.256E+04 | 1.266E+04 | 1.275E+04 | 1.284E+04 | 1.292E+04 |
| 22300 | 1.207E+04 | 1.222E+04 | 1.236E+04 | 1.248E+04 | 1.260E+04 | 1.271E+04 | 1.281E+04 | 1.290E+04 | 1.298E+04 |
| 22400 | 1.208E+04 | 1.224E+04 | 1.238E+04 | 1.252E+04 | 1.264E+04 | 1.276E+04 | 1.286E+04 | 1.296E+04 | 1.305E+04 |
| 22500 | 1.209E+04 | 1.226E+04 | 1.241E+04 | 1.255E+04 | 1.268E+04 | 1.280E+04 | 1.291E+04 | 1.302E+04 | 1.312E+04 |
| 22600 | 1.210E+04 | 1.227E+04 | 1.243E+04 | 1.258E+04 | 1.272E+04 | 1.285E+04 | 1.297E+04 | 1.308E+04 | 1.318E+04 |
| 22700 | 1.210E+04 | 1.228E+04 | 1.245E+04 | 1.261E+04 | 1.275E+04 | 1.289E+04 | 1.302E+04 | 1.314E+04 | 1.325E+04 |
| 22800 | 1.210E+04 | 1.229E+04 | 1.247E+04 | 1.263E+04 | 1.279E+04 | 1.293E+04 | 1.307E+04 | 1.320E+04 | 1.332E+04 |
| 22900 | 1.210E+04 | 1.230E+04 | 1.248E+04 | 1.266E+04 | 1.282E+04 | 1.298E+04 | 1.312E+04 | 1.325E+04 | 1.338E+04 |
| 23000 | 1.210E+04 | 1.230E+04 | 1.250E+04 | 1.268E+04 | 1.285E+04 | 1.302E+04 | 1.317E+04 | 1.331E+04 | 1.345E+04 |
| 23100 | 1.209E+04 | 1.231E+04 | 1.251E+04 | 1.271E+04 | 1.289E+04 | 1.306E+04 | 1.322E+04 | 1.337E+04 | 1.351E+04 |
| 23200 | 1.209E+04 | 1.231E+04 | 1.253E+04 | 1.273E+04 | 1.292E+04 | 1.310E+04 | 1.327E+04 | 1.343E+04 | 1.358E+04 |
| 23300 | 1.208E+04 | 1.232E+04 | 1.254E+04 | 1.275E+04 | 1.295E+04 | 1.313E+04 | 1.331E+04 | 1.348E+04 | 1.365E+04 |
| 23400 | 1.208E+04 | 1.232E+04 | 1.255E+04 | 1.277E+04 | 1.297E+04 | 1.317E+04 | 1.336E+04 | 1.354E+04 | 1.371E+04 |
| 23500 | 1.207E+04 | 1.232E+04 | 1.256E+04 | 1.278E+04 | 1.300E+04 | 1.321E+04 | 1.341E+04 | 1.359E+04 | 1.378E+04 |
| 23600 | 1.206E+04 | 1.232E+04 | 1.257E+04 | 1.280E+04 | 1.303E+04 | 1.324E+04 | 1.345E+04 | 1.365E+04 | 1.384E+04 |
| 23700 | 1.205E+04 | 1.232E+04 | 1.257E+04 | 1.282E+04 | 1.305E+04 | 1.328E+04 | 1.350E+04 | 1.371E+04 | 1.391E+04 |
| 23800 | 1.204E+04 | 1.232E+04 | 1.258E+04 | 1.283E+04 | 1.308E+04 | 1.331E+04 | 1.354E+04 | 1.376E+04 | 1.397E+04 |
| 23900 | 1.203E+04 | 1.231E+04 | 1.258E+04 | 1.285E+04 | 1.310E+04 | 1.334E+04 | 1.358E+04 | 1.381E+04 | 1.404E+04 |
| 24000 | 1.202E+04 | 1.231E+04 | 1.259E+04 | 1.286E+04 | 1.312E+04 | 1.338E+04 | 1.363E+04 | 1.387E+04 | 1.411E+04 |

Acknowledgments. The authors are particularly indebted to Mr. G. Delluc, Ms. M. F. Elchinger, and Dr. B. Pateyron of the Université de Limoges who performed the calculations of the tables. Thanks are also due to Dr. Z. Njah of the Université de Sherbrooke and to Prof. J. Heberlein and Mr. P. C. Huang of the University of Minnesota who have made many helpful suggestions and provided comparisons of these data with those derived at the University of Minnesota over the past 15 years, and a compilation of a large number of experimental and theoretical literature data carried out by Dr. J. Lesinski at the Université de Sherbrooke.

## REFERENCES

1. W. B. White, G. B. Dantzig, and S. M. Johnson, *J. Chem. Phys.* **28** (1958): 751.
2. ADEP—Banque de données de l'Université et du CNRS Ed. Direction des Bibliothèques des Musées et de l'Information Scientifique et Technique (1986). (a) B. Pateyron, Thèse de Doctorat es Sciences Physiques (Université de Limoges, France, 1987). (b) B. Pateyron, J. Aubreton, M. F. Elchinger, and G. Delluc, "Thermochemical Equilibria in Multicomponent Systems on Microcomputers," International Meetings on Phase Equilibrium Data, Paris, 5–13 September (1985). (c) B. Pateyron, J. Aubreton, M. F. Elchinger, and G. Delluc, "Thermodynamic and Transport Properties at High Temperature: Hydrogen Plasma and Water Plasma," International Meetings on Phase Equilibrium Data, Paris, 5–13 September (1985). (d) M. F. Elchinger, B. Pateyron, G. Delluc and P. Fauchais, "Radiative and Transport Properties of Some Nitrogen–Oxygen Mixtures Including Air," Proceeding of 9th International Symposium on Plasma Chemistry, Pugnochiuso, Italy, R. d'Agostino, ed., Vol. 1. p. 127 (1989).
3. C. E. Moore, *Atomic Energy Levels* (NBC Circ. 467, Vol. 3, 1958).
4. K. S. Drellishak, Ph.D. Thesis (Northwestern University, 1963).
5. B. J. McBride and S. Gordon, NASA TN-D-40976 (1967).
6. G. Herzberg, *Spectra of Diatomic Molecules,* 2nd ed. (New York: Van Nostrand, 1950).
7. J. O. Hirschfelder, C. F. Curtis, and R. B. Bird, *Molecular Theory of Gases and Liquids* (New York: Wiley, 1964).
8. S. Chapman and T. G. Cowling, *Mathematical Theory of Non-uniform Gases* (London: Cambridge University Press, 1964).
9. R. S. Devoto, Ph.D. Thesis (Stanford University, 1965).
10. C. Gorse, Thèse 3ième Cycle (Université de Limoges, France, 1975).
11. C. Bonnefoi, Thèse 3ième Cycle (Université de Lomoges, France, 1975).
12. C. Bonnefoi, Thèse de Docteur ès Sciences Physiques (Université de Limoges, France, 1983).
13. J. Aubreton, Thèse de Docteur ès Sciences Physique (Université de Limoges, France, 1985).

# Index

www.ingramcontent.com/pod-product-compliance
Ingram Content Group UK Ltd.
Pitfield, Milton Keynes, MK11 3LW, UK
UKHW022322190726
13856UKWH00001B/156

* 9 7 8 1 4 8 9 9 1 3 3 8 8 *